American Industrial Archaeology

In loving memory of
Francis Carleton McVarish
and
Charles Edgar Maynard

American Industrial Archaeology

A Field Guide

Douglas C. McVarish

Walnut Creek, California

LEFT COAST PRESS, INC.
1630 North Main Street, #400
Walnut Creek, CA 94596
http://www.LCoastPress.com

ISBN 978-1-59874-098-1 hardcover
ISBN 978-1-59874-099-8 paperback

Library of Congress Cataloging-in-Publication Data

McVarish, Douglas C.
American industrial archaeology : a field guide / Douglas C. McVarish.
p. cm.
Includes bibliographical references and index.
ISBN 978-1-59874-098-1 (hardcover : alk. paper) — ISBN 978-1-59874-099-8 (pbk. : alk. paper)
1. Industrial archaeology—United States. I. Title.
T21.M4588 2008
609.73—dc22
200801985

Printed in the United States of America

♾ The paper used in this publication meets the minimum requirements of American National Standard for Information Sciences-Permanence of Paper for Printed Library Materials, ANSI/NISO Z39.48-1992.

08 09 10 11 12 5 4 3 2 1

Contents

Tables and Illustrations

TABLES

FIGURES

FOREWORD

Reading Industrial Sites

Carolyn L. White, University of Nevada, Reno

When I first discussed this volume with Douglas McVarish, we at Left Coast Press anticipated bringing it out as part of the Guides to American Artifacts series edited by Timothy J. Scarlett and myself. The series consists of comprehensive guides to classes of historical artifacts commonly found in excavations, archives, museums, and private collections. These handbooks seek to aid in the identification and interpretation of the technical, temporal, and diagnostic significance of the objects described. In addition to facilitating the identification and analysis of classes of artifacts, the authors explore the interconnections between objects and social identity.

This volume meets many of the aims that we have for the series. Above all, its purpose is usefulness. As a general guide, however, it approaches a subject so broad in scope that it simply does not fit within the confines of the series, as it attempts to describe many different kinds of sites, rather than focusing on a single artifact type. So, while it is not officially part of the series, it is a closely allied work. We hope it will provide vocabulary, descriptions, technical terminology, and comparative examples for those engaged in industrial archaeology. It is organized to facilitate both basic and detailed interpretations of these sites. Further, the volume is heavily illustrated to aid in the identification of features, buildings, structures, and landscapes through characteristic examples found within different kinds of industry.

Industrial sites are complex spaces. To fully understand them, researchers must comprehend the particular mode of industry in terms of technology, processes, materials (raw and finished), innovations, and connections to related industries. In addition, aspects of landscape, history, and the social lives of the people who worked in these places are enmeshed with the industrial production itself. All of this information must be mapped onto the structure and layout of the site. The scale of industrial sites further complicates their study, and scholars rely on a large-scale perspective when studying and recording these built environments.

The complexity and scale of industrial sites, as well as the incredible variety of extant site types, make the study of industrial places a very challenging task. Until now, there were few single resources to which one could turn to examine the array of industrial sites found on the American landscape. *American Industrial Archaeology: A Field Guide* seeks to remedy this problem.

McVarish covers a broad array of industrial site types within the expansive categories of transportation, extraction, production, and power generation. He breaks each industry into their assorted components, elucidating the functions and processes of each element so that they can be understood separately and as parts of a connected whole. He describes the functional and technical changes of each industry through time, providing physical and temporal contexts of the separate pieces and the composite industry. McVarish provides concise descriptions of the actions and processes in each industry and creates an authoritative vocabulary for working with and understanding the sites in the field.

The straightforward compilation and presentation of the historical background and technical aspects of so many industries will meet the varied needs of archaeologists, architectural historians, historians, landscape planners, and others interested in the industrial past. This book will make it easier to identify and interpret industrial sites.

Preface

The idea for this book emerged from my work as an architectural historian for a cultural research management firm. As part of compliance with federal environmental legislation, particularly Section 106 of the National Historic Preservation Act of 1966, as amended, state and federal agencies are called upon to assess the significance and integrity of properties under the criteria of the National Register of Historic Places and often to prepare a narrative history, description, and photographic documentation of those significant properties that will be damaged or destroyed by a government-sponsored, -funded or -permitted undertaking.

In the course of assisting agencies in complying with their responsibilities under Section 106, I have prepared documentation of several airports, a shipyard, numerous bridges, several electrical generating plants, a commercial pottery, a mine fan, a barge canal, a forge, a textile plant, manufactured gas facilities, and two slackwater navigation corridors. My graduate education in planning and historic preservation had not provided me with knowledge in the aspects of technology represented by these resources. Each time I started a new project I wished there was a book that would provide me with basic knowledge of the particular property type. This book is an attempt to provide such a source.

The property types represented in the book were chosen both because of my experience in conducting research and preparing documentation concerning them and because of the number of remaining historic resources of the types selected. Most of the chapters contain a brief introduction to the technology or characteristics of the property type and, through illustrations, provide information to enable the historian or researcher to identify important elements of the property type in the field. Most chapters include a glossary with brief definitions of terms that the researcher may encounter in researching a property type. The introductory chapter includes general guidance on researching industrial properties, while the concluding chapter presents an introduction to some of the major types of industrial landscapes.

The model for the book concept has been *A Field Guide to American Houses* by Virginia and Lee McAlester (Alfred A. Knopf, 1984). The McAlesters' guide is not intended as a scholarly treatise discussing in detail the manifestations and development of architectural styles, but rather as a reference source for the layperson and professional called upon to survey historic houses. This guide to industrial archaeology does not delve deeply into industrial processes and industrial communities. Instead, it provides the reader with information about some of the industrial sites, structures, and objects that a survey may include and provides basic terminology to describe these industrial facilities. For those who choose to delve deeper, references are provided.

Acknowledgments

First, I want to thank the staff of John Milner Associates, Inc., for their support and advice, especially my colleagues in the Philadelphia office: Rebecca Yamin, Grace Ziesing, Lori Aument, Courtney Clark, Matt Harris, and Dawn Thomas. Courtney and her husband, Dave, also contributed several illustrations to the book. Other JMA colleagues whom I worked with on projects that form the core of some of the chapters include Michael Roberts, Rick Meyer, Joel Klein, and Charles Cheek. Dan Roberts, the president of JMA, has been a particular source of personal and professional encouragement. JMA consultant, photographer John Herr, is always able to show how industrial buildings and structures can be creatively documented. As mentioned, the book grew out of a series of industrial documentation projects that I have worked on while at JMA. Many people helped during the course of these projects, including Conrad Weiser of the U.S. Army Corps of Engineers (Pittsburgh); staff of the COE offices in Louisville, Huntington, West Virginia, and Buffalo; staff of the New Jersey Department of Transportation; the staff of DMJM Aviation; and Terry Sroka of the Reading Regional Airport. A particular thanks is due to Dr. Sydne Marshall of Tetra Tech FW for her support and friendship.

I also greatly appreciate those individual and organizations that allowed use of images reproduced in this volume. Among them are the American Railway Engineering and Maintenance Association, the American Association of State Highway and Transportation Officials, the Highway Research Board, and United States Steel Corporation.

At Left Coast Press, I am grateful to Mitch Allen for taking a chance on a first-time author, to Carolyn White for her careful editing, her support, and her incisive comments, to Jennifer Collier for guiding me through the final editing and production process, and to Carol Leyba and Leyba Associates for their work in producing this volume.

A book such as this would be impossible without archival collections. I owe a special debt to the Hagley Library in Wilmington, Delaware, particularly to archivists Marge McNinch and Christopher Baer. I also am grateful for the ingenuity and helpfulness of Richard Boardman, Map Librarian of the Free Library of Philadelphia.

Finally, I want to thank members of my McVarish and Maynard families, particularly my wife, Lois Maynard, Joanne Church, Lorraine Seddon, Ginny McVarish, and Mary McVarish.

CHAPTER 1

Industrial Archaeology Research

> *Industrial archaeology.* The study of sites and structures reflecting changing industrial processes and practices. (National Park Service)

Industrial archaeology is a subject with a variety of definitions. To some, it is the study and excavation of past industrial sites undertaken by those trained in the theory and practice of historical archaeology. To others, it is research, study, and recordation of industrial sites and structures undertaken not necessarily by archaeologists but often by industrial historians or architectural historians. This concept, embodied in the conferences and publications of the Society for Industrial Archeology, is the one that has guided preparation of this book.

The author has used a broad definition of industrial archaeological resources. Not only does it include properties specifically associated with a variety of industrial practices and processes, it includes other property types only peripherally associated with industry, such as airports, canals and slackwater navigation systems, and roads and highways.

This chapter introduces some of the major sources used in documenting historic industrial properties.

MAP RESEARCH

Maps are one of the most widely available historic sources of documentation of industrial sites. The best known, and still among the most useful sources of map depictions of historic industrial buildings, are the Sanborn Map Company's fire insurance maps. Sanborn maps were first issued in the late nineteenth century and periodically updated; they were compiled for some 12,000 cities and towns in the United States, Canada, and Mexico.

Sanborn maps were designed for insurance agents to determine the degree of hazard associated with a particular property, and as such they show the size, shape, and construction of buildings, locations of windows and doors, sprinkler systems, and types of roofs. The Sanborn collection includes some 50,000 editions of fire insurance maps comprising an estimated 700,000 individual sheets. The largest extant collection of Sanborn maps is held by the Library of Congress, and a checklist of these maps was compiled between 1974 and 1978 by the Geography and Map Division of the Library of Congress. This volume is available in many reference collections and is the best first step to determine whether a community has been mapped by Sanborn.[1]

Sanborn maps are available in several formats. Many libraries have original elephant folio versions of some maps, often in bound volumes. In most cases, these volumes were updated by pasting map updates over the original maps. The date of the last revision, indicated on a table at the beginning of each map series, is effectively the date of information in the volume. Because of the size of these volumes, they are rarely available for photocopying, though they may be digitally photographed.

Microfilm reels of the Library of Congress collection of Sanborn maps are available in many public and research libraries; most libraries hold reels of their own state and surrounding states. The Chadwyck-Healey Company microfilmed the Library of Congress's collection of Sanborn maps. Microfilm reels are arranged alphabetically by state, generally in two series. The first series is composed of older maps, arranged sequentially by compilation date; the second series contains the newest maps.

Sanborn maps are also available through an online subscription service offered by UMI.[2] Access is available through many larger public and research libraries. Again, many collections offer access only to maps of their own state or their own state and adjoining states. Sanborn maps, both in hard copy and digital format, and permission to reprint them are available from the current owner, Environmental Data Resources, Inc.[3]

Other Building Footprint Map Series

Although Sanborn became nationally dominant, other regional and local companies issued footprint maps. For example, ample Sanborns exist in Philadelphia, but in the nineteenth century, a local Philadelphia mapmaker, Ernest Hexamer, published a valuable series of insurance surveys of individual maps of industrial facilities in the tri-state area (Pennsylvania, New Jersey, and Delaware). For those facilities surveyed by Hexamer, the amount of information included and the building detail are greater than in the comparable Sanborn map. In addition to footprints of each building, at least one elevation drawing of the facility was also included in the later insurance surveys. The same company also published atlases of building footprints for Philadelphia and vicinity. In the twentieth century, Sanborn's Philadelphia-area competition was the Franklin Survey Company. This company, which exists today as a map publisher and dealer, issued a series of volumes covering Philadelphia and much of its Pennsylvania suburbs. The best collection of Philadelphia-area maps, including the original Hexamer surveys, is in the Map Collection of the Central Library of the Free Library of Philadelphia.

Other areas of the country have their own map series. For example, Chicago's growth, development, and redevelopment are chronicled in various editions of Elisha Robinson's Robinson's *Atlas of the City of Chicago*, available at the Chicago Historical Society and the University Library of the University of Illinois, Chicago, among other repositories. Cities in several midwestern states were included in maps published by Chicago's Rascher Insurance Map Publishing Company, while the Fire Underwriters Inspection Bureau compiled maps of many small cities and towns in Minnesota. Portions of Iowa were surveyed by the Bennett Company and the Iowa Insurance Bureau. The Bennett Company, based in Cedar Rapids, compiled maps between 1897 and 1915, while the Iowa Inspection Bureau was founded in Des Moines in 1915. These maps are found in the collections of the State Historical Society, the University of Iowa, and Iowa State Archives.

To locate Sanborn and other fire insurance maps, the best repositories are state historical societies, state archives, state universities, and larger public libraries. A useful, though incomplete national union list of fire insurance maps is available on the Western Association of Map Libraries website.[4]

Census Schedules and Products of Industry

For those researching industrial sites in operation during the second half of the nineteenth century, the products of industry Census schedules may be a valuable source of information. Manuscript schedules identifying individual enterprises are available for 1850, 1860, 1870, and 1880 and have been microfilmed by the National Archives. Entries include the name of the owner of the facility, number of employees, description of power source(s), raw materials used, and types and value of production.

The single complete collection of microfilmed schedules is available at the National Archives in Washington, DC. Regional National Archives facilities typically have microfilmed schedules for their home state.[5] These schedules may also be available at state universities, state libraries or archives, and some larger local historical societies. Industries are arranged by county and by municipality within the county.

Building Permits and Assessment Records

Building Permits

Building permit records are occasionally very useful, depending on the record retention and permit policies of the particular municipality.

Some municipalities—for instance, Fall River, Massachusetts—retain record cards indicating the dates of permits for a particular location, the applicant, the cost of the proposed improvement, and a brief description of the improvement, but have not retained the permits themselves. Other municipalities, including Philadelphia, have microfilmed cards that include dates of the permit but little other information.[6] Other jurisdictions retain complete historical files including, on occasion, blueprints of the original construction and later improvements. Still others retain only ten years or so of permit applications. Access to these records is informal, and researchers must negotiate the busy schedules and heavy workload of the staff.

Assessment Records

Assessment records vary greatly in availability, information, and location. The most useful records are record cards: 8½ by 11 inches, typed card stock records. Cards may include period photographs taken by the assessor, sketched floor plans, as well as information on heating and power systems. Such cards are available in the Trentoniana Collection of the Trenton Public Library for properties in New Jersey's capital and may be available in other municipal archives.

More typical are assessment books, some of which date back to the eighteenth century, in which the property owner is indicated along with the value of his real and personal property. In these series, there may be a specific delineation in a particular year of a building's materials, size, and use, but the inclusion of such information is unpredictable.

Road and Bridge Research

Several major categories of research materials are useful for researching roads and bridges. Road "returns" or road papers and construction records are among the most useful.

In many areas, the laying out of roads was a subject of jurisdiction of the local court. Those individuals who sought the road's construction would present a petition to the court. If construction was approved, the court would appoint "viewers" to supervise its laying out. In some cases, the alignment of the road was described in a document that could be accompanied by a map of the road, often illustrated with houses or industrial buildings as landmarks. These returns or papers can provide critical information concerning the date of construction or alteration of an older roadway.

In some eastern states, road papers are filed with county records and are among other underutilized and nearly forgotten resources. In some states, such as New Jersey, older returns have been microfilmed and are available at the state archives, and original copies are available either in books or in loose files at the county. In other states, such as Delaware, hard copies of the original returns are in the collection of the state archives.

Researchers seeking historical information about a state highway and its associated records can use two useful sources of information. For general information about the cost and specifics of construction, transportation or highway department annual reports are useful. State libraries may contain a complete run of these reports, often illustrated with notable construction projects of a particular year.

For details of road and bridge engineering, the best source is "as-built" drawings, which vary in availability and access by state. For example, in Delaware, older as-builts are available on microfilm at the Delaware Public Archives, while more recent, as well as older as-builts, have been digitized and are available through the Delaware Department of Transportation (DelDOT) computer system. As-builts are less accessible in other states. In one state, in recent years, many of these records were not copied and were stored with miscellaneous drawings in a remote warehouse.

Journal Articles

Period articles in journals are an immensely useful resource for all sorts of industrial structures and complexes. If a facility or bridge was notable in its time, its construction or planning may have been the subject of one or more journal articles. The best way to access these articles is through indices such as the *Engineering Index* or the *Industrial Arts Index*. Indices are available online at college and university libraries and large public libraries, but electronic versions of older compilations are rarely found. Fortunately, many college and university libraries have a complete run of both indices, although often in remote storage. *The Engineering Index*, which is still published, began publication in the late nineteenth century, while the *Industrial Arts Index* was published from 1913 to 1957.

Period journals are also invaluable for their inclusion of contemporaneous projects, the state of technology, and recent innovations, allowing a particular project and industry to be placed in a broader context. For example, a typical issue of *Power Plant Engineering* included a feature article on a recent construction project, as well as other articles discussing important aspects of technology—turbines, condensers, generators, switching equipment, and the like. This journal also had a year's-end issue with articles summarizing recent technological developments.

In rare cases, long runs of the periodicals are available in open stacks. In other cases, such as at the Free Library of Philadelphia, most industrial and engineering periodicals are available on microfilm. In still other places, such as the Universities of Delaware and Pennsylvania, the library system has retained the periodicals themselves but keeps them in remote storage due to their limited demand. Although most libraries will retrieve these volumes, such circumstances make browsing impossible.

A limited but still useful online collection of both periodical articles and books is *Making of America*, a cooperative project of Cornell University, the University of Michigan, and the Andrew W. Mellon Foundation, described as a digital library of primary sources in American social history from the antebellum period through Reconstruction.[7] Among its strengths are science and technology. The collection includes digitized pages of nineteenth-century texts that may be printed and searched by keyword.

The portion of the database maintained by Cornell includes full-text collections of *Manufacturer and Builder* (1869–1894) and *Scientific American* (1846–1869), as well as more general-interest magazines (e.g., *The Atlantic Monthly* [1857–1901], *Harper's New Monthly Magazine* [1850–1899], and *Scribner's Monthly* [1870–1881]) which may also have articles of interest to the industrial historian.

Newspaper Articles

Local newspapers also provide information on the initial and ongoing construction of industrial structures, as well as the place of buildings and structures in a community's esteem. If the approximate date of construction is known, the researcher may skim a year or so of the local newspaper on microfilm. In very rare cases, such as in Bangor, Maine, or Newark, New Jersey, this task has been made much easier by the presence of a printed or manuscript newspaper index. In other cases, such as in Temple University's Urban Archives in Philadelphia, a newspaper's morgue is available, arranged by subject.

Textbooks

Textbooks are another source of valuable information on science and engineering. For instance, the books in the International Textbook Company series were written by recognized authorities of the late nineteenth to mid-twentieth centuries and are still widely available in libraries and in online and brick-and-mortar bookstores. These volumes offer an excellent introduction to the engineering technology of their day.

Patents

Original patent drawings can be a useful means to explore an aspect of industrial technology. Patent drawings of all patents issued since 1790 are available at the United States Patent and Trademark Office website.[8] The accompanying narrative is sometimes present as well. The available information is limited by the fact that to find the patent drawings for earlier designs using this database, it is necessary to know the patent number.

Patent numbers are contained in numerous indices, and a reference librarian at a U.S. Government depository library may be the best guide. One index that is useful for earlier patents is the Patent Office's *Subject-Matter Index of Patents for Inventions Issued by the United States Patent Office from 1790 to 1873, Inclusive*. This somewhat idiosyncratic index was reprinted in 1976 by Arno Press.[9] Google also provides access to full-text patents and drawings on its web browser. Although this information is indexed, the index is not totally inclusive.

Trade Catalogs

Trade catalogs are a useful means of researching industries, industrial archaeological elements, and specific companies. Produced to sell a company's wares to prospective purchasers, most are in the form of pamphlets and booklets. Local historical societies and libraries may have catalogs for local companies. Research libraries, including the Science and Industry Reading Room of the Carnegie Library of Pittsburgh and the Hagley Library (Wilmington, Delaware), have comprehensive collections.

Corporate Archives

Many larger companies maintain corporate archives. Access policies differ and often are determined by the company's general counsel. A list of corporate archives with brief collection descriptions, access policies, and contact information is available online.[10] It may be useful to contact the company to determine whether it retains superseded records if such is not noted in this collection. Furthermore, current employees may have knowledge of existing architectural or engineering drawings. Vacant buildings may house forgotten records.

Occasionally businesses deposited records in curatorial facilities. Among major archival collections containing industrial records are the Hagley Library, the Kroch Library at Cornell University, the Archives of Industrial History of the University of Pittsburgh,[11] and the Youngstown (Ohio) Historical Center of Industry and Labor.[12] The easiest way to determine whether the records of a company in question are contained in a major public archives is by means of the National Union Catalog of Manuscript Collections (NUCMC), available online.[13]

The R. G. Dun & Company Collection in the Baker Library of the Harvard Business School is a useful source for the historian of the nineteenth century. This collection, available onsite only, contains 2,580 volumes of handwritten credit reports on firms from the United States and Canada from 1841 to 1892. Information on the duration of the business, net worth, sources of wealth, and the character and reputation of owners, their partners, and successors is included. Information about the collection and restrictions on its use are available online.[14]

Historic American Building Survey/ Historic American Engineering Record

While chances are slim that a given industrial site has been previously documented, a similar building, structure, or site may have been the subject of a HABS/HAER recordation project. Most, though not all, of the HABS/HAER collection has been digitized and indexed and is available for searching on the Library of Congress website.[15] All HABS/HAER documentation contains large-format black-and-white photographs, and many include descriptive and historical reports. Some are also illustrated with color slides and/or measured drawings. Most of this documentary information may be downloaded and reprinted.

NOTES

1 Library of Congress, Geography and Map Division, Reference and Bibliography Section, *Fire Insurance Maps in the Library of Congress: Plans of North American Cities and Towns Produced by the Sanborn Map Company: A Checklist* (Washington, DC: Library of Congress, 1981).

2 Information on Digital Sanborn Maps, 1867–1970, owned by ProQuest Information and Learning Company, is available on the UMI website (http://sanborn.umi.com).

3 Environmental Data Resources' website is www.edrnet.com/reports/historical.html.

4 The website for the Western Association of Map Libraries' fire insurance map union list is www.lib.berkeley.edu/EART/sanbul_libs.html. The use of fire insurance maps in research is discussed in Diane Oswald's book, *Fire Insurance Maps: Their History and Applications* (College Station, TX: Lacewing Press, 1997).

5 The locations of regional National Archives research collections are listed on the National Archives website: www. archives.gov.

6 Copies of some, but not all, older Philadelphia building permits are retained by the Philadelphia City Archives.

7 The two Making of America websites are: moa.umdl.umich.edu (Michigan) and cdl.library.cornell.edu/moa.

8 www.uspto.gov/patft/index.html.

9 United States Patent Office, *Subject-Matter Index of Patents for Inventions Issued by the United States Patent Office from 1790 to 1873, Inclusive* (New York: Arno Press, 1976), Library of Congress catalog number: T223 .D7 A45 1976.

10 www.hunterinformation.com/corporat.htm.

11 The website address of the Archives is www.library.pitt.edu/libraries/archives/ais.html.

12 The website of the Youngstown Historical Center of Industry & Labor is www.ohiohistory.org/textonly/places/younst/index.html.

13 www.loc.gov/coll/nucmc/nucmc.html.

14 http://library.hbs.edu/hc/collections/.

15 www.loc.gov.

CHAPTER 2

Bridges

Bridges are one of the most common structures surveyed and documented by historic resource researchers. They range from humble concrete-and-steel girder spans found in almost any community, to covered bridges of the Northeast and Midwest, suspension bridges that span major waterways, and modern long-span bridges, the cable-stayed bridge. The following typology is arranged roughly chronologically by bridge material and type. The typology is designed to enable the researcher to identify the type of bridge being surveyed and its major components. The chapter concludes with a glossary of bridge terms referenced in the text.

STRUCTURAL ENGINEERING: BRIDGE-RELATED CONCEPTS

A few basic principles of structural engineering are useful for understanding the parts and construction techniques of bridges. These principles govern the construction of most bridges. When a load is placed on a beam, it bends, or becomes curved. The curved shape causes stresses in the beam. The largest stresses are at the bottom and top of the beam. The greater the load placed on the beam, the more the beam will curve and, consequently, the bigger the stresses at the top and bottom will be. The top part of the beam is affected by compressive forces, while the bottom of the beam is affected by tension or tensile forces. The longer the span, the more pronounced the curvature is for any load.

Most trusses are intended to do the same work as a beam but with less material. Trusses are constructed of individual short, straight members that are arranged to form triangles. If the loads the truss must carry are placed only at joints, then each truss member will receive only axial loads. Connecting pins let each member rotate slightly so that none will bend.

Bridge members exhibit stress and fail in relation to their form and the type of stress they bear. Members along the top of a truss are in compression and have compressive stresses. Members along the bottom are in tension and have tensile stresses. Members between the top and bottom or between the two ends may be either in tension or compression, depending on how they are arranged and where the loads are located. The largest stresses will be in members proximate to the load placement. Structural members fail in compression by buckling, and long members buckle more easily than shorter ones. Members with small cross sections buckle more easily than those with larger cross sections. Diagonal bracing is required at the top of a truss to prevent the tops of the trusses from tipping or leaning. The weight of a bridge itself is termed the dead load, while the weight of traffic on the bridge is termed the live load.[1]

STONE BRIDGES

Stone Slabs

Stone slab bridges represent an early structural type, a simple beam. The only systematic survey in the United States of this rare structure type is that of the Massachusetts Highway Department, which identified 56 bridges in the state. The oldest identified Massachusetts stone slab bridges are three approach spans flanking the stone arches of the Adams Street Bridge in Dorchester Lower Mills; these were probably built in 1765 (Figure 2.1). Other Massachusetts stone slab bridges are located in Acushnet, Cohasset/Scituate, Leominster, Ludlow, Medfield, Middleborough, North Attleborough, Peabody, Plympton, Waltham, and West Bridgewater.[2]

Stone spans are necessarily short in length. The three Adams Street Bridge approach spans, made of roughly shaped blocks of gray granite, are 7.5 to 10 feet long. In their simplest form, stone spans are built by placing the spanning stone on top of walls at either end. These stones were presumably placed using derricks, blocks and tackles, and horse power.

When the central arch spans of the Adams Street Bridge were constructed, additional stonework was added atop the original slabs to permit the arch construction. This structure includes a fascia slab, a narrow course of irregular blocks above the slab stones, and a thick layer of large-scaled quarried granite rubble on top, resulting in a total overburden of about 10 feet.[3]

Stone Arch

Historic Overview

The oldest known stone arch bridge in the United States that carries a modern highway is the Frankford Avenue Bridge. This bridge, which crosses the Pennypack Creek in Philadelphia, was constructed in 1697 as part of the King's Road. Stone arch bridge construction was relatively labor-intensive, particularly for large structures, which made them ill suited for much of the sparsely populated United States in the seventeenth and eighteenth centuries. During the nineteenth century, stone arch bridges were never as popular in the United States as in Europe; stone rubble was the preferred material. The first stone bridge in the United States constructed with highly dressed stone and uniform mortar joints was the Baltimore and Ohio Railroad's Carrollton Viaduct in Baltimore.[4]

Figure 2.1. Adams Street Bridge, Dorchester Lower Mills, Massachusetts. Detail of south stone slab span. HAER MA-131-16, National Park Service. Photograph by Wayne Fleming, October 1995.

In general, stone arch road bridges of the United States were small-scale structures built by local masons. A number of large bridges were erected along the National Road between Maryland and Ohio and in the Philadelphia area. Stone spans were more frequently used in aqueduct and railroad construction. In the early twentieth century, stone bridge construction flourished in some rural areas, and in the 1930s a number of Civilian Conservation Corps (CCC) projects featured stone bridges. Notable examples of twentieth-century stone arch bridges are those associated with the carriage roads of Maine's Acadia National Park.

Stone Arch Bridge Technology and Arch Types

The typical components of a stone arch bridge are shown in Figure 2.2. Spandrel walls serve as retaining walls for fill material, usually stones, large rocks, and soil. The arch barrel surrounds the fill material. The parapet is an extension of a spandrel wall and is the outermost edge of the roadway. Stone arch bridges are held in equilibrium by the creation of an upward compressive force equal to the pull of gravity, a tensile force by the stones pushing against each other.

Decorative features may include a belt course, dressed voussoirs, a keystone, and a date stone. A belt course is a horizontal band of masonry that extends across the spandrel wall and may be flush with or projecting from this wall. Voussoirs are the wedge-shaped ring stones of the arch. The center voussoir at the crown of the arch is the keystone.

Stone masonry may be classified by the type of material used, the method of finishing the stone surfaces, and the method of construction. The type of material used in a masonry bridge is typically sandstone, limestone, gneiss, or marble, but may be brick, granite, or other building stones.

Several categories are used to describe the finishing of the stone surfaces. Rubble masonry employs rough stones, unsquared and unfinished. Squared-stone masonry consists of stones that are squared and roughly dressed or finished. Ashlar consists of stones that are precisely squared and carefully dressed. Rubble can be laid to approximate regular rows or courses (coursed rubble) or can be uncoursed (random rubble). Ashlar and squared stone can also be laid coursed or uncoursed.

Arch bridges may be classified by the curve of the intrados or lower surface. Arch types include semicircular, segmental, multi-centered, parabolic, and elliptical (Figure 2.3). When an arch is sprung from a horizontal bed, it is termed full-centered.[5]

WOOD BRIDGES

Two primary types of wood bridges have been built in the United States: the wood trestle and the wood truss. The structural evolution of the wooden bridge began with wood stringer construction and progressed to beams, trussed beams, trussed arches, arches, and

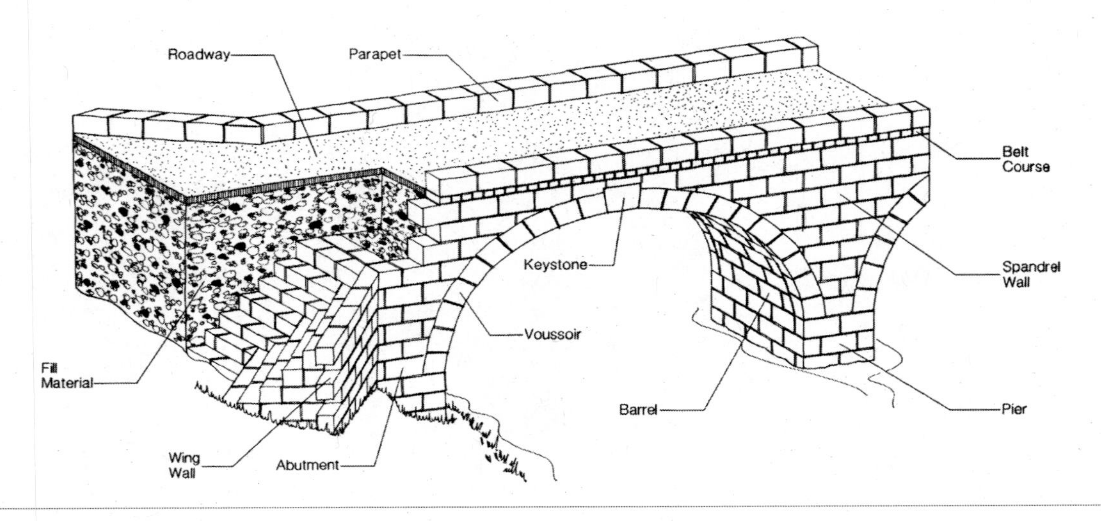

Figure 2.2. Typical stone arch bridge. Reprinted, by permission, from Commonwealth of Pennsylvania, *Historic Highway Bridges in Pennsylvania* (Harrisburg: Pennsylvania Historical and Museum Commission and Pennsylvania Department of Transportation, 1986), 28.

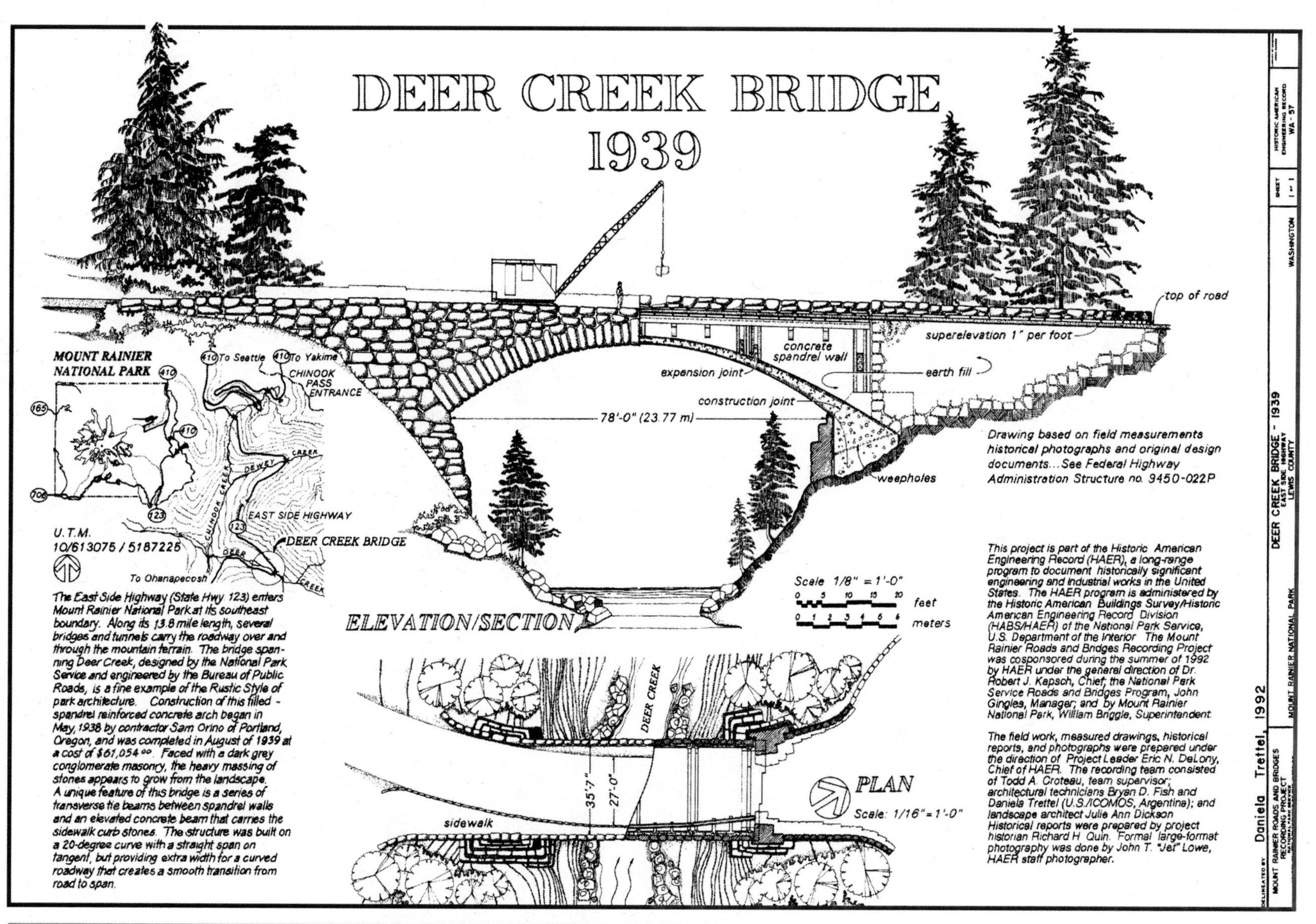

a

Figure 2.3. *(continues on next page):* Types of bridge arches. ***a***, Semi-elliptical arch. Deer Creek Bridge, Mount Ranier National Park, Washington. HAER WA-57, National Park Service. Delineated by Daniela Trettel, 1992.

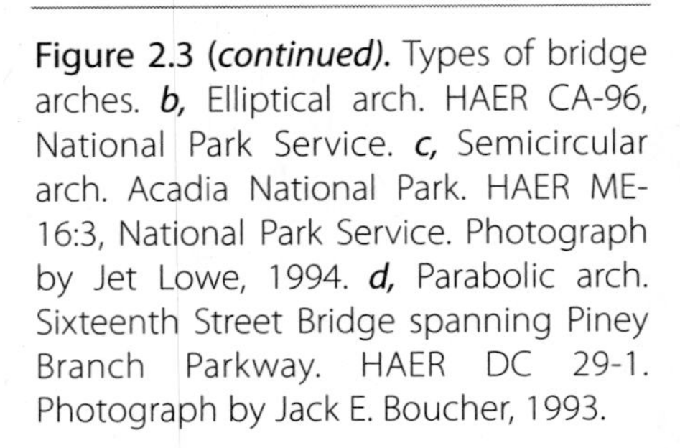

Figure 2.3 (*continued*). Types of bridge arches. ***b,*** Elliptical arch. HAER CA-96, National Park Service. ***c,*** Semicircular arch. Acadia National Park. HAER ME-16:3, National Park Service. Photograph by Jet Lowe, 1994. ***d,*** Parabolic arch. Sixteenth Street Bridge spanning Piney Branch Parkway. HAER DC 29-1. Photograph by Jack E. Boucher, 1993.

b

c

d

trusses. Wood trusses were then modified to include iron rods. Subsequently, iron posts were added, and the all-metal truss was introduced in the mid-nineteenth century. According to bridge historian Henry G. Tyrell, the majority of bridges built in the United States prior to 1860 were built of timber.[6]

Covered Bridges

The covered bridge is both the best known and most beloved early bridge type in the United States. Although most commonly found in the New England states, New York, and Pennsylvania, smaller numbers are found in midwestern and southern states. As late as 1930, from 450 to 500 covered bridges survived in the eastern United States.

Kingpost Truss

The kingpost is the oldest and simplest truss used in bridge construction (Figure 2.4a). It consists of a basic truss triangle with two timbers slanting down from the center to the ends of the lower chord. The kingpost extends vertically from the center of the upper chord to the center of the lower chord, forming two adjoining triangles. The post is a tension member, while the diagonals and lower chord are in compression. It is used primarily for short spans of approximately 20 to 30 feet, with few over 35 feet. A majority of remaining kingpost truss bridges are in Pennsylvania.[7]

Multiple Kingpost Truss

The multiple kingpost truss, an elaboration of the simple kingpost truss, was designed to span distances of as much as 100 feet (Figure 2.4b). It is the truss used in a Burr arch-truss bridge (see below). One unusual feature is the open center panel, often found when there is an odd number of panels. An X center panel is occasionally found in other examples. Most remaining covered bridges of this type are in Ohio and range in size from 50 to 100 feet long.[8]

Queenpost Truss

This truss succeeded the kingpost in the design chronology and was developed to span longer distances, though seldom more than 60 or 70 feet (Figure 2.4c). The truss is formed of a truncated triangle with a post at each end and a horizontal crosspiece located directly under but independent of the top chord of the bridge. Most remaining queenpost truss covered bridges are in Pennsylvania or Vermont.[9]

Town Truss

The Town truss or Town lattice, named for its originator, Connecticut native Ithiel Town, was first used in 1820 (Figure 2.4d). It became popular because it employed smaller lumber dimensions than other truss types, required a limited amount of framing and hardware, could be erected by unskilled laborers, and could span up to 200 feet. It consists of a web of light planks, 2 to 4 inches in thickness and 10 to 12 inches wide, latticed at an angle of 45 or 60 degrees and fastened with wooden pegs or trunnels at each intersection. This bridge design was used for highways and, later, railroads. Well over 100 examples of this bridge type remain in New England, Pennsylvania, Ohio, and the southern states.[10]

Burr Arch-Truss

Patented by Theodore Burr in 1804, the same year he constructed a bridge spanning the Hudson River at Waterford, New York, the Burr arch-truss combines reinforced arches with multiple kingpost trusses. The truss is of the multiple kingpost type, with panel posts generally spaced 10 feet apart (Figure 2.4e). Between each panel post is a diagonal timber sloping up toward the center of the bridge. The arch is pegged to each intersecting truss member throughout the truss, and the ends of the arch are anchored in the bridge piers or abutments. The arches stiffen the trusses and reduce deflection, allowing wider streams and rivers to be spanned. Though not a true arch, it relies on the interaction of arch segments with the truss members to carry the load. A majority of remaining Burr arch-truss bridges are in Pennsylvania, with lesser numbers in Indiana, Ohio, and New England.[11]

Wheeler Truss

The Wheeler truss was developed by Isaac H. Wheeler of Sciotoville, Ohio, for which he was awarded Patent 107,576 in 1870 (Figure 2.4f). Unlike other wooden trusses, the Wheeler truss employs a three-chord system: upper, middle, and lower. Diagonal braces are set at 22-degree angles with a center vertical. These diagonals pass through, and are bolted to, all three chords. The counterbraces are short timbers, notched into the center verticals and the diagonal braces above and below

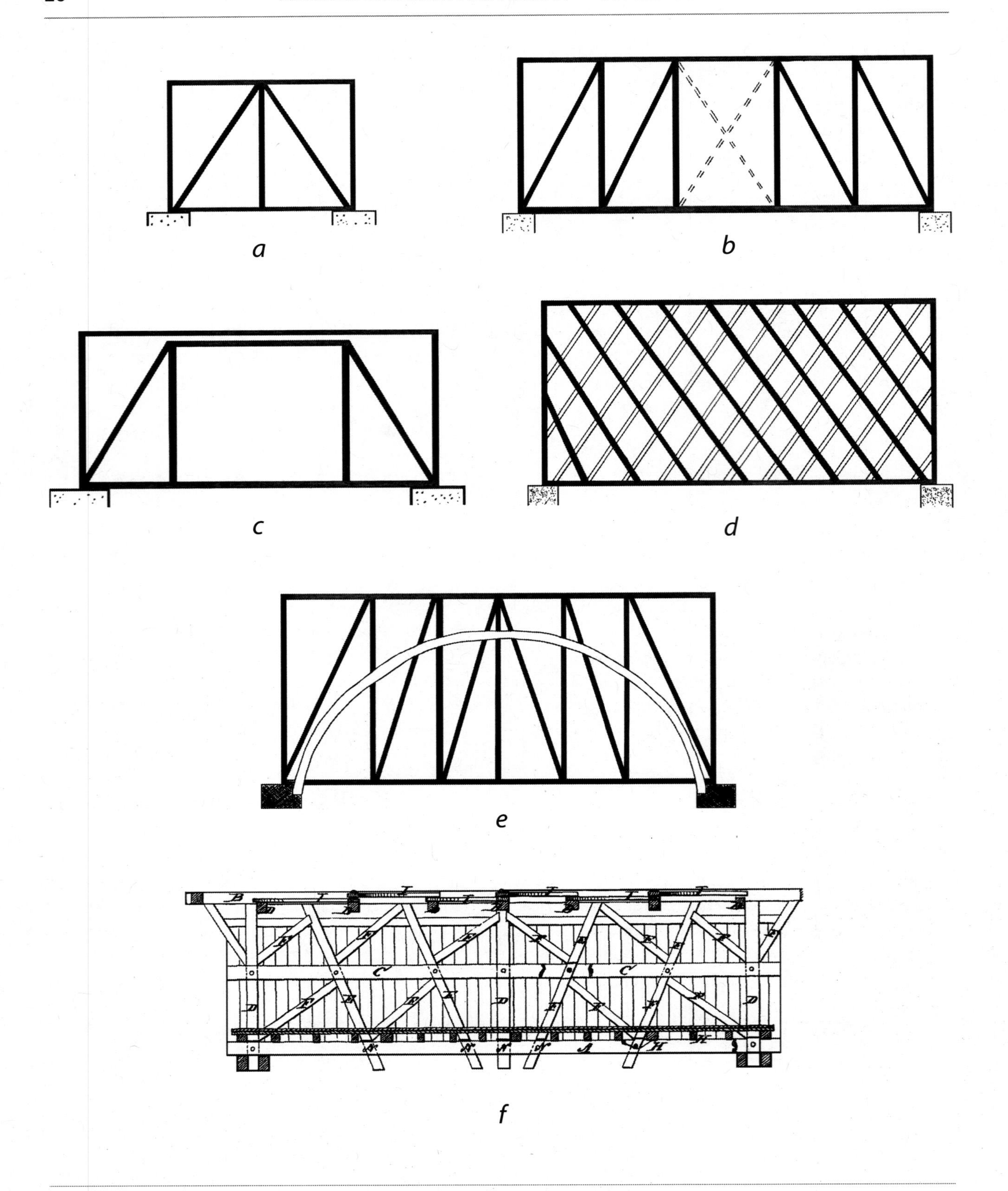

Figure 2.4. Types of wood bridge trusses: ***a,*** kingpost; ***b,*** multiple kingpost; ***c,*** queenpost; ***d,*** Town; ***e,*** Burr arch; ***f,*** Wheeler, Patent No. 107,576 (1870). Drawings by Courtney Lynch and David Clark.

the middle chord. Despite the economy of this design, few were built, predominantly in southern Ohio and northern Kentucky.

Wooden Arches

Wooden arches were occasionally used in the place of trusses, usually tied at the ends to the lower chord of the bridge. The live load was carried by iron suspension rods, extending from the arch to the lower chord using blocks, washers, and nuts for adjustment. Vertical timbers were placed close to each suspension rod, and auxiliary diagonal suspension rods might also be present. Only a few bridges of this type remain, mainly in Vermont.[12]

Other Truss Forms

Other, less common covered bridge truss types include the Long truss, the Paddleford truss, the Howe truss, the Haupt truss, the Warren truss, the Pratt truss, the Brown truss, the Smith truss, the Partridge truss, the Childs truss, the Brown truss, the bowstring, the post truss, and the McCallum truss. Each is pictured and described in Raymond E. Wilson's "Twenty Different Ways to Build a Covered Bridge," reprinted in *American Wooden Bridges* (New York: American Society of Civil Engineers, 1976).

Timber Bridges

Timber highway bridges were initially a permanent form, but were later used for temporary structures and in places where transportation was difficult and suitable timber was available. Timber truss bridges were usually made with the Howe type of truss, with timber top and bottom chords and diagonal braces and steel rods for vertical ties.

Wood Stringer Bridges

While covered bridges receive substantial attention due to their photogenic qualities and association with the country of the nineteenth century, wood stringer bridges (Figure 2.5) were more common. Initially, structures with

Figure 2.5. Wood stringer bridge. Houghton, Michigan. Photograph by the author, 2007.

log cribs or rough stone abutments and round logs hewn to receive a plank floor were used to span openings up to 40 feet. For wider streams, pile or frame trestle bents were used in quiet waters, and log cribs filled with stone were used where heavy freshets occurred.[13]

Pile bents were generally used where the ground was soft or covered with water. Among the varieties of lumber used for bents are red cedar, red cypress, pitch pine, yellow pine, and white pine.

Wood stringer bridges were built on both railroads and roads, especially where limited clearance was required beneath the bridge. Spans of plain stringer bridges were limited to about 15 feet for railroad construction. For spans between 15 and 30 feet, compound keyed stringers were used. The keys were generally cast iron and were proportioned for the longitudinal shear. The thin fin at the middle of the key was cut into the wood to hold it from working out of place. For more elaborate stringer bridges a pile cap was used. For lower-quality bridges and for framed bents, solid caps—either tenoned or drift-bolted to the pile or post—were used.[14]

Timber bridges remain in use in many areas, especially on lightly traveled roads in estuaries and tidal marshes where the crossing is of considerable length. Some modern timber bridges have been constructed on masonry substructures, and some have been constructed on timber pile bents or abutments. The bridge decks are usually constructed of timber or concrete.[15]

METAL BRIDGES

From the mid-nineteenth century to the advent of reinforced concrete construction in the early twentieth century, metal was the most frequently used material for bridges. Metal bridge types include beams, plate girders, arches, and trusses. The simplest type of metal bridge is the beam bridge. These bridges consist of I-beams spanning the opening, placed close together to carry the floor of the bridge. These bridges are the metal equivalent of wood stringer bridges.

Slab-Steel Beam Bridges

Slab-steel beam bridges (Figure 2.6) are constructed by placing steel beams side by side with the ends resting on the abutments. The majority of all extant bridges in the United States are the slab-and-beam type.[16] This construction consists of several beams that span in the direction of the roadway and are topped with a reinforced concrete deck. The longitudinal beams can be made of a variety of materials but are usually of concrete or steel.[17]

Reinforced concrete floors on steel highway bridges are supported on joists or stringers and by floor beams, or by the floor beams alone. Stringers are used for beam bridges and are commonly used for truss bridges, while the stringerless floor is commonly used on plate girder bridges.

Bridge engineers can choose to use a mechanical, shear-resistant device to connect the slab to the beams. This device ensures that the slab will act with the steel beams and assists the beams in carrying longitudinal bending moments. As a result, smaller steel beams can be used. The saving in beam size must be weighed against the cost of shear connectors. On beam spans of longer than 40 feet, the composite slab and beam bridge has proven economical.[18]

Plate Girder Bridges

As early as 1847, plate girder spans were built for small railroad bridges. The first such span was built by James Milholland for Maryland's Baltimore and Susquehanna Railroad. These girders were built-up members fabricated from riveted shapes. They competed with the metal truss and had the advantage of being able to be completely finished at the bridge shop and shipped to the site.[19]

As Frank W. Skinner noted in the early twentieth century, plate girders were among the simplest possible bridges and were extensively and increasingly used for both highway and railroad short spans.[20] They were solid and rigid, easily erected, economically manufactured, and did not require adjustment after completion. Most such bridges were less than 75 feet in length, although lengths of up to 100 feet were feasible. By the 1930s, plate girders were used for spans up to 150 feet and were considered to be the most durable metal bridge.

A plate girder is a built-up I-beam consisting of a single web plate and two flanges. The simplest plate girder span was constructed with a rectangular web, parallel pairs of top and bottom flange angles, with or without flange cover plates, vertical end, and intermediate web-stiffener angles. It rested on flat plates riveted to its lower chord flanges and bolted to the masonry. Plate girders were fastened together by lateral bracing.

Figure 2.6. Steel beam bridge. Prescott Bridge, Raymond, New Hampshire. HAER NH-16, National Park Service. Photograph by Bruce Alexander, 1988–1989.

The elementary plate girder span has two main longitudinal girders with solid continuous vertical web plates and T-shaped parallel horizontal flanges. They are either deck or through spans. The former supports the floor directly on the top flanges, the latter by floor beams connected to their webs. Deck span girders are generally braced together by transverse diagonals in vertical frames, called sway frames, and by horizontal lateral diagonals in the planes of the top and bottom flanges. Where the spans are long, two or more plates are spliced together to form the web plates, and horizontal plates are riveted to the flange angles to increase the flange area (Figure 2.7).[21]

Through plate girder spans are used when the clearance between the high water and the underside of the bridge is insufficient to permit the use of a relatively deep deck structure.

The floor system of highway bridges generally consists of a reinforced concrete slab supported on longitudinal steel stringers, which are in turn supported by transverse floor beams. Some designs eliminate the stringers.[22]

Encased Girder

Although undoubtedly functional, steel girders are often considered ugly. To improve the appearance, concrete is sometimes used to encase the I-beams that support the deck, fascia girders are concrete encased, and the parapet is constructed of paneled concrete. In other instances, the application of veneer stone yields the appearance of a stone bridge.

Continuous Steel Girder

Many late twentieth-century steel truss bridges are continuous spans in which the truss extends beyond two adjacent support piers. In the same fashion, steel girder spans were constructed in which a single girder extends over multiple piers. The first major use of continuous steel girders in American bridge construction occurred in the 200-foot main span of the Capital Memorial Bridge in Frankfort, Kentucky (1937–1938). By 1960, the longest steel girder continuous span bridge was the 387-foot main span of the Connecticut Turnpike Quinnipiac River Bridge (1958).

Evolution in steel girder bridges also led to innovations in pier construction, specifically the development of the hammerhead or T-shaped pier in which the deck is carried on heavy brackets of steel or concrete cantilevered from a central column. First used in Europe, the first American use of these piers occurred on the Housatonic River Bridge of Connecticut's Merritt Parkway, constructed in 1939–1940.[23]

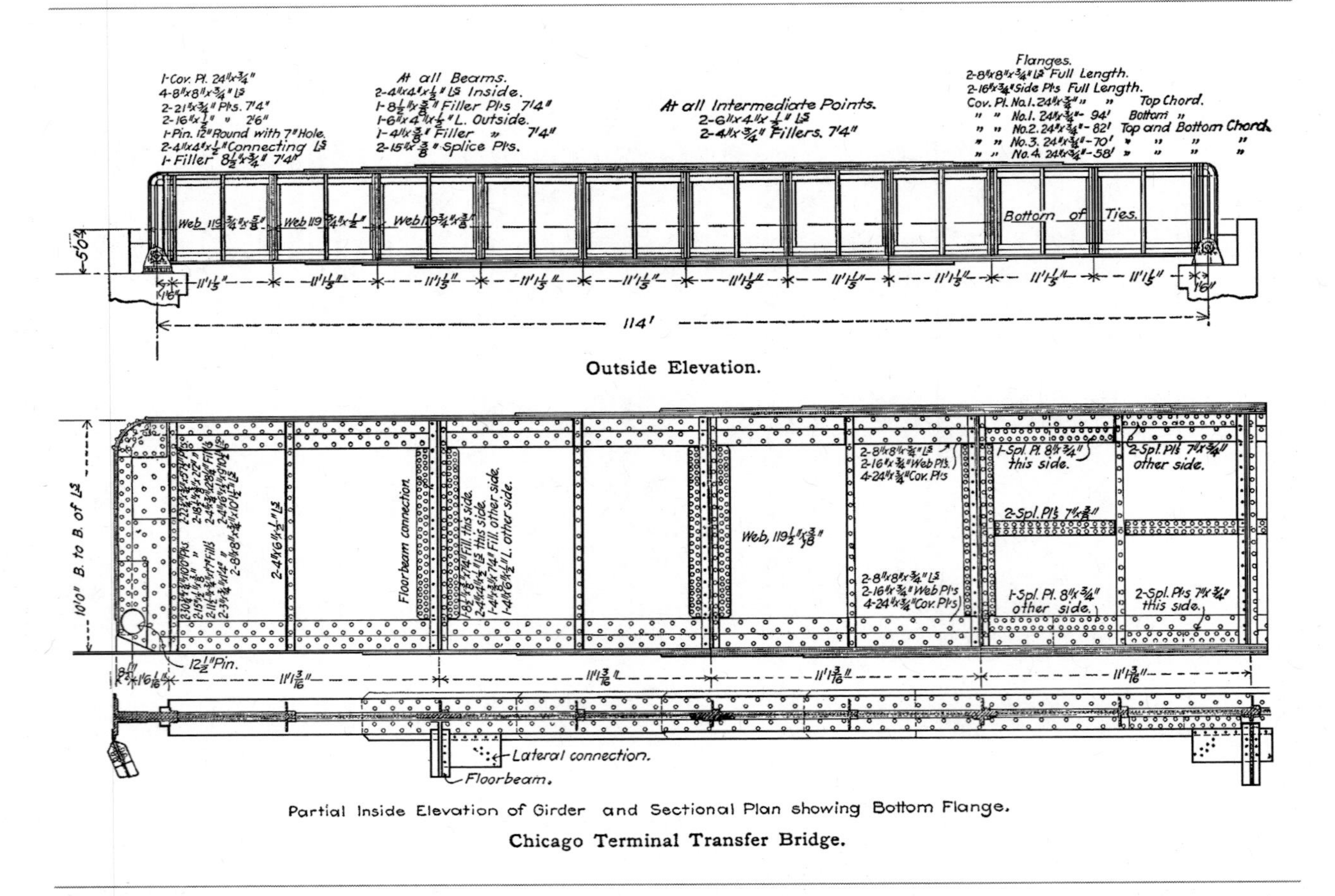

Figure 2.7. Details of plate girder bridge, Chicago Terminal Transfer Bridge. Reprinted from Frank W. Skinner, *Types and Details of Bridge Construction* (New York: McGraw Publishing Company, 1908: II), 104.

Rigid Frame Bridges

Rigid frame bridges were first introduced in the United States in 1922. The first bridges were designed by Arthur G. Hayden, chief designing engineer of the Westchester County Park Commission in New York, as an alternative to the more expansive arch structures. Arches required heavy abutments, along with extensive grading and/or excavation. Hayden's design was intended to resemble a true arch but eliminated the massive abutments and offered a broad arch-like opening with a relatively shallow superstructure and a minimal loss of headroom beneath the span. Its strength is attributable to the rigid connection between horizontal and vertical members that spread the load more evenly throughout the entire structure.

The steel structural members of rigid frame bridges are typically hidden by cast concrete or stone veneer arch and wing walls. First constructed on New York's Bronx River Parkway (Figure 2.8), these bridges were later used for grade-separated interchanges on other parkways, such as Connecticut's Merritt Parkway, Virginia's Mount Vernon Memorial Highway, and the Blue Ridge Parkway, as well as other general roadways.[24]

Metal Arch

The steel arch is distinguished from the truss or girder in that the reactions at the supports are inclined rather than vertical. Arches may be grouped in accordance with the methods by which stresses are distributed throughout the arch (Figure 2.9). These groupings include hingeless or fixed arches, single-hinged (or one-hinged) arches, two-hinged arches, and three-hinged arches. Metal arches permit bridges of greater span than metal trusses. Although the hinges in arch bridges are substantially larger, they are similar to pin connections used for trusses.

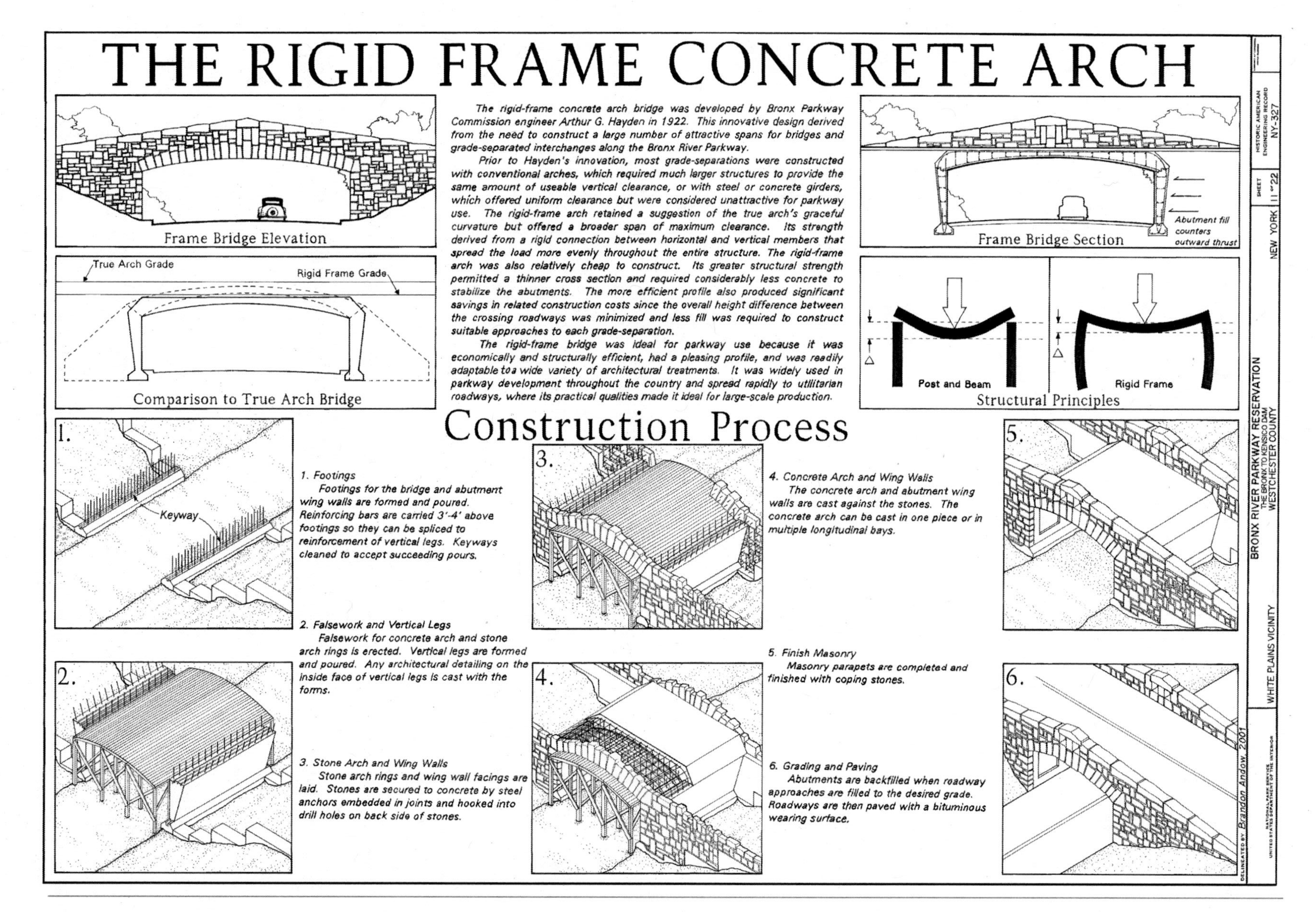

THE RIGID FRAME CONCRETE ARCH
Frame Bridge Elevation
True Arch Grade
Rigid Frame Grade
Comparison to True Arch Bridge
The rigid-frame concrete arch bridge was developed by Bronx Parkway Commission engineer Arthur G. Hayden in 1922. This innovative design derived from the need to construct a large number of attractive spans for bridges and grade-separated interchanges along the Bronx River Parkway.
Prior to Hayden's innovation, most grade-separations were constructed with conventional arches, which required much larger structures to provide the same amount of useable vertical clearance, or with steel or concrete girders, which offered uniform clearance but were considered unattractive for parkway use. The rigid-frame arch retained a suggestion of the true arch's graceful curvature but offered a broader span of maximum clearance. Its strength derived from a rigid connection between horizontal and vertical members that spread the load more evenly throughout the entire structure. The rigid-frame arch was also relatively cheap to construct. Its greater structural strength permitted a thinner cross section and required considerably less concrete to stabilize the abutments. The more efficient profile also produced significant savings in related construction costs since the overall height difference between the crossing roadways was minimized and less fill was required to construct suitable approaches to each grade-separation.
The rigid-frame bridge was ideal for parkway use because it was economically and structurally efficient, had a pleasing profile, and was readily adaptable to a wide variety of architectural treatments. It was widely used in parkway development throughout the country and spread rapidly to utilitarian roadways, where its practical qualities made it ideal for large-scale production.
Abutment fill counters outward thrust
Frame Bridge Section
Post and Beam
Rigid Frame
Structural Principles
Construction Process
1.
Keyway
1. Footings
Footings for the bridge and abutment wing walls are formed and poured. Reinforcing bars are carried 3'-4' above footings so they can be spliced to reinforcement of vertical legs. Keyways cleaned to accept succeeding pours.
2. Falsework and Vertical Legs
Falsework for concrete arch and stone arch rings is erected. Vertical legs are formed and poured. Any architectural detailing on the inside face of vertical legs is cast with the forms.
2.
3. Stone Arch and Wing Walls
Stone arch rings and wing wall facings are laid. Stones are secured to concrete by steel anchors embedded in joints and hooked into drill holes on back side of stones.
3.
4.
4. Concrete Arch and Wing Walls
The concrete arch and abutment wing walls are cast against the stones. The concrete arch can be cast in one piece or in multiple longitudinal bays.
5. Finish Masonry
Masonry parapets are completed and finished with coping stones.
6. Grading and Paving
Abutments are backfilled when roadway approaches are filled to the desired grade. Roadways are then paved with a bituminous wearing surface.
5.
6.
DELINEATED BY: Brandon Andow, 2001
BRONX RIVER PARKWAY RESERVATION
THE BRONX TO KENSICO DAM
WESTCHESTER COUNTY
WHITE PLAINS VICINITY
NEW YORK
SHEET 11 of 22
HISTORIC AMERICAN ENGINEERING RECORD
NY-327

Figure 2.8. The rigid frame concrete arch. Bronx River Parkway Reservation, White Plains vicinity, New York. HAER NY-327, National Park Service. Delineated by Brandon Andow, 2001.

Classified by the method in which the rib is fabricated and the deck carried by the rib, arch bridges may be grouped as solid rib arches, braced rib arches, and spandrel-braced arches. The ribs of metal arches may be formed of I-beams, plates and angles, cast iron or cast-steel segments, or riveted trusses.

Fixed (Hingeless) Arch

The principal advantage cited for this arch type is its rigidity.[25] The best-known example of a hingeless metal arch bridge in the United States is the 520-foot Eads Bridge in St. Louis. Other hingeless arch bridges include the double-deck Henry Hudson Parkway Bridge in New York City (1935–1936) and the 950-foot Rainbow Bridge over the Niagara River in western New York (1939–1941).[26]

One-Hinged Arch

This type of construction was rarely used and had no particular advantages.[27] In this arch, the hinge is placed at its crown. A hinge is formed by a bridge pin bearing upon either a structure or cast-iron pedestal.

Two-Hinged Arch

The two-hinged arch is more rigid than the three-hinged arch and provides adequate clearance above water as a through arch. The arch hinges are placed at the junction of the arch and the abutments. In two-hinged arch trusses, the chords are parallel or converge toward the skewbacks.

The best-known two-hinged arch bridge is Gustav Lindenthal's 1914–1916 Hell Gate Bridge, a railroad bridge in New York City. This bridge has a deck supported from the arch ribs by hangers and a curved upper chord connected to the arch rib by Pratt truss webbing. The bridge has a clear span of 977 feet. Originally a three-hinged span, the crown hinge was replaced by a rigid joint in April 1916 to increase the bridge's stability.[28] Other well-known two-hinged arch bridges were the 510-foot Washington Bridge (1898) over the Harlem River in New York City and braced and spandrel arched spans over the Niagara River.

Three-Hinged Arch

The three-hinged arch was introduced in the United States in 1869 by Joseph M. Wilson, a Pennsylvania Railroad engineer who later established the noted Philadelphia architecture firm, Wilson Brothers.

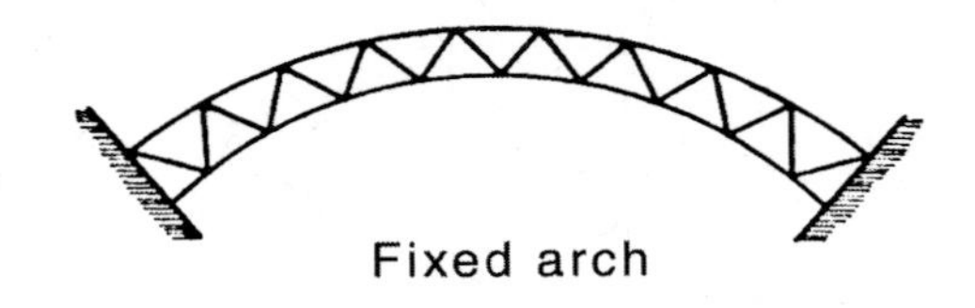

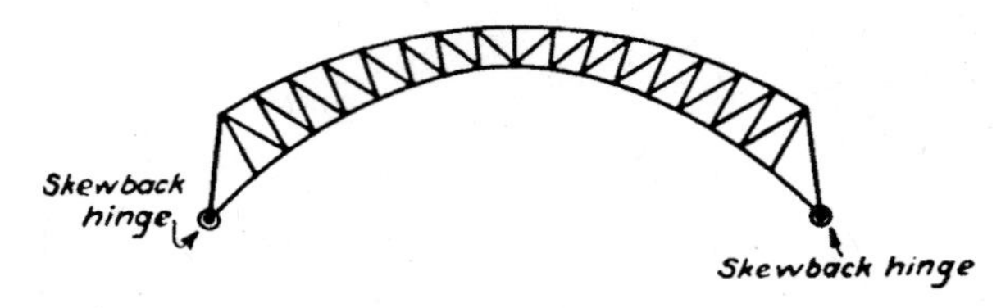

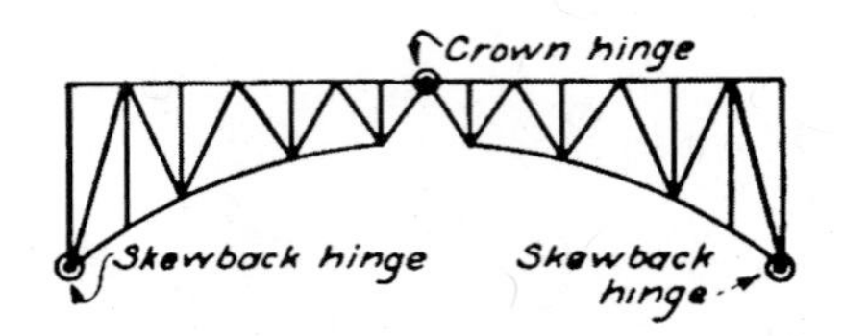

Figure 2.9. Types of bridge arches. Reprinted from George A. Hool and W. S. Kinne, editors-in-chief, *Movable and Long-Span Steel Bridges*, 2nd edition (New York: McGraw-Hill Book Company, Inc., 1943), 359.

Originally used in railroad bridges, by 1890 the three-hinged arch had become an accepted structural form for highway bridges. The three-hinged arch features hinges at the two end points and the crown. Of the hinged types, this is by far by the most common.

Prominent three-hinged arch spans included the 540-foot Bellows Falls Bridge in Vermont (1905, now demolished), the Fairfax Bridge on the Carbon River in Washington state, and the 456-foot wrought-iron Lake Street Bridge in Minneapolis.

Metal arched bridges can also be classified according to the method in which the rib is fabricated and the deck carried by the rib.

Solid-Ribbed Arch

Solid-ribbed arches (Figure 2.10a) are constructed of curved plate girder ribs. The roadway is carried on posts resting on top of the arches or suspenders hung from the bottom. These arches may be either fixed or possess one, two, or three hinges. The two-hinged type may be either parallel curved or of the crescent type. The three-hinged type may be either parallel curved or lenticular in section. The most common form of solid webbed arch rib is that of a parallel curved rib of constant depth throughout its length.[29]

Brace-Ribbed Arch

In the brace-ribbed arch (Figure 2.10b) the solid web is replaced by a system of diagonal bracing. Either a single or a double-intersection web system may be used. Brace-ribbed arches consist of two parallel or nearly parallel chords connected by open webbing. Brace-ribbed arches can be constructed as hingeless, one-hinged, two-hinged, or three-hinged arches.[30]

Spandrel-Braced Arch

Spandrel-braced arches (Figure 2.10c) are constructed only as deck structures, with the roadway carried on top of the arch. The main arch consists of the curved bottom members. The roadway is carried by the horizontal top chord. Web trussing, generally Pratt trusses, connects the horizontal top chords. Arches of this type are usually constructed with two or three hinges due to the difficulty of adequately anchoring to avoid movement at the abutments. This type of construction consists of a fixed horizontal top chord, a curved or arched bottom chord, and a system of diagonal bracing connecting the two.[31] Often this type of bridge could be erected as a cantilever, thus eliminating the need for falsework and saving money in erection.

Bowstring Arch

Bowstring arches were used for relatively short spans, generally 100 feet or less. The construction of most of these bridges was simple, requiring little skill or equipment. This type of bridge was developed and patented by Squire Whipple in 1841. Its distinguishing feature was a polygonal top chord that arched upward from the bottom chord at the abutments and was designed to function as an arch as well as a truss member. The majority of variations on Whipple's original design proposed improvements to the fabrication of the arch chord and methods to simplify the joints between the web members and the chords. Most relied on simple crossed-diagonal web panels. Such spans were marketed extensively throughout the Northeast and Midwest

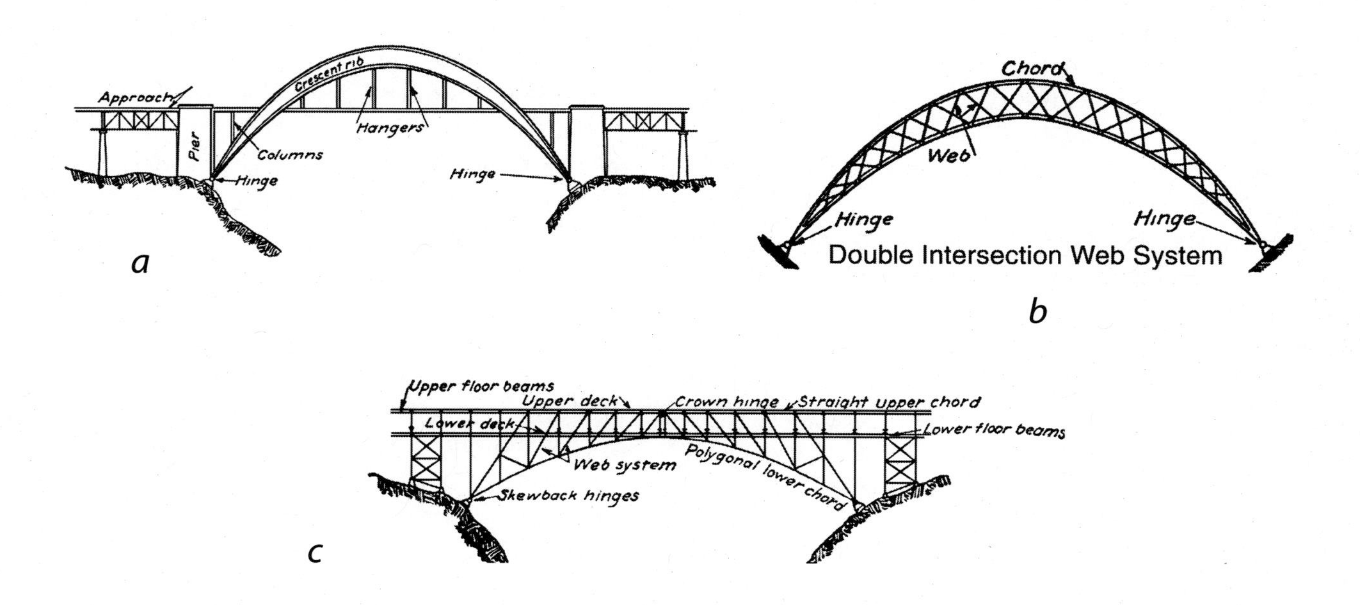

Figure 2.10. Configuration of bridge arches: *a,* deck span with solid-ribbed arch; *b,* braced-rib arch; and *c,* spandrel-braced arch (Hool and Kinne 1943: 360–361).

as pony truss spans by major fabricators such as the King Bridge Company and Wrought Iron Bridge Company.[32]

Metal Truss

The metal truss bridge was developed in response to the rapid growth of the country's road and rail network during the nineteenth century. This bridge form was adaptable to a wide variety of site conditions, its structural behavior could be analytically understood, and its prefabricated components made it simple to manufacture, ship, and erect.[33]

A truss is a framework composed of individual members configured such that loads applied at the joints produce only direct tension or compression. Every truss is a triangle or a combination of triangles. The members of the truss are fastened together either with pins or with plates and rivets.[34]

During most of the nineteenth century, American truss bridges used pin connections to hold the members together. This technology required that holes be drilled in the ends of the members, which were then aligned with one another. A cylindrical pin, similar to a metal dowel, was slipped through the opening to form a structural connection. Pin connections were popular because they allowed for speedy erection of trusses. They were, however, susceptible to loosening under shaking caused by fast-moving trains.

Riveted connections provided a solid, rigid means of joining together the truss members. Riveted connections could not be hand-driven easily in the field, thus limiting their use until the development of portable pneumatic riveting systems late in the nineteenth century. Warren trusses, rarely built with pin connections, became popular with the advent of field riveting.

Metal Bridge Truss Types

Metal bridge trusses may be divided into a number of standard types.

King Post Truss

The simplest truss type is the king post, a triangle with a vertical member extending from the center of the base to the upper vertex. First used to support bridges in the Middle Ages, the king post could be used to support through spans of up to 60 feet. The king post is also the germ of other, more complex truss types.

Weight placed on the horizontal beam causes it to deflect and pull down on the king post, placing it in tension. The post, in turn, pulls down on the two diagonals, placing them in compression and transmitting the force out to the abutments. These trusses were originally constructed of wood or of metal pipe.[35]

Queen Post Truss

When spaces to be spanned were too wide for the king post truss, a second parallel post was added, with a top chord connecting the upper ends of the two posts. The queen post was useful in spanning distances up to 80 feet.[36]

Waddell or A Truss

This truss design was patented by nineteenth- and twentieth-century bridge engineer, J. A. L. Waddell. It consisted of a large subdivided triangle with two, smaller, subdivided interior triangles (Figure 2.11a). Described by Waddell as the most rigid, pin-connected bridge ever built, it was frequently used prior to the development of the modern riveted Pratt truss bridge.[37]

Pratt Truss

The Pratt truss was patented by Thomas Pratt and Caleb Pratt in 1844 (Figure 2.11b). The top chord and verticals act in compression, while the diagonal members act in tension. Originally built with wood compression members and iron tension members, the type, later executed entirely in metal, continued to be widely constructed well into the twentieth century. It was typically constructed with pin-connected joints. Never a complicated structure, it was adaptable to a wide variety of situations.

The Baltimore and Ohio Railroad used Pratt trusses almost exclusively from 1880 to 1905. Bridge engineer Waddell claimed that the Pratt truss was the most commonly used truss type for spans under 250 feet. Many surviving truss highway bridges employ Pratt trusses; they are typically spans of 160 to 200 feet, using an inclined upper chord (camelback) truss. The Pratt truss is one of two truss types widely used in twentieth-century rail and highway spans in the United States.[38]

Double-Intersection Pratt Truss

The double-intersection Pratt truss, also known as the Whipple, Whipple-Murphy, or Linville truss, added more diagonals to the basic Pratt truss (Figure 2.11c). The diagonals extend across two panels but keep the parallel top and bottom chords of the simple Pratt profile. Squire Whipple's double-intersection truss was patented in 1847 and was first used several years later on the Saratoga and Rensselaer Railroad near Troy, New York.[39] The double-intersection Pratt truss was widely used for long-span railroad bridges. The truss was most commonly used in the trapezoidal form with straight top and bottom chords, although bowstring trusses were also built.[40]

Lowthrop Truss

This truss, developed by F. C. Lowthrop, suggests a cast-iron, pin-connected version of the Pratt truss (Figure 2.11d). The truss could be used in spans up to 150 feet and was used for the 11-span 1,122 foot Cataraugus & Fogelsville Railroad Bridge in Pennsylvania.[41]

Fink Combination Truss

Albert Fink and Wendell Bollman were assistants to Benjamin H. Latrobe, chief engineer for the Baltimore and Ohio Railroad. The patented Albert Fink design (Figure 2.11e) has wood compression members and wrought-iron web and lower chord members in tension. They were designed by railroads so that the iron members could be used again in the construction of new iron bridges. These bridges typically range from 75 to 150 feet in length. Like the Bollman, the Fink is formed by superimposing a number of inverted king-post trusses. Unlike the Bollman, these trusses are of varying spans and may be of varying depth.[42]

Fink Suspended Truss

The suspended (suspension) truss, shown in Figure 2.11f, was adopted by the B&O Railroad for bridges on new lines west of the Cumberland Gap. Fink's truss, patented in 1854, consists of seven king-post elements: one long king post, under which are two smaller king posts whose diagonals reach from each abutment to mid-span, and which segments are further divided into two smaller king posts. The Fink truss could be expanded with as many as 15 inverted king-post elements to span longer distances. The trusses were used in lengths from 75 to 206 feet. In 1980, two Fink truss bridges were known to survive: one near Zoarville, Ohio, and the other in West Lynchburg, Virginia.[43]

Bollman Truss

Wendell Bollman's patented"suspension truss bridge" is two bridges in one: a Pratt truss with counters and vertical end posts, upon which are superimposed a system of wrought-iron rods radiating diagonally from the vertical end posts to the foot of each king post (Figure 2.11g). The inverted king-post trusses are of uniform depth and span. These rods support the floor beams and deck much as the hangers of a suspension bridge do. The design was used extensively on the B&O Railroad from 1850 to 1875. Bridge length typically ranged from 75 to 100 feet.[44] The only known surviving Bollman truss bridge, formerly used to carry the B&O Railroad, now sits in a park in Savage, Maryland, between Washington, DC, and Baltimore.

Post Diagonal Truss

S. S. Post's initial patented diagonal-truss iron bridge was constructed on the Erie Railroad in 1865 (Figure 2.11h). Many other bridges of this type were erected between 1865 and 1880. The lower ends of the post are inclined half a panel toward the ends that form the characteristic single triangle at mid-span. The inclined iron posts form one system; diagonals of iron bars and rods with eyes at each end running counter to the posts formed two more systems. These bridges were typically constructed in lengths from 100 to 300 feet.[45]

Howe Truss

Developed by William Howe, this truss has its vertical members in tension and its inclined web members in compression (Figure 2.11i). The upper and lower chords were commonly constructed of timber, while the vertical tension members were either iron or steel rods. The diagonals were constructed of wood. The Howe truss was once popular in regions where timber was plentiful, but it has all but disappeared except in the Pacific Northwest.[46]

Parker Truss

The Parker truss, developed by C. H. Parker, is a Pratt truss with an inclined top chord (Figure2.11j). This chord resulted in a substantial reduction of material for long spans.[47]

Pennsylvania (Petit) Truss

The Pennsylvania (Petit) truss is a Parker truss modified by introducing sub-struts or sub-ties to resist or

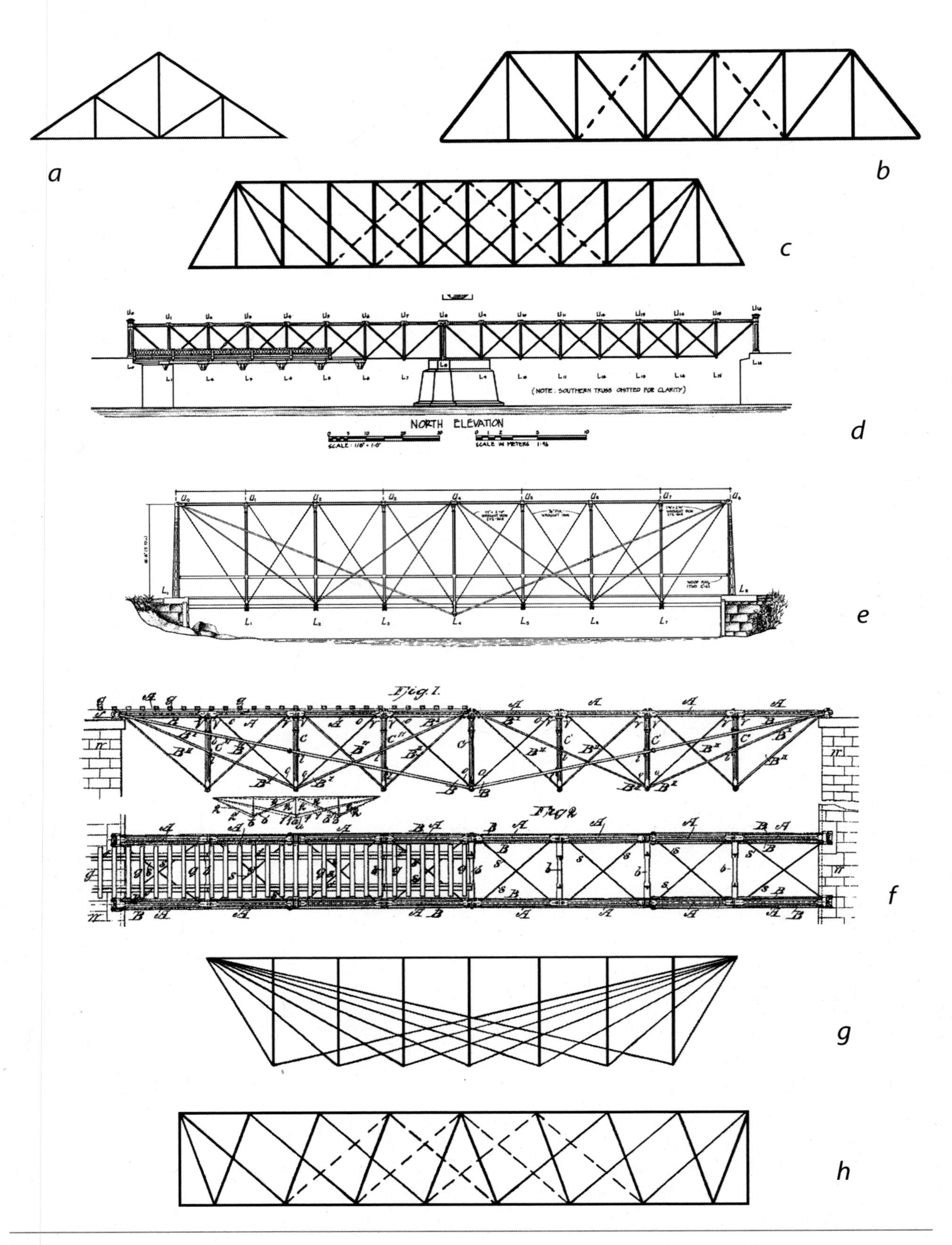

Figure 2.11. Metal bridge trusses. ***a,*** A truss. Reprinted from J. A. L. Waddell, *Bridge Engineering* (New York: John Wiley and Sons, Inc., 1916), 478. ***b,*** Pratt truss (Waddell 1916: 468). ***c,*** Double-intersection Pratt or Whipple truss (Waddell 1916: 472). ***d,*** Lowthrop truss. West Main Street Bridge, 1870. HAER NJ-19, National Park Service. Delineated by Carolyn Givens, 1985. ***e,*** Fink truss. Fink Through-Truss Bridge. Flemington vicinity, New Jersey. HAER NJ-18, National Park Service. Delineated by Lori A. Allen, 1985. ***f,*** Fink suspended truss (Patent No. 10,887). ***g,*** Bollman truss (Waddell 1916: 473). ***h,*** Post diagonal truss (Waddell 1916: 473).

transmit stress (Figure 2.11k). In basic form, it is a Baltimore truss with inclined upper chords. This truss type was developed in the 1870s by engineers of the Pennsylvania Railroad.[48]

Baltimore Truss

The Baltimore truss modifies the basic Pratt configuration by adding additional, auxiliary members, like the Pennsylvania truss but without the inclined upper chord (Figure 2.11l). It employs parallel chords in which the main panels have been subdivided by an auxiliary framework. Like the Pennsylvania truss, the Baltimore truss was developed by engineers for the Pennsylvania Railroad in the 1870s.[49]

K-Truss

The K-truss employs a web system in which the diagonal members intersect the vertical members at or near mid-height (Figure 2.11m). The assembly in each panel forms a letter "K." The K-truss, more economical than the Pennsylvania truss, has smaller secondary stresses and was relatively simple to erect the field. Though used infrequently, it was first employed in the cantilever and anchor arms of the Quebec Bridge over the St. Lawrence River, completed in 1917, designed by prominent bridge engineer Ralph Modjeski. Riveted and with a curved top chord, the K through truss became a familiar sight on the primary roads of Oklahoma in the 1930s, with spans of 140 to 210 feet.[50]

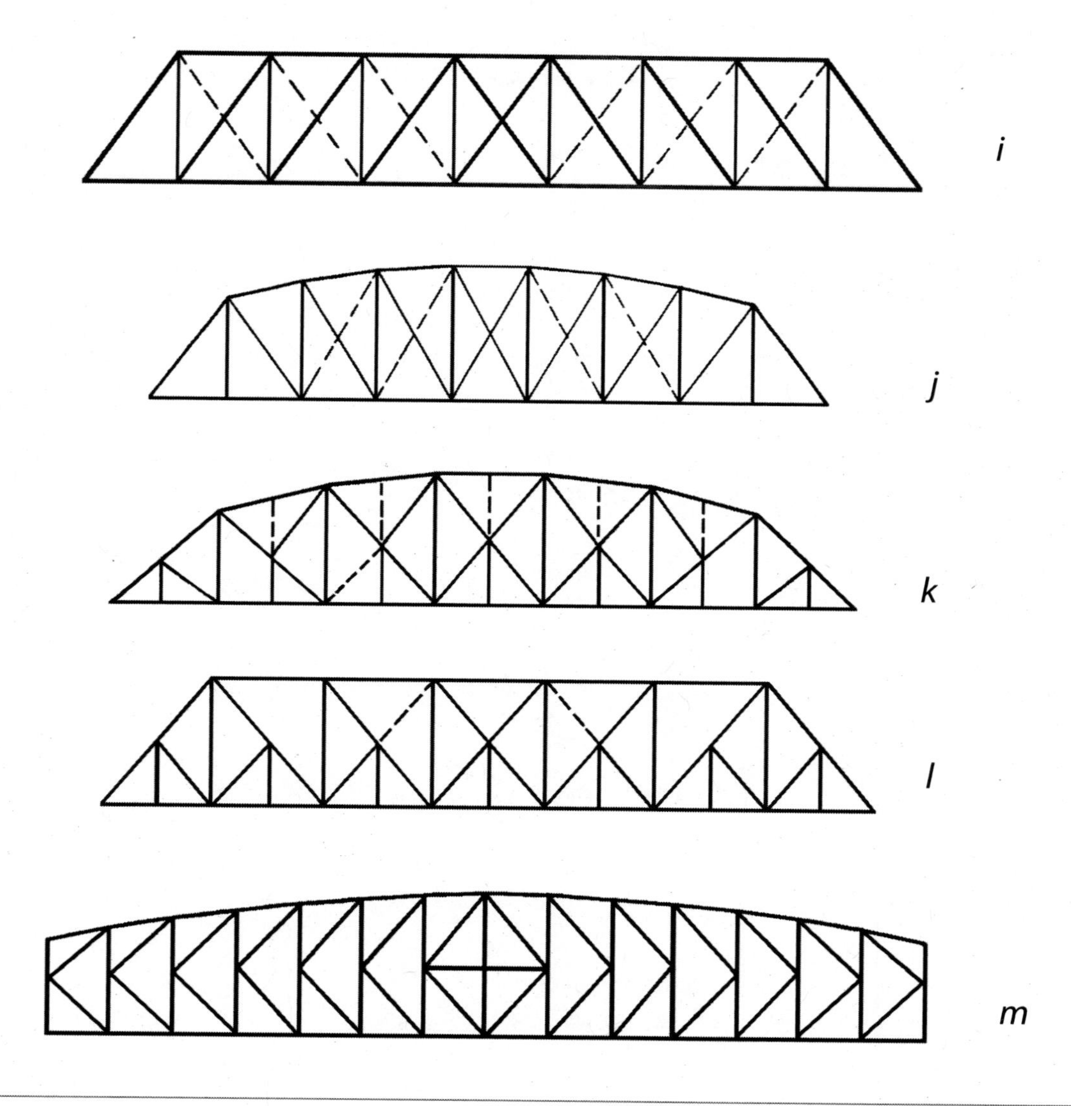

Figure 2.11 ***(continued)***. Metal bridge trusses. ***i,*** Howe truss (Waddell 1916: 473). ***j,*** Parker truss (Waddell 1916: 469). ***k,*** Pennsylvania (Petit) truss (Waddell 1916: 470). ***l,*** Baltimore truss (Waddell 1916:469). ***m,*** K-truss (Waddell 1916: 479).

Lenticular Truss

This truss derives its name from the curved upper and lower chords (Figure 2.11n). The basic truss form is like that of a Pratt truss. This truss was developed in part by Friedrich August von Pauli, who published details of his truss design in 1865, and in part by William O. Douglas of Binghamton, New York, who patented his initial design in 1878. Pauli's most famous American design was Pittsburgh's Smithfield Street Bridge. In the latter part of the nineteenth century, the Berlin Iron Bridge Company of East Berlin, Connecticut, manufactured and erected almost 800 lenticular truss bridges in the United States. These bridges are sometimes referred to as pumpkin-seed bridges, elliptical truss bridges, or parabolic truss bridges because of their lens shape.

The use of riveted plates, angles, and channel sections to build the top chords and vertical posts is characteristic of these bridges. The bridges were designed as either through or pony trusses. In comparison with typical modern bridge sections, the components of a lenticular pony truss are relatively light and could have been handled by workers at the site. For example, a 14.75-foot-long upper chord would weigh about 640 pounds, and an 8- foot-long vertical post would weigh about 130 pounds. [51]

Truss-Leg Bedstead

This truss type circumvents the need for hard rock abutments, which were not found in the open prairies and plains reached by the railroads in the 1880s (Figure 2.11o). This truss extends the vertical end posts of a Pratt truss down below grade where they are attached to a "deadman," a heavy timber buried in the ground. The weight of the timber, combined with the weight of the soil and roadbed, is sufficient to keep the bridge from moving.[52]

Warren Truss

The Warren truss was patented in 1848 by two British engineers, James Warren and Theobald Monzoni. The truss was originally formed by equilateral or isosceles triangles (Figure 2.11p). Later modifications included subdivision by verticals (subdivided Warren truss) and the addition of alternate diagonals (double Warren truss). The Warren truss was widely used throughout much of the United States from about 1860 to the early twentieth century. It was typically constructed with riveted joints and was used to span distances up to 400 feet.

Warren trusses of riveted construction, modified with vertical members, became the standard long-span bridges on many railroads during the early 1900s and replaced the Pratt truss for shorter spans. Deck Warrens with vertical end posts were popularly used on elevated railroads in cities. The Warren truss was never as widely used for road bridges, as were the Pratt and Whipple double-intersection trusses. More modern forms survive in bascule bridges. The Warren truss is one of two truss types widely found in twentieth-century rail and highway spans.[53]

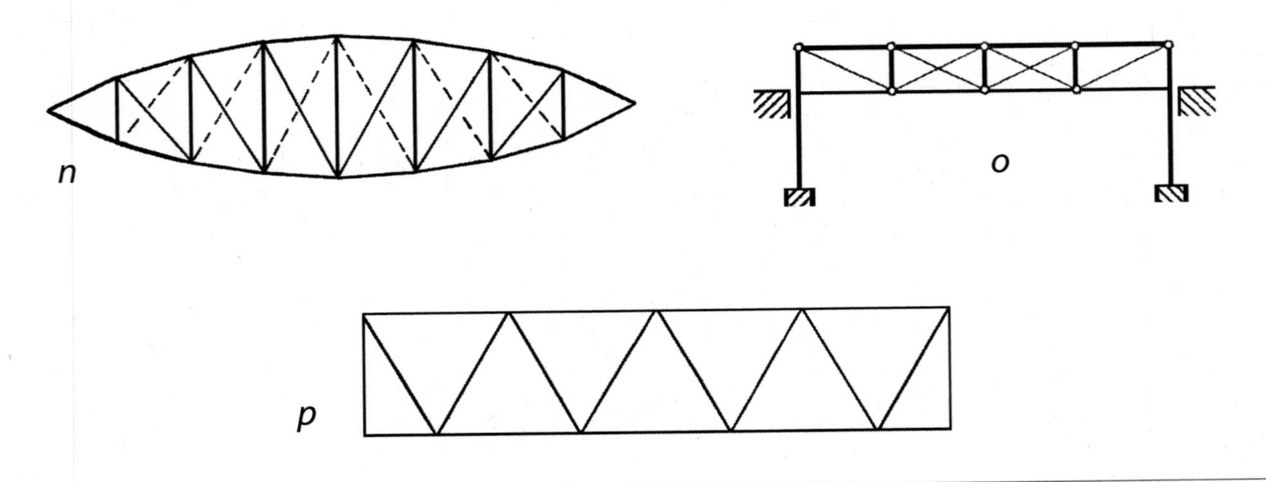

Figure 2.11 (*continued*). Metal bridge trusses. ***n,*** Lenticular truss (Waddell 1916: 474). ***o,*** Truss-leg bedstead. From Milo S. Ketchum, *The Design of Highway Bridges of Steel, Timber and Concrete* (New York: McGraw-Hill Book Company, 1920), 106. ***p,*** Warren truss (Waddell 1916: 472).

Warren Quadrangular Truss
This truss is a double-intersection Warren truss in which two series of equilateral triangles intersect to form a lattice-like configuration (Figure 2.11q).[54]

Wichert Truss
This truss type was developed by E. M. Wichert of Pittsburgh, Pennsylvania, in 1930 and was called the Automatically Adjustable Continuous Bridge. It is a continuous truss with a chord and web configuration that continues uninterrupted over one or more intermediate supports (Figure 2.11r), as compared with simple supported trusses, which are supported only at each end. Wichert's first major truss bridge was the 1937 Homestead High Level Bridge over the Monongahela River near Pittsburgh.[55]

Thacher Truss
The Thacher truss was patented in 1884 by Edwin Thacher, a pioneer in the development of steel bridge reinforcement and prolific designer of reinforced concrete bridges. Thacher served as an engineer for Andrew Carnegie's Keystone Bridge Company in the 1870s. The Thacher truss represents a hybrid between the double-intersection Pratt and Warren designs (Figure 2.11s). It is distinguished by diagonal compression members located at the center of each span and diagonal tension members that extend from the connection between the top chord and the inclined end post. By the 1980s, only two single-span examples and one two-span example were known to remain.[56]

Metal Truss Bridge Forms

Through Truss
In most truss bridges, the deck is located inside the main trusses, with floor beams attached to the bottom of the verticals. When the deck is located near the bottom chord, the truss is called a through. In a through truss bridge (Figure 2.12a), entry to the bridge is generally by means of a portal; the truss is visible to either side of the vehicle or pedestrian. A through truss bridge generally measures from 80 to 200 feet or more.

Pony Truss
A through truss bridge with insufficient height for upper lateral bracing is termed a pony truss (Figure

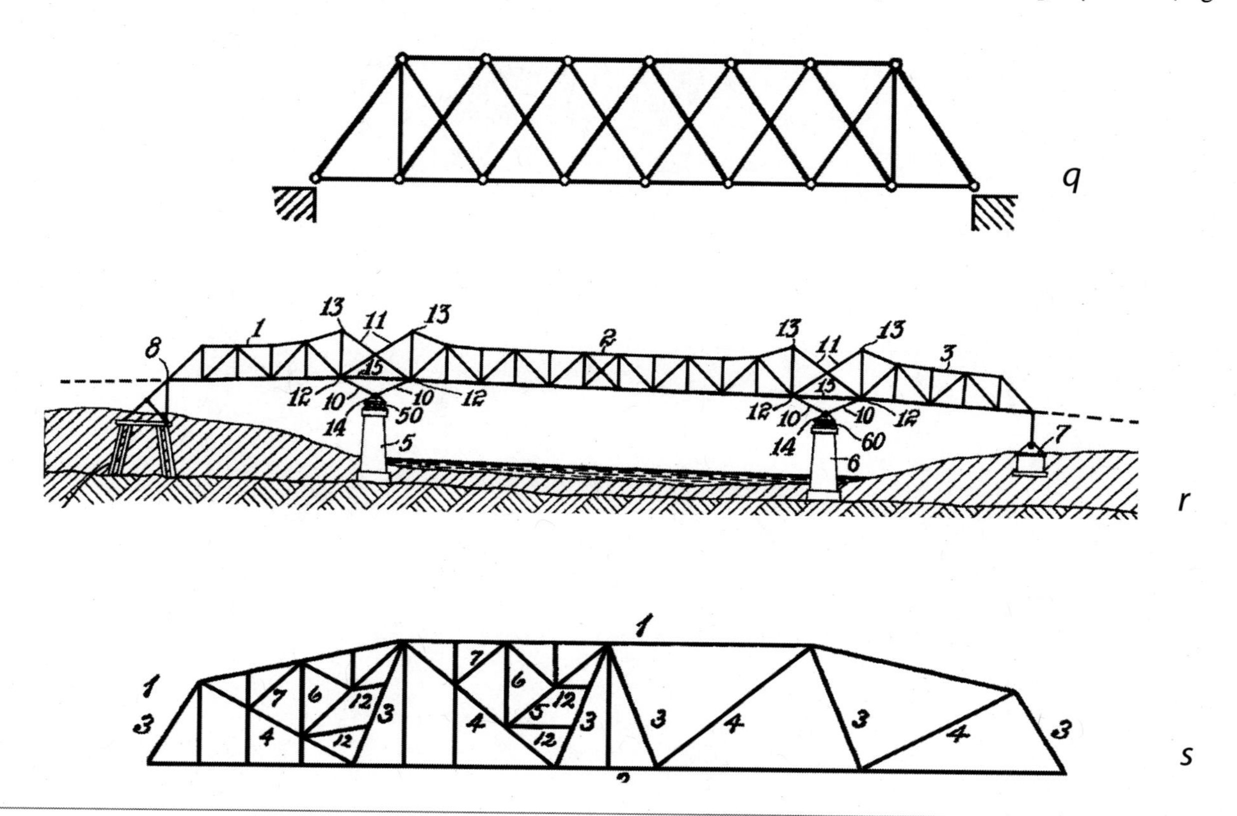

Figure 2.11 (*continued*). Metal bridge trusses ***q,*** Warren quadrangular truss (Ketchum 1920). ***r,*** Wichert (Patent No. 2,079,095); ***s,*** Thacher (Patent No. 310,747).

a

b

c

Figure 2.12. Truss configurations. ***a,*** Through truss. Richmond Bridge, Richmond, Texas. HAER TX 11-7, National Park Service. Photograph by Jay Storr, May 1989. ***b,*** Pony truss. Fannin County Road 222 Bridge, Dial, Georgia. HAER GA 58:2, National Park Service. Photograph by Sammy Fowler, June 1985. ***c,*** Deck truss. Jefferson Street Viaduct over Des Moines River, Ottumwa, Iowa. HAER IA-86:7, National Park Service. Photograph by Bruce A. Harms, 1996.

2.12b) or half-through truss. Pony truss bridges generally range from 30 feet or less to 80 feet.

Deck Truss
A truss bridge that carries the deck on its upper chord is termed a deck truss (Figure 2.12c). The deck truss is the least common type of older metal truss bridge.

Steel Trestles

A steel trestle is used to carry roadways or railroad lines (Figure 2.13) at a considerable distance above the ground. The tower and intermediate spans are commonly built of plate girders, whether the trestle carries a railroad or a highway roadway. The tower consists of two trestle bents braced together by longitudinal bracing.

CONCRETE BRIDGES

Concrete Arch Bridges

One of the first concrete arch bridges in the United States was built of unreinforced concrete in Prospect Park, Brooklyn, New York, in 1871. The first reinforced concrete arch bridge was built in San Francisco's Golden Gate Park in 1889.[57] The first parabolic concrete arch bridge in the United States was W. J. Douglas's unreinforced arch to carry 16th Street over Piney Branch in Washington, DC.[58] Concrete arch bridges in parks or cities were typically designed with voussoirs of molded concrete and rusticated or brush-hammered spandrel walls.

With the introduction of metal reinforcement in concrete and the lightening of arch ribs, concrete arches became flatter and longer spans were possible. Multi-arch bridges could employ smaller piers between arches of the same span. The equal horizontal thrust of the arches allowed them to act as buttresses, canceling much of the other's load.

Most large reinforced-concrete bridges in the United States are arch structures (Figure 2.14). Arch bridges may be divided into two types: deck arches, where the roadway lies atop the arch, and rainbow (through) arches, where the arch extends above the roadway. Deck arches were first popularized in the United States in the 1890s by Fritz von Emberger and Edwin Thacher. Later Daniel Luten constructed numerous examples of the type, reportedly as many as 17,000. Luten's bridges are characterized by curved, inscribed solid parapets.[59]

Open Spandrel Arch
The spandrel is the area between the bottom of the arch and the roadway deck. Open spandrel bridges have pierced spandrel walls. In its lightest structural shape, the open spandrel arch does not contain fill material. For spans greater than about 100 feet, open spandrel arches were often substantially less expensive to construct than closed spandrels.

Closed Spandrel Arch
A closed spandrel bridge is one in which the spandrel area is completely filled. The filled spandrel arch consists of a barrel arch that contains rubble or other fill

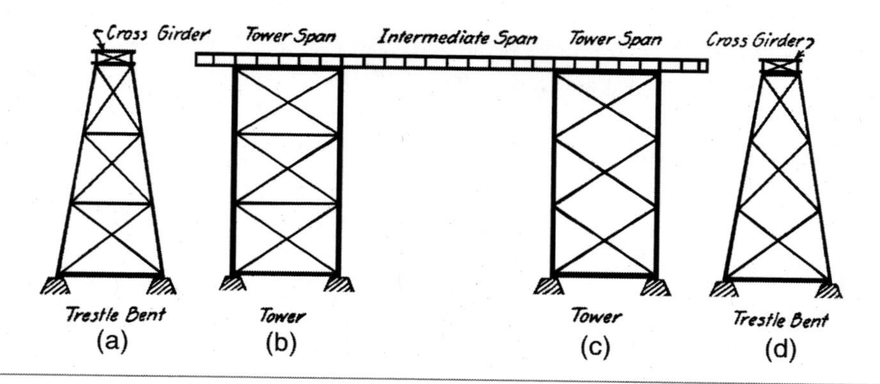

Figure 2.13. Railway steel trestle. Reprinted from Milo S. Ketcham, *The Design of Highway Bridges of Steel, Timber and Concrete*, 2nd edition (New York: McGraw-Hill Book Company, 1920), 110.

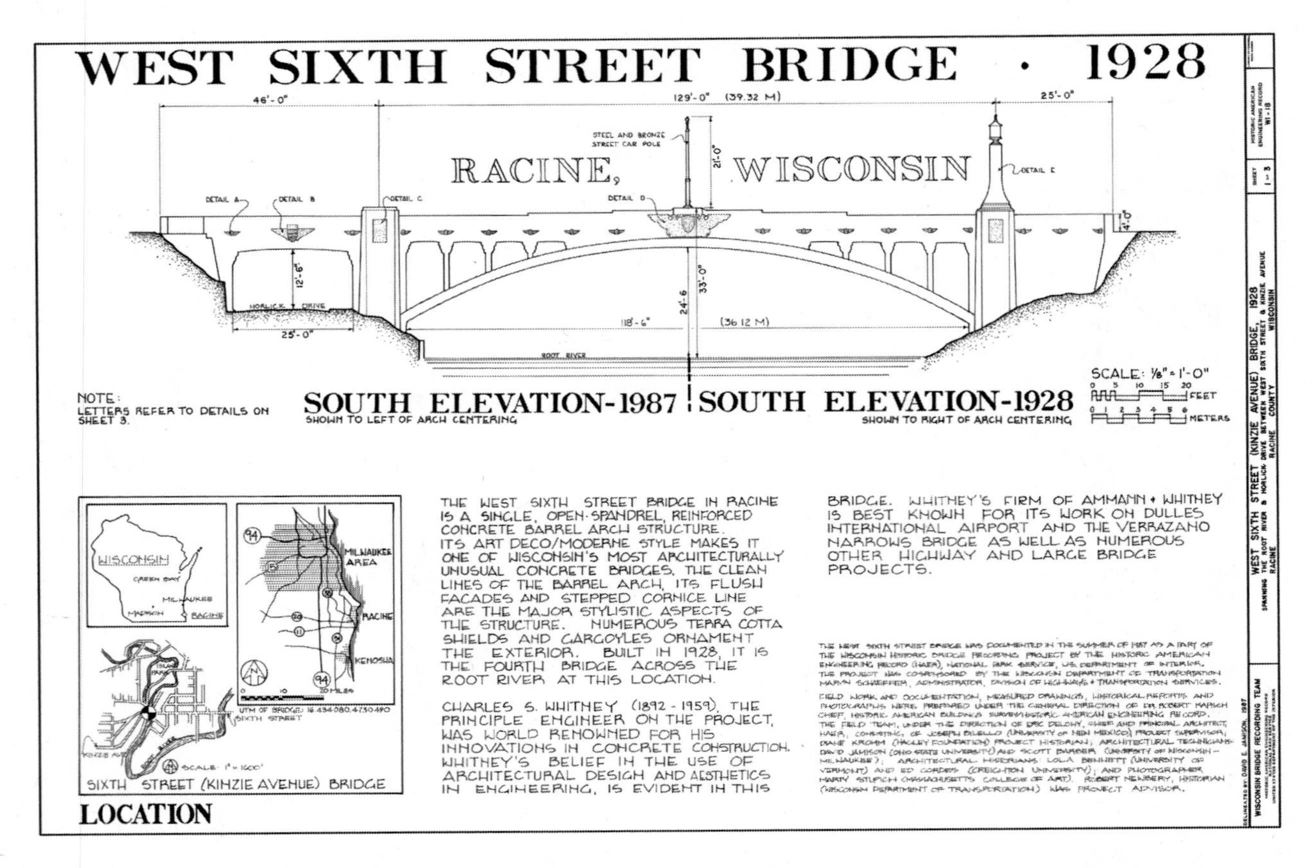

Figure 2.14. Example of reinforced concrete arch bridge. West Sixth Street Bridge, Racine, Wisconsin. HAER WI-18, National Park Service. Delineated by David E. Jamison, 1987.

material and terminates in walls that act as retaining walls for the fill.[60]

Ribbed Arch

One method of reducing the amount of concrete in arch bridges involves the form of the arch itself. Instead of building a solid arch extending the width of the bridge, the arch can be divided into a series of parallel ribs that function as separate arches.

Prominent examples include the three-ribbed Key Bridge between Washington, DC, and Virginia, the two-ribbed Tunkhannnock Creek (Nicholson) and Hop Bottom viaducts (Figure 2.15) in Pennsylvania, and the two-ribbed George Westinghouse Memorial Bridge in western Pennsylvania.

Rainbow Arch

The best known type of concrete through arch bridge, commonly known as a rainbow arch, was patented by James B. Marsh (1856–1936), owner of the Marsh Engineering Company in Des Moines, Iowa, in 1912. Most rainbow arch bridges were constructed in the Midwest, although examples exist in the eastern states. In this bridge type, the deck is suspended by vertical suspenders from arches that extend above the roadway. The arches act in compression, but the hangers carry the floor in tension, necessitating significant amounts of reinforcement. However, the design reduced the amount of concrete needed as compared with that of a closed spandrel arch.[61]

Concrete Frame, Slab, and Girder Bridges

The concrete slab (Figure 2.16) or concrete girder bridge is one of the most common bridge designs in the United States. The simplest design consists of a single reinforced concrete slab with vertical supports. Some such concrete slab bridges, such as those on farm or ranch roads in the Texas Hill Country, lack guard rails, parapets, or railings, while more heavily

Figure 2.15. Detail of Hop Bottom Viaduct, Susquehanna County, Pennsylvania, showing ribbed arches. Photograph by the author, 2005.

Figure 2.16. Concrete slab bridge. Escalante River Bridge, Garfield County, Utah. HAER UT-80, National Park Service. Photograph by Michael R. Polk, February 1994.

used bridges were usually built with railings or parapets. The slab is anchored to the supports by vertical steel dowels. To counteract tension in the slab, longitudinal reinforcement is added to the main reinforcement.

The slab bridge's advantage over other concrete bridges is simplicity in design and reduced material and labor costs in construction and reinforcement. Slab bridges require more concrete than steel and girder bridges of the same span, but because of simple and inexpensive formwork, they could be erected at a relatively low cost. In the 1920s, concrete slab bridges were found economical for spans up to 30 feet, with the majority from 10 to 25 feet.[62]

Continuous Slab Bridges

In the early 1900s, contemporary with advances in concrete slab building construction, the continuous slab carried on mushroom columns was used in bridge construction. The first span of this type was built in 1909 in St. Paul, Minnesota, to carry Lafayette Avenue over the tracks of the Soo Line. A mushroom column is a cylindrical column with a funnel-shaped capital. A drop slab is a flat slab with a perpendicular panel projecting from its underside. Later, where moderate loads were expected, the mushroom capital and drop slab were discarded, leaving the cylindrical column and a slab of uniform depth.[63] Among the largest bridge structures to use the mushroom capital and drop slab is the upper deck of Chicago's Wacker Drive.

Concrete Box Girder Bridges

The first reinforced concrete box girder bridge in the United States was constructed in 1937. Four states constructed them before 1950, and twenty-six states had constructed them by 1960. This type of construction was particularly popular in California, where 3,100 reinforced concrete box girder bridges and 1,100 prestressed concrete box girder bridges had been designed and built by 1977.

By 1977, about 90 percent of all reinforced concrete box girder spans in the United States on state highways were less than 100 feet in length. Approximately 40 percent of California's prestressed concrete box girder bridges had span lengths exceeding 150 feet.

Reinforced concrete box girders have been built with spans ranging from 30 to 235 feet, with the typical range from 50 to 150 feet. Simple spans are limited to a length of about 110 feet, while longer spans are prestressed. The most economical girder spacing for ordinary box girder bridges ranges from about 8 to 12 feet.

The deck of a concrete girder bridge has two primary functions: to support the live load on the bridge and to act as the top flange of the longitudinal girders. Deck reinforcement ordinarily requires four layers of reinforcing steel. Most decks range from 6 to 9 inches in thickness.

Box Girder Bridge Elements

The primary elements of a box girder bridge include the soffit, the end diaphragm, the bent cap, and the hinge. The soffit functions as a compression flange for negative girder movement; it contains the positive girder reinforcement and also constitutes an architectural feature. The end diaphragm is typically used between the girder stems at abutments and piers. Bent caps are used to connect the tops of the supporting bents and are typically constructed of precast concrete. To improve the appearance of the bridge, it is preferable to keep the bent caps within the limits of the box girder superstructure so that the bridge is a single plain surface interrupted only by the columns. In very long bridges, hinges are required to accommodate temperature fluctuations. One of the most common types of box girder hinges uses two steel angles, placed back to back, as the horizontal sliding surface. A thin sheet of asbestos, lead, or other suitable material is used to lubricate the surface and facilitate sliding.

T-Beam Bridges

T-beam bridges were constructed of cast-in-place reinforced concrete beams with flanking integral monolithic deck sections. The main reinforcing steel was placed longitudinally at the bottom of the beam stem and deck; reinforcing was placed transversely to the stem. The bridges feature a series of reinforced concrete beams integrated into the concrete slab to form a monolithic mass that appears in cross section like a series of uppercase *T*s connected at the top. The T-beam bridge was introduced nationally around 1905 as a more efficient use of material than a slab bridge for spans over about 25 feet. In this design the deck thickness, longitudinal beam size, and spacing are proportioned to produce strong, economical sections. T-beam bridges were commonly constructed until the late 1960s. In numerous states, including Pennsylvania and New Jersey, the T-beam bridge is ubiquitous, particularly on state highways.[64]

Deck Girder Bridges

Concrete deck girder bridges are adapted to spans between 25 feet to 60 feet where headroom is not limited (Figure 2.17). They are especially suited to viaduct construction.

Three types of floor arrangements are used in deck girder bridges: girder and slabs, where concrete slabs span between longitudinal girders; girder, floor beam, and slab, in which the slab is supported by floor beams spanning between the longitudinal girders; and girder, floor beam, and two-way slab, where the panels of the floor slab are supported along four edges by the floor beams and longitudinal girders and are reinforced by bars placed in two directions at right angles to each other.

Since the exterior girders generally carry appreciably smaller loads than the interior girders, these outer members can be narrower than the interior members. In the simplest arrangement, pipes and ducts are either suspended or bracketed under the slab. In other bridges, pipes and ducts are carried in a cavity beneath a thin sidewalk slab.

Lateral spacing of the longitudinal girders dictates the cost of the bridge. Closely spaced girders mean thinner slabs and a larger number of main girders; wide spacing means thicker slabs but a smaller number of girders.

Through Girder Bridges

Through girder bridges are structures in which the main longitudinal girders extend above the roadway. The through girder bridge is well adapted to spans of 25 to 65 feet with a roadway of not more than 20 feet in width. They are used mainly where the headroom is too small for deck girder designs.

The floor construction may consist of a solid slab spanning between the main girders; closely spaced floor beams spanning between the girders with the slab supported by the floor beams; or widely spaced floor beams with one or more lines of longitudinal beams spanning between the floor beams and a slab divided into panels

Figure 2.17. Detail of La Verkin Creek Bridge, La Verkin vicinity, Utah, showing concrete deck girder construction. HAER UT-81, National Park Service. Photograph by Kim A. Hyatt, 1994.

supported on four sides and reinforced in two directions. Main reinforcement of the slab consists of bars placed near the bottom at right angles to the girders.[65]

Prestressed Concrete Girder Bridges

In reinforced concrete girder bridges, the steel reinforcing bars are placed in the lower part of the girders in order to withstand tensile stresses. Usually reinforcing bars are simply embedded in the structure during construction and placed in tension. It became possible to reduce the amount of material in a girder by "prestressing" (stretching) the steel reinforcing members so that they were in tension before placement of loads on the girder. The first such bridge was Philadelphia's Walnut Lane Bridge (1950), constructed in consultation with Belgian engineer Gustav Magnel. Steel reinforcement for each girder consists of four prestressed steel cables. Because of prestressing, the main girders of this bridge require a depth of only 6 feet, 7 inches.[66]

Flat-Slab Bridges

In flat-slab bridges the floor construction consists of a reinforced-concrete slab extending in four directions and supported directly by isolated reinforced-concrete columns without the aid of beams. In their heyday, flat-slab design was suited both for highway and railway bridges. For multi-span structures of spans up to 40 feet, flat-span offered the most economical solution to the problem. In all cases, the cost of formwork and vertical supports was appreciably lower than that of any other type. Flat slabs are monolithic, with the supporting columns eliminating expensive expansion bearings.

A flat-slab bridge consists of some or all of the following elements: continuous slab; drop panels at columns; column heads; columns; spandrel beams; and footings for columns. A drop panel is a thickening of the slab at the column, used in most flat-slab bridges for reasons of economy. A column head is the flaring out at the top of the column in the shape of a truncated cone or pyramid. Interior columns may be round, square, or octagonal in cross section. Exterior columns usually have square or rectangular cross sections. The strength of a flat-slab construction depends upon the rigidity of the columns, particularly the exterior columns. Spandrel beams are used to carry the railing and to strengthen the exterior of the slab.

Precast Concrete Short Span Bridges

Precast concrete manufacturers offer a variety of short-span bridge systems (Figure 2.18). Designs include single-piece arches, two-piece arches, and three-sided boxes. Each system has its own limitation on span

Figure 2.18. Precast concrete short-span bridge. Photograph from Terre Hill Concrete, Lancaster, Pennsylvania. Reprinted with permission.

length, span height, and load capacity. Finishes commonly available are colored smooth-as-cast, textured formliner, exposed aggregate, acid etch, brick, and sand blasts. These bridges are manufactured well in advance of installation and can be set onto a bridge foundation in a matter of hours using a small crew and crane.[67]

MOVABLE SPAN BRIDGES

Movable bridges can be divided into several major types: swing or revolving; folding or jack-knife; bascule; semi-lift bascule; and vertical lift.

Swing or Revolving Bridges

Swing bridges were the most common movable spans in use in the nineteenth and early twentieth centuries (Figure 2.19). The earliest swing bridges were constructed of wood and were put in motion by the approaching vehicle, usually a boat traversing a narrow waterway.

Swing bridges turned about a vertical axis. They were generally constructed of steel plate girders, open-webbed, riveted girders, riveted trusses, or pin-connected trusses. Deck, pony, and through trusses were all constructed. By the early twentieth century, riveted trusses predominated, eliminating the problem of wear on the pins and pinholes due to bridge movement. The bridge could be supported on a simple pivot, a drum turning on rollers on a circular track, or a combination of the two.[68]

A swing span bridge rotated on a central pier and rested in a position perpendicular to the roadway, thus opening two channels for passing marine traffic. The operator's house was usually placed above the roadway at the center of the span. Disadvantages included the time needed to open and close the bridge and the obstruction of the pivot pier.

Swing Bridge, Center-Bearing

The earliest swing bridges were center-bearing, wooden structures (Figure 2.19a). The centers were usually constructed of cast iron and in some cases were fitted with steel disks. In a center-bearing bridge the span's weight is supported on the center pivot. Since the center pivot wears with use and is expensive and difficult to replace, steadying elements were frequently overloaded.

As the technology improved, center-bearing bridges became a preferred type. Their advantages over rim-bearing bridges were that they approximated a fixed span; the main trusses or girders were ordinary two-span continuous structures; they were simpler; they were easier to erect and adjust; and their center pier could be smaller.

When the bridge was open, each truss was supported at the end by a cross girder that rested on a center bearing. This bearing was usually a phosphor bronze disk, 1 to 3 feet in diameter, between two hardened steel disks. To prevent the bridge from tipping, six to eight balance wheels (18 inches in diameter) were fastened to the trusses and the floor system.

Swing Bridge, Rim-Bearing, or Turntable-Bearing

Early in the nineteenth century, English engineers built the first rim-bearing swing bridges. Among the earliest rim-bearing bridges in the United States was the original Rush Street Bridge across the Chicago River built in 1856. In a rim-bearing bridge the span's weight is supported on small roller bearings or wheels that run on tracks a short distance from the center (Figure 2.19b). Solely rim-bearing bridges had the disadvantage of requiring skill in construction and delicate adjustments during erection. Repair work was costly.

The trusses in a rim-bearing bridge are supported by a large circular girder, which rotates with the span. The girder rests on conical rollers. Long, heavy spans were designed as either rim-bearing or combination rim-bearing and center-bearing. In the rim-bearing bridge type, both dead and live loads at the center are carried by the drum, the roller upon which it turns, the track on which the rollers run, and by the masonry near its outer circumference. The result is that the mechanical parts carry a much larger portion of the total load than the pivot of a center-bearing bridge.

Shear-Pole Swing Spans

The least expensive and least common type of swing span is the shear-pole swing span (Figure 2.19c). When navigation traffic was small and opening infrequent, particularly in locations where navigation was closed in winter, the design proved satisfactory. Most shear-pole bridges were erected on small, single-track railroads, such as the one that stood at the opening of the Barge Canal in Burlington, Vermont.

The shear poles are usually constructed of steel. The centers are of the simple-pivot, disk type. Dead and

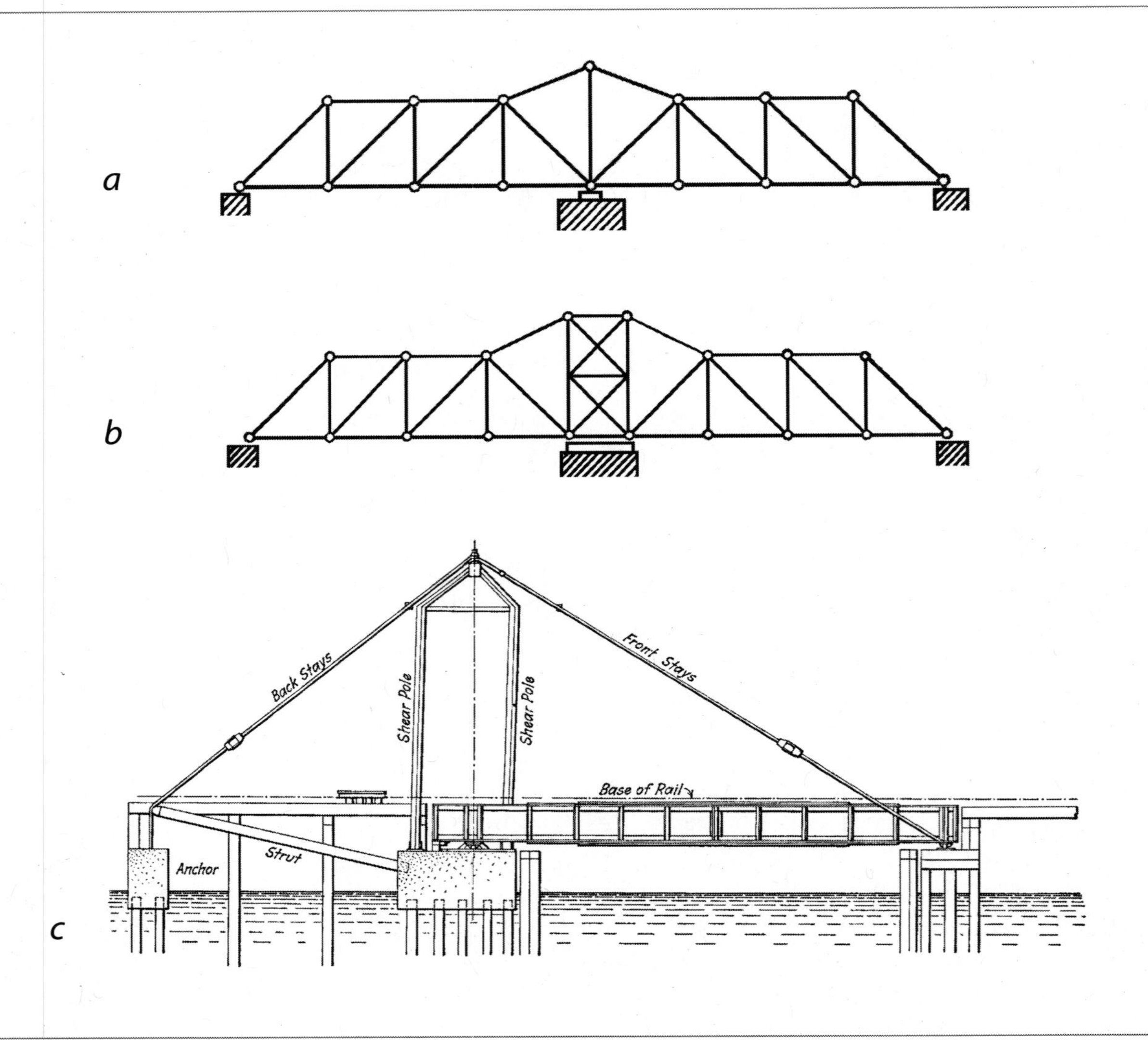

Figure 2.19. Types of movable spans: ***a,*** center-bearing swing bridge (Ketcham 1920: 110); ***b,*** rim-bearing swing bridge (Ketcham 1920: 110); ***c,*** shear-pole swing span. Reprinted from Otis Ellis Hovey, *Movable Bridges* (New York: John Wiley & Sons, Inc., 1926).

live loads are delivered to the pivot by cross girders. Two transverse balance wheels steady the span when swinging. The free ends of light bridges are supported by rollers, which run on an inclined track until they register against stops on the pedestals. The center of the pivot on the shear pole is set slightly eccentric with respect to the pivot on the pier so that the free end of the bridge is lifted a little when it moves open. The most common method of operation is by hand winches.

Folding or Jack-Knife Bridge

The jack-knife bridge was used for railroad bridges. It consists of a deck girder under each rail, one or more needle beams under the free end, and a gallows frame over the pivots, braced and anchored back to the shore; with ropes, or rods, extending from the needle beams to the tops of the gallows frame. The last few remaining examples of this type were used on railroads in New England.[69]

Bascule Bridges

The earliest type of bascule bridge was a single span, hinged or trunnioned at one end, moving in a vertical plane about the trunnion by means of an outhaul line attached to the free end and running upward and inward to the source of power. The ancestry of this bridge can be traced back to the medieval drawbridge. The earliest examples were small, constructed of timber, and hand-operated.

During the late nineteenth and early twentieth centuries, these bridges were developed in numerous patented types, characterized broadly as roller lift, trunnion, and pivot. The bascule bridge, because it rises from one end, was desirable when a large, clear channel was necessary. Bascule bridges can be single leaf or double leaf. In the double leaf, the moving leaf is shorter and lighter, and the counterweight arm may be shorter and the size of the counterweight reduced.

Bascule-type vertical-lift bridges were generally more expensive than other types. They had the advantage of requiring no cables and sheaves and being able to be used to retrofit existing immovable span bridges.[70]

Roller Lift Bascules: Scherzer and Rall Types

The principal commercial examples of the roller lift bascule are the Scherzer and Rall types, the former designed by the Scherzer Rolling Lift Bridge Company and the latter under patents controlled by the Strobel Steel Construction Company.[71]

Between 1893 and 1921, the Scherzer Rolling Lift Company was granted twelve patents for variations in their rolling lift bascule design. A Scherzer bridge is characterized by its large concrete counterweight and segmental circular moving girder (Figure 2.20a).[72] In the Scherzer bascule the leaf rotates on the quadrant which rolls along horizontal track girders. The center of gravity of the leaf is at the center of this quadrant and moves in a horizontal line as the bridge opens. A counterweight is attached to the short arm projecting shoreward so that the leaf is maintained in balance at all positions. The operating machinery has only to overcome inertia and the friction of the moving parts. The span is operated by a pivot working in a rack pivoted to the upper part of the quadrant.

The Rall type, invented by Theodor Rall of Chicago, is the subject of patents 817,516 and 1,094,473 (Figure 2.20b). The bridge superstructure rotates about the center of gravity of the leaf where a pivot or trunnion rests in a roller carried by a horizontal track girder. When the leaf is closed, the main girder or truss bears on the pin fixed to the pier, and the roller is slightly raised off the track girder. As a result, the load on the bridge is carried directly by the pin to the pier. The swing strut is connected at one end to the movable girder by a second pin and at the other end to the first pin. When the leaf rises, it first revolves around the first pin until the roller is in full bearing with the track girder. Then, as the opening continues, the roller moves horizontally on the track girder, while the second pin of the main girder describes an arc with the first pin as the center. The leaf is operated by the main pinion engaging a rack fixed to the strut which is pivoted to the girder.[73]

Trunnion Bascules: Brown, American Bridge, Page, Chicago, and Strauss

Among the leading types of trunnion bascule bridges were the Strauss (Strauss Bascule Bridge Company), Brown, American Bridge, Page, Chicago (Chicago Bascule Bridge Company), and the Waddell and Harrington types. The most common were the simple trunnion or "Chicago" type and the multiple trunnion or "Strauss" type.

Brown

In the late nineteenth century, the method of balancing a bascule leaf by means of a counterweight moving vertically was introduced. Thomas Ellis Brown, Jr., of New York used this idea to develop a patented type. The first Brown bridge constructed was that at Ohio Street, Buffalo, across the Buffalo River. The Brown type of bascule differs from the others primarily in the method of operation and in the application of its counterweights (Figure 2.20c).[74]

American Bridge (Abt)

The American Bridge bascule (Figure 2.20d) was developed in 1920 and was patented by Hugo A. F. Abt, assistant engineer for the American Bridge Company. The first bridge using this design was for the Wabash Railway across the River Rouge in Detroit, Michigan.

The counterweight of the bridge is in a steel box with arms hinged to the top of the tower. A compression link, pin-connected to the counterweight box, is articulated to a tension-operated link hinged to the rocker-link joint. A track in the tower, inclined about 30

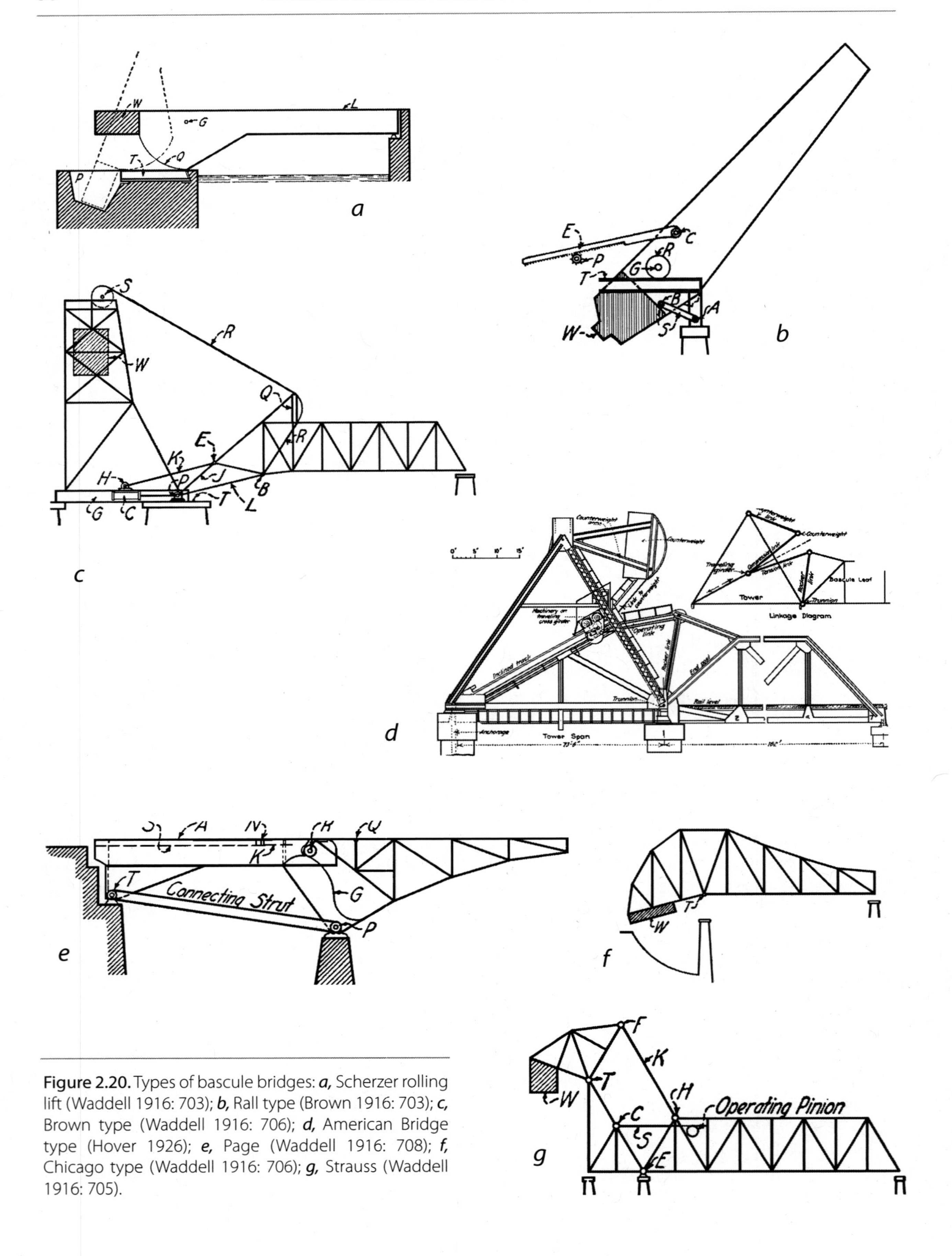

Figure 2.20. Types of bascule bridges: ***a,*** Scherzer rolling lift (Waddell 1916: 703); ***b,*** Rall type (Brown 1916: 703); ***c,*** Brown type (Waddell 1916: 706); ***d,*** American Bridge type (Hover 1926); ***e,*** Page (Waddell 1916: 708); ***f,*** Chicago type (Waddell 1916: 706); ***g,*** Strauss (Waddell 1916: 705).

degrees with a horizontal plane, bisects the angle between the compression and tension links forming a scissor joint. Power is applied by pinions engaging racks on the inclined track girders. As the counterweight and the operating-link joint rotate in arcs of circles, and power is applied along the bisector of the scissors-joint angle, an almost exact balance is attained in all positions of the moving leaf.[75]

Page
The Page bascule (Figure 2.20e) has the unusual feature of a tilting approach span designed for highway bridges. This approach span is used as a counterweight. In through railroad bridges the approach span is fixed, and a tilting counterweight is placed overhead.

The approach span pivots on trunnions at the shore end, while the free end is carried by rollers resting on specially curved track girders fixed to the main trusses of the bascule. As the leaf rises, the track girder rotates with it about the pin and causes the end of the approach span to drop also. This approach span is loaded so that it balances the weight of the leaf in all positions.[76]

Chicago
The Chicago bascule (Figure 2.20f) employs trusses supported on trunnions in line with the lower chord placed a short distance back from the center of gravity of the span. Counterweights are rigidly attached to the shore arm, and a pit is provided in the pier for reception when the bridge is opened. The leaf is operated by a pinion and segmental rack attached at the end of the short arm.

The entire weight of the leaf and counterweight is carried by the trunnions located approximately at the center of gravity of the mass. The first example of this type was the Clyburn Avenue Bridge in Chicago erected about 1899.[77]

Strauss
There were several bascule bridge designs developed by the Strauss Bascule Bridge Company, the most distinctive being the overhead counterweight type and the heel trunnion type. The Strauss Company also issued a design known as the Strauss Underneath Counterweight type, in which the principle is the same as that of the overhead but with the counterweight and link located underneath the roadway.[78]

In the overhead counterweight type, the counterweight is elevated at the rear of the short end of the leaf above the tail trunnion. The bascule leaf pivots upon the main trunnion placed to the rear of the center of gravity. In the heel trunnion type (Figure 2.20g), awarded Patent 1,211,639, the counterweight is connected with the counterweight frame which is pivot-mounted upon the fixed tower or supporting frame. The counterweight trunnion is placed forward of the counterweight in the bridge framework, while the main leaf trunnion is placed in the forward of two piers.

Bascule Bridge Machinery
A typical machinery layout for a simple trunnion, double-leaf bascule bridge was employed in the Michigan Boulevard bridge in Chicago. Each leaf was operated by two 100-horsepower motors located in a machinery room under the lower street level and behind the anchor pier. Pinions on extensions of the armature shafts engaged a master gear containing a bevel-gear equalizer on a transverse shaft, extending the entire width of the bridge. At each end a miter gear drove a longitudinal shaft, extending to the grouped reduction gearing, the first shaft of which was driven by a pair of miter gears. The bevel gear shafts turned in unison, bearing at their ends, and were supported at intermediate points by pedestal bearings. These gears were mounted on cast-steel union frames. The first and second shafts were each fitted with double-block brakes. The movements of the bridge were controlled from operating rooms at each end of the bridge.[79]

Direct-Lift Bascule Span

The direct-lift bascule is not properly a bascule, but rather a type of vertical-lift bridge. This design was patented in 1915 by Theodor Rall and was awarded Patent 1,140,316. These bridges are divided into two general types. The first type employs counterweighted lifting girders rotating about fixed trunnions supported by stationary towers (Figure 2.21a). The second employs counterweighted lifting girders attached to lift spans that roll on stationary girders (Figure 2.21b). These two types of bridges were manufactured by the Strobel Steel Construction Company and the Strauss Bascule Bridge Company.

The semi-lift bridge was used for spans requiring a headroom, when open, of 150 feet or less. This

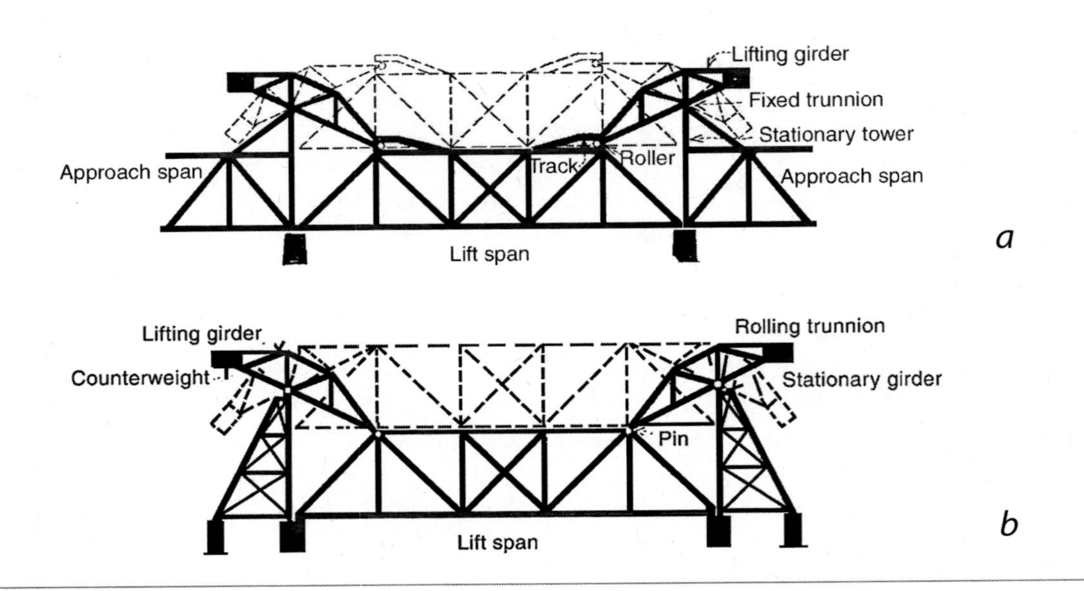

Figure 2.21. Rall direct-lift span: ***a,*** Type A; ***b,*** Type B (Hool and Kinne 1943: 26).

technology could also be applicable to modifications of existing fixed spans.[80]

Vertical-Lift Bridge

Vertical-lift bridges were developed much later than bascule and swing bridges. While a few were built in the early nineteenth century as canal crossings, widespread use of this type did not occur until the 1890s.

Hool and Kinne divided vertical-lift spans into three basic types: (1) bridges with a lifting span and no lifting deck; (2) bridges with a lifting deck and a fixed or a lifting overhead span; and (3) bridges of the bascule type.[81]

Vertical-lift bridges with a lifting span and no lifting deck can be subdivided into two subtypes: those with two towers comprised of four columns braced together in both directions (further subdivided into those with no overhead span and with an overhead span between the tops of the towers), and those with two towers of two vertical columns, with cross bracing in one direction and overhead trusses between the tops of the columns.

The majority of vertical-lift bridges have a lifting span and two towers of four columns each and lack a lifting deck. Bridges with two-column towers and overhead spans are more adaptable for short lift spans and lifts of moderate heights. Towers with inclined rear columns were preferred to those with all vertical columns due to increased structural stability. Vertical-lift bridges have been constructed with spans up to 425 feet in length, with lifts up to 140 feet, and with weights over 3 million pounds.

Very few vertical-lift bridges have been constructed with lifting decks. This type of bridge was used when two decks were required for traffic. The upper deck was usually designated for highway traffic and the other for railroad trains. In the case of a lifting deck and a lifting span, the lifting span had to be raised only for the passage of larger traffic, and thus traffic on the upper deck was infrequently interrupted.[82]

Waddell Type

The most common type of vertical-lift bridge was exemplified by the South Halstead Street Bridge in Chicago, completed in 1892, typical of those designed by J. A. L. Waddell. It was the first vertical-lift bridge of substantial size and importance to be constructed in the United States and had a span of 130 feet with a vertical lift of 140 feet. The bridge had a clearance of 15 feet in its lowered position.

Each tower had four sheaves at the top which turned on axles (Figure 2.22a). The counterweights were composed of cast-iron blocks and were held by steel cables, eight of which were fastened at each corner of the truss. Wrought-iron chains were used to counterbalance the cables. Eight up-haul and eight downhaul cables were employed. Pneumatic cylinders were used to stop the bridge in both the open and closed positions.[83]

Strauss Type

The main alternative to the Waddell type of lift bridge was developed by the Strauss Bascule Bridge Company. In this design, wire ropes and sheaves were avoided, but twenty auxiliary pin joints were introduced in addition to the four main trunnions (Figure 2.22b). This type of bridge was suitable only where the vertical lift for clearance required above the water was relatively small. The lift span was connected to a counterbalancing mechanism mounted on the two towers. The counterbalancing device at each end comprised a truss mounted on the main trunnion at the top of post, one end being pivotally connected to the hip of the main span through the hanger and carrying at its other extremity two independent concrete counterweights. The larger counterweight was pivot-connected to the truss at the base,

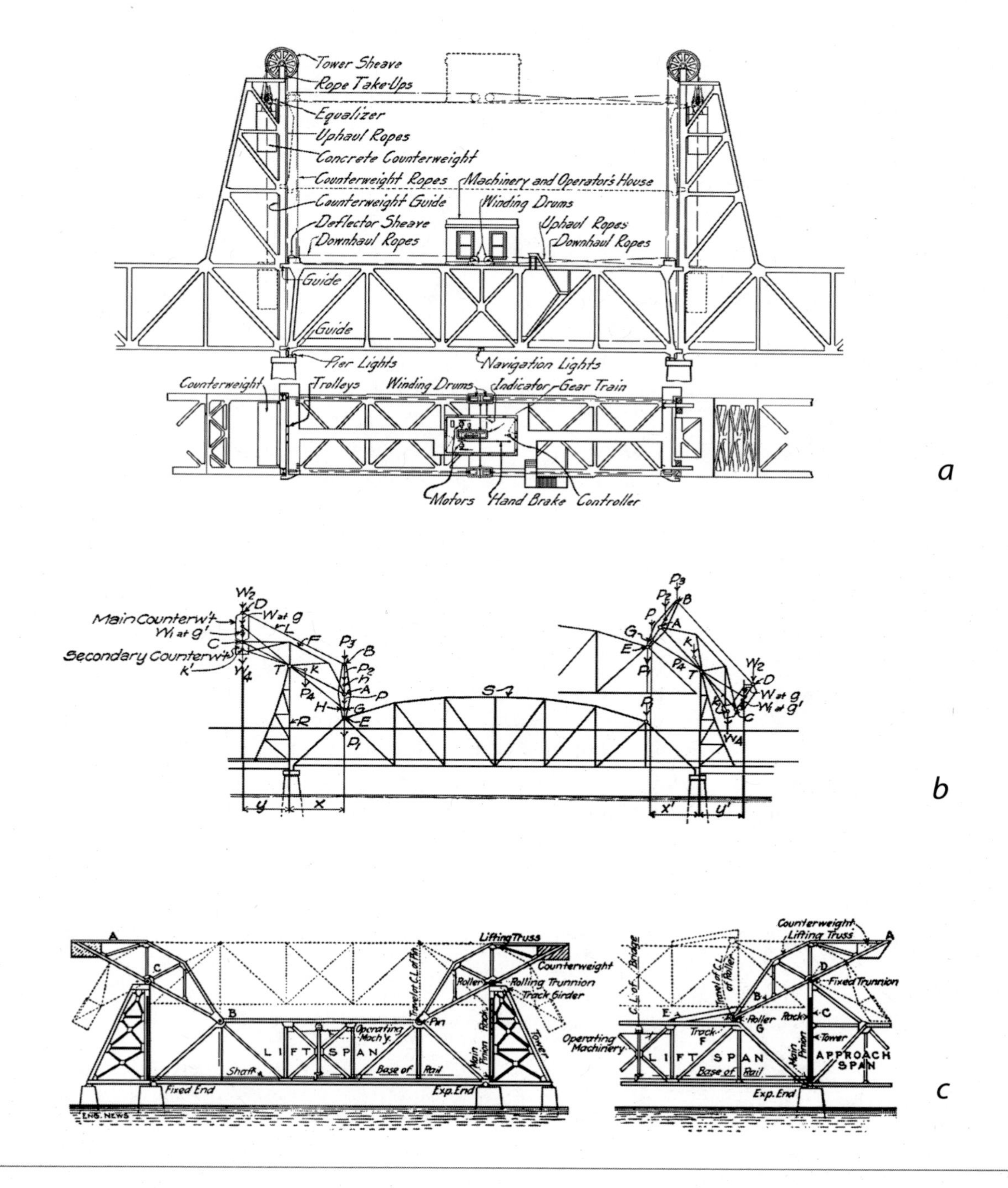

Figure 2.22. Types of vertical-lift bridges: ***a,*** Waddell type; ***b,*** Strauss type; ***c,*** Strobel type (Hovey 1926).

and at the top was pivotally connected to the upper extremity by means of the member, termed the counterweight link. The second counterweight was rigidly connected to the truss, and its only function was to bring the center of gravity of the truss into the fulcrum.

Strobel Type

The Strobel type (Figure 2.22c), designed by the Strobel Steel Construction Company of Chicago, was a method of balancing a vertical-lift-drawn span without cables, using principles similar to that of the Rall bascule design. A pair of counterweight trusses was mounted on each of the two towers. Each counterweight was proportioned to balance one-half of the weight of the lift span, when in closed position. At the end opposite the counterweight, the truss was articulated to the top chord at a panel point of the bridge truss. The trunnion turned in a roller adapted to move along a track on top of the tower. As the draw span was lifted, the trunnion rolled back, permitting point B to move in a vertical line. As a result, the counterweight and the pin each moved in an arc of a circle about the movable trunnion.

Components of Vertical-Lift Span Bridges

The major components of a vertical-lift span bridge include the truss, towers, guides and centering blocks, counterweights, sheaves, and machinery house and operator's house.

The lift-span truss was generally designed in the same way as an ordinary fixed-span truss, with the exception that suitable seating devices were needed at the ends of the lift span, as were means devised for fastening the cables. In addition, the trusses had to accommodate machinery, the machinery house, and the operator's house. The towers were generally composed of two vertical front and two inclined rear columns, well braced in both directions. Provision was made for fastening the sheaves on the vertical columns by suitable sub-posts or by a sheave girder. In this type of tower, the counterweights moved up and down inside the tower.[84]

Guides, usually roller guides, were attached at the eight corners of the lift span to keep it in line while it was being raised and lowered. Centering blocks were attached to the four lower corners to hold this span in place when it was in the lowered position. Counterweights for vertical-lift bridges were typically made of concrete cast on a steel framework and weighed about 5 percent less than the weights to be balanced. Movable weights equaling about 10 percent of the balanced weight were also provided. The inside face of the counterweight was provided with guides that engaged tracks on the tower. Counterweight cables consisted of steel strands around a hemp center.

The pitch diameter of the sheaves was equal to at least 60 times the diameter of the cable. Sheaves were constructed of either cast steel or were built up from structural steel and were fastened to its shaft by at least three keys.

The machinery for a vertical-lift span was usually placed in a machinery house atop the center of the lift span. In deck girder spans, the machinery was placed beneath the bridge floor and between the girders. If an operator did not stay in the machinery house, he usually had a small house with an unobstructed view of the water and the bridge traffic.[85]

CONTINUOUS SPAN

A continuous span bridge consists of a single truss extending over several piers. Conditions favorable to the use of the continuous type are long spans, good foundations, piers of moderate height, moderate truss depth, spans of approximately equal length, and cantilever erection. Advantages include economy of material, suitability for erection of one or more spans without falsework, rigidity under traffic, elimination of expensive and troublesome hinge details, and safety of the completed structure.

The continuous span was extremely rare until the last quarter of the nineteenth century.[86] By the 1930s, prominent continuous spans included the Queensboro Bridge in New York (1909), the Ohio River Bridge of the Chesapeake and Ohio Railroad at Sciotoville, Ohio (1914), the Chesapeake and Ohio Railroad Ohio River Bridge at Cincinnati (1928– 1929), the Lake Champlain Bridge (1931), the Cape Cod Highway Bridges in Bourne, Massachusetts (1935), and the Clinch River Bridge near Knoxville, Tennessee (1935).[87]

CANTILEVER BRIDGES

Most nineteenth-century American bridges were designed so that each span rested independently on its piers or abutments. Known as simple trusses, these bridges did not extend continuously over the piers. For bridges with long distances between piers, it was desirable to design the truss continuously over a pier, constructing a bridge that would cantilever beyond the piers. Following World War I, this type of bridge was widely used for highway and railroad crossings. Cantilever bridges were always inferior in rigidity to simple truss spans and were nearly always more expensive.[88] Although the cantilever principle is very old, the modern cantilever bridge is a development of the continuous bridge.

Examples of cantilever bridges include the Passaic and Hackensack River spans of New Jersey's Pulaski Skyway, the Commodore Barry Bridge between New Jersey and Pennsylvania at Chester, Pennsylvania, the Lewis and Clark Bridge in Longview, Washington, Pennsylvania's Beaver Bridge over the Ohio River, the Thebes Bridge (Illinois), the Monongahela Bridge in Pittsburgh, Pennsylvania, the Huey P. Long Bridge in New Orleans, the Grace Memorial Bridge in Charleston, South Carolina, the Carquinez Strait Bridge in Vallejo, California, the Bridge of the Gods in Cascade Locks, Oregon, and the Coos Bay Bridge in North Bend, Oregon.

Cantilever truss bridges may be erected with falsework under the main spans only. Such bridges may be made as through, deck, or half-through. By the early twentieth century, the practical length of such bridges was up to 2,000 feet in length. Anchorages at the shore ends of the anchor arms usually consist of eyebars extending down into abutments and attached to girder platforms embedded in the masonry. Some cantilever bridges, such as the San Francisco-Oakland Bay Bridge and the Pulaski Skyway, employ a swinging truss, an anchor truss with curved chords that provides a transition between the deck and through truss or the approach and main span.[89]

FLOATING BRIDGES

Floating bridges, or pontoon bridges, have been used since antiquity. More permanent floating bridges are largely a product of modern times. Permanent floating bridges have been employed when the water is so deep that bridges supported on piers are infeasible.[90] Modern U.S. permanent floating bridges include the Governor Albert D. Rosellini-Evergreen Point Bridge, completed in 1963, which spans Lake Washington in Washington state, carrying Route 520 from Seattle to Medina. At 7,578 feet in length, it is the longest floating bridge in the world. Other U.S. floating bridges include the Homer M. Hadley Memorial Bridge, Mercer Island-Seattle, Washington (completed 1989), the Hood Canal bridge in Washington (completed 1961; rebuilt 1982); the Lacey V. Murrow Memorial Bridge, Seattle-Mercer Island, Washington (completed 1993), the Eastbank Esplanade pedestrian bridge, Portland, Oregon (2001), and the wood-framed Floating Bridge, Brookfield, Vermont (originally built 1820; rebuilt 1884; pontoons replaced 1978).

The design of the original Lacey V. Murrow Bridge was typical of floating bridges. The structure of the bridge consisted of the pontoons, the cables securing the pontoons to the underwater anchors; the anchors themselves, of which there were three types; the sliding draw pontoon near the center of the pontoons; and the transition span between the approaches and the pontoons. The pontoons were constructed of reinforced concrete.[91]

SUSPENSION BRIDGES

The first true stiffened suspension bridge in the United States was the Niagara wire-cable bridge, built by John A. Roebling between 1851 and 1855. The stiffening trusses were 18 feet deep and designed to carry a single-track railway on the top of the structure. It spanned 821 feet between the centers of the towers. The railroad floor was suspended from two cables, and the highway floors from two other cables. The cables were each composed of 3,640 iron wires. The towers, originally of masonry, were 60 feet high, 15 feet square at the base, and 8 feet square on top.

Cable Systems

The earliest modern suspension bridges consisted of cables anchored at each end with vertical tension members, called suspendors, attached to the roadway deck for support (Figure 2.23). As the suspension bridge evolved, several types of cable systems were developed, most notably chain links, twisted wire ropes, and twisted strand cables.

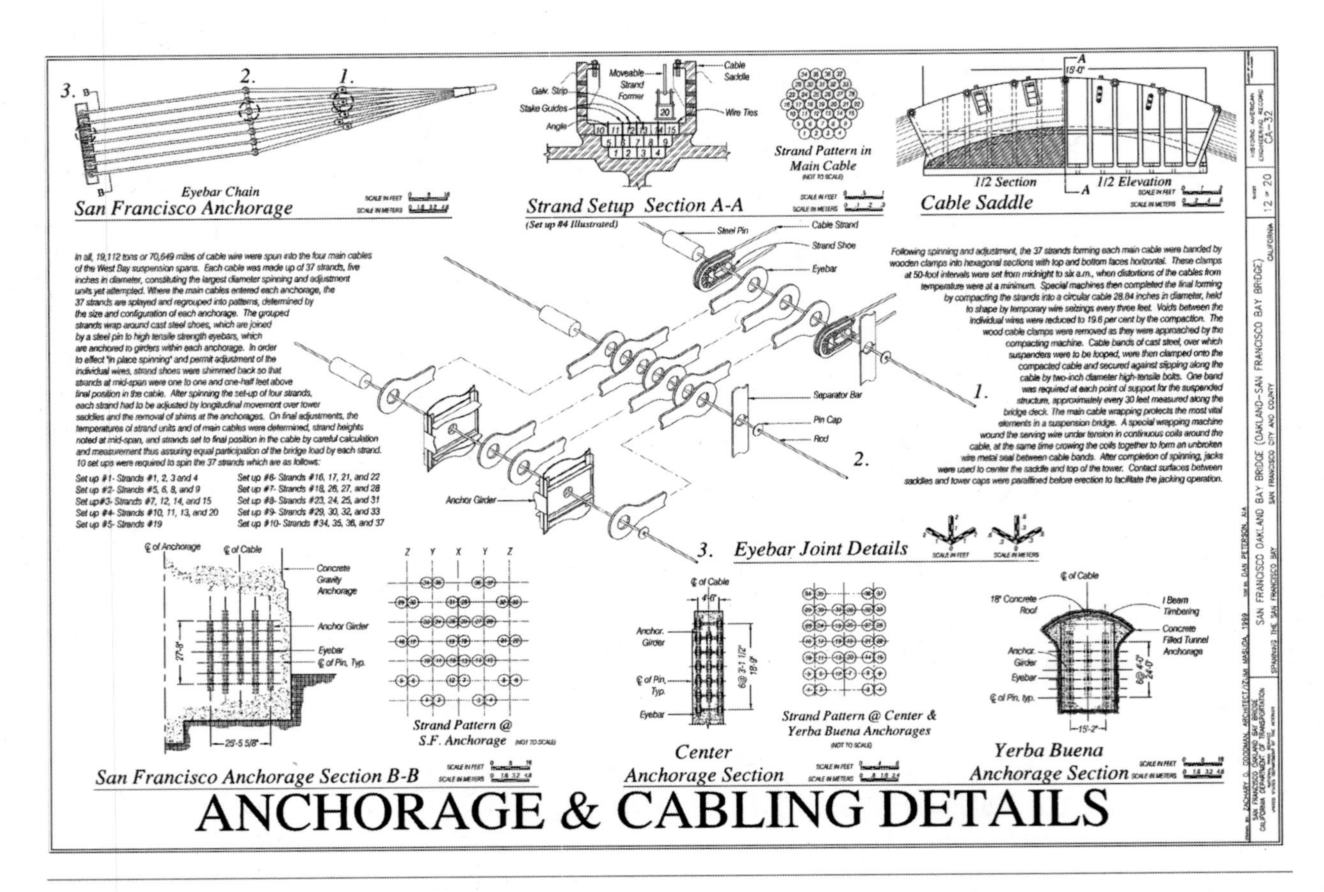

Figure 2.23. Saddle, cable, and anchorage details, suspension bridge. From San Francisco-Oakland Bay Bridge, San Francisco, California. HAER CA-32, National Park Service. Delineated by Zachary D. Goodman, Zumi Masada, 1999.

A suspension chain consists of pin-connected forged eyebars. Earlier designs used structural steel, while later designs used nickel steel.[92]

Twisted wire ropes were an alternative to parallel wire cables for spans up to 600 to 700 feet.[93] These ropes were prepared by mills (such as New Jersey's Roebling Mill) and were shipped on drums. They were hauled across the spans by mean of light temporary carrier cables and secured at the anchorages. Each rope end was expanded and fixed in a steel socket. Twisted wire ropes generally possessed considerable flexibility, particularly when successive layers of wires had alternate directions of twist.[94] By the 1920s, parallel wire cables had been made in diameters up to 21 inches.

In erecting parallel wire cables, the individual wires are strung in place between the anchorages. When the desired number is reached, they are bundled together in a strand. Cables generally consist of 7, 19, or 37 strands, separated into component strands at the anchorage. The strands diverge slightly to make room for the "strand shoes," grooved castings of horseshoe shape around which the respective strands are looped. The shoes are pin-connected to the anchorage eyebars.

Towers

Suspension bridge towers are constructed of masonry or, more frequently, steel. In masonry towers, such as those on the Brooklyn Bridge, the tower may consist of shafts rising from a common base beneath the roadway and connected together at the top with Gothic arches.

In steel towers, the tower generally consists of a column or leg for each suspension system. For lateral stability, the tower legs are braced together by cross-girders and cross-bracing or by arched portals. Steel tower columns are constructed of plates or angles, either open or closed in cross section. Horizontal diaphragms placed at intervals stiffen the section. The cross section grows toward the base. With high towers, individual legs may be made of braced tower construction, with each leg consisting of four columns spreading apart toward the base and connected with cross-bracing.[95]

Saddles

Saddles, located on top of the towers, serve as resting places for the cables. The saddle may rest on rollers to permit longitudinal movement in response to changes in cable length, or it may be bolted to the tower. In the latter case, the tower takes up the movement, either by bending, if fixed at the base, or by pivoting, if hinged at the base.

Tower saddles of long-span bridges, such as New York's Brooklyn and Williamsburg bridges, are generally provided with rollers, although New York's Manhattan Bridge uses fixed saddles. Objections to movable saddles are uncertainty in roller operation, tendency to clog or rust, and need for frequent maintenance.

Where chains are used instead of cables, the saddle support is generally of the rocker type: the entire tower acts as a rocker or else is anchored with a smaller rocker on its top. Such a rocker is pin-connected to the tower at its lower end and to the eyebar chains at its upper end.[96]

Most suspension bridges consist of two sets of towers over which the main cables are draped. This configuration divides the bridge into a main span and two side or anchor spans.[97]

Anchorages

At the anchorages, cable strands loop around shoes that are pin-connected to the anchor chains. The chains extend in straight, broken, or curved lines to their final pin-connection with anchor plates, girders, or grillage. Twisted wire rope cables can be anchored directly without the use of eyebar chains. The ends are secured in sockets that bear against the anchor girders. The anchorage masonry serves the function of taking up the pull of the cable or chain and transmitting it to the foundation.[98]

CABLE-STAYED BRIDGES

Cable-stayed bridges (Figure 2.24) have largely replaced traditional suspension bridges for the spanning of wide bodies of water. In this bridge type, tension members extend from one or more towers at varying angles to join with the abutment and piers to carry the deck. An advantage of the cable-stayed bridge is that it can be erected with no falsework for the main span.

The first modern cable-stayed bridge, designed by West German engineers for a bridge at Stromsund, Sweden, was completed in 1955. The earliest modern cable-stayed bridge in the United States was Alaska's Sitka Harbor Bridge, with a maximum span length of 450 feet.[99] Among recently completed cable-stayed bridges in the United States are the Delaware Route 1 Chesapeake and Delaware Canal Bridge; the 1,546-foot-long Cooper River Bridge in Charleston, South Carolina; the 1,378-foot U.S. 82 Bridge in Greenville, Mississippi; the 1,300-foot Dame Point Bridge in Jacksonville, Florida; the 1,250-foot Sidney Lanier Bridge in Brunswick, Georgia; and the 1,250-foot Houston Ship Channel Bridge in Baytown, Texas.[100]

Cables are composed of steel wire strands that are generally anchored at an anchorage head by button-heads bearing on a steel plate. The space in front of the anchor plate is filled by a mixture of small steel balls and an epoxy material. The number of cables used has varied from one to twenty. When a single cable is used, the bridge girder must be much larger than when multiple cables are used.

Several different cable patterns have been used in cable-stayed bridges. The radiating pattern consists of cables radiating from top or near the top of the tower, and the harp pattern (Figure 2.24a:2) consists of evenly spaced cables running up the tower. Other patterns include the fan (Figure 2.24a:3) and the star (Figure 2.24a:4).[101]

Bridges have been built with both single and double vertical planes of cables. Single-plane cables lie on the bridge center line and connect to a single pylon tower. Such bridges must have a torsionally rigid box girder. Double-plane cables connect to paired, load-bearing pylons.

The form of the tower is another identifying element of cable-stayed bridge types. The choice of single-plane or double-plane cable systems dictates the use of a single-pylon tower (Figure 2.24b:1), a double- pylon tower (Figure 2.24b:2), or an A-frame tower (Figure 2.24b:3). In a double-pylon tower, the pylons may be tied together at the top to form a portal (Figure 2.24b:4). Cables can be terminated at the tower or can be carried over a tower saddle and terminated on the

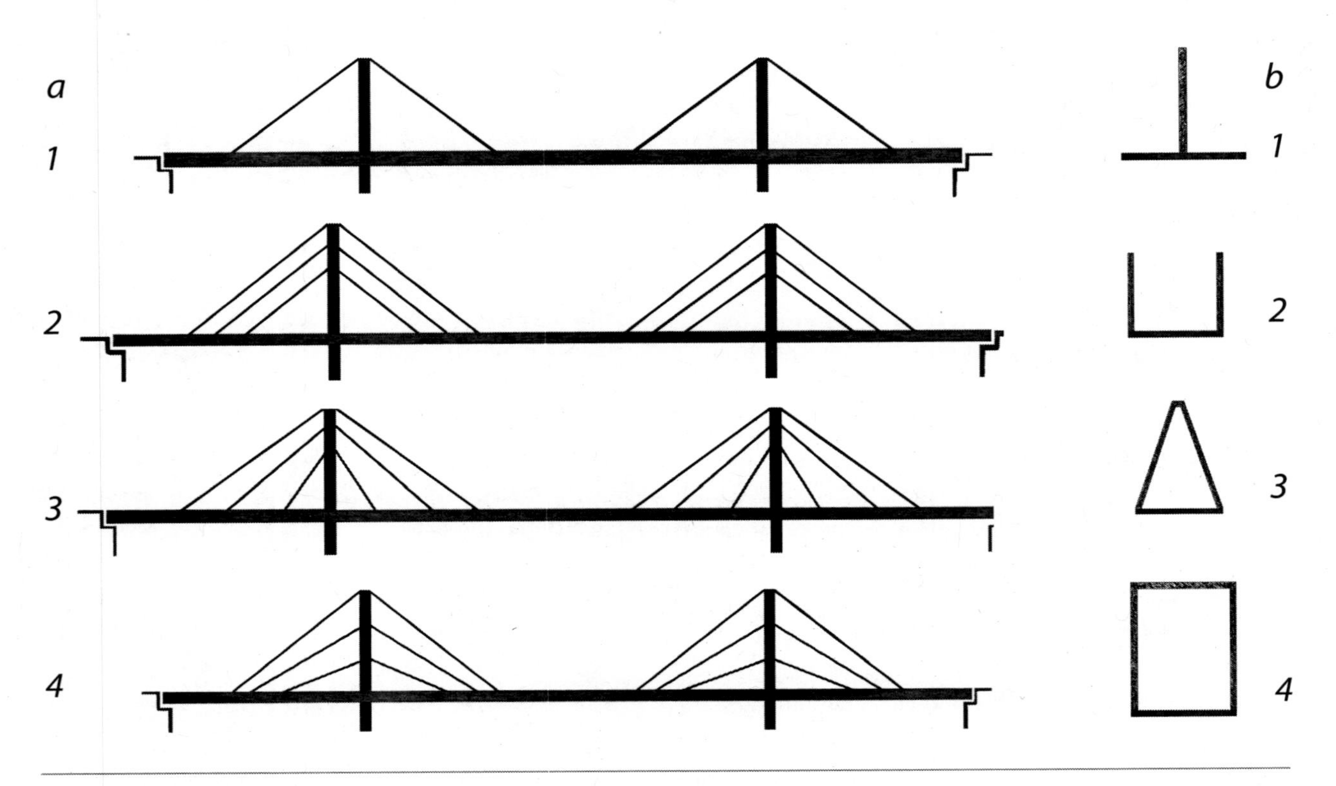

Figure 2.24. *a,* Patterns of cables: (1) mono; (2) harp; (3) fan; and (4) star. *b,* Types of cable-stayed bridges: (1) single; (2) double; (3) portal; and (4) A-shaped. Drawn by David Clark.

girder in the end span. Towers may be constructed of steel, reinforced concrete, or a steel shell filled with concrete. A single tower, resulting in two spans, is quite common in modern cable-stayed bridges.[102]

The use of cable stays greatly reduces the size of the girders required for the span. Girders used are made of prestressed concrete or steel of various sizes and shapes. Some bridges, including the Sitka Harbor span, employ plate girders.[103]

The bridge infrastructure of the United States represents a wide variety of types, as described above. Although specific truss types have become extinct, and numerous types have not been constructed for a half-century or more, the researcher often does not have to look far to see a preserved wood covered bridge or an iron or steel truss span carrying a lightly traveled road. The entries above permit a basic identification of bridge type, and reference to some of the sources indicated below will permit a more detailed description of the elements of a particular bridge.

GLOSSARY

The glossary is primarily compiled from definitions included in *Bridges and Tunnels of Allegheny County, Pennsylvania,*[104] J. A. L. Waddell's *Bridge Engineering* (1913), and the Glossary of Bridge Terms on the Ohio DOT Preventive Maintenance website.[105]

Abutment. The portion of a bridge structure that supports the end of a span or accepts the thrust of an arch.

Anchorage. The portion of a suspension bridge to which the cables are attached. Anchorages are generally located at the outermost ends and are constructed as a solid mass.

Approach. The road surface leading to a bridge.

Arch barrel. The curved inner surface of an arch.

Arch ring. The outer course of stone forming the arch. This course is constructed of a series of voussoirs.

Ashlar. Cut, squared, building stone dressed on all four sides.

Back. The upper or convex side of an arch.

Balustrade. A decorative railing, particularly one constructed of concrete or stone, consisting of a top and bottom rail and vertical supports, the balusters.

Beam. A horizontal structural member supporting vertical loads by resisting bending. A girder is a larger beam, often made of multiple plates.

Bearing. A support element that transfers loads from the superstructure to the substructure while permitting limited movement.

Bearing seat. The top of a pier or abutment that supports a bridge bearing.

Bent. The group of members forming a single vertical support of a trestle, designated as a pile bent where the principal members are piles, and as a framed bent where of framed timbers.

Bobtail. An asymmetrical configuration of a swing bridge in which the pier is located off-center and the shorter end of the span requires a counterweight to balance it.[106]

Bolster beam. A timber placed between the abutment and truss of a covered bridge that extends. This beam is commonly found in Town trusses.

Bottom chord. The bottom, usually horizontal, supporting members of a truss.

Box girder. A steel beam built up from shapes to form a hollow cross section.

Brace. In a covered bridge, a diagonal timber that slants upward toward the center of the bridge. In a steel or iron bridge, a beam or girder placed at an angle to an upright structure for support.

Brace-ribbed arch. An arch with parallel chords connected by open webbing.

Cable. The portion of a suspension bridge extending from one anchorage to the other over the tops of the towers.

Camber. A built-in, upward curve of a bridge. Also, a positive curve built into a beam that compensates for vertical load and anticipated deflection.

Cantilever. A structural member that projects beyond a supporting column or wall and is counterbalanced and/or supported on only one end.

Cap. A horizontal member upon the top of piles or posts, connecting them in the form of a bent.

Catenary. The curve formed by a rope or chain hanging freely between two supports. The curved cables or chains of suspension bridges may be referred to as catenaries.

Centering. Temporary structure or falsework, generally of wood, erected to support an arch during construction.

Chord. The upper and lower horizontal members of a truss system.

Closed spandrel wall. A masonry or concrete wall constructed solidly to enclose fill material.

Column. A vertical structural member used to support compressive loads.

Continuous span. A superstructure that extends as a single piece over multiple supports.

Corbels. In timber bent bridges, pieces of timber placed lengthwise of the stringers, between the stringers and the caps.

Corbelled arch. Masonry constructed over an opening by progressively overlapping the courses from each side until they meet at the top center.

Counterbrace. On a covered bridge, a diagonal timber or rod that slants upward away from the center of the bridge.

Counter-posts. Bracing members used to stiffen taller framed bents.

Cradle. The part of a suspension bridge that carries the cable over the top of a tower.

Crown. The highest part of the arch ring.

Dead load. The weight of a bridge itself without any traffic moving over it.

Deck. The surface of the bridge that carries vehicular traffic. In a covered bridge, the floor.

Deck truss. A truss that carries its deck on the top chord

Dressed stone. A stone block that has been squared and shaped to form a precise fit with other stones.

Elliptical arch. An arch formed by multiple arcs, each of which is drawn from its own center.

Embankment. An angled grading of the ground often used to support bridge approach roads.

End post. The outermost vertical or angled compression member of a truss.

Expansion joint. A meeting place between two parts of a structure designed to allow for movement of the parts due to heat or cold or moisture levels. It commonly takes the form of a hinged or movable connection.

Extrado. The outer exposed curve of an arch.

Eyebars. A structural member of a bridge, usually formed of steel or iron, with an oblong body and enlarged round head at each end. Each head is pierced with a hole through which a pin is inserted to connect it with other members. Eyebars are used for the main tension members of pin-connected trusses.

False arch. Masonry built over a wall opening by uniformly advancing courses from each side until they meet at midpoint. Also known as a *corbelled arch*.

Falsework. Temporary structures used in construction to support spanning or arched structures in order to hold the component in place until the structure is able to support itself.

Fill. Earth or stone used to raise the ground level, form an embankment, or fill the interior of an abutment, pier, or closed spandrel.

Fixed arch. An arched structure anchored in its location.

Flange. Used in a plate girder bridge, the simplest form consists of a pair of unequal-legged angles, with the long legs placed out and riveted to the web plate.

Flared post. In a covered bridge, a post whose top slants toward the end of the bridge.

Floor beams. Horizontal, transverse members between the trusses that support the deck.

Footing. The enlarged bottom portion of the substructure or foundation that rests directly on soil, bedrock, or piles.

Footpiece. A specially designed casting usually placed between the diagonals and chords of a covered bridge for rigid assembly of members.

Frame trestle. A bridge structure in which the upright members or supports are framed timbers.

Girder. A horizontal structural member that supports vertical loads by resisting bending. It is generally constructed of multiple metal plates riveted or welded together.

Gusset plate. A metal plate used to unite multiple members of a truss.

Hinged arch. An arch supported by pinned connections at either end and/or in the middle. Compare to fixed arch.

Humpback. A bridge having relatively steep approach embankments leading to the bridge deck.

I-beam. A steel structural member with a cross-sectional shape similar to the capital *I*. The top and bottom portions are the flanges, and the central portion is the web.

Impost. The surface that receives the vertical weight at the bottom of an arch.

Intrado. The interior arc of an arch.

Joist. A beam, usually one that supports a brick floor or deck.

King post. Two triangular panels with a common vertical. The simplest triangular truss.

Lacing. An assemblage of bars, channels, or angles on a truss composed of diagonal members. Lacing bars are used to join the parts of the member together and make it act as a solid member.

Laminated arch. In a covered bridge, a series of planks bolted together with staggered joints to form an arch. Used in the place of a solid timber.

Lateral bracing. The bracing assemblage that engages a member perpendicular to the plane of the member; intended to resist lateral movement and deformation and to provide resistance against raking of parallel elements in truss and girder bridges.

Lattice. An assemblage of smaller pieces arranged in a grid-like pattern, often used to reduce the weight of a girder.

Lift span. The movable portion of a vertical-lift bridge.

Panel. The portion of a truss between two vertical posts.

Parapet. A low wall along the outside edge of a bridge deck used to protect vehicles and pedestrians.

Phoenix column. A bridge member developed and patented by the Phoenix Bridge Company of Phoenixville, Pennsylvania. Hollow and circular, it consists of four, six, or eight wrought-iron segments flanged and riveted together.[107]

Pier. A structure, usually concrete or stone, located between the abutments to support a multi-span structure.

Pile. (1) A long column driven into the ground to form part of a bridge foundation. (2) A member usually driven or jetted into the ground and deriving its support from the underlying strata and by the friction of the ground on its surface.

Pin. A cylindrical bar used to connect various members of a truss; typically inserted through the holes of a meeting pair of eyebars.

Plate girder. A structural member consisting of a vertical steel or iron web plate to whose top and bottom edges are riveted horizontal pairs of angles to form flanges, and to whose ends are attached vertical angles which transmit the load to the supports.

Portal. The opening at the ends of a through truss that forms the entrance.

Post. A vertical member of a truss that is perpendicular or near perpendicular to a chord.

Pylon. A monumental vertical structure marking the entrance to a bridge.

Rib. One of a series of arched members parallel to the length of a bridge.

Rim bearing. A term applied to swing spans to indicate that the dead load is supported by a circular girder near the periphery of the pivot pier instead of near its axis.

Rocker. A cast or built-up steel frame fastened to the end of a truss or column to permit a slight rotation.

Rocker bearing. A bearing or support for solitary trestle bents or cantilever spans which permits a slight rocking with changing position of the live load and with variations in temperature.

Roller bearing. A single roller or group of rollers so housed as to permit longitudinal movement of the structure.

Runners. Lengthwise planks laid over crosswise planks of a covered bridge deck, probably added following the invention of the automobile to reduce noise from loose planks.

Saddle. A U-shaped element placed atop a suspension bridge tower into which cables are laid.

Sash brace. A horizontal member secured to the posts or piles of a bent.

Segmental arch. An arch formed from an arch drawn from a point below the spring line, forming a less than semicircular arch.

Semi-elliptical arch. See elliptical arch.

Shoe. A pedestal-shaped member beneath the superstructure bearing that transmits and distributes loads to the bearing area.

Sill. A lower horizontal member of a framed bent.

Simple span. A span in which the effective length is the same as the length of the spanning structure.

Skew angle. An acute angle between the alignment of the superstructure and the alignment of the substructure.

Skewback. The inclined surface upon which an arch rests.

Soffit. The under or concave side of an arch.

Span. The horizontal distance between bridge abutments or piers.

Spandrel. The roughly triangular area above an arch and below a horizontal bridge deck,

Stanchion. One of the larger vertical posts supporting a railing.

Strand shoe. A portion of an eyebar embedded in the anchorage onto which the cable wires are looped.

Stringer. (1) A beam parallel to the length of a span that supports the deck. (2) A longitudinal member of a timber bent span extending from bent to bent and supporting the ties.

Strut. A compressive member of a truss bridge.

Subsill. A timber embedded in the ground to support a framed bent.

Substructure. The portion of the bridge, including abutments and piers, that supports the superstructure.

Superstructure. The portion of a bridge that carries the traffic load.

Suspendors. Tension members of a suspension bridge that hang from the main cables to carry the deck. Also, similar members of an arched bridge that features a suspended deck.

Sway brace. A member bolted or spiked to the bent of a timber bridge and extending diagonally across its face. Sway braces are generally needed only for pile or framed bents exceeding 10 feet in height.

Swing span. A movable bridge span that opens by rotating horizontally on axis.

Tie. (1) A tension member of a truss. (2) A transverse timber of a timber bent span resting on the stringers and supporting the rails.

Tied arch. An arch with a tension member across its base to connect one end to another.

Tower. A tall pier or frame supporting the cable of a suspension or cable-stayed bridge.

Trussed arch. A metal arch bridge that features a curved truss.

Upper chord. The top horizontal member of a truss.

Voussoirs. Wedge-shaped blocks used to form an arch.

Web. The system of members connecting the top and bottom chords of a truss.

Wing walls. Extensions of a retaining wall as part of an abutment, used to contain the fill of an approach embankment.

REFERENCES

General Sources

American Society of Civil Engineers. *American Wooden Bridges*. New York: ASCE, 1976.

Boothby, Thomas. "Designing American Lenticular Truss Bridges, 1878–1900." *IA: The Journal of the Society for Industrial Archeology* 30:1 (2004): 5–18.

Burr, William H., and Myron S. Falk. *The Design and Construction of Metallic Bridges*. New York: John Wiley and Sons, 1905.

Chatterjee, Sukhen. *The Design of Modern Steel Bridges*. Oxford, England: BSP Professional Books, 1991.

Comp, T. Allen, and Donald C. Jackson. *Bridge Truss Types: A Guide to Dating and Identifying*. Nashville: American Association for State and Local History, Technical Leaflet No. 95, 1977.

Condit, Carl W. *American Building Art: The Nineteenth Century*. New York: Oxford University Press, 1960.

—. *American Building Art: The Twentieth Century*. New York: Oxford University Press, 1961.

Darnell, Victor C. *Directory of American Bridge-Building Companies*. Washington, DC: Society for Industrial Archaeology, Occasional Publication No. 4, 1984.

Degenkolb, Oris H. *Concrete Box Girder Bridges*. Ames: The Iowa State University Press, 1977.

DeLony, Eric. "Surviving Cast- and Wrought-Iron Bridges in America." *IA: The Journal of the Society for Industrial Archeology* 19:2 (1993): 17–47.

Federal Highway Administration. *Covered Bridge Manual*. Publication No. FHWA-HRT-04-098. McLean, VA: Turner-Fairbank Highway Research Center, FHWA, 2005.

Fletcher, Robert, and Jonathan Parker Snow. "A History of the Development of Wooden Bridges." *American Society of Civil Engineering Paper no. 1864. Proceedings ASCE 1932*. Volume LVIII, pp. 1455–1498. Republished No. 4, New York, 1976.

Foster, Wolcott C. *A Treatise on Wooden Trestle Bridges According to the Present Practice on American Railroads*. New York: John Wiley and Sons, 1891.

Heins, C. P., and D. A. Firmage. *Design of Modern Steel Highway Bridges*. New York: John Wiley & Sons, Inc., 1979.

Hool, George A., and W. S. Kinne. *Movable and Long-Span Steel Bridges*. New York: McGraw-Hill Book Company, 1923. (2nd edition: New York: McGraw-Hill Book Company, Inc., 1943.)

—. *Reinforced Concrete and Masonry Structures*. New York: McGraw-Hill Book Company, Inc., 1924.

__. *Steel and Timber Structures*. New York: McGraw-Hill Book Company, 1942.

Hovey, Otis Ells. *Movable Bridges*, volume I: *Superstructure*; volume II: *Substructure*. New York: John Wiley & Sons, Inc., 1926.

Jackson, Donald C. *Great American Bridges and Dams*. Washington, DC: Preservation Press, 1988.

Kemp, Emory, editor. *America Bridge Patents: The First Century (1790–1890)*. Morgantown: West Virginia University Press, 2005.

—."The Introduction of Cast and Wrought Iron in Bridge Building." *IA: The Journal of the Society for Industrial Archeology* 19:2 (1993): 5–16.

Ketchum, Milo S. *The Design of Highway Bridges of Steel, Timber and Concrete*. 2nd edition. New York: McGraw-Hill Book Company, 1920.

Kirkham, John Edward. *Highway Bridges: Design and Cost*. New York: McGraw-Hill Book Company, 1932.

Nelson, Lee H. *The Colossus of 1812: An American Engineering Superlative*. Somerset, NJ: American Society of Civil Engineers, 1990.

Parsons Brinkerhoff and Engineering & Industrial Heritage. *A Context for Common Historic Bridge Types*. National Cooperative Highway Research Program, Project 25-25, Task 15, 2005. Online at www4.trb.org/trb/crp.nsf/reference/boilerplate/attachments/$file/25-25(15)_FR.pdf.

Rastorfer, Darl. *Six Bridges: The Legacy of Othmar H. Amman*. New Haven: Yale University Press, 2000.

Robb, Frances C. "Cast Aside: The First Cast-Iron Bridge in the United States." *IA: The Journal of the Society for Industrial Archeology* 19:2 (1993): 48–62.

John A. Roebling's Sons Company. *Suspension Bridges: A Century of Progress*. Trenton, NJ: John A. Roebling's Sons Company, 1935.

Simmons, David A. "Bridges and Boilers: Americans Discover the Wrought-Iron Tubular Bowstring Bridge." *IA: The Journal of the Society for Industrial Archeology* 19:2 (1993): 63–79.

Skinner, Frank W. *Types and Details of Bridge Construction*. 3 volumes. New York: McGraw Publishing Company, 1904–1908.

Taylor, Frederick W., Sanford E. Thompson, and Edward Smulski. *Reinforced-Concrete Bridges*. New York: John Wiley & Sons, Inc., 1939.

Tyrell, Henry G. *History of Bridge Engineering*. Chicago: G. B. Williams Company, 1911.

Waddell, J. A. L. *Bridge Engineering*. 2 volumes. New York: John Wiley and Sons, Inc., 1916.

—. *De Pontibus: A Pocket Book for Bridge Engineers*. New York: John Wiley and Sons, 1914.

—. *Economics of Bridgework: A Sequel to Bridge Engineering*. New York: John Wiley and Sons, 1921.

Winpenny, Thomas. *Without Fitting, Filing or Chipping: An Illustrated History of the Phoenix Bridge Company.* Easton, PA: Canal History and Technology Press, 1996.

State and Regional Guides

Alabama

Prince, A. G. *Alabama's Covered Bridges: Past and Present.* Birmingham, AL: the author, 1972.

Arkansas

Arkansas State Highway and Transportation Dept. *Arkansas Historic Bridge Inventory Review and Evaluation.* 2 volumes. Little Rock: Arkansas State Highway and Transportation Department, 1996.

California

Mikesell, Stephen D. *Historic Highway Bridges of California.* Sacramento: California Department of Transportation, 1990.

Colorado

Fraser, Clayton. *Historic Bridges of Colorado.* Denver: Colorado Department of Highways, 1987.

—. Highway Bridges in Colorado. National Register of Historic Places Multiple Property Documentation Form. Denver: Fraserdesign, 2000.

Connecticut

Clouette, Bruce. *Where Water Meets Land: Historic Movable Bridges of Connecticut.* Hartford: Connecticut Department of Transportation, 2004.

Clouette, Bruce, and Matthew Ross. *Connecticut's Historic Highway Bridges.* Hartford: Connecticut Department of Transportation, 1991.

Howard, Andrew R. *Covered Bridges of Connecticut: A Guide.* Unionville: Village Press, 1985.

Delaware

Lichtenstein Consulting Engineers. *Delaware's Historic Bridges.* Paramus, NJ: Lichtenstein Consulting Engineers, 2000.

Spero, P. A. C. and Company. *Delaware Historic Bridge Survey and Evaluation.* Baltimore: P. A. C. Spero and Company, 1991.

Florida

Florida Department of Transportation Environmental Management Office. *The Historic Highway Bridges of Florida.* Tallahassee: Florida Department of Transportation, 1992.

Georgia

Smith, Frank D. *Georgia's Historic Bridges.* Edited and typed by Mary E. Waters. Conyers, GA: F. D. Smith, 2000.

Hawaii

Schmitt, Robert C. Historic Hawaiian Bridges. Photocopy of typescript in the University of Hawaii Manoa Library, 1987.

Idaho

Herbst, Rebecca. *Idaho Bridge Inventory: Including a History of Bridge-Building in Idaho.* Boise: Idaho Transportation Department, 1983.

Illinois

Chicago Department of Public Works, Bureau of Engineering. *The Movable Bridges of Chicago: A Brief History.* Chicago: Department of Public Works, 1979.

Draper, Joan E. *Chicago Bridges.* Chicago: Chicago Department of Public Works, 1984.

Eaton, Thelma. *The Covered Bridges of Illinois.* Ann Arbor, MI: Edwards Brothers, 1968.

Indiana

Cooper, James L. *Industry and Ingenuity in Artificial Stone: Indiana's Concrete Bridges.* Greencastle, IN: J. L. Cooper, 1997.

—. *Iron Monuments to Distant Posterity: Indiana's Metal Bridges, 1870–1930.* 1987.

Gould, George E. *Indiana Covered Bridges thru the Years.* Indianapolis: Indiana Covered Bridge Society, 1988.

Sinclair, Stan. *Illustrated Guide to Parke County Covered Bridges.* Rockville, IN: Stan Sinclair, 1991.

Iowa

Gasparini, Dario, Eugene M. Farrelly, and Dawn M. Harrison. *Structural Studies of Iowa Historic Bridges.* Ames: Iowa Department of Transportation, 1996.

Finn, Michael R. *Bowstring Arch Bridges of Iowa.* Anamosa, IA: Jones County, 2004.

Hippen, James. *Marsh Rainbow Arch Bridges in Iowa.* Boone, IA: Boone County, 1997.

Kentucky

White, Vernon. *Covered Bridges: Focus on Kentucky.* Berea, KY: Kentucky Imprints, 1985.

Maine

Jakeman, Adelbert M. *Old Covered Bridges of Maine.* Ocean Park, ME: Adelbert M. Jakeman, 1980.

Maryland

Legler, Dixie, and Carol Highsmith. *Historic Bridges of Maryland*. Crownsville: Maryland Historical Trust, 2002.

Spero, P. A. C. & Company and Louis Berger & Associates. *Historic Highway Bridges in Maryland: 1631–1960*. Baltimore: P. A. C. Spero & Company, 1995.

Massachusetts

Bennett, Lola, and Richard Kaminski. "Lower Merrimack River Bridges." *Civil Engineering Practice: Journal of the Boston Society of Civil Engineers Section/ASCE* 15:2 (Fall–Winter 2000): 43–62.

Lutenegger, Alan J., and Amy B. Cerato. *Lenticular Truss Bridges of Massachusetts*. www.ecs.umass.edu/cee/cee_web/bridge/1.html.

Massachusetts Highway Department. Historic Bridge Inventory. http://www.mhd.state.ma.us/default.asp?pgid=content/brhist&sid=about.

Michigan

Michigan Department of Transportation. *Michigan Historic Bridges*. www.mdot.state.mi.us/environmental/historic/bridges.

Minnesota

Hess, Jeffrey A. *Final Report of the Minnesota Historic Bridge Survey*. 2 volumes. St. Paul: Minnesota Department of Transportation, 1988.

Montana

Axline, Jon. *Conveniences Sorely Needed: Montana's Historic Highway Bridges, 1860–1956*. Helena: Montana Historical Society, 2005.

Quivik, Frederic L. *Historic Bridges in Montana*. Washington: National Park Service, 1982.

Nebraska

Nebraska State Historical Society. *Spans in Time: A History of Nebraska Bridges*. Lincoln: State Historical Society and Department of Roads, 1999.

Nevada

Knight, Kenneth C., and T. H. Turner. *An Inventory of Nevada's Historic Bridges*. Carson City: Nevada Department of Transportation, 1988.

New Hampshire

Kenyon, Thedia Cox. *New Hampshire's Covered Bridges*. Sanbornville, NH: Wake-Brook House, 1957.

New Jersey

Dale, Frank. *Bridges over the Delaware River: A History of Crossings*. New Brunswick: Rutgers University Press, 2003.

New Jersey Department of Transportation. *Historic Bridge Inventory*. www.state.nj.us/transportation/works/environment/HistBrIntro.htm.

Richman, Steven M. *The Bridges of New Jersey: Portraits of Garden State Crossings*. New Brunswick: Rutgers University Press, 2005.

New Mexico

Rae, Steven R., Joseph E. King, and Donald R. Abbe. *New Mexico Historic Bridge Survey*. Santa Fe: New Mexico State Highway and Transportation Department, 1987.

New York

Berfield, Rick. *Covered Bridges of New York State: A Guide*. Syracuse: Syracuse University Press, 2003.

Duensing, Dawn, historian. Bronx River Parkway Reservation. HAER No. NY-327, 2001. Westchester Archives, www.westchesterarchives.com/BRPR/report/BRPR_opening.html.

New York State Department of Transportation. *Historic Bridge Inventory/Management Plan*. www.dot.state.ny.us/eab/hbridge.html.

Reier, Sharon. *The Bridges of New York*. Originally published 1977. Mineola, NY: Dover Publications, Inc., 2000.

Ohio

Ketcham, Bryan E. *Covered Bridges on the Byways of Ohio*. Cincinnati: Bryan E. Ketcham, 1969.

Lichtenstein Consulting Engineers. *Third Ohio Historic Bridge Inventory for Bridges Constructed 1951–1960*. Columbus: Ohio Department of Transportation, Office of Environmental Services, 2004. www.dot.state.oh.us/oes/hist_bridges.htm.

Ohio Department of Transportation. *Second Ohio Historic Bridge Inventory*. Columbus: ODOT, 1990.

—. *The Concrete Arch Supplement to the Ohio Historic Bridge Inventory*. Columbus: ODOT, 1994.

Watson, Sara Ruth, and John R. Wolfs. *Bridges of Metropolitan Cleveland: Past and Present*. Cleveland: no publisher, 1981.

Wood, Miriam. *The Covered Bridges of Ohio: An Atlas and History*. Columbus: the author, n.d.

Oklahoma

Oklahoma Department of Transportation. *Spans of Time*. www.okladot.state.ok.us/hqdiv/p-r-div/spansoftime.

Oregon

Smith, Dwight A., James B. Norman, and Pieter T. Dykman. *Historic Highway Bridges of Oregon*. Salem: Oregon Department of Transportation, 1985.

Pennsylvania

Commonwealth of Pennsylvania. *Historic Highway Bridges in Pennsylvania*. Harrisburg: Pennsylvania Historical and Museum Commission and Pennsylvania Department of Transportation, 1986.

Cridlebaugh, Bruce S. *Bridges and Tunnels of Allegheny County, Pennsylvania*. http://pghbridges.com.

Dale, Frank. *Bridges over the Delaware River: A History of Crossings*. New Brunswick: Rutgers University Press, 2003.

Evans, Benjamin D., and June R. Evans. *Pennsylvania's Covered Bridges: A Complete Guide*. Pittsburgh: University of Pittsburgh Press, 2001.

Kidney, Walter C. *Pittsburgh's Bridges: Architecture and Engineering*. Pittsburgh: Pittsburgh History and Landmarks Foundation, 1999.

Shank, William H. *Historic Bridges in Pennsylvania*. York: American Canal and Transportation Center, 1980.

Theodore Burr Covered Bridge Society of Pennsylvania, Inc. www.tbcbspa.com.

Zacher, Susan. *The Covered Bridges of Pennsylvania: A Guide*. Harrisburg: Pennsylvania Historical and Museum Commission, 1982.

Tennessee

Hufstetler, Mark L. Survey of Older Beam and Slab Bridges of Tennessee. Ph.D. dissertation, University of Tennessee, 1999.

Vermont

McCullough, Robert. *Crossings: A History of Vermont Bridges*. Norwich, VT: Vermont Historical Society, 2005.

Virginia

Miller, Ann Brush, and Kenneth M. Clark. *Survey of Metal Truss Bridges in Virginia*. Charlottesville: Virginia Transportation Research Council, 1997.

—. *A Survey of Movable Span Bridges in Virginia*. Charlottesville: Virginia Transportation Research Council, 1998.

Miller, Ann Brush, Kenneth M. Clark, and Matt Grimes. *A Survey of Masonry and Concrete Arch Bridges in Virginia*. Charlottesville: Virginia Transportation Research Council, 2000.

Miller, Ann Brush, Daniel D. McGeehan, and Kenneth M. Clark. *A Survey of Non-Arched Historic Concrete Bridges in Virginia Constructed Prior to 1950: Final Report*. Charlottesville: Virginia Transportation Research Council, 1996.

Virginia Department of Transportation. *Virginia's Covered Bridges*. Richmond: Virginia Department of Transportation, 2000.

Washington

Holstine, Craig E., and Richard Hobbs. *Spanning Washington: Historic Highway Bridges of the Evergreen State*. Pullman: Washington State University Press, 2005.

Wisconsin

Wisconsin Department of Transportation. *Historic Highway Bridges in Wisconsin: Movable Bridges*. Volume III, part 1. Madison: Wisconsin Department of Transportation, 1996.

Notes

1 This structural engineering discussion is paraphrased from Civil Engineering Bridge Building Contest (http://campuspages.cvcc.vccs.edu/TechPrep/Bridge_building.htm), and Early Evolution of Trusses (www.du.edu/~jcalvert/tech/ truss.htm).

2 Stephen J. Roper, Structural Historian, Massachusetts Highway Department, personal communication, October 5, 2005.

3 Bruce Clouette, Adams Street Bridge, Historic American Engineering Record MA-131, 1996, 2–3.

4 P. A. C. Spero and Company, *Delaware Historic Bridges Survey and Evaluation* (Baltimore: P. A. C. Spero and Company, 1991), 13.

5 George A. Hool and W. S. Kinne, *Reinforced Concrete and Masonry Structures* (New York: McGraw-Hill Book Company, 1924), 435–436.

6 Spero 1991, 22.

7 Raymond E. Wilson, "Twenty Different Ways to Build a Covered Bridge," in American Society of Civil Engineers, *American Wooden Bridges* (New York: American Society of Civil Engineers, 1976), 130; Commonwealth of Pennsylvania, *Historic Highway Bridges in Pennsylvania* (Harrisburg: Commonwealth of Pennsylvania, 1986), 59.

8 Wilson, 131.

9 Wilson, 130; Commonwealth of Pennsylvania, 59.

10 Wilson, 133.

11 Wilson, 130–131; Commonwealth of Pennsylvania, 59.

12 Wilson, 132.

13 Robert Fletcher and J. P. Snow, "A History of the Development of Wooden Bridges," in American Society of Civil Engineers, eds., *American Wooden Bridges* (New York: American Society of Civil Engineers, 1976), 54.

14 Fletcher and Snow, 54–55.

15 Spero 1991, 23.

16 C. P. Heins and D. A. Firmage, *Design of Modern Steel Highway Bridges* (New York: John Wiley & Sons, Inc., 1979), 55.

17 Milo Ketchum, *The Design of Highway Bridges of Steel, Timber and Concrete*, 2nd edition (New York: McGraw-Hill Book Company, 1920), 145.

18 Heins and Firmage, 55–56.

19 Spero 1991, 14; Carl W. Condit, *American Building Art: The Nineteenth Century* (New York: Oxford University Press, 1960), 106.

20 Frank W. Skinner, *Type and Details of Bridge Construction*, Part II: *Plate Girders* (New York: McGraw Publishing Company, 1906), 3.

21 Skinner, 3–5.

22 George A. Hool and W. S. Kinne, *Steel and Timber Structures* (New York: McGraw-Hill Book Company, 1942), 361–362.

23 Carl W. Condit, *American Building Art: The Twentieth Century* (New York: Oxford University Press, 1961), 100.

24 Dawn Duensing, Bronx River Parkway Reservation, HAER NY-327, 2001; Michigan Department of Transportation, "Gillespie Ave./Clinton River [Bridge]," www.michigan.gov/mdot.

25 George A. Hool and W. S. Kinne, *Movable and Long-Span Steel Bridges*, 1st edition (New York: McGraw-Hill Boon Company, 1923), 362.

26 Condit 1961, 128.

27 Hool and Kinne 1923, 362.

28 Condit 1961, 121.

29 Hool and Kinne 1923, 360.

30 Hool and Kinne 1923, 360–361.

31 Hool and Kinne 1923, 361.

32 David Guise, "Horton's Bowstring Truss," *Society for Industrial Archeology Newsletter* 31:2 (2002): 14–15; Condit 1960, 113.

33 Spero 1991, 42.

34 Ketchum, 103.

35 David Weitzman, *Traces of the Past: A Field Guide to Industrial Archaeology* (New York: Charles Scribner's Sons, 1980), 61; Historic Bridges of Iowa-Construction Types (www.ole.dot.state.ia.us/historicbridge/construction.asp).

36 Weitzman, 88.

37 J. A. L. Waddell, *Bridge Engineering*, volume I (New York: John Wiley and Sons, Inc., 1916), 477–478.

38 Weitzman, 110; Condit, 1961, 82; Commonwealth of Pennsylvania, 113.

39 Anthony J. Bianculli, *Trains and Technology: The American Railroad in the Nineteenth Century*, volume 4 (Newark: University of Delaware Press, 2002), 48.

40 Commonwealth of Pennsylvania, 115.

41 Weitzman, 108–109

42 Weitzman, 94–95.

43 Weitzman, 96–97.

44 Weitzman, 100–101.

45 Weitzman, 102–103.

46 Weitzman, 92–93.

47 Waddell, 24.

48 Commonwealth of Pennsylvania, 117.

49 Commonwealth of Pennsylvania, 118.

50 Condit 1960, 90.

51 Alan J. Lutenegger and Amy B. Cerato, Lenticular Truss Bridges of Massachusetts, www.ecs.umass.edu/cee/cee_web/bridge/1.html.

52 Weitzman, 68.

53 Weitzman, 111; Condit 1960, 82; Commonwealth of Pennsylvania, 120.

54 Commonwealth of Pennsylvania, 121.

55 Pittsburgh Bridges: A Spotter's Guide to Bridge Design, http://pghbridge.com/basics.htm; Condit 1961, 98.

56 Spero 1991, 88–89; David Guise, "Early American Truss Bridges: Thacher's Truss," *Society for Industrial Archeology Newsletter* 30:2 (2001): 14–15.

57 Spero 1991, 87.

58 Condit 1961, 197.

59 Spero 1991, 93.

60 Spero 1991, 90.

61 Virginia M. Price, Lake City Bridge 1914, Lake City, Iowa, HAER IA-46, 1995.

62 Ketchum, 273; John Edward Kirkham, *Highway Bridges: Design and Cost* (New York: McGraw-Hill Book Company, 1932), 114.

63 Condit 1961, 211–213.

64 Spero 1991, 215–216; P. A. C. Spero & Company, *Historic Highway Bridges in Maryland: 1631–1960, Historic Context Report* (Baltimore: P. A. C. Spero & Company, 1995), 148; Mead and Hunt, *Contextual Study of New York State's Pre-1961 Bridges* (Cincinnati: Mead and Hunt, 1999), 27.

65 Frederick W. Taylor, Sanford E. Thompson, and Edward Smulski, *Reinforced-Concrete Bridges* (New York: John Wiley & Sons, Inc., 1939), 93.

66 Commonwealth of Pennsylvania, 13.

67 National Precast Concrete Association, Precast Concrete Short Span Bridges. On the NPCA website, www.precast.org.

68 Spero 1991, 78–79.

69 Otis Ellis Hovey, *Movable Bridges*, volume II (New York: John Wiley & Sons, Inc., 1927), 18.

70 Hool and Kinne, *Movable and Long-Span Steel Bridges*, 2nd edition (New York: McGraw-Hill Book Company, Inc., 1943), 5.

71 Hool and Kinne 1943, 15.

72 Spero 1991, 59; Waddell, 702.

73 Waddell, 703.

74 Waddell, 706.

75 Hovey, 135.

76 Waddell, 708.

77 Hool and Kinne 1943, 20.

78 Hool and Kinne 1943, 24–25.

79 Hovey, 63.

80 Hool and Kinne 1923, 26.

81 Hool and Kinne 1943, 162–163.

82 Hool and Kinne 1943, 159–163.

83 Hool and Kinne 1943, 163.

84 Hool and Kinne 1943, 172–173.

85 Hool and Kinne 1943, 173–175.

86 Condit 1961, 91.

87 Condit 1961, 93–94; Hool and Kinne, 1943, 201.

[88] J. A. L. Waddell, *De Pontibus: A Pocket Book for Bridge Engineers* (New York: John Wiley and Sons, Inc., 1914), 55.
[89] Condit 1961, 109–110.
[90] William Michael Lawrence, Lacey V. Murrow Memorial Bridge, Historic American Engineering Record, WA-2, 1993: 2.
[91] Lawrence, 7.
[92] Hool and Kinne 1943, 327.
[93] Hool and Kinne 1943, 330.
[94] Hool and Kinne 1943, 330.
[95] Hool and Kinne 1943, 332.
[96] Hool and Kinne 1943, 333.
[97] Hool and Kinne 1923, 335.
[98] Hool and Kinne 1923. 335.
[99] Heins and Firmage, 421–422.
[100] U.S. 82 Greenville Bridge website: www.greenvillebridge.com/2b1_ranking.htm.
[101] Heins and Firmage, 422–429.
[102] Heins and Firmage, 430–431.
[103] Heins and Firmage, 433.
[104] www.pghbridges.com.
[105] www.dot.state.oh.us/preventivemaintenance/glossary/glossmenu.htm.
[106] Spero 1991, 82.
[107] Thomas R. Winpenny, *Without Fitting, Filing or Chipping: An Illustrated History of the Phoenix Bridge Company* (Easton, PA: Canal History and Technology Press, 1996), 6.

Chapter 3

Railroads

Although railroads remain a dominant presence on the industrial landscape, they have undergone substantial evolution in recent years. Semaphore signals are gone, wooden railroad ties have vanished in many places. Myriad buildings and structures have been demolished as railroad lines become increasingly automated and switching is done from central control locations on many lines. The older features of the railroad landscape may still be found both adjacent to abandoned lines and on lines outside of the principal rail corridors. This chapter contains a guide to some of the buildings, structures, and objects that may be encountered in studies of rail corridors.

Railroad researchers can locate information in a wealth of sources. Numerous railroad company records exist, containing varying amounts of information about physical facilities. Such records are dispersed among historical societies, state archives, research libraries, and other collections throughout the country. Among particularly notable collections are the Railroad Museum of Pennsylvania, the Hagley Library in Wilmington, Delaware, and the Minnesota Historical Society in St. Paul. A comprehensive list of railroad museums with brief descriptions and links to websites is located at www.railmuseums.com.

Other potentially useful sources of information include railroad historic and technical society websites and railroad enthusiast websites. Among these are several associated with the Pennsylvania Railroad Historical and Technical Society, and other sites that include inventories of passenger stations and other railroad buildings and structures.

Of particular interest to the East Coast researcher seeking information on railroad buildings and structures are the Railroad Valuation Records of the Interstate Commerce Commission, included in RG 134 in the Textual Reference Division of the National Archives in College Park, Maryland. Most valuation records were created between 1915 and 1920 by the ICC and railroad engineers who undertook a massive project to inventory nearly every aspect of the U.S. railroad system to determine the net worth for each railroad. Included in these reports are descriptions of buildings and structures, and, in some cases, photographs as well.[1]

Railbeds

Railroad tracks sit on top of a bed composed of a cambered base, designed to allow water to run to drains on either side of the line. The track is supported on ballast, which provides support, load transfer, and drainage to the track and keeps water away from the rails and ties. Ballast is typically laid to a depth of 9 to 12 inches and is usually made of irregularly shaped basalt stones laid on a layer of sand.

Tracks consist of two rails secured on ties to keep the rails at the correct spacing, ordinarily at 25- to 30-inch intervals. Around 1890, about 60 percent of the ties in the United States were made of oak, but regional variations were common in the twentieth century. Oak was the preferred material in much of the East and the Mississippi Valley. Yellow pine was used extensively in the southern Atlantic Coast and the Gulf States. Mountain pine was widely used by railroads in west Texas, New Mexico, and Arizona. Cedar was used in Maine, Michigan, Wisconsin, and Washington. Other woods, including tamarack, hemlock, spruce, wild cherry, honey locust, and black walnut, were also used. In recent years, some railroads have begun to use composite ties, made of old tires and recycled plastic.[2]

RAILS

The first rails laid on American railroads were wooden stringers used for very short tram roads around coal mines. With the invention of the locomotive, rails were initially replaced by the cast-iron "fishbelly" rail, cast in lengths of about 3 feet. A fishbelly rail is a short-span rail with an undulating lower surface that provides greater thickness between the ties than above them, for greater strength.

Various forms of wrought-iron strap rails followed. Later rails were either "T" rails or double-headed or "bull-headed" rails carried in chairs. The double-headed rail was designed with a symmetrical form so that when one head was worn down by traffic, the rail could be reversed, though this proved impractical. The bull-headed rail, used only on United Kingdom railways and a few U.K.-designed railways, had the lower head only large enough to properly hold the wooden key with which the rail was secured to the chairs.[3]

Standardization of rail sections in the United States began in the late nineteenth century after an investigation by the American Society of Civil Engineers. Subsequent standards were devised by the American Railway Association and the American Railway Engineering Association.[4] The heaviest rails weighed 120 to 150 pounds per yard, with the Pennsylvania Railroad using 155-pound rails, the heaviest in the country, on its main line. The standard for main line tracks in the United States is presently 141 pounds per yard. Until the mid-to-late twentieth century, U.S. railroads used sections of rail that measured 39 feet so that they could be carried to and from the site on 40-foot gondolas.

Dating Rails

By dating rails, it is possible to determine the date of abandonment of a line, approximate the maximum weight of locomotives on the rails, and estimate the traffic. When heavier locomotives were acquired by a railroad, heavier rails were laid to accommodate it. These changes occurred mostly on the main track, while lighter rails continued in service in yards and on sidings.

As a first step, the rail's composition must be determined. Iron rails were used before 1884 and steel rails followed. Generally, iron rails had a pear-shaped head, while steel rails had a roughly rectangular head. The width of the base of the rail, then, correlates closely to rail weight. The rail weight enables an estimate to be made of the date and capacity of older rails (Tables 1 and 2).[5]

An important element of rail construction is the joint between rail sections, a joint that requires the same strength and stiffness as the rails that it joins. Commonly used early twentieth-century joint types include the Bonzano, Wolhaupter, Weber, atlas suspended, 100 per cent, and continuous rail joint. In the United States early joints were staggered, while in Europe joints were parallel. Most modern railroads use continuous welded rail (CWR) in which a mile or more of rail is welded together, making a stronger joint, a smooth ride, and less required maintenance. Welded rail is usually laid on concrete ties.

Spikes

Rails needed to be held to the ties by a fastening that provided sufficient resistance and was cheap and easily applied. The track spike fulfilled the last two requirements but had comparatively small resisting power. The Goldie spike aimed to improve the form by reducing the destruction of wood fibers. The sides were smooth, the edges clean-cut, and the point was ground down to a pyramidal rather than chisel-shaped form.

TABLE 1. IRON RAIL WEIGHTS AND DIMENSIONS

Pounds/ Yard	Height of Rail	Width of Base	Width of Head
70	3 ¾	3 7/8	2 ¼
65	3 7/8	4	2 ¼
60	3 3/8	3 7/8	2 1/8
56	3 ½	3 3/8	2 3/8

TABLE 2. STEEL RAIL WEIGHTS AND DIMENSIONS (ASCE STANDARDS, 1895)

Pounds/Yard	Height of Rail	Width of Base	Width of Head	Depth of Head
152	8	6 ¾	3	1 27/32
131	7 1/8	6	3	1 ¾
100	5 ¾	5 ¾	2 ¾	1 45/64
95	5 9/16	5 9/16	2 11/16	1 41/64
90	5 3/8	5 3/8	2 5/8	1 19/32
85	5 3/16	5 3/16	2 9/16	1 35/64
80	5	5	2 ½	1 ½
75	4 13/16	4 13/16	2 15/32	1 27/64
70	4 5/8	4 5/8	2 7/16	1 11/32
65	4 7/16	4 7/16	2 13/32	1 9/32
60	4 ¼	4 ¼	2 3/8	1 7/32
55	4 1/16	4 1/16	2 ¼	1 11/64
50	3 7/8	3 7/8	2 1/8	1 1/8
45	3 11/16	3 11/16	2	1 1/16
40	3 ½	3 ½	1 7/8	1 1/64
30–35	3 ½	3 ½	1 ¾	

Switches, Frogs, and Derails

Connections between two or more tracks require individual structures of definite design for their effective functioning. These usually were a switch, a frog, derails, and connecting or lead rails.

Switch

A railroad switch is a mechanical device that guides trains from one line of rail tracks to another (Figure 3.1). A switch consists of a pair of switch rails, which are designated as right-hand or left-hand switch points; one or more rods to hold the points in correct relation to each other; and gauge and switch plates to support the switch rails at the proper elevation with reference to the stock rails. The gauge and switch plates, in connection with plain or adjustable rail braces, also maintain the correct position of the stock rails. Necessary accessories include heel blocks, for effecting a rigid joint at the heel of the switch, jaw or transit clips for uniting the rods with the switch points, and metal guards for foot protection.[6]

Each switch contains paired and linked tapering rails that can be moved laterally into one of two positions that route the train to each track. Usually the position of the switch is changed electrically on main lines and controlled from a remote control center. On rarely used sidings, low-traffic branch lines, marshalling yards, or historic railroads, a switch may be manually operated from a switch stand using a lever and accompanying linkages. Prior to the widespread availability of electricity, switches at heavily traveled junctions were operated from a switch tower near the tracks by an elaborate system of rods and levers.[7]

Frog

A simple turnout frog is a device to enable the wheels running on one track to cross the rail of a diverging track. It provides continuous channels for the wheel flanges and supports the intersection of the flangeways. Turnout frogs are divided into two principal types: rigid frogs without movable parts, and movable wing frogs, in which one or both wings move outward to create a flangeway for the wheels. The latter type includes the

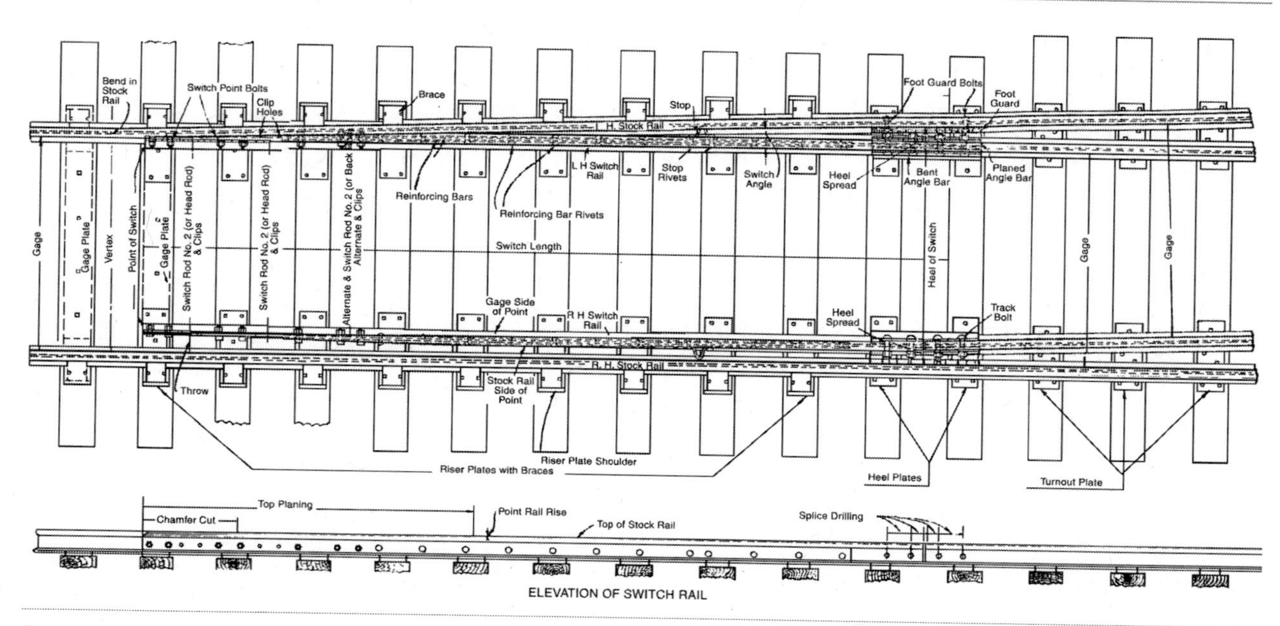

Figure 3.1. Diagram of railroad switch. Reprinted from Elmer T. Howson, *Railway Engineering and Maintenance Cyclopedia*, 2nd edition (New York: Simmons-Boardman Publishing Company, 1926), opposite page 258.

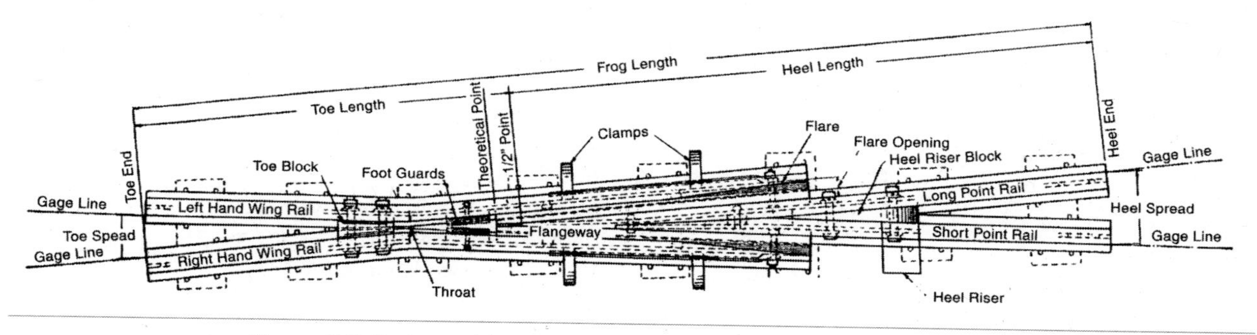

Figure 3.2. Spring rail frog showing names of parts (Howson 1926: 261).

spring rail pattern (Figure 3.2), in which only one wing is movable, and the double-spring rail, or sliding pattern where both wings move in unison to open a flangeway on one side of the point while it is closed on the other. In the early twentieth century, the double-spring rail frog and the sliding rail frog were used only when speeds were slow.[8]

Derails

Derails were commonly placed on passing sidings to prevent the slack of a train from overrunning the clearance point, and they were also used on outlying sidings to prevent cars moved by the wind or by trainman error from overrunning the clearance point. Siding with grades descending toward the main track where gravity could cause a car to foul the main track was almost always protected by derails. Derails were also frequently found on car repair tracks to protect workers.

The common types of derails are the single- or double-point derail; the lifting derail, in which one point deflects the wheels while the other lifts the flanges over and off the rail; and the block derail. The block derail comes in two forms, the revolving (or hinged) derail and the sliding derail. Derails were thrown mechanically from a central plant; manually, in union with switches through connections; or independently by a separate stand.[9]

SIGNALING

The first signaling systems used telegraphs to transmit orders between stations along the line. Upon receipt, the station would display a signal to the engineman of the approaching train.

Two types of signals are used on railroads in the United States: permissive and absolute. These are visually distinguishable. Permissive signals usually have a number plate on the base of the mast on which they are mounted, or the letter *I*. Absolute signals lack a number plate, or show the letter *A*. A red permissive signal means stop and proceed at a restricted speed until a more favorable signal is reached. A red absolute signal means stop and stay until the signal changes or verbal permission is received from the dispatcher.[10]

The lights displayed by a signal have three properties: aspect, name, and indication. The aspect describes what is displayed—such as red over green. The name is the formal name classification for the signal. The indication is the instruction conveyed by the signal.

The earliest railroad signal was a cloth flag. This was followed by ball signals introduced on the New Castle & Frenchtown Railroad in 1832, which conveyed information by their position and color to trains and stations. In the 1860s, the semaphore signal was introduced (Figure 3.3). These signals employed a pointed blade narrower at the inner end that was moved mechanically. Initially a "lower quadrant" semaphore was used: the blade moved in the lower right portion of an imaginary circle whose center was the blade's pivot point. Shortly after 1900, the three-position (vertical, diagonal, horizontal) upper-quadrant semaphore was introduced. With the advent of light signals, semaphores were seen as inferior since they used two different aspects, blade position by day and light by night, to convey their message.

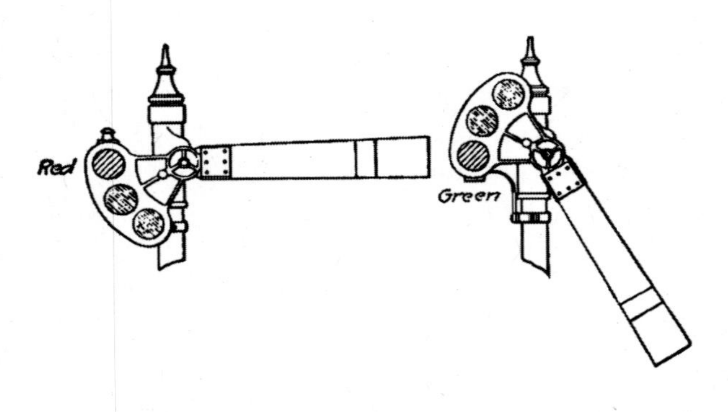

Figure 3.3. Semaphore signal. Reprinted from Edgar Everett King, *Railway Signaling* (New York: McGraw-Hill, 1921), 9.

Color Light Signals

Color light signals were introduced in the first decade of the twentieth century to provide signals in subway tunnels. These lights consist of three bulbs shining through three lenses or "roundels," of red, yellow, and green. Typically, the lights are grouped vertically, with green at the top. Some use a triangular arrangement, while some, such as the Chicago and North Western Railroad, used a horizontal configuration.

Searchlight Signals

Searchlight signals use a single lamp with a series of movable lenses to project three colors, and they first appeared in 1920. A relay-type mechanism moved colored lenses in front of a white light. In recent years, the low power consumption of the single light has been offset by the high maintenance cost of the movable light system, and tri-light signals are preferred for new installations.

Position-Light Signals

Developed by the Pennsylvania Railroad and introduced in 1915, the position-light signal was used only on this railroad and its affiliated lines (Figure 3.4). High-intensity lights of fog-penetrating yellow were illuminated in rows of three. A fully equipped signal head could display lights in horizontal, vertical, or two diagonal rows, but partial units were common. Position-lights were phased out on account of high maintenance costs and the lack of standard colors.

Color Position-Light Signals

First used in 1921, color position-lights were originally identified with the Baltimore and Ohio Railroad. Unlike the position-light, these signals have no center lamp and use larger lenses of three colors. The two lamps in each row are green (vertical), yellow (diagonal), and red (horizontal). Although more costly than other types, it has the advantage of redundancy. Its message is conveyed two ways: the position of the lights and the color of the lights. Amtrak has converted its position-light signals on the Northeast Corridor by removing the center lamp and changing the remaining lenses to color.[11]

Figure 3.4. Position-light signal. New York, New Haven and Hartford Railroad, Automatic Signalization System, Connecticut. HAER CT-8, National Park Service. Photograph by Thomas Brown, 1980.

Signal Bridges

Signal bridges were developed to improve signal visibility and to accommodate multiple tracks. These bridges, constructed of steel or iron, consist of paired legs with diagonal bracing on either side of the right-of-way and a horizontal bridge with bracing extended across the tracks. Most were equipped with ladders and a top walkway to permit maintenance on the signal equipment. Signal equipment was frequently mounted on the top of the bridge.

RAIL YARDS

Rail yards are critical parts of the nation's rail system. In some rail yards, freight trains are assembled, so that cargo will reach the proper destination, while in others rail cars are stored and cargo is transferred to other motor vehicles for delivery to destinations. Rail yards are divided into two types: classification yards and team-delivery yards

Classification Yards

A classification yard, also known as a marshalling yard, is a machine for separating trains, or drafts of cars, according to destinations, routes, commodities, or traffic requirements. Upon arrival in the yard, cars are taken to a lead or drill track. From there the cars are sent through a series of switches (called a ladder) onto the classification tracks. Larger yards tend to put the lead on the slope of an artificially built hill (called a hump) so that the force of gravity propels the cars through the ladder.[12]

Classification yards may be divided into three types: flat-shunted yards, hump yards, and gravity yards. In flat-shunted yards, the tracks lead into a flat shunting neck at one or both ends of the yard, where the cars are pushed with a shunting engineer to sort them into the right track. Among American flat-shunted yards were those in Houston, Texas, Decatur, Illinois, and East Joliet, Illinois.

Hump yards have the largest shunting capacity, capable of serving several thousand cars a day. Cars are pushed to the lead track by an engine, and single cars, or coupled cars in a block, are uncoupled just before the crest of the hump and roll by gravity to destination tracks in the classification bowl. The world's largest classification yard is the North Platte, Nebraska, hump yard.

Gravity yards, in which the whole yard is arranged on a slope, require less use of shunting engines. Never as widely used in the United States as in Europe, few, if any, remain in active use in this country.[13]

Team-Delivery Yards

Team yards, located as close as possible to the industrial center of the city, serve as a storage place for railroad equipment and a transfer location for freight. Consisting of a number of tracks each holding 10 to 15 cars, the yard is equipped with track scales, transfer cranes, gantry cranes, and paved driveways.

Transfer cranes are usually located at the entrance of the yard and at the point most accessible to shippers and receivers of freight. Typically, such a crane is of the bridge type, with a trolley on top covering one track at the adjacent railway. The gantry crane can transfer not only to or from the shipper or receiver of freight, but from one car or truck to another.

RAILROAD BUILDINGS

Railroad Shops

Railroad shops were constructed by railroads to provide central maintenance locations for locomotives and cars. The components of railroad shops usually include a locomotive shop, a freight car shop, passenger car shops, storehouses, a roundhouse, a blacksmith shop, a foundry, a planning mill, a lumberyard, and a scrap department.[14]

Two examples of twentieth-century shop complexes documented by the Historic American Engineering Record are the Pennsylvania Railroad's facilities in Juniata and Altoona, Pennsylvania, and the Southern Railroad's shops in Spencer, North Carolina. The Juniata Locomotive Shops had a paint shop, an electric and hydraulic house, a boiler shop, a blacksmith shop, a boiler house, office and storeroom, hydraulic transfer table and pit, erecting shops, and a machine shop, totaling 118,986 square feet of floor area. The Altoona shops included locomotive and car shops totaling 854,980 square feet of floor area.[15]

Major buildings and structures of the Spencer Shops (Figure 3.5) include the turntable, roundhouse, the electric shop and storehouse, the machine and erecting shop, the smith and boiler shop, the woodworking shop, the paint shop, the freight car repair shed, and the car repair shed.

Locomotive Shops

Locomotive shops were usually arranged parallel with the general line of yard tracks. Locomotives entered and left the shop on the center of the three working tracks. No turntable or transfer table was necessary, as the locomotive was transferred from the entering track to the working spaces on other tracks by traveling cranes.

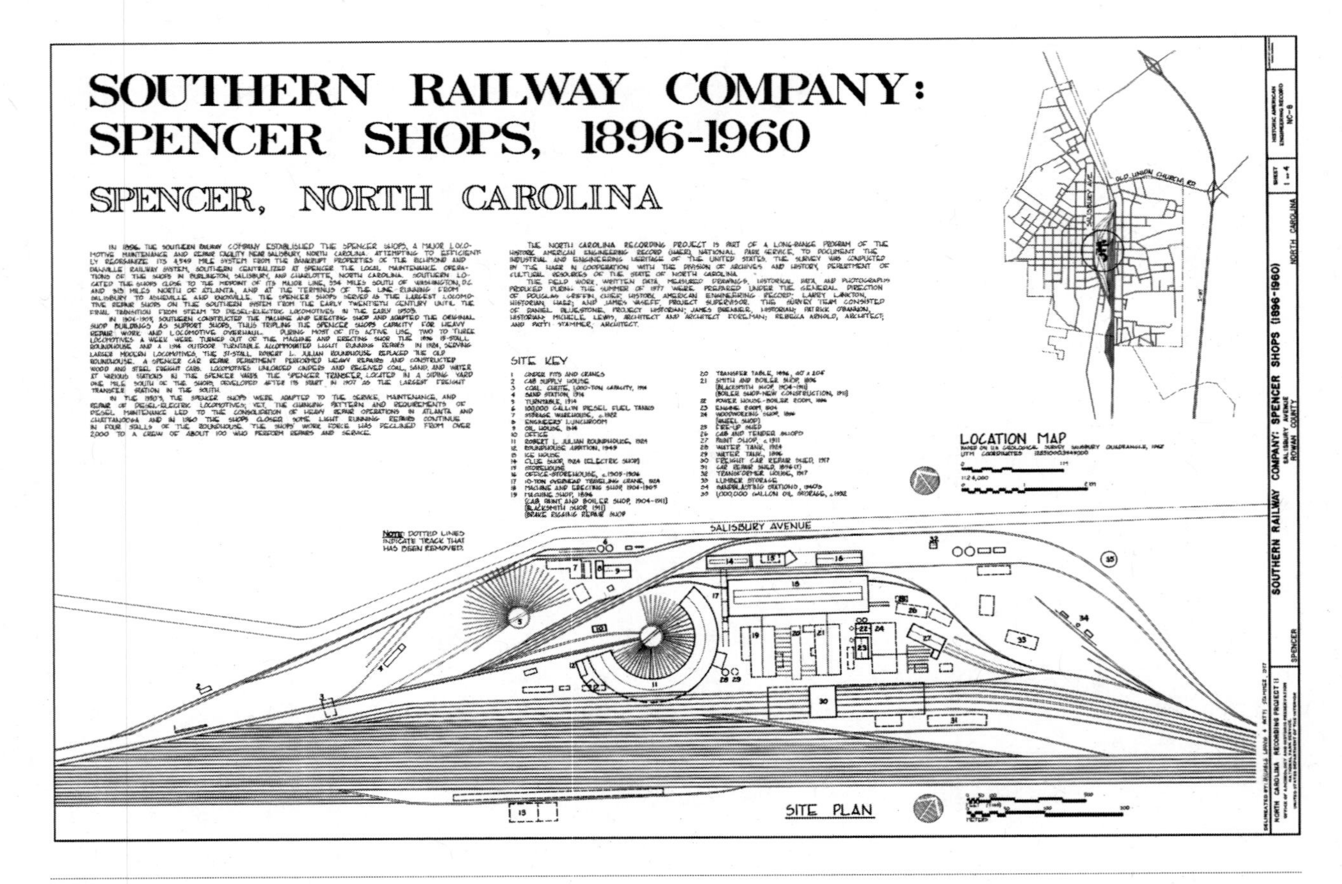

Figure 3.5. Southern Railway Company: Spencer Shops, Spencer, North Carolina. HAER NC-8, National Park Service. Delineated by Michelle Lewis and Patti Stammer, 1977.

These buildings were long and narrow and had at least three bays. In the early twentieth century, most such shops were constructed with a steel frame with clear height from floor to roof trusses and brick exterior walls tied to the structural frame. To provide illumination, the side walls often had windows extending nearly to the roof. Other shops employed saw-tooth monitors to increase light levels.[16]

Freight Car Shop

Car shops were erected by railroads to provide minor and major repairs to freight cars. Minor repairs include changing wheels, repairing brake rigging, and replacing draft rigging couplers. Major repairs include complete overhaul and part replacement and repair of cars damaged in service. An example of a layout of car shops is the D, L & W Shops in Keyser Valley, Pennsylvania.

Similar to the locomotive shop, the freight car shop was typically constructed with a steel skeleton and brick walls. A common design includes four longitudinal tracks on 20-foot centers, with cars moved from track to track by overhead cranes.[17]

Passenger Car Shops

The typical passenger car shop includes the following: coach shop, paint shop, wheel shop, truck shop, inspection facilities, storehouse facilities, toilet and washroom, wood mill, and powerhouse. The 1926 American Railway Engineering Association layout for a small passenger car shop illustrates typical building sections (Figure 3.6).[18]

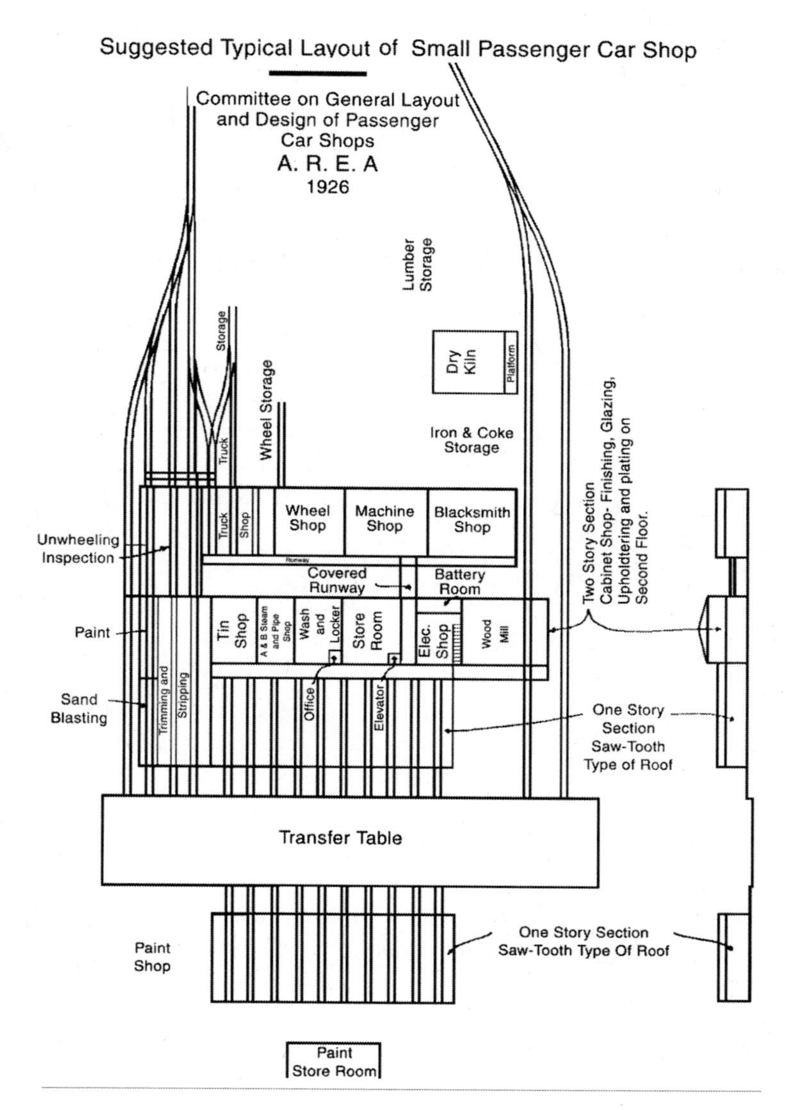

Figure 3.6. Suggested typical layout of small passenger car shop. Reprinted, with permission, from American Railway Engineering Association, *Proceedings of the Annual Convention of the American Railway Engineering Association* (Chicago: American Railway Engineering Association), AREA 1926:306.

These shops required ample natural light to be admitted through the roof, as well as through windows in the walls. Most shops had brick walls with a large window area and used either wood or steel roof trusses and supports. The width of early twentieth-century car shops ranged from 90 feet to a maximum of 225 feet. Clear height to the roof trusses was generally about 20 feet.

In some yards the coach repair and paint shops were placed on the same side of the transfer table pit to conserve space, while in other yards the two shops were on opposite sides of the pit.[19]

Storehouses

As the supply base for an entire railway system, the storehouses were located to provide for good shipping connections. The chosen location allowed direct access to the general yard system and to the outside lead. Tracks extended to either side of the building, and loading docks permitted a number of cars to be loaded or unloaded simultaneously. To facilitate loading, the long narrow building usually had a first floor 4 feet above grade or at the level of an ordinary boxcar.[20]

Engine Houses or Roundhouses

Engine houses were used on railroads for housing engines when not in use and for cleaning engines, making light repairs, and washing out; they were located in terminal or division yards or at points where engines were changed or held in reserve. The buildings had to be sited to offer easy ingress and egress from the main tracks of the railroad.

Engine house designs can be divided into square houses and polygonal houses or roundhouses. Square houses were smaller, typically, and roundhouses were used almost universally for larger houses. Smaller buildings generally had one or more tracks at one gable end, the length of the building exceeding the longest engine used. The roundhouse is a house built in a circular form around a turntable with tracks leading from the turntable radially into the house. Closed roundhouses were constructed in a full circle, while open or segmental roundhouses formed a segment of a circle. It was customary to provide two passageways into a closed roundhouse.

Blacksmith Shop

The blacksmith shop was placed in an isolated location to provide light and air on all sides. Because a large proportion of the material that passed between the locomotive and blacksmith shops was heavy and bulky, the blacksmith shop was situated to provide the shortest and most direct route to the locomotive shop. The blacksmith shop included large hammers and furnaces, as well as bolt headers, shears, forging machines, and other metal-handling equipment.[21]

Foundry

The foundry was located so that castings could be conveniently delivered to the storehouse for line shipments and to the shops where castings were machined or assembled. In the early twentieth century, many shops had brass and iron foundries. A typical foundry had brick walls and was amply provided with natural light by large windows and a roof monitor. The interior was divided into three bays, with the central bay the widest. Foundries ranged in length to 300 feet or more.[22]

Planing Mill

The planing mill was located to one side of the shop buildings adjacent to the lumberyard. Because much of its output was used by the freight car shop, a logical position was between the lumberyard and the freight car shop. Its typical form was that of a long, narrow building with a large floor area to accommodate the milling equipment. Roof trusses were of wood or steel. Early twentieth-century planing mills ranged in size from 70 to 125 feet wide and from 230 to 500 feet long.[23]

Freight House

Freight houses can be divided into several types: outbound houses, inbound houses, transfer houses, and combination houses. In outbound houses, freight is received by truck and forwarded by rail. In inbound houses, the freight is received by rail and held for the consignee. In transfer houses, freight is transferred from one line to another; combination houses combine two or three of the types in one building.[24]

Local freight houses were frequently single-story, wood-framed buildings, surrounded by high platforms on several or all sides. If tracks ran along only one side of a house, it was termed a side house or station, whereas an island house or station had tracks on both sides. Freight doors were usually sliding doors from 7 to 10 feet wide and from 7 to 12 feet high, either single or in pairs.

In urban areas, multiple-level freight houses and warehouses were deemed most efficient due to high property values and space restrictions. An example of such a freight station with warehouse space on the upper floors is shown in cross section in Figure 3.7.[25]

Signal Towers

Signal towers were used on railroads to allow a watchman, signalman, gateman, switch-tender, or operator to be stationed at a sufficient elevation above the railroad to give a view of the tracks and surroundings, or to permit the visibility of signals or signalmen by approaching trains, vehicles, or other signal stations. There are two classes of signal towers: those intended to protect exposed points on the line and those forming part of a block-signaling system. At a maximum, there were an estimated 4,400 signal towers nationwide.

Signal towers that protected exposed points on the line were simply watchmen's houses set on trestles, and were used to afford protection at railroad and highway

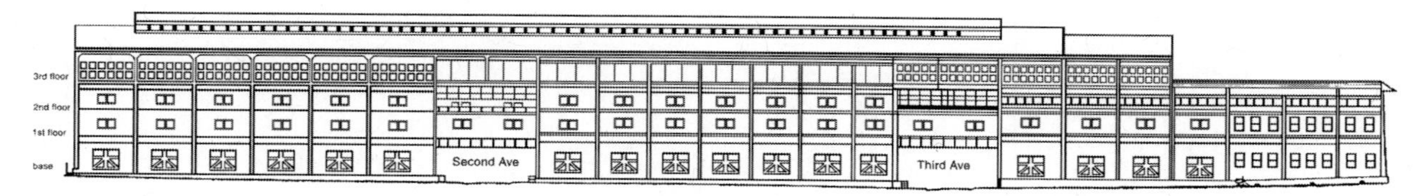

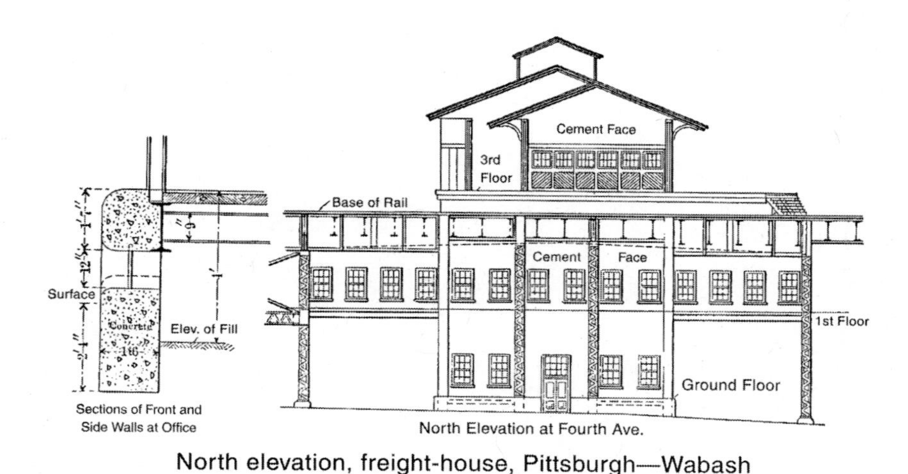

Figure 3.7. Pittsburgh-Wabash Freight House. Reprinted from John A. Droege, *Freight Terminals and Trains*, 2nd edition (New York: McGraw-Hill Book Company, Inc., 1925).

grade crossings, tunnels, sharp curves, points where the view was obstructed, and at the head of switch and yard systems. The other towers formed part of an extensive signaling system by which the road was divided into sections or "blocks." A signal tower was located at the end of each block or section.

With interlocking switch systems, it was customary to locate the working levers in the signal room of a signal tower so that a single individual could control the switches and movement of trains. Such signal towers were present at terminal yards, stations, junction points, and crossover systems. One example is the Delaware, Lackawanna & Western Railroad's 1908 tower in Scranton, Pennsylvania (Figure 3.8).

Interlocking Tower

An interlocking tower is a subcategory of switch tower (Figure 3.9). The function of an interlocking tower is to provide control over a junction of track switches and lineside signals, the former controlling course and direction, and the latter controlling speed. The term "interlocking" derives from a safety feature where line signals and track switches are linked to lock out conflicting routing choices.

The earliest of these switching mechanisms consisted of shoulder-high levers that moved long rods connected to the track switches, physically governing the position of rails. Later systems were power-assisted, where small levers caused electric motors to perform the same function. These power-assisted systems allowed control of larger areas.

The lower story of the buildings housed track and repeater relays, generator sets, terminal boards, and power boards. A critical element of the second story was the view of the tracks, provided by either double-hung windows or single, fixed lights. Towers ranged in size from 12 feet square to 16 by 28 feet, with the largest up to 50 or 100 feet long.[26]

Head or Stub Stations

The head or stub station is an end-of-the-line station, the tracks of which end in bumpers or in loops where the runs of trains within or part of a system terminate or continue after reverse movement. Most of the largest

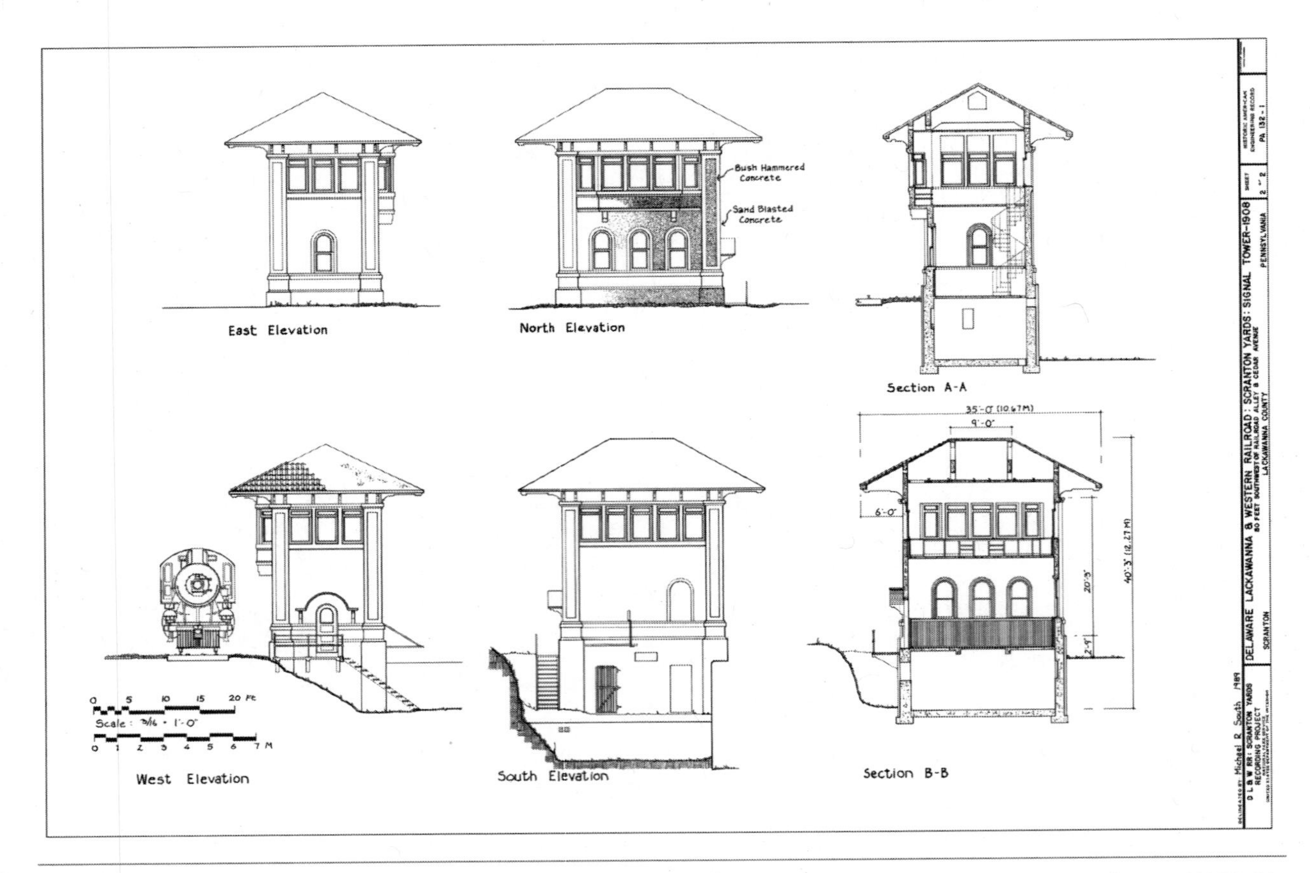

Figure 3.8. Delaware, Lackawanna & Western Railroad: Scranton Yard Signal Tower. Scranton, Pennsylvania. HAER PA-132, National Park Service. Delineated by Michael R. South, 1989.

Figure 3.9. South Station Interlocking Tower and Signal Bridge #7, HAER MA-58, National Park Service. Photograph by William Edmund Barrett, 1983.

and busiest stations in the country were head stations, including Grand Central Station in New York, Boston's North and South Stations, St. Louis's Union Station, Washington, DC's Union Station, Chicago's and Philadelphia's former Reading Terminal and Broad Street Station (now demolished). Such stations feature elaborate entrance halls, waiting rooms, and ticket counters, making them prominent civic spaces.

These large stations were erected in cities to accommodate passengers at a railroad terminus. Frequently several railroads entering the same city would unite and jointly use a "union depot." Facilities included those for passenger service, such as waiting rooms, restrooms, parcel and coat rooms, newsstands, and ticket offices; those for baggage, express and mail service; those for station service, including offices, storerooms, and shops; those for depot service, such as janitors' rooms and HVAC facilities; hotel accommodations; and general railroad offices.

Train Sheds

A hallmark of the nineteenth- and early twentieth-century urban train station was the arched train shed (Figure 3.10). In some instances, an arch of a single span covered from 25 to 32 tracks. These sheds proved expensive to both build and maintain, and were dark, dirty, cold in winter, and hot in summer. They were rarely lighted adequately.

In the early twentieth century, new types of sheds were introduced to improve upon the arched shed: the umbrella, the butterfly, and the Bush shed. The umbrella shed is a flat-roofed canopy with a central support that covers the platform itself but not the track area. The butterfly shed is similar to the umbrella shed except that its roofline takes the form of a shallow V, resembling a butterfly's wings. The Bush shed was the invention of Lincoln Bush and was first used in Delaware, Lackawanna, and Western's Hoboken Terminal. It differs from the umbrella and butterfly sheds

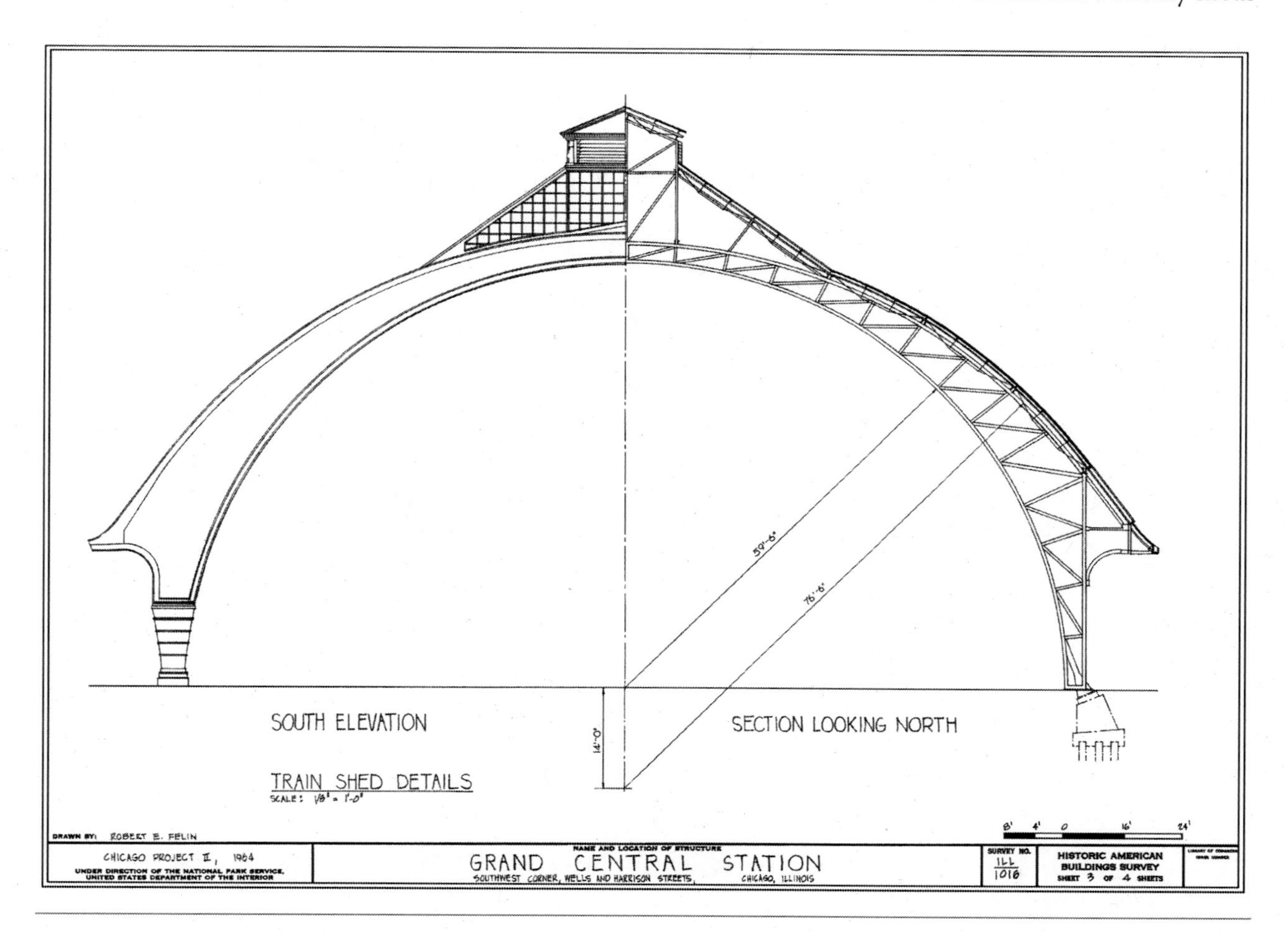

Figure 3.10. Grand Central Station, Chicago. Section of train shed. HABS IL-1016, National Park Service. Delineated by Robert E. Felin, 1964.

in that it is continuous except for the slot or gutter through which locomotive smoke and steam are discharged. The first of these sheds consisted of longitudinal sections constructed of structural steel, reinforced concrete, and wire glass. There were three continuous wire glass sidelights, 6 feet wide, placed longitudinally in the roof of each section. Early Bush sheds were erected in Scranton, Oakland, Chicago, Baltimore, Detroit, Jersey City, and Hoboken.[27]

Passenger Depots

Depots dedicated to passenger accommodation were constructed at local stations where the passenger business warranted a separate building or where the freight business was conducted separately.

Railroads often adopted standard depot plans, of which there are several types, flag depots being the simplest. A flag depot was constructed where there was no scheduled train stop, but the train would stop upon a signal. The smallest such buildings include a waiting room, a ticket office, and a baggage room. Other types of stations, characterized by the relationship between the building and the tracks, include the one-sided station, the two-sided station, and the head station. The two-sided station comprises separate buildings for each of the two directions of travel, while the head station, located at the end of a rail line, became the predominant form for city railroad terminals.[28]

Larger depots are two-story buildings that contain large waiting rooms, restrooms, smoking rooms, dining rooms and kitchens, baggage rooms, express rooms, mail rooms, telegraph offices, parcel rooms, newsstands, supply rooms, employee rooms, and offices. These larger buildings approach terminals in size and elaboration.

In the early twentieth century, the American Railway Engineering Association recommended that small passenger stations include a waiting room, a women's retiring room, toilets, ticket and telegraph offices, and a baggage room (Figure 3.11).[29]

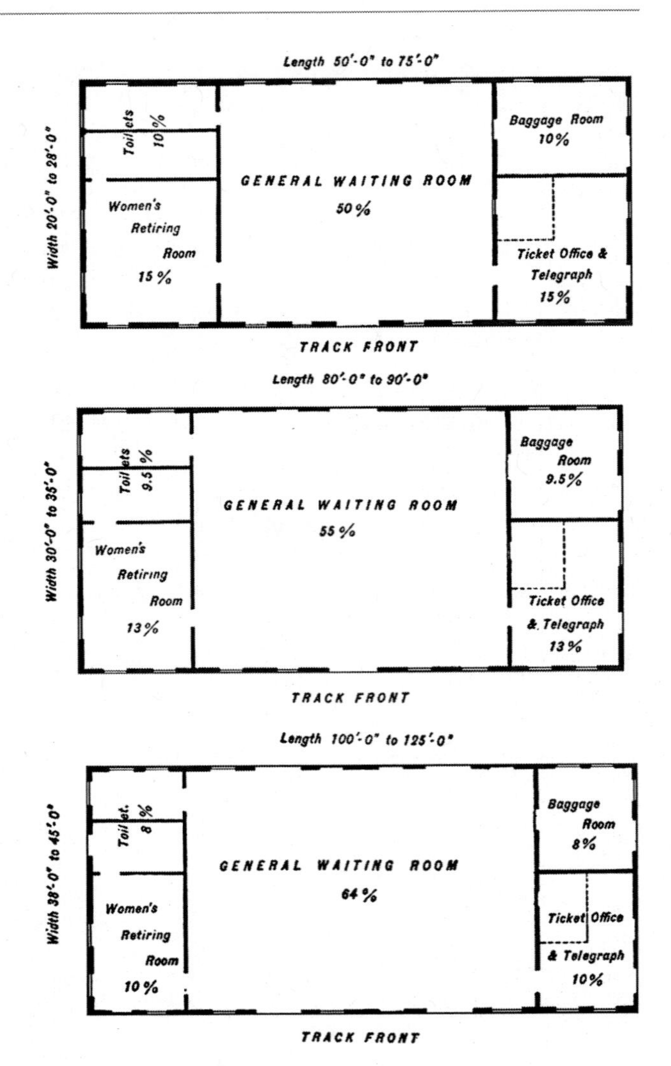

Figure 3.11. A.R.E.A. Standard Plans for Small Stations. Reprinted from John A. Droege, *Passenger Terminals and Trains.* Reprinted in 1969 by Kalmbach Publishing Company, Milwaukee (New York: McGraw-Hill Book Company, 1916), 254.

Combination Stations

Combination passenger and freight stations were erected at points where the freight and passenger business was small, and economics favored all station work under a single roof (Figure 3.12).[30]

Sand Houses

Sand houses or plants were required on railroads to store sand for use on engines to increase the friction of the driving wheels on the rails on heavy grades or when the rails were slippery. Sand houses were provided at all points on railroads where engines were changed or in connection with engine houses and coaling stations.

Freshly dug sand is always moist and absorbs moisture from the air, making artificial drying necessary.

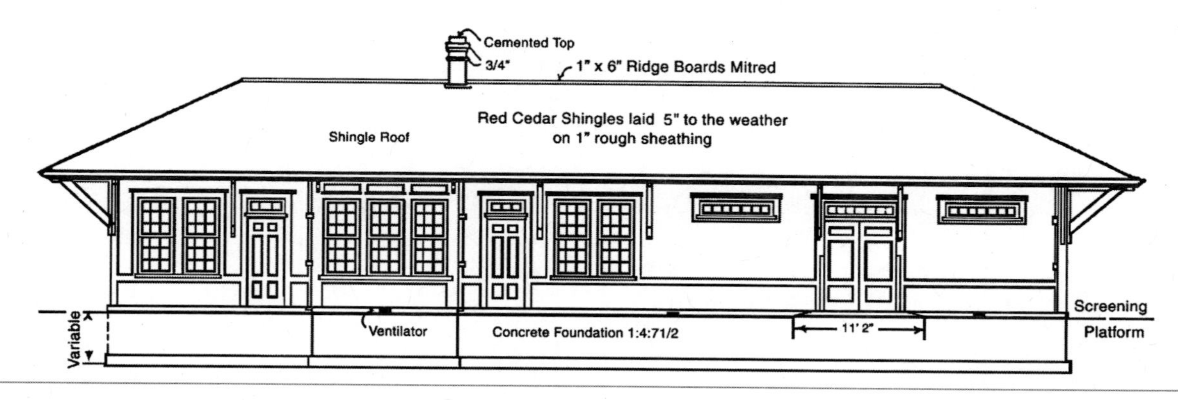

Figure 3.12. Elevation of combination station, Virginia (Droege 1916: 267).

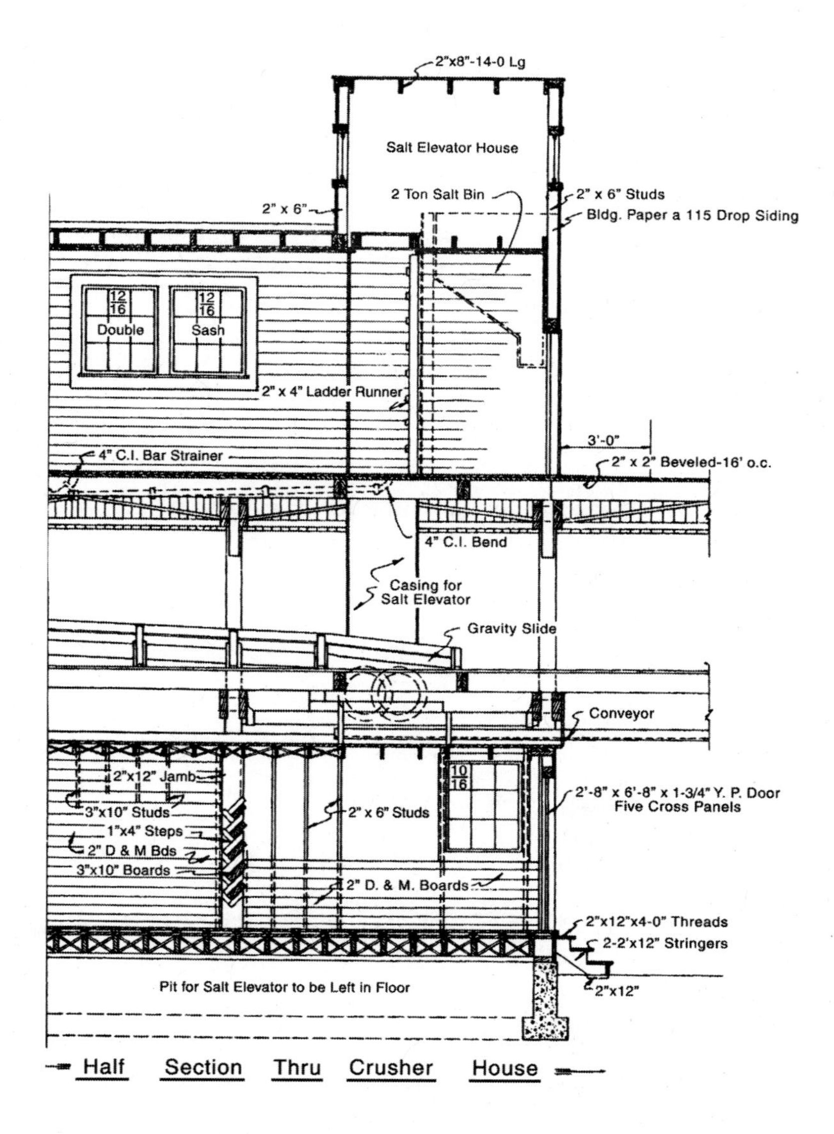

Figure 3.13. Section through a crusher house, Jackson, Mississippi. Reprinted, with permission, from American Railway Engineering Association (AREA 1922: 862).

Sand houses had three distinct functions: storing wet sand, drying sand, and storing dry sand. The structures can be divided into various types and combinations, the most common being a ground storage plant; in combination with a coaling station, a complete gravity plant; or a complete mechanical sand plant.[31]

Ice Houses

Prior to mechanical railroad car refrigeration, ice houses were used to handle natural and artificial ice for railroad cars. A simple icing station was a storage facility with an elevator at one end through the anteroom annex to the storage room. The ice house included a crusher room where block ice was delivered. These storage buildings were generally wood framed. A cross section of the crusher house erected in Jackson, Mississippi, is shown in Figure 3.13. The Jackson ice house was used primarily for icing northbound shipments of fruits and vegetables.[32]

Section Houses

On a railroad the term section house is used to designate a dwelling provided by the railroad company for the foreman or track hands employed on a particular section of track. Section houses were cheaply constructed and were of one of two types: for the accommodation of one or more families or for the accommodation of a number of men.

These houses were usually of wood-framed construction, roofed in shingles, and sheathed on the exterior with clapboards or vertical boards. Typically, a section house was a one-story, four- or five-room wood-framed house (Figure 3.14), though sometimes these houses were one and one-half or two stories in height. By the early twentieth century, many railroads provided structures with a front porch, a living room, two bedrooms, a bathroom, a combined or separate dining room and kitchen, a small pantry, and a complete basement.[33]

Section Tool Houses

Section tool houses were used to store hand cars, tools, and supplies required for track and roadbed construction or maintenance. Generally placed from 3 to 10 miles apart, they were usually of wood-framed construction, sheathed only on the outside, and roofed in tin, shingles, or corrugated iron. A large door enabled the entrance and egress of a hand-car truck. These houses typically ranged from 12 by 14 feet to 16 by 30 feet in footprint.

Watchman's Shanties

Watchman's, flagman's, or switch tender's shanties, often called flag houses, switch houses, or watch boxes, were used along railroads at exposed points or at yard systems, crossovers, or leaders, where regular switch tenders were required. Railroad companies often used standard designs because of the required number of such buildings. These buildings were wood-framed, sheathed on the outside either with vertical boards and battens or with weatherboarding, tongue-and-groove boards, or corrugated iron, and were roofed in tar, shingles, slate, or corrugated iron. Their size was limited owing to their location between tracks or on the edge of a right-of-way. The interior contained sufficient space for a small stove, a bench, a locker, and storage space. Windows were arranged to command a good view of the tracks. Buildings could be square, octagonal, or oblong in footprint.

Oil Houses

Oil houses were constructed as fireproof storage facilities for oil delivered from tank cars. Openings for ventilation were provided above the level of the top of the tanks. Typically, such houses were constructed of reinforced concrete and/or brick.

RAILROAD STRUCTURES

In addition to a variety of buildings, the historic railroad corridor also includes structures. Among these structures are platforms, water stations, coaling stations, ash and cinder handling facilities, waterfront

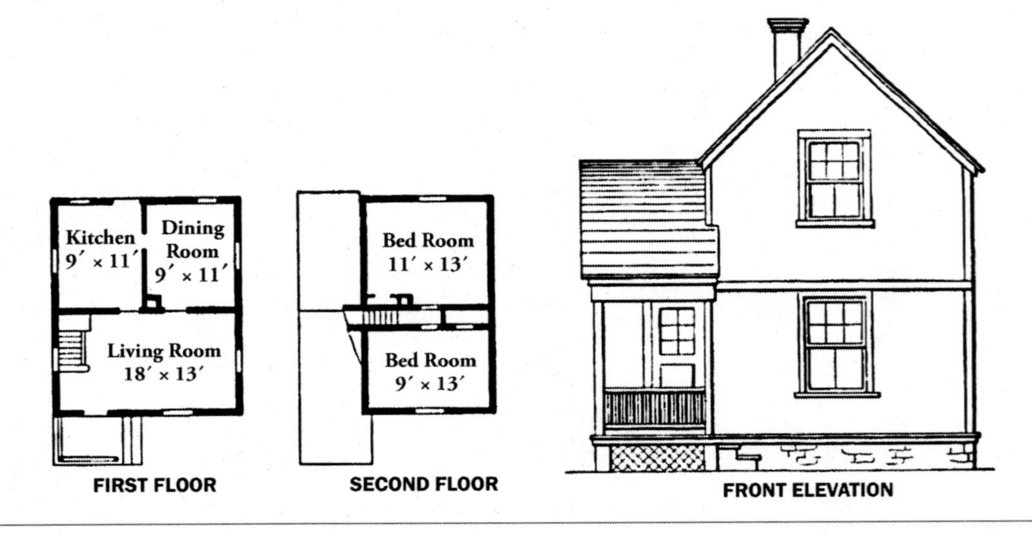

Figure 3.14. Standard Four-Room Section House of the Missouri-Kansas-Texas (Howson 1926: 597).

piers and docks and other terminal facilities, track scales, trestles and bridges, tunnels, and culverts.

Platforms

Platforms were necessary along tracks at passenger and freight depots to accommodate passengers and to facilitate baggage and freight transfer to and from train cars. Low platforms were used generally for passengers, while high platforms were used for freight depots. Passenger platforms were typically constructed of wood, while some freight platforms were constructed of more permanent materials such as reinforced concrete.

Water Stations

Water stations were required on a railroad to supply water to steam locomotives and were generally located from 5 to 20 miles apart. Water was pumped from ponds, lakes, springs, wells, or streams below the level of the railroad by suction pipe.

Wooden water tanks were universally used in the United States. These tanks were generally cylindrical in shape and were formed from 14 to 16 feet long staves. Diameters were 16, 18, 20, 22, 24, and 30 feet, and capacities ranged from about 20,000 to 80,000 gallons. The floor of the tank was generally set about 12 to 15 feet above the track. Foundations were usually wooden trestle bents on mud sills or on small stone foundation walls. In warmer climates, the pipe ran up the outside and discharged over the top of the tub, whereas in colder climates, the pipe entered the floor of the tank with proper protection against freezing. The discharge or delivery pipe was connected with a gooseneck spout attachment at the face of the tank or with a standpipe located some distance from the tank.

Coaling Stations

At one time railroads were a major consumer of coal, using from 20 to 25 percent of the coal produced in the United States. At a coaling station, coal was dumped into the locomotive tender. Wooden coaling stations, placed alongside tracks, resembled coal tipples. A large holding tank was located above or alongside the tracks. Rectangular chutes positioned over the locomotive filled the tender. By the 1930s, most coaling stations were made of steel and concrete. With the conversion of the railroads to diesel fuel in the 1950s, coaling stations and water tanks became obsolete and were demolished, though some examples still survive in rural areas.[34]

Coaling stations can be classified by their number of pockets (Figure 3.15). In two- or four-pocket plants,

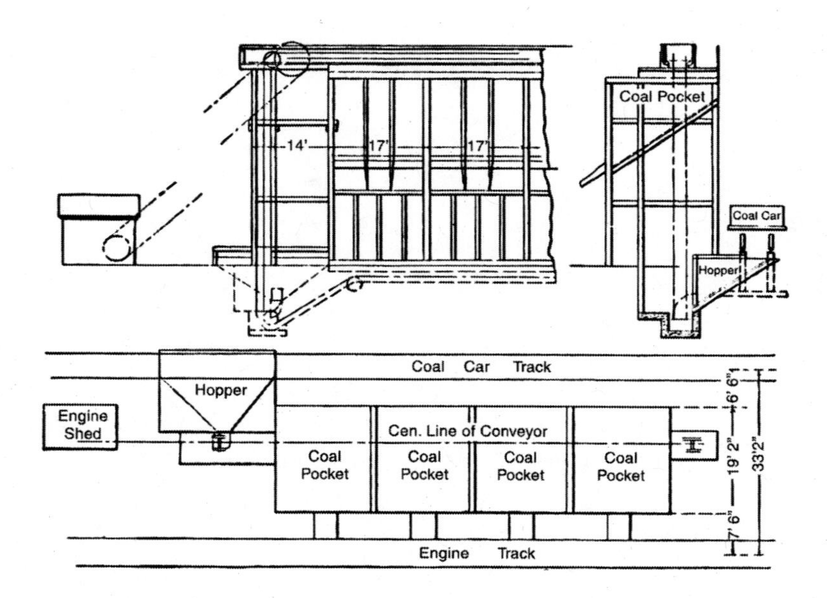

Figure 3.15. Four-pocket McHenry coaling plant. Reprinted from Orrock 1909: 146.

the coal car was spotted over the hopper and dumped, the coal running by gravity into the boot, where it was hoisted by endless chain and buckets to the pockets. Gates were provided to each pocket so that the coal could be dumped into the one desired by leaving the gate open. In these plants the coal car proceeded on one track and the engine on a second track.

A second type of coaling station was the four-pocket, three-track plant. It had two double elevated coal pockets, located between three tracks and connected together on top by a house spanning two tracks. The elevating mechanism consisted of an endless chain and buckets and a steel boot. Coal cars ran along the outside tracks while locomotives ran along the two inner tracks.[35]

Ash Pile and Cinder Handling

A by-product of the burning of coal in engines was the production of great amounts of ash. To dispose of this waste, railroads provided pits into which ash was deposited for later removal. The pit was constructed by excavating beneath the level of the track and building an open-top masonry enclosure with a suitably drained masonry floor and walls on which a line of pedestals was placed to support the girders that carried the track rails over the pit.

The walls and floors of ashpits were usually made of concrete, while the pedestals were usually made of iron. Several methods were developed for removing ashes. One method used a narrow-gauge steel tram car to carry the ashes away, while another used a steam jet. Dump buckets were also employed.[36]

Waterfront Terminals

The transfer of ordinary freight between rail and water was accomplished in various ways: first, by car floats or ferries, using float bridges to adjust for varying depths of water; second, by lightering, transferring between car and vessel or pier by use of a barge or self-propelled vessel; or third, by car and vessel direct, either trucking freight to vessel or vice versa.

The float was brought up to a transfer bridge, a hinged bridge with means of adjusting the free end to match the level of the track on floats of different heights and at different stages of the tide on tidal waterways. A late nineteenth-century car transfer bridge located on the Atlantic & Danville Railroad in West Norfolk, Virginia, made this adjustment via chains passing over a pulley in the top of a gallows frame over the free end of the bridge.[37]

Coal Piers

Piers where coal was transferred from cars to boats or from boats to cars were located at many freshwater and tidewater ports and were equipped with mechanical apparatuses to facilitate rapid, and economical, movement (Figure 3.16).

A typical coal pier is the Pennsylvania Railroad's Canton Coal Pier in the Canton area of Baltimore, built in 1916–1917. This pier carried coal from railroad cars out to the pier by 4-ton cable-drawn dump cars that transferred the coal into the ship by traveling loaders. The railroad hopper or "road" cars were loaded and were run one at a time down to the foot of the incline. A cable-propelled "barney," a narrow-gauge electric locomotive, then pushed them up to the car dumper. The 100-ton hopper under the car dumper was fed by gravity into cable cars that traveled on an endless, reversing track. The cars were hauled up a grade to the steel superstructure on the pier where the coal was delivered to traveling loaders. The loaders moved along the pier on a two-rail track, carrying a hopper and a chute for emptying the coal onto the barges. The emptied cars then moved by the force of gravity to the empty yard.[38]

Ore and Lumber Docks

Because water transportation was substantially more economical than rail for bulky commodities, often freight was transported from cars or vessels or vice versa, particularly on the Great Lakes. In many instances, handling included storage in addition to transfer. Waterfront facilities required machinery for transferring cargo between cars and vessels, as well as additional machinery for unloading for storage.

For bulk loading and unloading of ore, Hulett unloaders were employed. The prototype of the unloader was developed for Andrew Carnegie in 1899. The unloader consisted of two parallel girders at right angles to the length of the dock, mounted on trucks that supported the trolley or carriage which, in turn, carried the walking beam, the end of which supported a vertical leg with a bucket at its lower end. The bucket leg was suspended in a vertical position, and the operator rode

Figure 3.16. High-lift dump. Port Covington Terminal. Coal Pier #4. Baltimore, Maryland. HAER MD-75, National Park Service. Photograph by Philip Szczeranski, 1988.

in it just over the bucket. By means of a hoisting mechanism, the beam oscillated up and down, carrying the bucket up over the hatch or to the bottom of the hold. When the bucket reached the pile of ore, it was closed and filled. Then the leg was raised and trolleyed back over the hopper on the dock, into which the contents of the bucket were discharged. From the hopper, the ore was dumped into an auxiliary bucket car, which, in turn, transferred the ore to the cars. The bucket had a capacity of about 10 gross tons.[39] While about 80 unloaders once operated in the United States and Canada, in 2000, only two sets remained, four in Cleveland and a pair at the LTV Steel Coke Plant in South Chicago. The advent of the self-unloading freighter doomed the Hulett unloaders.[40]

In the 1920s, the world's largest ore-shipping dock was that of the Duluth, Missabe & Northern on Duluth-Superior Harbor, Minnesota-Wisconsin. The 2,304-foot-long, 76-foot-wide, and 84-foot-tall superstructure had a structural steel frame with concrete partition wall, pocket walls, and sidewalk slabs. The entire pocket structure was carried by two lines of steel columns, arranged in pairs to form bents. There were 384 pockets with 6,540 cubic foot capacity. Eight standard 50-ton ore cars could be dumped into each pocket. The top of the dock was 76 feet 5 inches wide.[41] Few ore docks survive in the Great Lakes region. In addition to those in Duluth, two ore docks remain on the Superior lakefront in Marquette, Michigan (Figure 3.17), one of which still loads iron ore pellets into freighters.

Scales

In the 1920s, track scales for freight service ranged in capacity from 100 to 150 tons, with a platform length of 36 to 60 feet. Scales were often placed on or just beyond the hump in a classification yard. Several railroads had standard track scales. The 50-foot DL&W track scale,

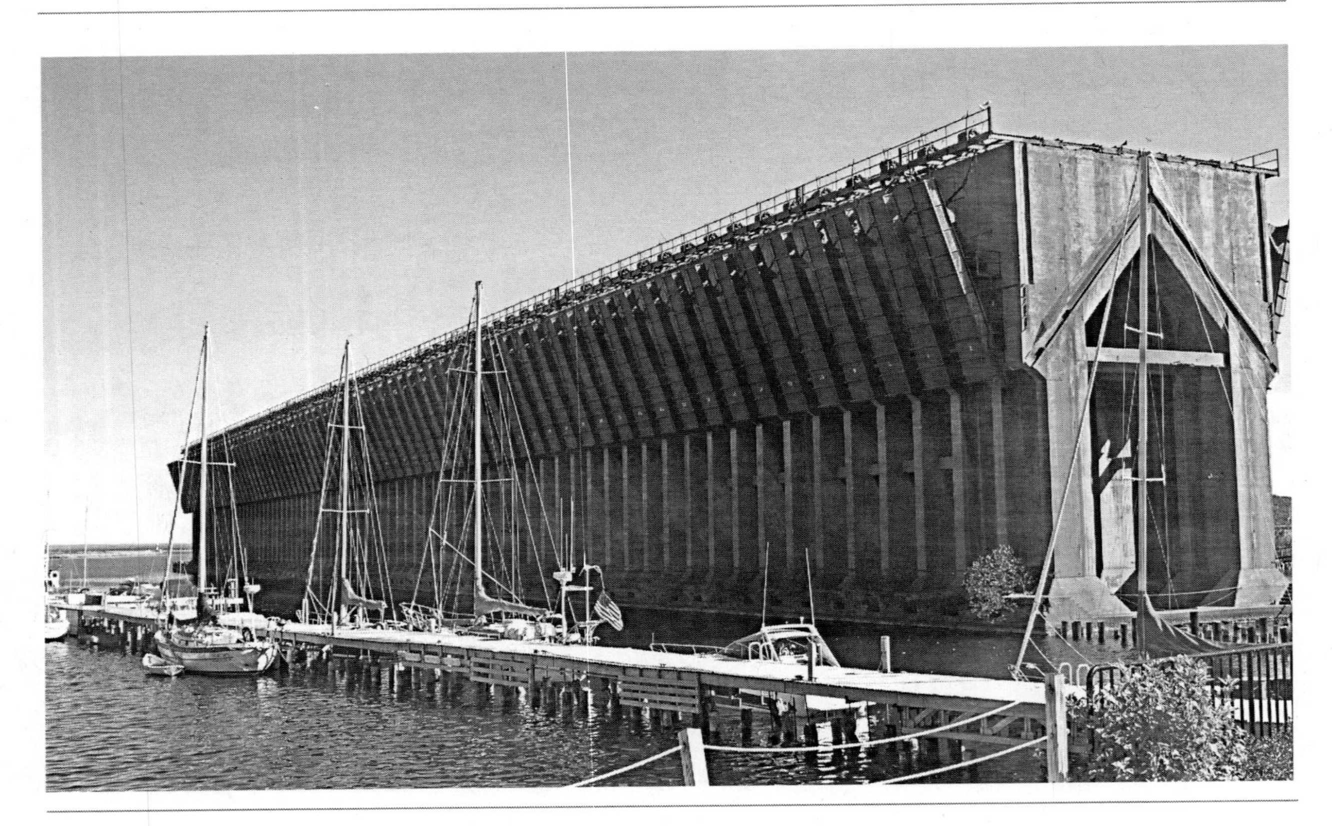

Figure 3.17. Ore Dock #6. Marquette, Michigan. Photograph by the author, 2007.

built by the Fairbanks Scale Company in 1926, had a capacity of 167 tons.[42]

Railroad Trestles and Pile Bridges

Trestle bridges were erected to carry railroads across deep ravines. Wood-framed trestles, usually made from yellow pine or Douglas fir, were cheaper in initial cost than earth embankments when the height exceeded 15 to 20 feet (Figure 3.18). Steel trestles were used for permanent structures across deep ravines or gorges in mountainous areas. In the twentieth century, concrete trestles also began to be used.

A trestle bridge consists of a floor system like that of a stringer bridge resting on trestle bents instead of retaining wall abutments and piers. The bent for a single track consists of a sill on which rest four posts, two vertical or plumb posts, and two batter posts. The posts carry a cap. The sill rests on a foundation consisting of one or more mud sills or masonry piers or piles. Bents are strengthened by sway bracing, one plank diagonally on each side of the bent.[43]

Ballast Deck Trestles

Although timber trestles were usually constructed with an open deck, ballast decks were sometimes used to reduce the hazard of fire. The simplest ballast deck is an open box formed by placing timber flooring over the stringers.

The concrete trestle usually employs a ballast deck. This trestle sometimes rests upon open concrete bents, each composed of four concrete columns connected by transverse struts or caps, footed on concrete pedestals or piles. In other cases, narrow concrete piers are used. The bent spacing generally ranges from 15 to 20 feet. Usually the floor slab is poured in place after the bents are constructed.[44]

Pile bridges are erected as temporary or permanent structures across tidal estuaries, marshes, lakes, and moderate-sized, soft-bottomed, sluggish streams.[45] Pile bridges employ the same floor system as trestle bridges and differ only in that pile bents are used instead of framed trestle bents. As with trestle bents, sway braces are employed.

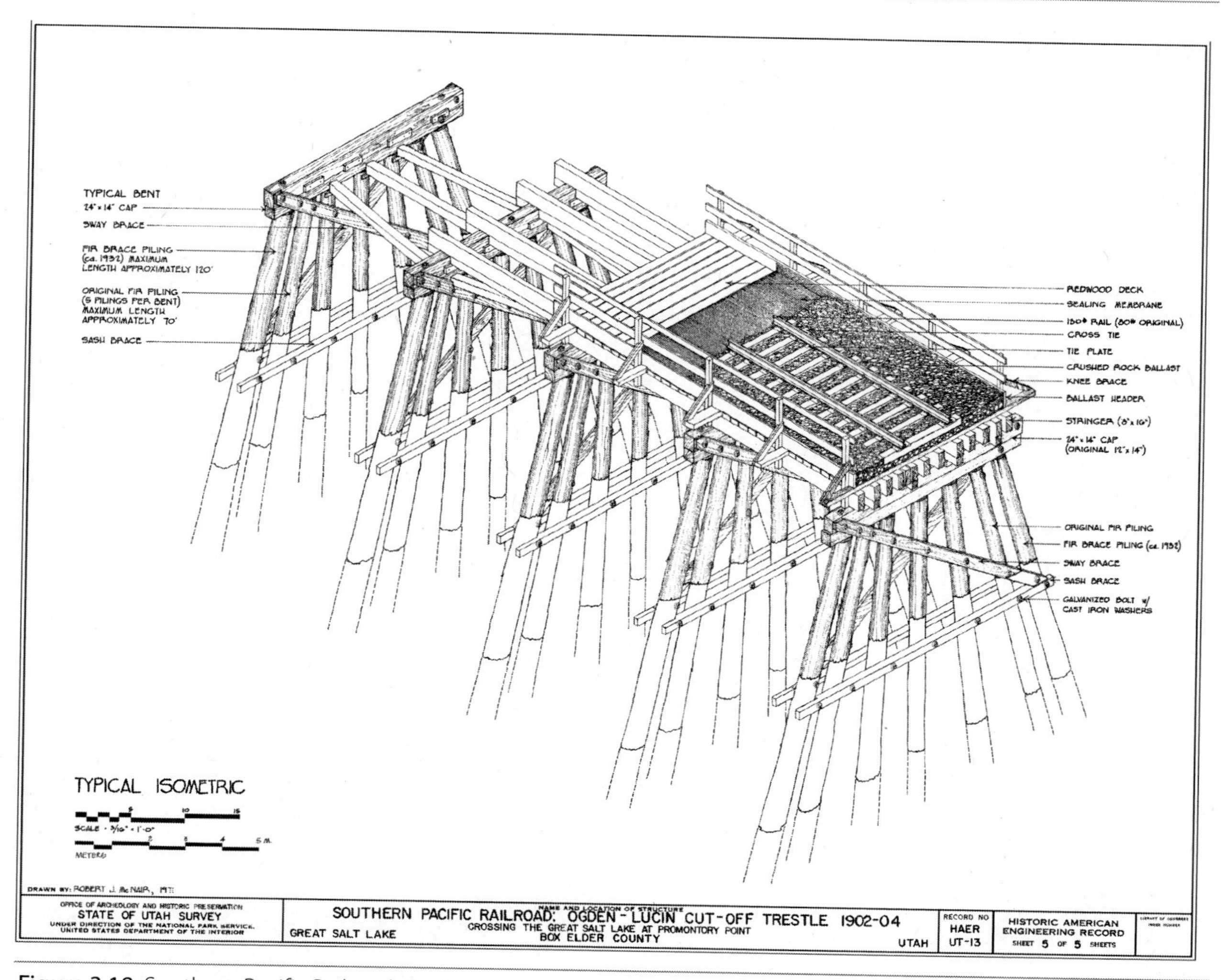

Figure 3.18. Southern Pacific Railroad: Ogden Cut-off Trestle, 1902–1904. HAER UT-13, National Park Service. Delineated by Robert J. McNair, 1971.

Tunnels

The oldest railroad tunnel in the United States is the Staples Bend Tunnel near Johnstown, Pennsylvania, a 901-foot tunnel with a 19-foot-high and 20-foot-wide arch section constructed during 1831–1832. In service only a short time, it survives as part of the Allegheny Portage National Historical Site. Other early tunnels include the Norwich and Worcester Railroad's tunnel in Tafts, Connecticut, and the Philadelphia and Reading's Black Rock Tunnel near Phoenixville, Pennsylvania. These early tunnels were generally constructed by drilling holes through rock with a hammer and chisel and then blasting with black powder. Sometimes the surrounding rock was sufficiently firm and stable that no additional support was required, but frequently a timber or masonry lining was used.

The technology of tunnel construction changed in the 1860s with the introduction of compressed air drills. In the late nineteenth century, advances were made in underwater tunneling with the introduction of the hydraulic cast-iron shield. As the shield was moved ahead by hydraulic jacks, material from the river bottom entered the shield through doors and was removed by tunnel workers. During the twentieth century, advances in tunneling technology included the introduction of high-speed drills that could excavate the entire tunnel cross section at once. Prefabricated steel sections have largely taken the place of temporary timber supports, while precast concrete sections have been used in place of cast-iron tunnel liners. Major tunnel components include the portals, the liner, and, in many cases, one or more ventilation shafts.[46]

Among the longest railroad tunnels in the United States are the 7.8-mile Cascade Tunnel near Everett, Washington, completed in 1929, and the 6-mile Moffatt Tunnel in Colorado, completed in 1928.

Culverts

When a waterway to be spanned is not wide enough to justify a bridge, a culvert, usually a box, pipe, or arch culvert, is used. Wood box culverts consist of rectangular boxes as long as the width of the bottom of the embankment and are usually built of heavy timbers. A stone box culvert is built of rough, flat stone. The walls or ends are laid in cement mortar. Stone culverts are generally not built where the span is more than 4 feet. The pipe culvert consists of one or more lines of vitrified earthen, cast-iron, or concrete pipe with head walls of rubble or concrete masonry at the ends.[47]

RAILROAD SIGNS AND STRUCTURES

Information content of railroad signs can be divided into three types: for train operation, for employees, and for informing or warning the public.

Signs for Train Operation

Among signs for train operation are signs of uniform design: the milepost signs indicating the distance of a crossing, junction, station, yard or draw; whistle posts that warn of highways or stations; flanger signs; yard limit boards; and stop, reduce speed, or resume speed boards. The recommended color scheme for these signs was black on white with 5- or 6-inch letters.

Mileposts

Mileposts are used to show distance from a terminal or junction point. Designs of mileposts are generally railroad-specific. Pennsylvania Railroad mileposts are typically placed on the north side of the track and are numbered from major terminals such as New York, Philadelphia, Harrisburg, and Pittsburgh.[48]

Early twentieth-century mileposts display "1 M" for 1 mile at the top, and below it a cross board identifying the site or structure in that distance, such as a crossing, draw, junction, or station. A Camden and Amboy Railroad milepost in Burlington County, New Jersey, a cast-iron obelisk, is shown in Figure 3.19. Modern mileposts on the same rail line are simple metal sheets, black numbers on a white background, mounted vertically on a metal stake.

Whistle Posts

Early whistle post signs consist of a triangular-topped board carrying the letter *W* attached to a post (Figure 3.20). Whistle posts are located in advance of places where whistle signals are required by operating rules, such as highway grade crossings. The posts are located a sufficient distance from the crossing, often one-quarter mile, for the train to be able to sound the required grade crossing signal, two long–one short–one long, while traveling at maximum authorized speed, with the last long held until the train is on the crossing. Design of whistle posts is railroad-specific. Typically such posts are concrete posts marked with *W*. In earlier years, the post could be a wood, metal, or concrete board or a section of a telephone pole with the letter painted, cast, or indented in it. Modern whistle posts on the same line are a black *W* with a white background mounted on a metal post.

Figure 3.19. Camden and Amboy Railroad milepost, Burlington County, New Jersey. Photograph by author, 2007.

Figure 3.20. Camden and Amboy Railroad whistle post, Burlington County, New Jersey. Photograph by author, 2006.

Flanger Signs

A flanger in railroad terminology is a snowplow blade. In the 1920s, a flanger sign consisted of a black dot on a triangular sign, indicating when the flanger operator should lift the plow blade before a switch, a grade crossing, or anything between the rails that could snag the plow blade.

Subsequently railroads introduced signs resembling upside-down hockey sticks. A upward angling blade indicated that the plow blade should be raised for an obstruction, a downward angling blade placed after an intersection indicated that it was safe to lower the blade, while an upward angling blade with a horizontal blade meant to raise the flanger due to multiple obstructions.

Yard Limit

One type of yard limit sign is a board shaped like a carpenter's square carrying the words "YARD LIMIT" on the two angles. Another version of the yard limit sign consists of a cast-iron sign, measuring approximately 1 foot 6 inches high and 3 feet 6 inches wide mounted 8 feet above the ground, also imprinted with "YARD LIMIT." The sign was used to mark the transition from a rail yard to travel tracks.

Stop, Reduce, or Resume Speed

These signs consist of an angle board attached to a post with either the speed reduction or the new speed indicated.

Signs for Employees

These signs include bridge numbering board; section, subdivision, and valuation section posts; property posts; and curve and elevation posts. The bridge sign, section post, and combined subdivision and section post consist of a metal plate mounted on a wrought-iron post.

The property line was marked by a 5-foot-long post of concrete, 2 feet of which projected above the ground.

Public Warning Signs

Railroad crossing signs are now standardized. In the early twentieth century, these signs differed depending on the railroad and the location. Nearly all such signs were white with black letters. The American Railway Engineering Association recommended a highway crossing sign, mounted on a 16-foot pole with 8-foot-long cross buck members.

Trespass Signs

Early twentieth-century trespass signs are made of cast iron with raised borders and letters and a wrought-iron post. The signs, typically rectangular with rounded corners, usually measure 2.5 feet wide by 1.5 feet tall.

Approach Warning Signs

In the 1920s the standard approach warning sign consisted of a metal disk painted black on white with a painted black diameter central cross and "RR" in the two upper quadrants. The total height of the pole and sign was 4 to 5 feet above grade.

Crossing Gates

Crossing gates were widely introduced in the 1920s in response to substantial increases in automobiles. Early gates consist of a long, tapered wooden arm pivoted to a bearing near the counterweighted end. The gate was raised or lowered by means of a hand or power rotating device, with an associated mechanical bell or gong. The arms were painted with alternating black and white diagonal strips. Later, the gates could be wire-connected or pipe-connected and hand operated, electrically operated, or pneumatically operated. The gate shanty was usually located in a small elevated building near the edge of the roadway with a view of the tracks.[49] Many gates were supplemented by cross buck signs with attached flashing lights.[50]

Bumpers and Car Stops

Stopping railroad cars could be effected in several ways (Figure 3.21): through a structure sufficiently rigid to destroy the car if necessary rather than incur the greater damage of a runaway car; through a bumper designed to receive the force and provide protection against the shocks of fairly severe impacts; by a skate device which retarded and eventually stopped the car by friction; or by a device secured to the rails.[51]

The concrete bumper consisted of a foundation of mass concrete with a bumper impact wall. The bumper was equipped with steel draft gear cylinders that received the impact. A variation of this design was employed at the station end of a terminal track in numerous urban stations.

A bumping post was usually a structure in which the tension members were extensions of the track rails or special rails bolted to the track rails with a flangeway between to permit the car to run beyond the bumper without derailment should the bumper be dislodged by the impact. The shock was received on a bumper face at the height of the couplers and was transferred downward through suitably designed wooden posts on a substantial foundation. Commer-

Figure 3.21. Bumpers: ***a,*** Buda All-Steel Bumping Post; ***b,*** Ellis Freight Bumping Post; ***c,*** Hercules No. 1 Freight Bumping Post (Howson 1926: 308).

cial bumpers were of all-metal construction and were specially designed to occupy a minimum of track room and to be easily dismantled.[52]

GLOSSARY[53]

Absolute signal. A signal whose "stop" indication means "stop and stay."

Ballast. Material placed on the roadbed to hold the track in line.

Ballast curb. A longitudinal timber placed along the outer edge of the floor on ballast deck bridges to retain the ballast.

Bench wall. The abutment or side wall of a culvert or tunnel.

Block. A length of track of defined limits, with block signals or cab signals, or both, to govern its use by trains and engines.

Block signal. A fixed signal at the entrance of a block to govern trains and engines entering and using that block.

Body track. Each of parallel tracks of a yard, upon which cars are switched or stored.

Bridge tie. A transverse timber resting on the stringers and supporting the rails on a bridge.

Bumper. A permanent device used to stop rolling stock at an established limiting point.

Car ferry apron. A bridge structure supporting tracks connecting the car deck of a car ferry with the tracks extending to land, hinged at the shore end so that it is free to move vertically.

Car retarder. A braking device, usually power-operated, built into a railway track to reduce the speed of cars by brake shoes which press against the sides of the lower portions of the wheels.

Car stop. A temporarily installed device used to stop rolling stock at a particular point.

Catenary. An overhead conductor contacted by a pantograph or trolley and in which the trolley wire is suspended by hangers from a naturally sagging upper wire.

Classification yard. A yard in which cars are classified or grouped in accordance with requirements.

Closure rails. The rails between the parts of any special trackwork layout such as between the switch and the frog in a turnout.

Crossover. Two turnouts with the tracks between the frogs arranged to form a continuous passage between two nearby and generally parallel tracks.

Derail. A track structure for derailing rolling stock in case of an emergency.

Drill track. A track connecting with the ladder track, over which locomotives and cars move back and forth in switching.

Electrification. The installation of overhead wire or third-rail power distribution facilities to enable operation of trains hauled by electric locomotives.

Flag station. A station at which a train would not stop to take passengers unless it had been "flagged" by a hand or lantern signal. The simplest form of flag station was a landing that often consisted of a gravel or cinder platform 60 to 100 feet in length.

Flanger. A train-mounted snowplow blade.

Flat yard. A yard in which movement of cars is accomplished by a locomotive without assistance of gravity.

Frog. A track structure used at the intersection of two running rails to provide support for wheels and passageways for their flanges, thus permitting wheels on either rail to cross the other.

Gravity yard. A yard in which classification of cars is accomplished with the assistance of gravity.

Guard rail. A rail or other structure laid parallel with the running rails of a track to prevent wheels from being derailed.

House track. A track alongside of or entering a freight house.

Hump yard. A yard in which the classification of cars is accomplished by pushing them over a summit beyond which they run by gravity.

Ladder track. An extended track connecting either end of a yard with the main track.

Main track. A track extending through yards and between stations, upon which trains are operated by timetable or train order.

Point rail. The tapered rail of a split switch, also known as a *switch rail* or *switch point.*

Rail. A rolled steel shape, commonly a T-section, designed to be laid end to end in two parallel lines on cross ties to form a track for railway rolling stock.

Receiving yard. A yard for receiving trains.

Running rail. The rail or surface on which the tread of the wheel runs.

Running track. A track reserved for movement through a yard.

Split switch. A switch consisting essentially of two movable point rails with the necessary fixtures.

Spur track. A stub track of indefinite length diverging from a main line or track.

Stringer. A longitudinal member extending from bent to bent and supporting the track.

Stub track. A track connected with another at one end only.

Switch. A track structure used to divert rolling stock from one track to another.

Switcher. A switching locomotive, used for shifting or switching cars in yards and terminals.

Through (side) station. A station whose principal tracks are continued through the structure so that a train proceeding in either direction can continue its run in the same direction.

Track apron. Railroad track along the waterfront edge of a wharf or pier for transfer of cargo between ship and car.

Wye. A triangular arrangements of tracks on which locomotives, cars, and trains may be turned.

Yard. A system of tracks within defined limits provided for making up trains, storing cars, and other purposes.

BIBLIOGRAPHY

American Railway Engineering Association. *Proceedings of theAnnual Convention of the American Railway Engineering Association.* Chicago: American Railway Engineering Association, various years.

Berg, Walter G. *Buildings and Structures of American Railroads: A Reference Book for Railroad Managers, Superintendents, Master Mechanics, Engineers, Architects, and Students.* New York: John Wiley and Sons, 1893.

Bryan, Frank W., and Robert S. McGonigal. Railroad Signals: What They Do, What They Mean. *Trains Magazine*, January 2003. Online at www.trainsmag.com/Content/Dynamic/Articles/000/000/003/0351sexb.asp.

Carpenter, Richard C. *A Railroad Atlas of the United States in 1946*, volume 2: *New York and New England.* Baltimore: John Hopkins University Press, 2004.

Clemensen, A. Berle. *Delaware, Lackawanna and Western Railroad Line, Scranton to Slateford, Pennsylvania: Historic Resource Study.* Washington, DC: National Park Service, 1991.

Droege, John A. *Passenger Terminals and Trains.* Reprinted in 1969 by Kalmbach Publishing Company, Milwaukee. New York: McGraw-Hill Book Company, 1916.

—. *Freight Terminals and Trains.* 2nd edition. New York: McGraw-Hill Book Company, Inc., 1925.

Fairbanks-Morse & Company. *Fairbanks-Morse Locomotive Coaling Stations.* New York: Fairbanks-Morse, n.d.

Howson, Elmer T. *Railway Engineering and Maintenance Cyclopedia.* 2nd edition. New York: Simmons-Boardman Publishing Company, 1926.

Jonnes, Jill. *Conquering Gotham: A Gilded Age Epic: The Construction of Penn Station and Its Tunnels.* New York: Penguin, 2005.

King, Edgar Everett. *Railway Signaling.* New York: McGraw-Hill, 1921.

Middleton, William D. *Landmarks on the Iron Road: Two Centuries of North American Railroad Engineering.* Bloomington: Indiana University Press, 1999.

Orrock, John Wilson. *Railroad Structures and Estimates.* New York: J. Wiley & Sons, 1909.

Railroad Master Mechanic Editorial Staff. *Railroad Shop Up to Date: A Reference Book of Up to Date American Railway Shop Practice.* Chicago: Crandall Publishing Company, 1907.

Raymond, William Galt. *The Elements of Railroad Engineering.* New York: J. Wiley & Sons, 1909.

NOTES

1 These and other records are described in David A. Pfeiffer's three-part article on the National Archives website, "Riding the Rails Up Paper Mountain: Researching Railroad Records in the National Archives, " 1997. The URL for the first part (which includes a discussion of railroad valuation records) is www.archives.gov/publications/prologue/1997/spring/railroad-records-1.html?temp.

2 Track Page. Railway Technical Web Pages, www.railway-technical.com/track.html; David Weitzman, *Traces of the Past: A Field Guide to Industrial Archaeology* (New York: Charles Scribner's Sons, 1980), 13.

3 Track Page. Railroad Technical Web Pages.

4 Elmer T. Howson, editor, *Railway Engineering and Maintenance Cyclopedia*, 2nd edition (New York: Simmons-Boardman Publishing Company Co., 1926), 306.

5 Weitzman, 24–25.

6 Howson, 257.

7 William G. Raymond, *The Elements of Railroad Engineering*, 3rd edition, revised (New York: John Wiley & Sons, Inc., 1917), 82–83.

8 Howson, 261.

9 Howson, 304–305.

10 Frank W. Bryan and Robert S. McGonigal, "Railroad Signals: What They Do, What They Mean," *Trains Magazine*, January 2003. Online at www.trainsmag.com/Content/Dynamic/Articles/000/000/003/0351sexb.asp, 2.

11 Bryan and McGonigal, 6–7.

12 John A. Droege, *Freight Terminals of Trains*, 2nd edition (New York: McGraw-Hill Book Company, Inc., 1925), 63.

[13] John A. Droege, *Yards and Terminals and Their Operations* (New York: Railroad Gazette, 1906), 82–83.

[14] B. W. Benedict, editor, *Railroad Shop Up to Date: A Reference Book of Up to Date American Railway Shop Practice* (Chicago: Crandall Publishing Company, 1907), 9–11.

[15] Pennsylvania Railroad Company, *Pennsylvania Railroad Company, Altoona Shops and Motive Power Statistics* (Altoona: Office of the General Superintendent, Motive Power, 1890) as cited in *Pennsylvania Railroad Shops and Works: Special History Study* (Appendix B). www.cr.nps.gov/online_books/railroad/shsab.htm.

[16] Benedict, 36–37.

[17] Benedict, 92.

[18] American Railway Engineering Association, "General Layout and Design of Passenger Car Shops," in *Proceedings of the 27nd Annual Convention of the American Railway Engineering Association* (Chicago: American Railway Engineering Association, 1926), 294–307.

[19] Benedict. 107.

[20] Benedict, 171.

[21] Benedict, 74.

[22] Benedict, 136.

[23] Benedict, 127.

[24] Droege 1925, 299.

[25] American Railway Engineering Association, "Freight House Design," in *Proceedings of the 24th Annual Convention of the American Railway Engineering Association* (Chicago: American Railway Engineering Association, 1923), 691, 704.

[26] Howson, 792–793.

[27] John A. Droege, *Passenger Terminals and Trains* (New York: McGraw-Hill Book Company, 1916), 41–42; reprinted in 1969 by Kalmbach Publishing Company, Milwaukee.

[28] Lawrence Grow, *Waiting for the 5:05: Terminal, Station and Depot in America* (New York: Main Street/Universe Books, 1977).

[29] Droege 1916, 253–254.

[30] American Railway Engineering Association, *Manual of Recommended Practice, American Railway Engineering Association, C &M Section-Engineering Division—AAR* (Chicago: American Railway Engineering Association, 1954), 6-23-1.

[31] American Railway Engineering Association, "Ice Houses and Icing Stations," in *Proceedings of the 23rd Annual Convention of the American Railway Engineering Association* (Chicago: American Railway Engineering Association, 1922), 842–843.

[32] Howson, 597.

[33] George D. Torok, *A Guide to Historic Coal Towns of the Big Sandy River Valley* (Knoxville: University of Tennessee Press, 2004), 103.

[34] John Wilson Orrock, *Railroad Structures and Estimates* (New York: John Wiley and Sons, 1909), 145–148.

[35] Howson, 581.

[36] Anonymous, "Car Transfer Bridge at West Nofolk, Va. Atlantic & Danville RR," *Engineering News*, September 19, 1895.

[37] C. Crawford, Pennsylvania Railroad, Canton Coal Pier, Historic American Engineering Record, MD-34, 1988.

[38] Droege 1925, 279.

[39] Hardlines Design Company, Hulett Ore Unloaders, 1912–1992, OH-18, 2000.

[40] Droege 1925, 282–283; a detailed description of Ore Dock #6 is contained in William D. Middleton, *Landmarks on the Iron Road: Two Centuries of North American Railroad Engineering* (Bloomington: Indiana University Press, 1999), 154–157.

[41] Droege 1925, 219–220.

[42] William G. Raymond, *The Elements of Railroad Engineering* (New York: J. Wiley & Sons, 1909), 67–68.

[43] Howson, 532.

[44] Raymond, 58.

[45] Middleton, 103–108.

[46] Raymond, 59–60.

[47] Al Buchan, Right-of-Way and Track Details. New Jersey Division Meet National Model Railroad Association Clinic Handout. http://mywebpages.comcast.net/njdivnmra/boxcar/alb9901.html.

[48] Howson, 868.

[49] American Railway Engineering Association, "Signs," in *Proceedings of the 25th Annual Convention of the American Railway Engineering Association* (Chicago: American Railway Engineering Association, 1924), 625, 630.

[50] Howson, 308–309.

[51] Howson, 309.

[52] Howson, 309.

[53] Selected definitions were taken or paraphrased from the American Railway Engineering Association's *Glossary* (1953).

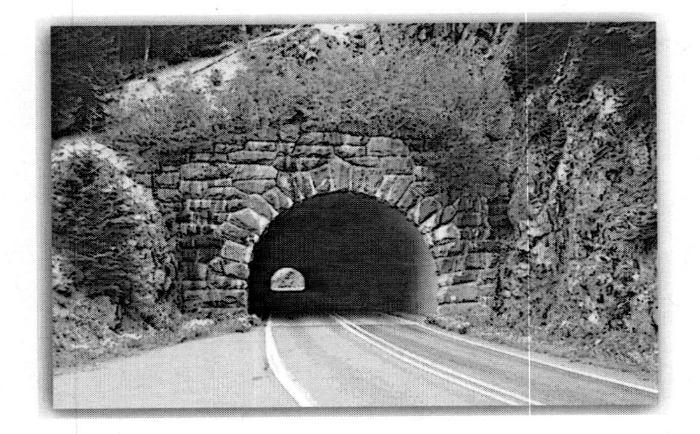

CHAPTER 4

Roads and Highways

In recent years, many cultural resource researchers have spent an increasing amount of their time involved with the survey of historic road corridors. While roads were at one point considered merely the spine on which buildings were located, they are now considered as historic resources in their own right. This chapter describes some of the elements of road and highway materials and the configurations that may be found in road corridor surveys.

HIGHWAYS IN THE UNITED STATES

One of the first surfaced highways in the United States was the Lancaster Turnpike from Lancaster, Pennsylvania, to Philadelphia, begun in 1792 and completed two years later. In 1811, construction was begun on the National Road, the first federally financed highway, later known as U.S. Route 40. This road started in Maryland and extended west through southern Pennsylvania, central Ohio, and Indiana to the Illinois state line.

Beginning in the 1840s, the plank road came to prominence in Canada and in some states such as New York and Indiana. This road surface, constructed of longitudinal logs, soon faded from popularity.

Around 1900, bituminous macadam began to be widely used on rural roads of the United States. The first concrete rural road in the United States was constructed in 1909 in Wayne County, Michigan, and concrete soon became the predominant material for rural highways.

The first limited access highway in the United States was the Long Island Motor Parkway, a private toll road that eventually stretched 45 miles from Queens to Lake Ronkonkoma. It opened its first section in 1908. The first modern superhighway in the United States, the Pennsylvania Turnpike, began construction in October 1938. Touted as an "all-weather" highway because of its tunnels and uniform road surface, access was limited to a small number of interchanges, curves were banks, and grades were shallow.[1]

By 1940, road mileage in the United States totaled about 3 million. Of this total, approximately 1.15 million was surfaced in some manner.

REMNANTS OF EARLY ROADS

The most obvious remnants of historic road corridors are bridges (see Chapter 2). Less imposing but equally diagnostic of a historic road corridor are mileposts.[2] Typically of stone construction, these markers indicate the distance from the site to nearby towns or cities. Many historic roadways still feature standing milestones, including portions of the Post Road, the National Road, King's Highway, and locally significant roads such as Delaware's Concord Pike and Massachusetts's Bay Road.

Several websites include limited inventories of identified milestones. Several National Road websites list standing milestones on that road, a few of which are listed in the National Register. Adam Paul and

Steve Okonski report milestones on the historic roads of the Baltimore area, including York Road, the Frederick (National) Road, Falls Road, and Philadelphia Road.[3]

Typical milestones were stone shafts, often about 5 feet tall (with the top 3 feet exposed). Most were square in cross section and some had curved heads. The material used depended on the availability of stone. In the Baltimore area, nearly all mileposts were constructed of granite, while the earliest stones on the Ohio section of the National Road were initially constructed in the 1830s of reinforced cementitious materials. The National Road markers weathered poorly, and many were replaced in the 1850s with sandstone markers. As of 2006, 83 existing Ohio National Road stones had been documented.[4]

Many local or regional roads have yet to be comprehensively surveyed for milestones and may yield numerous surviving stones. For example, in the Philadelphia area, Germantown Avenue-Germantown Pike, an eighteenth-century road, may well be marked in places by milestones. Two illustrated milestones sit beside the former Haddonfield Turnpike (present Haddon Avenue) in Camden County, New Jersey (Figure 4.1).

ROAD SURFACES

A wide variety of road surfaces have been used and continue to be used, depending on traffic volume, location, and budget. These range from untreated surfaces, such as gravel, to prepared surfaces of concrete, asphalt, and stone.

The earliest known paved American road was constructed in Pemaquid, Maine, in 1625. In 1825, the first macadam road in the United States was constructed on the Boonsborough Turnpike Road between Hagerstown and Boonsboro, Maryland. Macadam, named for the Scotsman John Loudon McAdam, consists of compacted layers of small stones cemented into a hard surface by means of stone dust and water. In 1877, the first asphalt paving in the United States

Figure 4.1. Mileposts. ***a,*** Haddon Avenue, Haddonfield, New Jersey; ***b,*** Haddon Avenue, Westmont, New Jersey. Photographs by the author, 2007.

occurred on Pennsylvania Avenue in Washington, DC. In 1906, the first bituminous macadam road in the United States was constructed in Rhode Island. In this road type, the foundation is macadam, upon which a bituminous material is poured, penetrating at least 2 inches into the foundation.[5]

Earth Roads

The lightest duty surface used on vehicular roads is well-drained natural earth bladed to a true cross section by a road grader. This surface is often prepared to accommodate a heavier traffic volume by the application of oil.

Gravel Roads

Gravel roads typically consist of a base course of local materials such as mud shells, scoria, or lime rock. Atop the base course is laid gravel, often to a depth of 2 to 4 inches. The gravel comes from natural deposits or is obtained by crushing large pebbles. The top course is spread, harrowed, and smoothed to correct irregularities. Soil mortar may be added to natural gravel to improve binding.

Portland Cement Concrete Pavements

The first all-concrete pavement in the United States was laid in Bellefontaine, Ohio, in 1893. This pavement was laid in blocks about 5 feet square. The resulting large number of joints was objectionable, and later installations employed larger slabs. The real growth in the use of concrete for road surfaces was precipitated by the increasing number of automobiles, which caused the rapid destruction of the then-predominant water-bound macadam roads.

Concrete pavements are usually constructed in one layer, with no distinct line between the foundation and wearing course. When two courses are used, a concrete base is laid first, followed by a concrete or cement-mortar wearing surface. In two-course construction it is permissible to use an aggregate that is less tough and hard for the base course.

According to the American Society for Testing Materials, Portland cement "is the product obtained by pulverizing clinker consisting essentially of hydraulic calcium silicates to which no additions have been made subsequent to calcinations other than water and/or untreated calcium sulfate."[6]

In the mid-twentieth century, three methods were used to prepare concrete for pavement construction. In one method, the aggregates and cement were proportioned at a central plant and the material was hauled in batches to the mixer at the job. The concrete was then deposited in place by means of a bucket traveling over a boom. In the second method, the concrete was mixed at a central plant and transported to the site in watertight trucks. In the third method, cement, aggregates, and water were proportioned at a central plant and mixed while being hauled to the job.

Hot-Mix Bituminous Pavement

In the mid-twentieth century, three basic types of hot-mix pavement were employed according to the grading of the aggregate: sheet asphalt, in which all the aggregate is very fine; fine-graded bituminous concrete, in which the maximum size of the aggregate is ½ inch; and coarse-graded bituminous concrete, in which aggregate as large as 1¼ to 1½ inches is used (Figure 4.2). Among the advantages of bituminous-concrete pavement are that it may be opened to traffic a few hours after it is completed, and it resists the impact and abrasive action of traffic.

A bituminous-concrete wearing course has a substantial base. Most surfaces are laid on a Portland cement concrete base or a "black base" of bituminous concrete. When the wearing course is of sheet asphalt, it was customary to provide a binder course of bituminous concrete and to make each of these courses 1½ inches thick. The purpose of the binder course is to even up the irregularities in the surface of the base course and to prevent shoving or creeping of the wearing course. By the mid-twentieth century, a binder course was rarely employed.[7]

Paving Brick

Once largely superseded, the use of paving brick has increased in recent years as the result of urban revitalization projects. The first brick pavements in the United States were laid in Charleston, West Virginia, in 1871.

Typical paving bricks (Figure 4.3) include those with lugs on one side and one end, those with one side with beveled ends, those with six plane faces, and those

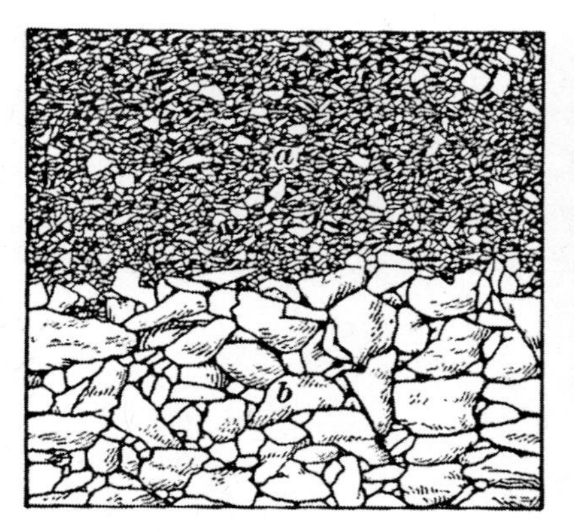

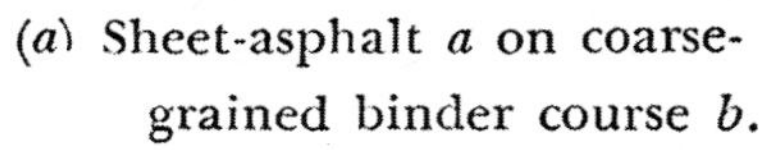

(a) Sheet-asphalt *a* on coarse-grained binder course *b*.

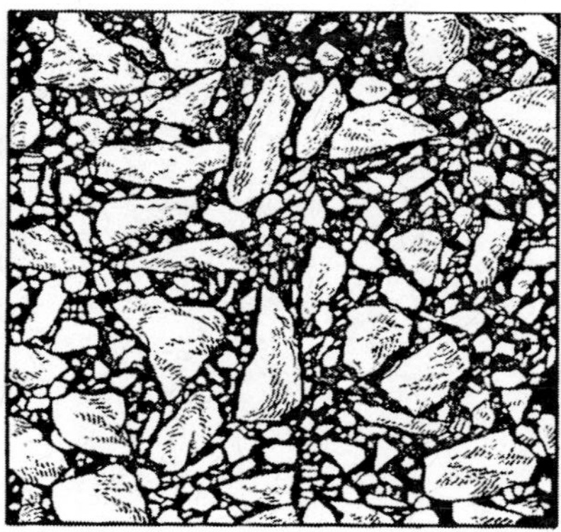

(b) Fine-grained bituminous concrete.

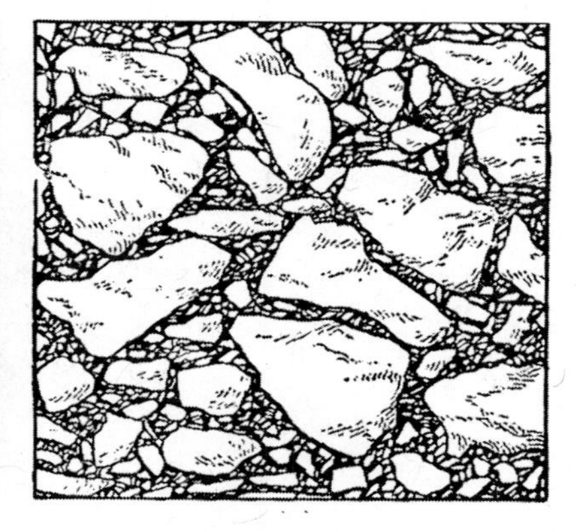

(c) Coarse-grained bituminous concrete.

Figure 4.2. Types of bituminous pavement . Reprinted, by permission of Thomson Education Direct, from Arthur G. Bruce and John Clarkeson, *Highway Design and Construction*, 3rd edition (Scranton, PA: International Textbook Company, 1956), 444.

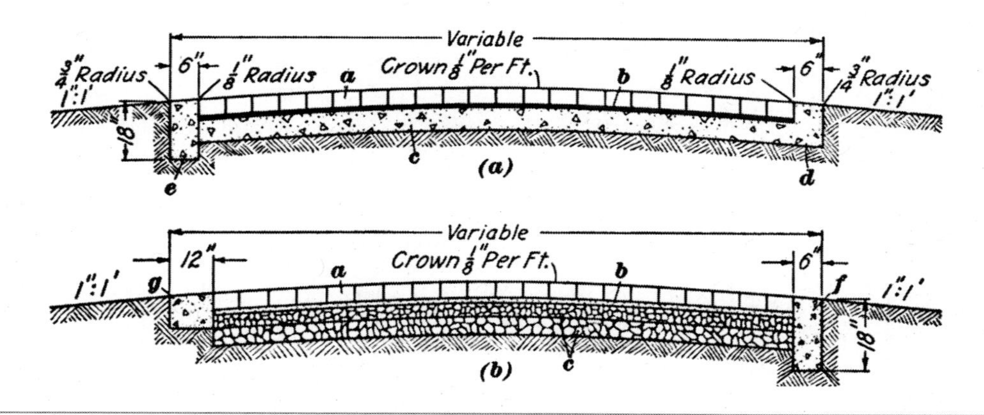

Figure 4.3. Typical cross sections of brick pavements (Bruce and Clarkeson 1956: 563). Reprinted, by permission of Thomson Education Direct.

with repressed lugs. Lug bricks are laid so that plane faces are at the top and bottom. The lugs serve to keep the bricks a sufficient distance apart.

Brick pavement is typically laid with a base course consisting of Portland cement concrete 6 to 8 inches thick. A bedding course is employed to take up any irregularities that may occur in the surface of the base course or any difference in the depths of the individual bricks. The bedding course consists of sand, granulated slag, limestone screenings, cement, or sand.[8]

Stone Block

Stone blocks have been used for street paving for over 2,000 years. In the mid-twentieth century, they were primarily used in pavements subjected to especially severe traffic conditions, such as around dock and warehouse districts or on steep hillsides. Granite is the predominant material used for stone blocks (Figure 4.4), although sandstone and trap rock have also been used. Generally, the length of the blocks ranges from 6 to 12

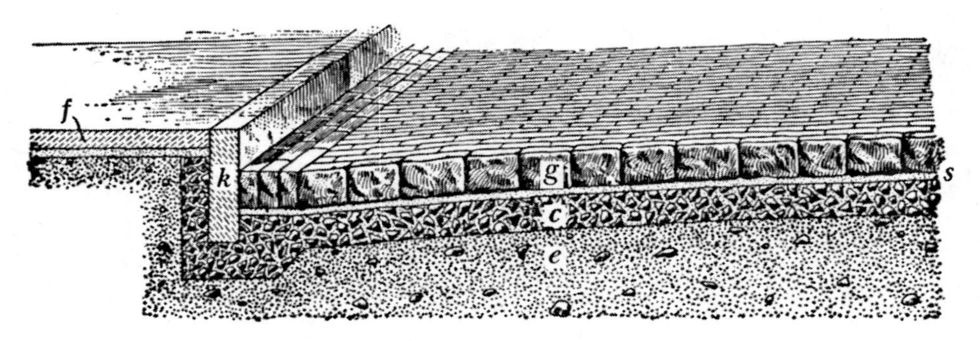

Figure 4.4. Typical section of granite block pavement (Bruce and Clarkeson 1956: 569). Reprinted, by permission of Thomson Education Direct.

inches, the width from 3½ to 4½ inches, and the depth from 4 to 5 inches.

The sequence of materials in stone paving typically includes a natural earth subgrade, a cement-concrete subcourse 8 to 10 inches thick, and a bedding course of sand or sand-cement usually 1 to 2 inches thick. Blocks are laid with the longest dimension at right angles to the curb in close contact so that the joints are narrow. The blocks are rammed into place by heavy rammers or tampers. The joints between the stone blocks are filled with cement grout, tar, pitch, asphalt, or mastic.[9]

Superseded Pavement Materials

Highway department construction reports and accounts of older highways may contain unfamiliar terms that represent discontinued methods of road construction. For example, one of the oldest and best known of the cold-laid types of bituminous surfaces is the Amiesite mixture, composed of crushed stone, liquefier, and asphalt cement. It was developed by Joseph Hay Amies, a Lutheran clergyman who undertook experiments with various paving materials. The stone was dried at the mixing plant where it was run into the mixer and was coated with a liquefier of crude naphtha, kerosene, and/or gasoline; it was then mixed thoroughly with hot asphalt cement and a small amount of powdered hydrate of lime. After the mixing was completed, the coated stone was loaded into cars or trucks and shipped to the job. It was customary to lay Amiesite in two courses, a binder course of about 1½ inches thick and a wearing course ½ to ¾ inch thick.[10]

Metal was the term used for stone chips used on roads. In roads sometimes referred to as tarmac, these stone chips were mixed with tar to form tarmac. Macadam or tarmac pavements can sometimes be uncovered beneath asphalt concrete or Portland cement concrete pavements.

CURBS

In cities and suburban areas, pavement edges are typically defined by curbs formerly constructed of granite, sandstone, or limestone and now typically of concrete. Curbing at parking areas and sidewalks is usually 6 to 8 inches in height with a nearly vertical outer face. Sloping curbs or island curbs are used around islands in intersections, where traffic is heavy, for protection of a raised median strip, or to make islands or the outsides of sharp curves more visible to the driver.[11]

Stone Curbs

Curbstones vary from 4 inches to a foot in width, from 8 to 24 inches in depth, and from 3 to 8 feet in length. The front face of the curb is typically hammer-dressed with a slight batter to a depth somewhat greater than that exposed above the roadway pavement. Where the sidewalk extends to the curbing, the back of the curb is dressed to a sufficient depth to allow the sidewalk pavement to fit closely against it. The top surface of the curb is dressed to a bevel corresponding to the slope of the adjoining sidewalk.[12]

Concrete Curbs

Concrete curbs were initially introduced in areas of lighter duty traffic, such as residential districts, parkways, and private drives. In areas with heavy truck traffic, the concrete curb can be strengthened by embedding a metal protector or a steel angle in the exposed corner of the curb. Where the face of the curb is curved or sloped so that wheels strike the face and not the edge, protective bars are not necessary.

Three general types of concrete curbs have been employed: the separate curb, the combined curb and gutter, and the integral curb. Separate curbs are often built after the pavement is completed. In the combined curb and gutter, the curb and gutter are cast in one piece, a construction commonly used for streets where the wearing course is other than concrete. In the integral curb, the curb is built monolithically with concrete pavement. Integral curbs have the advantage of acting as thickened edges for the concrete pavement, providing strength.

INTERSECTIONS

At-Grade Intersections

The American Association of State Highway and Transportation Officials divides at-grade intersections into three basic geometric types: three-leg, four-leg, and multi-leg intersections. These intersections can be further divided into unchannelized, flared, and channelized intersections (Figure 4.5a).[13] Channelization is defined as the separation or regulation of conflicting traffic movements into definite paths of travel by traffic island or pavement marking to facilitate the safe and orderly movement of vehicles and pedestrians. Channelization is employed to increase capacity, improve safety, increase convenience, and instill driver confidence. Channelization is frequently used to provide for left-turning traffic.[14]

Flared Safety Intersections

The George Washington Memorial Parkway is among the highways that have intersections with teardrop-shaped safety islands where residential or recreational development require access but cross traffic is expected to be light. The construction of short median strips that expand to a car-length width at the center of the intersection enable turning or entering traffic to safely cross one lane of traffic at a time. An extra lane width is provided in either direction for added safety. The configuration works best on slight curves where the widening required by the central islands does not appear to disrupt the continuous curvature.[15]

Three-Leg Intersections

The most common type of T-intersection involves the intersection of the terminus of one road and the continuation of a second road at a 90-degree angle (Figure 4.5b). Junctions of minor or local roads with more important highways can generally occur at an intersection ranging from 60 to 120 degrees. Such intersections may be channelized or unchannelized, depending on traffic volumes and configurations.[16]

Four-Leg Intersections

Four-leg intersections may range from the common crossroads, where traffic may be controlled by two or more stoplights or traffic signals, to more complicated intersections with channelization, islands, and auxiliary pavements. Examples of unchannelized and channnelized four-leg intersections are shown in Figure 4.5c, d.

Grade-Separated Interchanges

As with at-grade intersections, grade-separated interchanges are commonly divided into three- and four-leg intersections. An intersection with three intersecting legs consists of one or more highway grade separations and one-way roadways for all traffic movements. When two of the three intersection legs form a through road and the angle of intersection is not acute, the term used is a T-interchange. When all intersection legs form a through road or the intersection angle with the third intersection is small, the interchange may be considered a Y-type.[17] Examples of three-leg intersections with a single structure are diagrammatically depicted in Figure 4.6a. More complicated three-leg intersections with multiple structures are depicted in Figure 4.6b. A single quadrant trumpet interchange is shown in Figure 4.6c.

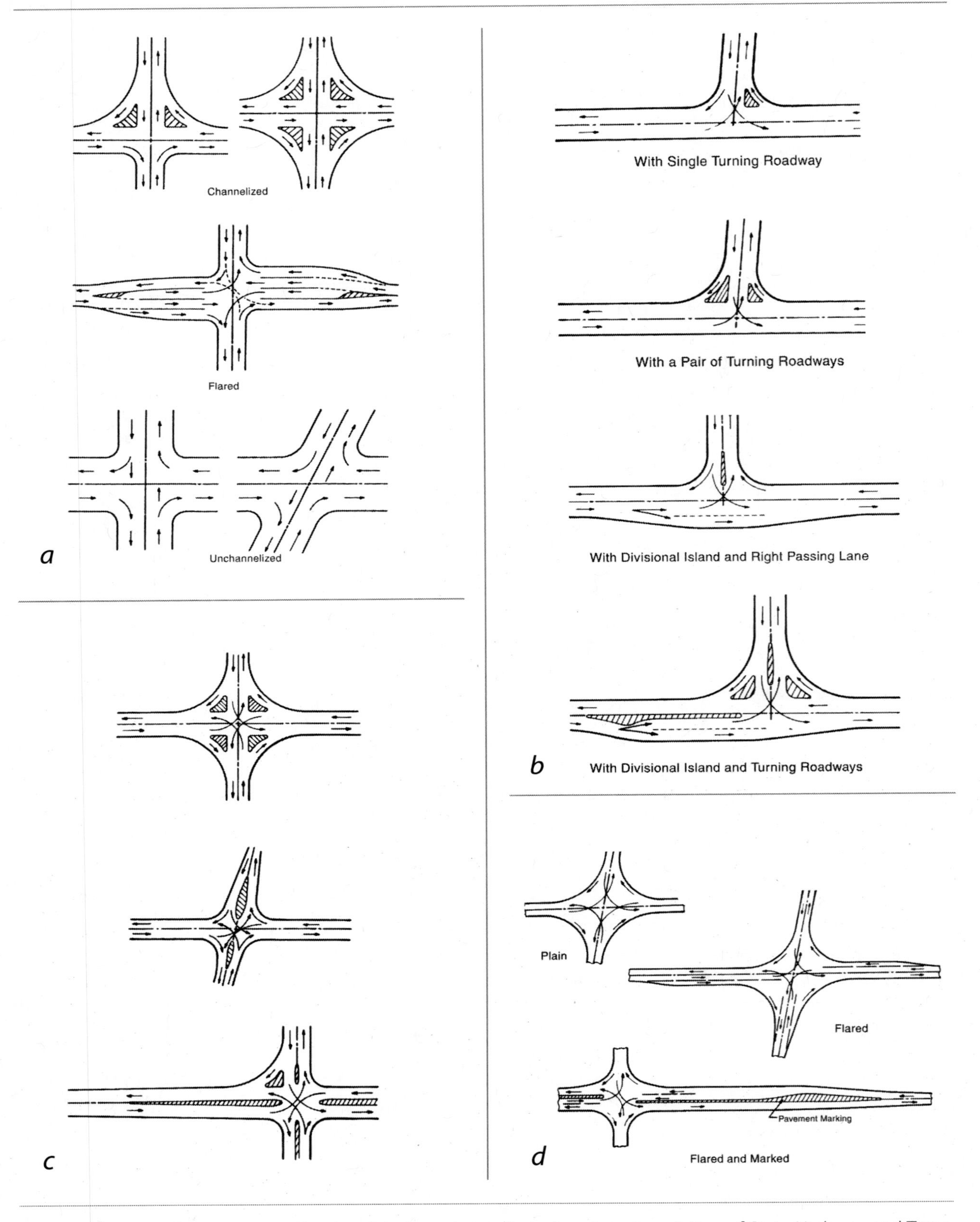

Figure 4.5. Types of intersections. Reprinted, by permission, from American Association of State Highway and Transportation Officials, *A Policy on Design of Urban Highways and Arterial Streets* (Washington: AASHTO, 1984). ***a***, general types of at-grade intersections; ***b***, three-leg channelized intersections; ***c***, four-leg intersections, channelized "T"; ***d***, unchannelized four-leg intersections, plain and flared.

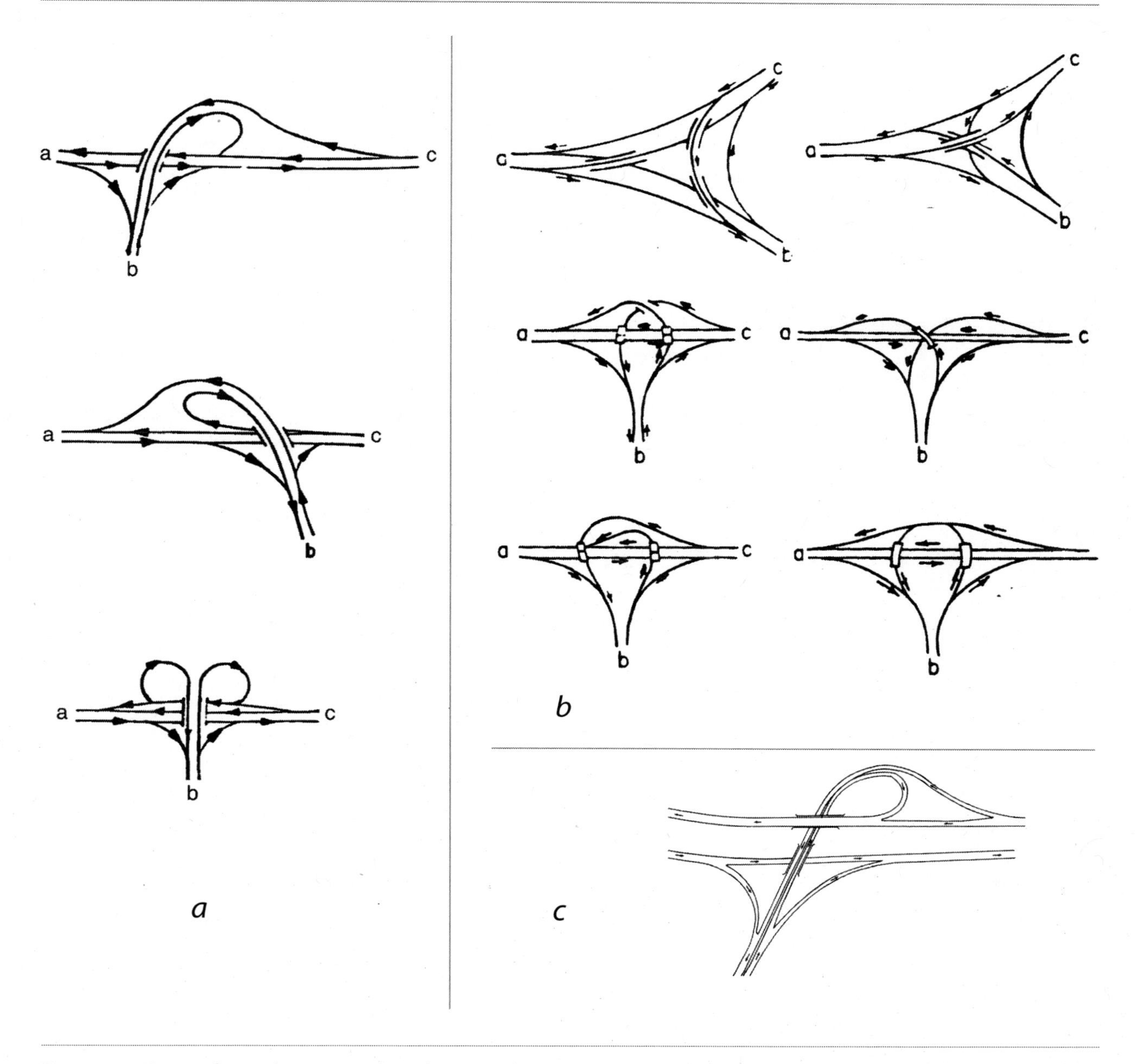

Figure 4.6. Types of interchanges: ***a,*** three-leg interchange with one structure; ***b,*** three-leg interchange with multiple structures; ***c,*** three-leg interchange (T-type or trumpet). (AASHTO 1984)

Four-leg interchanges may be divided into five general types: ramps in one quadrant; diamond interchanges; partial cloverleafs; full cloverleafs; and intersections with direct and semi-direct connections.

Ramps in One Quadrant

Interchanges with a single ramp are generally used for the intersection of roads with low traffic volumes. When a grade separation is provided at an intersection because of topography, a single two-way ramp will be adequate for all turning traffic.[18]

Diamond Interchanges

A full-diamond interchange is formed when a one-way diagonal-type ramp is provided in each quadrant. The ramps are aligned with free-flow terminals on the major highway, and left turns at grade are confined to the crossroad. The diamond interchange allows traffic to enter and leave the major road at relatively high speeds, and left-turning maneuvers require little extra travel.[19]

Cloverleafs

Cloverleafs are four-leg intersections that employ loop ramps to accommodate left-turning movements (Figure 4.7). Interchanges with loops in all four quadrants are referred to as full cloverleafs, and all others are termed partial cloverleafs.[20]

Virginia's George Washington Memorial Parkway included the first cloverleaf built by the federal government. The structure, built in 1930, employs a grade-separation structure to allow traffic on the parkway and 14th Street to flow unimpeded while spiral connecting ramps enable motorists to change direction without stopping or making dangerous left turns across traffic.[21]

The first cloverleaf in the United States was built a year earlier in Avenel, New Jersey, at the intersections of State Routes 4 and 25, the present intersection of U.S. 1 and 9 and NJ 35. The basic idea of a cloverleaf is that each traffic movement—right, left, and straight across from each of the incoming roads—could be made in the interchange without stopping for cross traffic.[22]

RAMPS

Ramps are defined as all types, arrangements, and sizes of turning roadways that connect two or more legs at an interchange. Ramp components include a terminal at each leg, and a connecting road, usually with some curvature, on a grade. Five basic ramp types are commonly employed: the diagonal, the one quadrant, the loop and semi-direct (or flyover), the outer connection, and the directional (Figure 4.8).[23]

TRAFFIC CIRCLES OR ROTARIES

The first one-way rotary system in the United States was Columbus Circle in New York City, completed in 1904. Columbus Circle was an early "nonconforming" traffic circle, along with other well-known urban land-

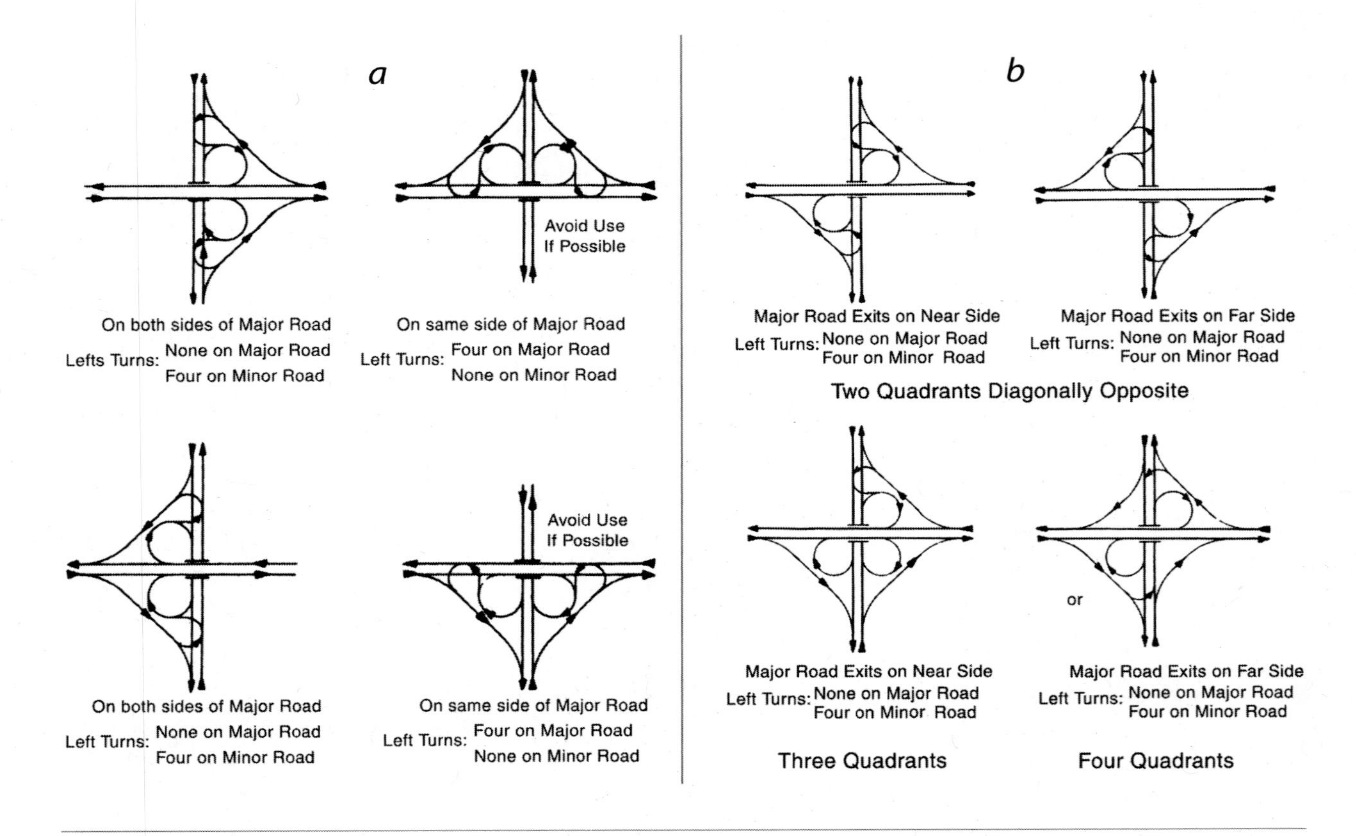

Figure 4.7. Types of cloverleafs: ***a,*** partial cloverleafs; ***b,*** whole cloverleafs (AASHTO 1984).

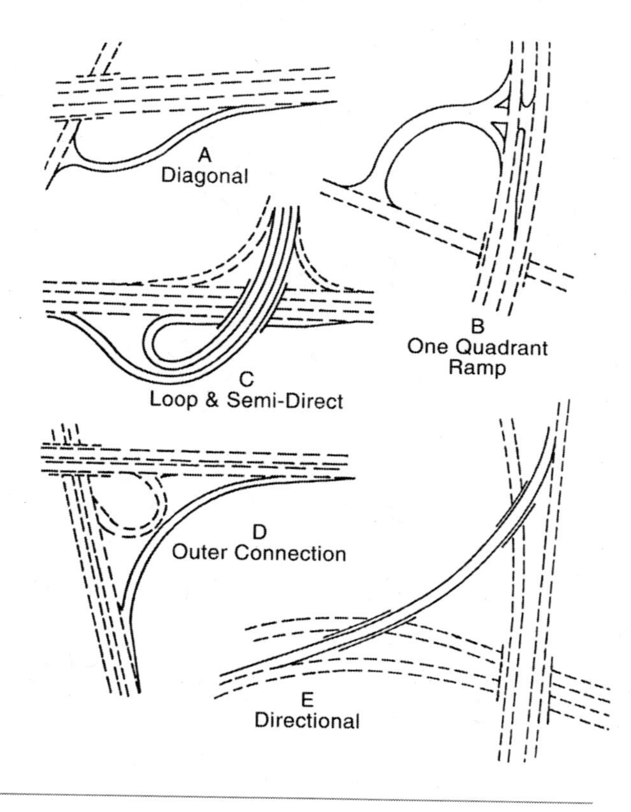

Figure 4.8. Types of highway ramps (AASHTO 1984).

marks such as Washington's Dupont Circle and Philadelphia's Logan Circle. Characteristics of these circles include these anomalies: entering traffic has the right-of-way; entries are regulated by stop signs or traffic lights; entries are tangential to the circle;[24] pedestrians cross onto the center island; the through road cuts through the circle; and the circulating traffic is controlled by traffic signals or stop signs (Figure 4.9).

The modern traffic circle is one of New Jersey's contributions to the history of American transportation. Although it did not originate there, New Jersey had more traffic circles than any other state. The state's first modern traffic circle was constructed in 1925 in the Philadelphia suburb of Pennsauken at the intersection of present U.S. Route 30 and U.S. Route 130. (Coincidently, it was located within sight of another New Jersey first, the first drive-in movie theater.)

The principle of the traffic circle is that it allows for greater volumes of traffic to pass through the intersection of two or more highways without having to stop. By 1932, traffic circles had been constructed in numerous large cities, including New Orleans (Lee), Baltimore (Druid Hill Park), Buffalo (Niagara), Washington, DC. (numerous circles include DuPont, Sheridan, Scott, and Chevy Chase), Portland, Oregon (Laurelhurst and Central Park), Indianapolis (Monument Place), and Boston (Columbia Road).

In traffic circles, all traffic merges and emerges from a one-way road around a central island. Because of the relatively large area required for the development of a rotary, the extra travel distance within it, the necessary speed reduction on the part of all entering vehicles, and the limited capacity of the weaving sections, rotaries soon fell out of favor.

A total of 67 traffic circles were eventually constructed in New Jersey, most of the construction taking place in the 1920s and 1930s. Traffic circles served the state until increases in vehicle speeds and traffic volumes reduced their efficiency and safety. Beginning in the 1970s, the New Jersey Department of Transportation began a concerted effort to remove traffic circles, using two methods. The first and more common approach was to cut through the center of the circle and install a traffic light. The less common and more expensive approach was to grade-separate the roadways.

In addition to the standard rotary, traffic engineers have developed several variations. In some rotaries,

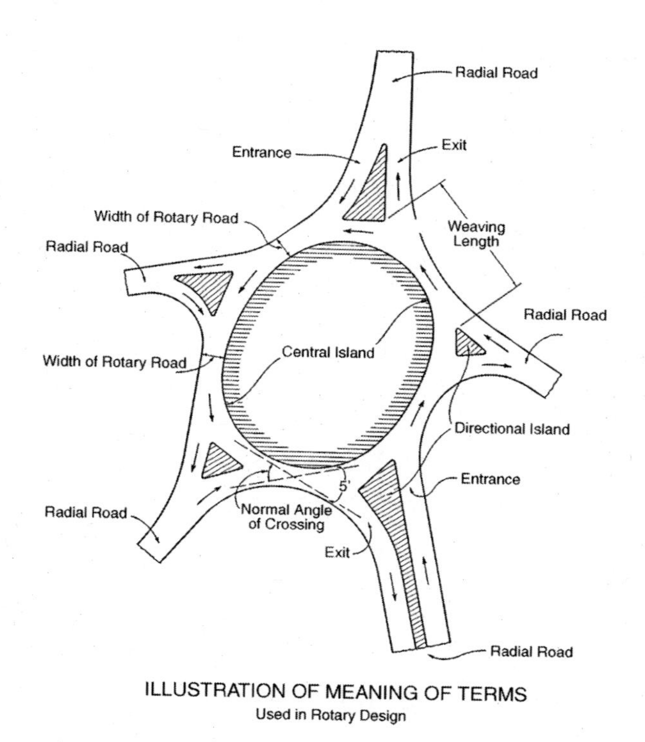

Figure 4.9. Terminology of traffic circles (AASHO 1942: 2).

traffic on one or two highways carrying the greater portion of the traffic has been facilitated by making the rotary tangent to the road or roads. Another variation to the standard rotary is to split the central island to reduce the distance that some of the entering traffic is required to travel. Another is the turbine-type rotary, in which most of the change of direction is made upon entering, and a near tangent path is used upon leaving. Yet another variation is the bridged rotary, which combines two or more grade separations to facilitate traffic flow on the major highways. A double rotary is a rare type found in a small number of locations, including Brooklawn, Camden County, New Jersey. A final variation, used at Washington's DuPont Circle, is the use of a rotary with an underpass carrying through traffic beneath the rotary.[25] Roundabouts—smaller and more controlled versions of traffic circles—are presently being built in California, Oregon, and Europe.

JUGHANDLES

New Jersey road engineers developed the jughandle in part as a means to replace traffic circles. As defined in the New Jersey Department of Transportation design manual, a jughandle is "an at-grade ramp provided at or between intersections to permit the motorists to make indirect left turns and/or U-turns." The jughandle was developed to minimize left-turn conflicts at intersections, and New Jersey has used them for years. Other states that have used jughandles to a lesser degree include Connecticut, Delaware, Oregon, and Pennsylvania.

Jughandles are one-way roadways in two quadrants of the intersection that allow for removal of left-turning traffic from the through stream without providing left-turn lanes. All turns—right, left, and U-turns—are made from the right side of the roadway. Drivers wishing to turn left exit the major roadway at a ramp on the right and turn left onto the major road at a terminus separated from the main intersection. Less right-of-way is needed along the roadway because left-turn lanes are unnecessary. However, more right-of-way is needed at the intersection to accommodate the jughandles.

HIGHWAY DRAINAGE

Highways, railways, streets, and similar constructions use two basic types of construction to prevent water from accumulating on the surface: surface and subsurface. Crowning or sloping the surface to shed the water is the first step. The water is concentrated in a depression, ditch, or gutter, parallel to the road centerline and thence into the nearest cross stream or other outlet. Usually the parallel ditches are adequately designed so that no water will remain on the roadway surface.

Open ditches are subject to erosion, sedimentation, and overflowing, possibilities that may be overcome by paving, riprapping, sodding, spillways, or by enclosing the stream. Culverts and culvert bridges are used to enclose streams and support the roadway above. They are typically found when natural or artificial streams intersect a roadway, at the bottom of depressions, or under intersecting roads and farm entrances when they form a part of the side ditch.

Culverts

A culvert is an enclosed channel serving as a continuation of and a substitute for an open stream when that stream meets an artificial barrier such as a roadway embankment. It differs from a bridge in that fill is usually used between it and the road surface. The most common types are concrete box or arch culverts, metal plate arch culverts, and pipe culverts of cast iron, corrugated metal, vitrified clay, and concrete. Arch culverts are used in high embankments, particularly in parks or residential areas where appearance is a particular concern.

Headwalls

To reduce the required length of pipe culverts and to protect the embankment from scour by water entering and leaving the culvert, it is common to construct headwalls of concrete or rubble masonry at the ends of the pipe. Headwalls serve the practical purpose of anchoring and preventing the displacement of short sectional culvert pipe and of protecting the embankment against erosion at the inlet and outlet. Alternatively, the pipe can be extended beyond the edge of the fill at the outlet end, eliminating the need for the headwall. Sometimes, riprap is used to protect the slopes from scour.

Among the most commonly used structures for road drainage is the reinforced concrete box culvert (Figure 4.10). Compared with the arch type (see below), the formwork for the box-type culvert is simpler and cheaper, and the width of excavation needed is

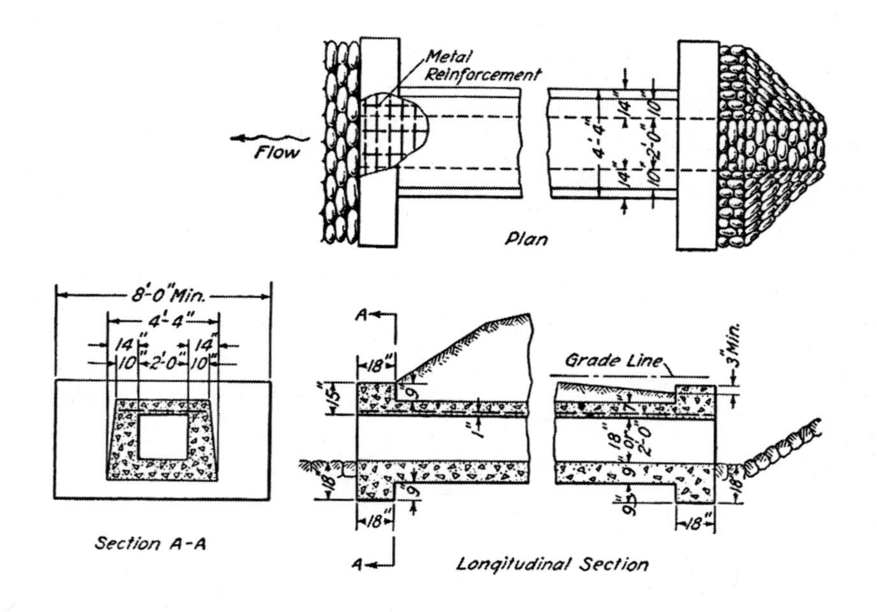

Figure 4.10. Reinforced concrete box culvert (Bruce and Clarkeson 1956: 330). Reprinted, by permission of Thomson Education Direct.

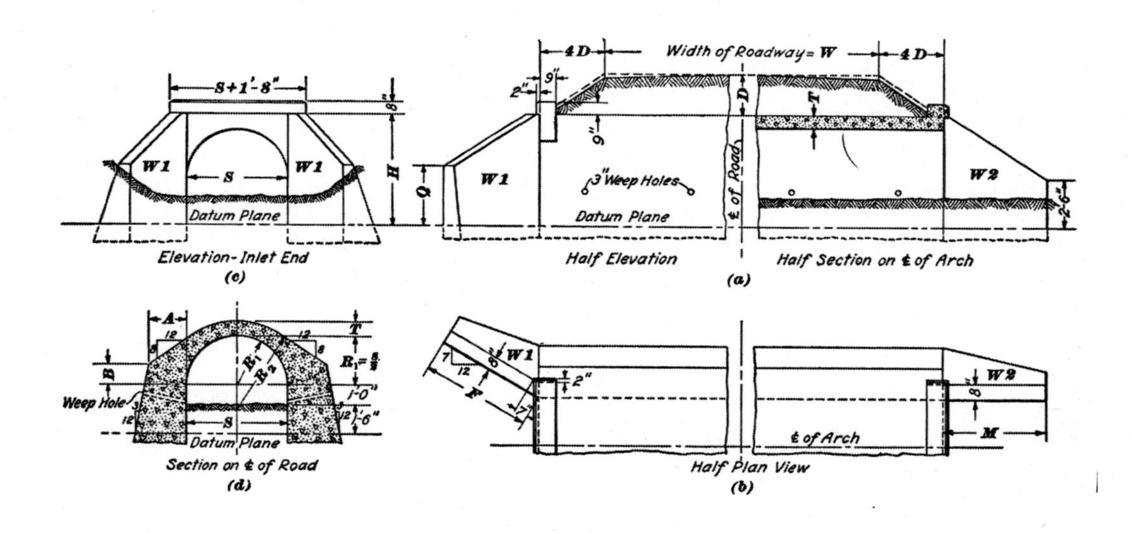

Figure 4.11. Design for a small concrete arch culvert (Bruce and Clarkeson 1956: 332). Reprinted, by permission of Thomson Education Direct.

smaller. These culverts generally employ spans from 2 to 5 feet.

Concrete Arch Culverts

When used with fills of considerable height, the arch culvert (Figure 4.11) may be more economical than the box type, because the increase in depth of fill does not require a proportional increase in the thickness of the arch ring.

Concrete Pipe Culverts

For small waterways and for cross drainage, pipe culverts are typically used. The average pipe culvert diameter is from 30 to 48 inches. The gradient of the pipe approximates that of the natural waterway. Headwalls are normally constructed of stone, concrete, or brick.[26]

Metal Plate Arch Culverts

By the mid-twentieth century, metal plate arch culverts had become increasingly common. The sectional bolted construction of metal plate arches enable them to be rapidly erected by unskilled labor. The principal time-consuming elements are concrete footings and headwalls, if used.

Multi-Plate Metal Culverts

In many locations culverts with cross sections larger than those provided by pipe have been needed. This led to the development in 1931 of plates with larger corrugations, 6 inches wide by 1½ inches deep. These plates, known as multi-plate, were corrugated, curved, and galvanized at the factory and shipped in nested, knocked-down form for bolted assembly in the field. Multi-plate culverts consist of plate sections bolted together in the field in the same manner as metal plate arches. These have been used for diameters up to 135 inches.

TUNNELS

Tunnels can be divided into two major categories: tunnels constructed by mining methods and tunnels constructed from the surface. The former are typically constructed through hillsides and mountains (Figure 4.12), while the latter are typically constructed in urban locations. Perhaps the best-known recent "cut and cover" tunnel is Boston's infamous "Big Dig." This method consists of excavating an open cut, building the tunnel within the cut, and backfilling over the completed structure.

In excavated tunnels, ventilation and structural support are two primary construction issues. By the 1930s, several means of ventilating tunnels had been introduced. For short tunnels, natural ventilation is deemed sufficient, but mechanical ventilation is required for tunnels 1,500 feet or longer or with vertical curves. In transverse ventilation, air is blown in through ducts placed at regular intervals along the walls of the tunnel and exhausted through a crown vent. In the longitudinal ventilation system, the contaminated air is exhausted through shafts or adits, while fresh air enters through portals or through adits or shafts. In tunnels over 1,500 feet long, mechanical ventilation is generally needed, with a fan located near the apex of the tunnel.

The type of tunnel lining used is dependent on budget and geological conditions. In some tunnels, the stability of the surrounding rock eliminates the need for a liner. In other tunnels, timber lining sets are used. Others use masonry linings or steel-plate linings.[27]

The five longest land vehicular tunnels in the United States are the Anton Anderson Memorial Tunnel, Whittier, Alaska (13,300 feet), the E. Johnson Memorial and Eisenhower Memorial, both on I-70 in Colorado (about 8,950 feet), the Allegheny Twin Tunnel on the Pennsylvania Turnpike (6,072 feet), and the Liberty Tubes in Pittsburgh (5,920 feet). The longest un-

Figure 4.12. Examples of highway tunnel portals. ***a,*** East Side Highway Tunnel, Packwood, Washington. HAER WA-75, National Park Service. Photograph by Jet Lowe, August 1992. ***b,*** Mosier Twin Tunnels, Troutdale vicinity, Oregon. HAER OR-36, National Park Service. Photograph by Roger Keiffer, September 1995.

derwater vehicular tunnels in the United States are the 1950 Brooklyn-Battery Tunnel (New York; 9,117 feet); the 1927 Holland Tunnel (New York; 8,557 feet); the 1995 Ted Williams Tunnel (Boston; 8,448 feet); the 1937 Lincoln Tunnel (New York; 8,216 feet); and the 1964 Thimble Shoals Tunnel of the Chesapeake Bay Bridge-Tunnel (Virginia; 8,187 feet).[28]

GUARDRAILS

Guardrails are employed where a road or highway is on an embankment and the side slopes are steeper than a 33-degree slope, or where motorists need protection from careening off the road surface. The ideal guardrail is designed to deflect the vehicle back toward the roadway rather than check its speed or stop it. Guardrails are divided into three broad categories according to the stiffness of the barrier and the relative amount of deflection that results from vehicle impact: rigid barriers, semi-rigid barriers, and flexible barriers.

Rigid Barriers

Rigid barriers are usually used only where the space available for deflection is limited, as on very narrow medians and bridge structures. Because they must be made unyielding, these barriers are usually constructed of reinforced concrete.[29]

According to a Transportation Research Board report, among the first concrete median barriers to be constructed were those erected in the mid-1940s on the U.S. 99 descent from the Tehachapi Mountains to California's Central Valley south of Bakersfield. The major purposes of median barriers are to prevent vehicle crossovers into oncoming traffic in narrow highway medians and to eliminate the need for costly and dangerous steel guardrail median barrier maintenance in high accident locations.

New Jersey introduced median barriers in the mid-1950s. The first ones, installed in 1955, were only 18 inches tall and resembled a low vertical wall with a curb on each side. As a result of operational problems, the shape was changed, and the height increased to 24 inches and then to 32 inches in 1959. By that time, the classic Jersey barrier shape had come into use, its first 2 inches from the pavement rising vertically, the next 10 inches rising at a 55-degree angle, and the remainder at an 84-degree angle (Figure 4.13a).

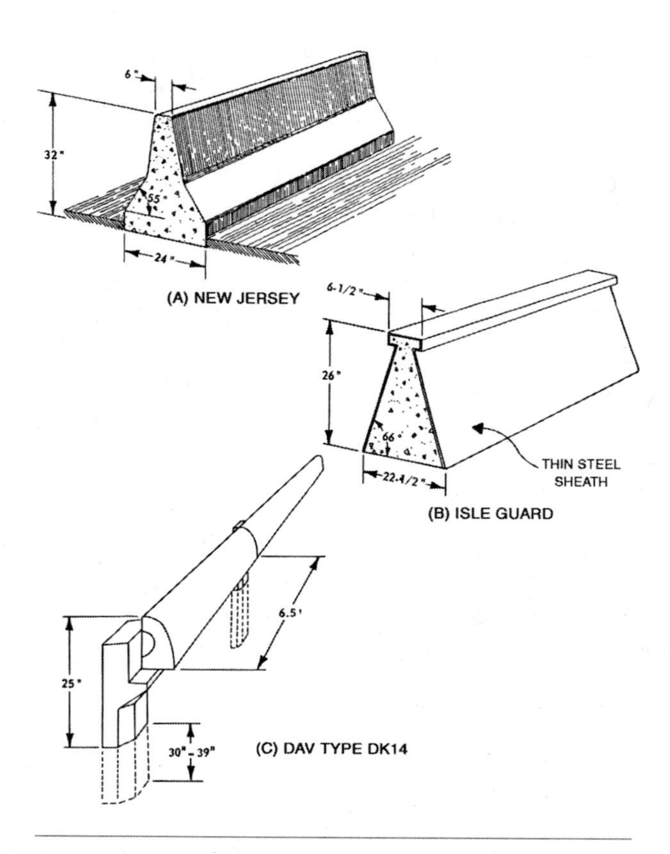

Figure 4.13. Types of rigid barriers. Reprinted, with permission, from *Highway Research Board—A Review of Current Practice,* National Cooperative Highway Research Program Report 36 (Washington: Highway Research Board 1967), 10.

Several other median barrier types are used on highways in addition to the Jersey barrier. The Ontario tall wall barrier has the same slopes as the Jersey barrier but is 42 inches tall. The F-shape barrier, similar to the Jersey barrier, has a lower bottom section and a flatter slope. This barrier is now generally preferred for both portable concrete barriers and permanent barriers. The constant slope barrier is 42 inches high, and its sides have a single slope of 79 degrees. Developed as the result of crash testing, it has been installed on I-95 and I-295 in Virginia. The GM barrier was developed in the early 1970s by General Motors in conjunction with the Texas Transportation Institute. It is similar to the Jersey barrier only thicker, and its breakpoint is about 3 inches higher. The shape was developed by crash testing during a time in which the average car was large. It was used in Virginia until the late 1970s.

Semi-Rigid Barriers

In the mid-twentieth century, the most commonly used types of guardrails were low, heavy, wooden guardrails; woven wire or steel cable and wooden, concrete, or steel posts; and wire mesh (Figure 4.14). These are the predecessors of today's common steel guardrails.

Post Types

In the mid-twentieth century, the prevailing material used for guardrail posts was wood. Wooden posts are resilient, so that when struck by a vehicle, they are often knocked entirely out of the ground, causing little damage either to them or to the vehicle. Posts are often treated with creosote or zinc chloride to prevent decay. The desired distance between wooden posts depends on the type of guardrail, generally 8 to 10 feet for wooden rail, 10 feet for steel cable rail, and 16 feet for steel-plate rails.

Concrete posts are durable and readily visible but are not resilient. If struck, they are likely either to break off at the ground line or to badly damage the vehicle. Concrete posts are square, round, or triangular in cross section, generally 6 to 8 inches across.

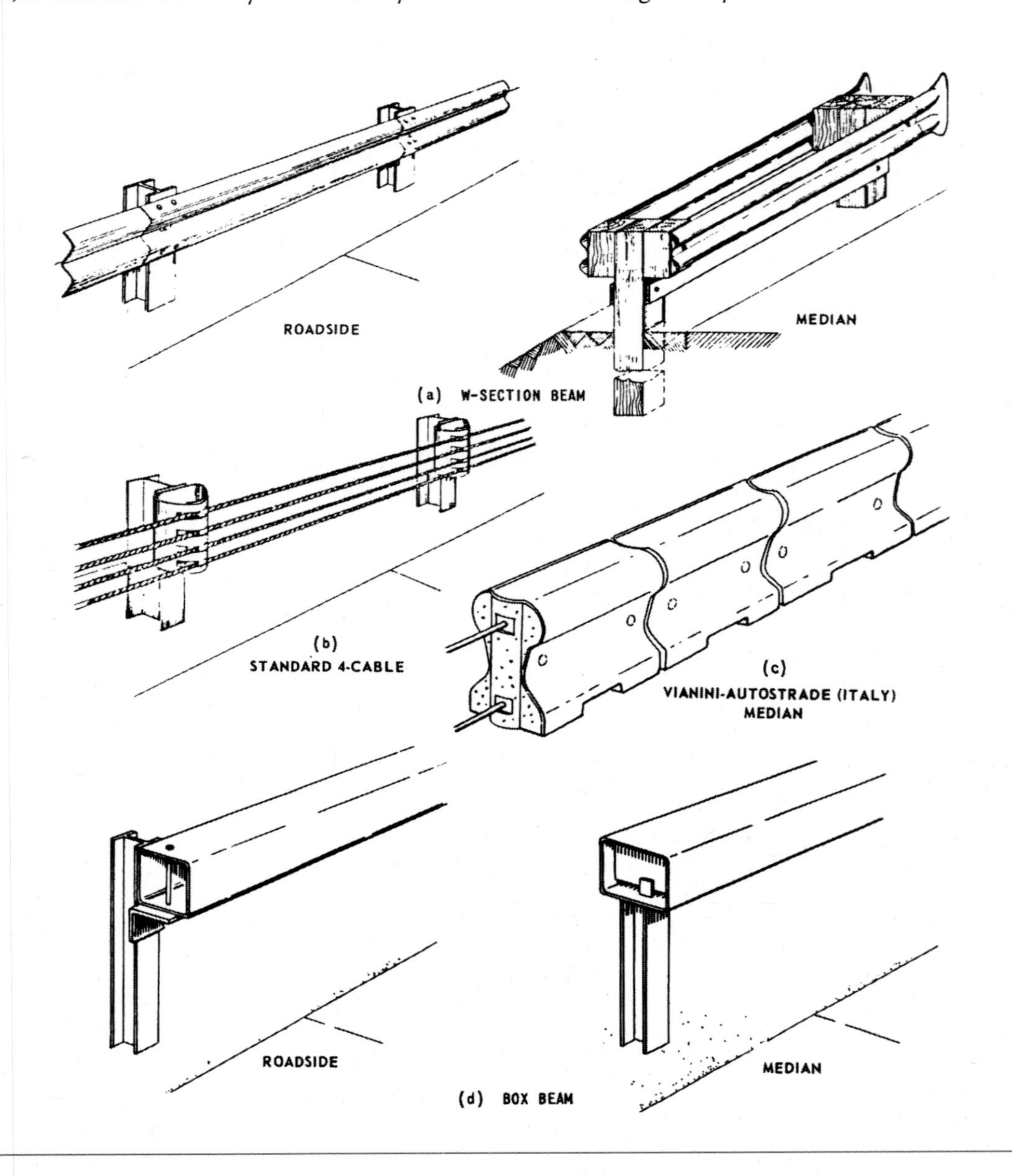

Figure 4. 14. Types of semi-rigid barriers (Washington, DC: Highway Research Board 1967), 11.

Steel posts are usually formed from channel or angle shapes. These posts are quite resilient and may be bent without breaking.

Rails

To be effective, a guardrail must be strong enough to resist the impact of a vehicle striking it and must be able to transmit the force of the impact to several posts. Major types of rails used in mid-twentieth-century guardrails include cable rails, wire-mesh guardrails, metal-plate rails, and wire tape.

Cable rails consist of galvanized drawn-steel wires that are twisted or woven in the form of a rope. These cables are generally ¾ inch in diameter. They are attached to the end post by means of wedge-locked cable ends and steel rods. The end post is anchored by means of the rod, provided with a turnbuckle and attached either to a reinforced-concrete or galvanized metal deadman buried in the ground. The cable is generally attached to intermediate posts by means of J-shaped bolts. Modern cable rails are usually ¾-inch-diameter wire rope with a minimum tensile strength of 25,000 pounds. The number of cables varies between two and four, and they are usually mounted on offset spring brackets that hold the cables at a separation of 4 to 6 inches.[30]

Early metal-plate guardrails are of various types. One type, called the resiliflex, consists of a continuous galvanized steel plate of no. 16 gauge and 12 inches wide. The plate is attached to the posts by means of steel springs. The flexing of the rail and the supporting springs cushion the blow of the vehicle and deflect it back toward the road. A second type, the Tuthill guardrail, consists of a convex steel plate mounted on spring steel supports attached to wooden posts just above the ground line.

Wire-tape rails were first introduced in about 1940. They consist of the same number and size of wire used in a standard ¾-inch cable, but the wires are woven into a flat band 2½ inches wide. The wires are held in place by half-round clips woven through them at 6-inch intervals.

W Beam

The most common present guardrails are the "W" beam and the blocked-out "W" beam. These may be installed on either steel, wood, or concrete posts but are most typically mounted on steel posts. Most are of the weak post variety. When hit, the posts collapse and the rail absorbs the energy. The mounting height of the top of the rail usually varies between 24 and 27 inches above the ground. The standard post spacing is 12.5 feet.

Box Beam

The box beam was introduced in New York State. The barrier consists of a 6 by 6 by 3/16-inch hollow, steel-tube horizontal railing weakly attached or supported on 3-inch steel posts spaced 6 feet apart. The operating principle of the barrier is that forces are resisted by the beam strength of the rail and are distributed over a large number of relatively weak posts.[31]

Flexible Barriers

By allowing large deflections, flexible barriers redirect or stop offending vehicles more gradually and subject occupants to lesser forces than do semi-rigid barriers. One such barrier, developed by the California Division of Highways, is the cable-chain-link fence consisting of two wire-rope cables fastened by U-bolts to fence posts at a height between 27 and 30 inches. A 48-inch chain-link fence is attached to the posts by steel wire ties. The posts are spaced on 8-foot centers.[32]

LIMITED-ACCESS HIGHWAYS

The genesis of the interstate highway and other limited-access highways that extend the length and width of many states was in the motor parkways of the northeastern states, particularly the New York City metropolitan area. The first U.S. parkway designed for automobiles was the Bronx River Parkway, completed in 1923, a precursor of later limited-access highways. This road traveled an elongated park, abutting property owners had no right of access, intersecting streets crossed the parkway on bridges, special ramps were provided for exits and entrances, and a winding free-flowing alignment was fitted to the topography. Additional parkways followed in the 1920 and 1930s, including the Hutchinson, Saw Mill River, Henry Hudson, and Taconic parkways, north of the city, and the Meadowbrook, Northern and Southern State and Wantagh on Long Island. By 1934, about 134 miles of parkways had been completed in the counties surrounding New York City.[33] In the same era, New Jersey's Palisades Parkway and Connecticut's Merritt Parkway were under construction.

Interchanges

Interchanges are necessary to provide for the transition and smooth flow of traffic between the parkway and local roads. In early parkway interchanges, short acceleration and deceleration lanes were provided to allow access and egress from the parkway at high speed. A study on the Merritt Parkway, however, showed that the lanes were insufficient in length for their intended use. Automobiles were unable to reach cruising speed before entering the parkway, and exiting motorists were forced to slow down below cruising, thus increasing the risk of rear-end collisions.[34]

Service Plazas and Waysides

Service plazas were part of the original design of the Merritt and other parkways, providing amenities ranging from restrooms and gasoline stations to full-service garages and restaurants. Typically, early service plazas were designed to harmonize with the architecture of the surrounding area. Those of the Taconic State Parkway called on traditional Hudson River Valley building styles and local building materials to give stations a rustic feel. The original signage and gasoline pump surrounds continued the rural character. An example of an early Taconic Parkway service area is the 1932 Briarcliff Wells Station near the south end of the highway.[35]

The original plans for the Merritt Parkway called for three pairs of service stations, set back from the highway reached by at-grade exits from the right traffic lane. The original block of each station was a rectangular, one-story structure with side-gabled, shallow pitch roofs clad in slate. Two sets were faced in brick, one with random ashlar stone masonry. In style and massing, the Merritt's service stations are similar to those originally constructed on many parkways in neighboring Westchester County. This represents the regionalization of architectural roadside improvements advocated at the time.[36]

The next generation of service areas is represented by the original plazas on the Pennsylvania Turnpike, designed to accommodate long-distance travelers. The most elaborate of these is the Midway Plaza, located a short distance east of the Bedford interchange and halfway between Pittsburgh and Harrisburg. Designed as the welcoming showplace of the turnpike, it is much more elaborate than the highway's other original service plazas. The main Midway building on the south side of the turnpike consists of a central two-story, five-bay section flanked by two large wings, each with a stone-end chimney. The plaza originally offered services not found in most other plazas, including full-course meals and recreational lounges and dormitory lodging for truckers. The Colonial Revival design of the buildings presaged that of numerous later rest areas, such as those on the Connecticut Turnpike.[37]

A next generation of rest areas or service plazas is represented by those constructed on the Ohio Turnpike in the 1950s. These plazas separated automotive and truck parking, included a service and restaurant building, and contained multiple fueling lanes.[38]

Toll Barriers and Booths

The design of toll barriers and booths on early parkways and turnpikes vary by highway. Early parkways, such as the Merritt Parkway, feature toll booths in keeping with the overall character of the road. The toll canopies are simple wood-framed structures. Pairs of de-barked tree-trunk columns resting on log plinths stand at each corner of a concrete base and support log trusses. The trusses in turn support beams at the eaves, ridge, and halfway up the roof, which in turn support closely spaced log rafters. The hipped roof is covered with wood shingles. As originally constructed, the plaza canopy had four interior toll booths, spaced evenly under the canopy. Clad in diagonal wood siding, the upper walls contained sliding windows (Figure 4.15).[39]

Figure 4.15. Former Merritt Parkway toll plaza, Housatonic Bridge Station, Connecticut. Relocated to Boothe Memorial Park, Stratford, Connecticut. HAER CT-138-1. Photograph by Jet Lowe, 1992.

While the rustic style of these booths fits the character of the road, other roads have sleek, modern, metal and glass toll plazas. Among the structures with such plazas are the Holland Tunnel between New York and New Jersey, New York's Triborough Bridge, and the Pennsylvania Turnpike.[40] These toll plazas are supported by steel columns, have steel-framed and glass booths, and feature a canopy with steel signboard. Because of the greatly increased traffic demand and larger average vehicle size, most older toll barriers have been either modified or replaced.[41] Components of later toll plazas include markers for channelization of traffic into proper lanes, barricades and protective walls, and a service/utility building. In the case of the Ohio Turnpike, this building houses automatic remote control counters, money vaults, ticket storage, locker room, lavatory, offices, heating plant, and automatic electric generator unit.[42]

GLOSSARY

Acceleration lane. An area of partial or full lane width of sufficient length to enable an entering vehicle to increase speed to the rate at which it is convenient or safe to merge with through traffic.

Asphalt filler. An asphaltic product used for filling cracks and joins in pavement structure.

Bituminous concrete. A roadway surface consisting of graded mineral aggregate and bituminous cement, the ingredients being proportioned to meet the requirements for stability under service conditions.

Cloverleaf. A grade separation with ramps for both directions of travel in each of the quadrants.

Deceleration lane. An added area of partial or full lane width of sufficient length to enable a vehicle that is to turn to slow down to the safe speed as it approaches the turn.

Flared intersection. One in which the number of traffic lanes or the pavement width exceeds the normal number of lanes or the normal width of intersecting highways.

Overcrossing. A grade separation in which the highway being discussed crosses over the intersecting road.

Ramp. A connecting roadway between two intersecting highways at a grade separation.

Separate turning lanes. Added traffic lanes separated from the intersection by an island.

T-grade separation. A grade separation at a T-intersection, generally with three direct connections and one inner loop.

Undercrossing. A grade separation in which the highway being discussed crosses beneath the intersecting road.

BIBLIOGRAPHY

American Association of State Highway Officials. *A Policy on Rotary Intersections.* Washington, DC: AASHO, 1942.

—. *A Policy on Grade Separations for Intersecting Highways.* Washington, DC: AASHO, 1944.

American Association of State Highway and Transportation Officials. *A Policy on Design of Urban Highways and Arterial Streets.* Washington, DC: AASHTO, 1973.

—. *A Policy on Geometric Design of Highways and Streets.* Washington, DC: AASHTO, 1984.

Bruce, Arthur G., and John Clarkeson. *Highway Design and Construction.* 3rd edition. Scranton, PA: International Textbook Company, 1956.

DeLony, Eric, and Sara Amy Leach, project leaders. Merritt Parkway. HAER CT-63. 1992.

Hewes, Lawrence Ilsley. *American Highway Practice.* 2 volumes. New York: John Wiley & Sons, Inc., 1942.

Highway Research Board. *Highway Guardrails—A Review of Current Practice.* National Cooperative Highway Research Program Report 36. Washington, DC: Highway Research Board, 1967.

Holley, I. B., Jr. "Blacktop: How Asphalt Paving Came to the Urban United States." *Technology and Culture* 44:4 (October 2003): 703–743.

Kauer, Theodore J. *Ohio Turnpike Report: An Engineering Report of the Planning and Construction of the Ohio Turnpike, Project No. 1.* Columbus: The Ohio State University Engineering Experiment Station, 1956.

Tuttle, L. S., and E. H. Holmes. "The Design of Street and Highway Intersections." *Public Roads: A Journal of Highway Research* 13:5 (July 1932): 73–88.

Western Regional Office, United States Bureau of Public Roads. "Highway Tunnels in Western States." *Public Roads: A Journal of Highway Research* 19:7 (September 1938): 125–133.

NOTES

[1] Kim C. Wallace, Pennsylvania Turnpike, Breezewood Interchange, HAER PA-349, 1994.

[2] Frank X. Brusca is a leading avocational historian of milestones and mileposts. His partial list of United States highways with

historic mile markers is on his Route 40 website: www.route40.net/history/milestones/inventory.shtml.

3 www.btco.net/ghosts/streets/milestones/milestone.html.

4 www.ohionationalroad.org/mile_markers.htm.

5 "Curbstone Presents-the American Road," curbstone.com/_macadam.htm.

6 Arthur G. Bruce and John Clarkeson, *Highway Design and Construction* (Scranton, PA: International Textbook Company, 1956), 501.

7 Bruce and Clarkeson, 440–443.

8 Bruce and Clarkeson, 560–562.

9 Bruce and Clarkeson, 569–570.

10 "A brief history of Amiesite Asphalt Company of America," www.amiesite.com/history.html.

11 Bruce and Clarkeson, 160–162.

12 Bruce and Clarkeson.

13 American Association of State Highway and Transportation Officials, *A Policy on Geometric Design of Highways and Streets* (Washington, DC: AASHTO, 1984), 819.

14 AASHTO, 819–821.

15 Ed Lupyak, George Washington Memorial Highway, Mount Vernon, Virginia, Intersections, HAER VA-69.

16 AASHTO, 831–832.

17 AASHTO, 931.

18 AASHTO, 939.

19 AASHTO, 942.

20 AASHTO, 949.

21 Ed Lupyak, George Washington Memorial Parkway, Intersections, Virginia, HAER VA-69, 1994.

22 The Cloverleaf Interchange, http://whereroadsmeet.8k.com/article/clover.htm.

23 AASHTO, 541.

24 American Association of State Highway Officials, *A Policy on Rotary Intersections* (Washington, DC: AASHO, 1942), 40–53.

25 Laurence Ilsley Hewes, *American Highway Practice*, volume II (New York: John Wiley & Sons, Inc, 1942), 396–397.

26 Western Regional Office, U.S. Bureau of Public Roads, "Highway Tunnels in Western States," *Public Roads: A Journal of Highway Research* 19:7 (Summer 1938): 125–133.

27 World Almanac Books, *The World Almanac and Book of Facts* (New York: World Almanac Books, 2006), 744.

28 Highway Research Board, *Highway Guardrails—A Review of Current Practice*, National Cooperative Highway Research Program Report 36 (Washington, DC: Highway Research Board, 1967), 8.

29 Highway Research Board, 12.

30 Highway Research Board, 13.

31 Highway Research Board, 14.

32 Antony Walmsley, "The Landscape," 2003, Henry Hudson Parkway website: www.henryhudsonparkway.org/hhp/history2.htm.

33 Eric DeLony and Sara Ann Leach, Merritt Parkway, HAER CT-63, 72.

34 Lauren Bostic, Taconic State Parkway 1931–1963, Service Areas, HAER NY-316, 1999.

35 DeLony and Leach, 94–95.

36 Kim C. Wallace, Pennsylvania Turnpike, Midway Plaza, HAER PA-347, 1994, 2–3.

37 Theodore J. Kauer, *Ohio Turnpike Report: An Engineering Report of the Planning and Construction of the Ohio Turnpike, Project No. 1* (Columbus: The Ohio State University Engineering Experiment Station, 1956).

38 DeLony and Leach, 97.

39 DeLony and Leach, 98.

40 In the case of the Merritt Parkway, tolls were eliminated in 1988 (as they were on the Connecticut Turnpike). The Greenwich toll plaza was removed to Greenfield Village in Dearborn, Michigan, while the Milford toll plaza was relocated to Boothe Memorial Park in Stratford, Connecticut (DeLony and Leach, 97).

41 Kauer, 57.

CHAPTER 5

Waterways

Engineered waterways may be divided into two basic categories: canals and slackwater navigation systems. Canals are artificial waterways used for inland navigation, often constructed parallel or in close proximity to rivers, and equipped with locks and, more rarely, inclined planes, to facilitate descents and ascents. Canals reached their heyday in the United States in the first half of the nineteenth century, beginning with New York State's Erie Canal. A slackwater navigation system is a series of locks and dams constructed in an existing or widened river channel to permit protected navigation. This development, often termed canalization, transforms a free-flowing waterway to a series of slackwater pools. First extensively developed in the later nineteenth century in the United States, construction of slackwater navigation systems continues today. Among prominent slackwater systems are the Upper Mississippi, the Ohio, the Kanawha, the Tennessee, the Monongahela, and the Allegheny.

THE CANAL

The first historically documented canal was one constructed in Upper Egypt ca. 4000 B.C.E. Numerous other canals are known to have been constructed in the ancient Middle East, including Sumerian canals in the valleys of the Tigris and the Euphrates.

In China, canal construction culminated with the completion of the last segment of the 650-mile-long Grand Canal in ca. 1293. Additional canals were built in Greece and in the Roman Empire. Canal building in the remainder of Europe began as early as the eleventh century. The most significant early canal was the French Languedoc Canal linking the Atlantic Ocean and Mediterranean Sea, which opened in 1681. The first modern British Canal, the Sankey Canal in northern England, was built ca. 1757–1761. The first extensive British canal system, the 93-mile, 74-lock Trent and Merset Canal, began construction in 1766.[1]

Early U.S. canals include Virginia's Patowmack Canal begun in 1785 and opened in 1802. The first 15-mile stretch of the Schuylkill and Susquehanna Canal of Pennsylvania was constructed in 1792–1794. Virginia's Dismal Swamp Canal was opened for small boats in 1784 and finally opened for full navigation in 1812. New York's first canal, the Little Falls Canal, was constructed between 1783 and 1795. This canal's first locks were 10 feet wide and 70 feet long.[2]

The canal era in the United States reached its heyday with the construction of New York's Erie Canal from the Hudson River to Buffalo, opened to traffic in 1825. It included 84 locks, each with a width of 15 feet and a length of 90 feet. Buoyed by the financial success of that waterway, backers financed the construction of networks of canals in New York, Massachusetts, Pennsylvania, Maryland, New Jersey, Virginia, North Carolina, and Indiana.[3]

The most recent large canal project in the United States is the Tennessee-Tombigbee Waterway constructed to connect the two rivers. The construction work began in December 1972, and the project opened

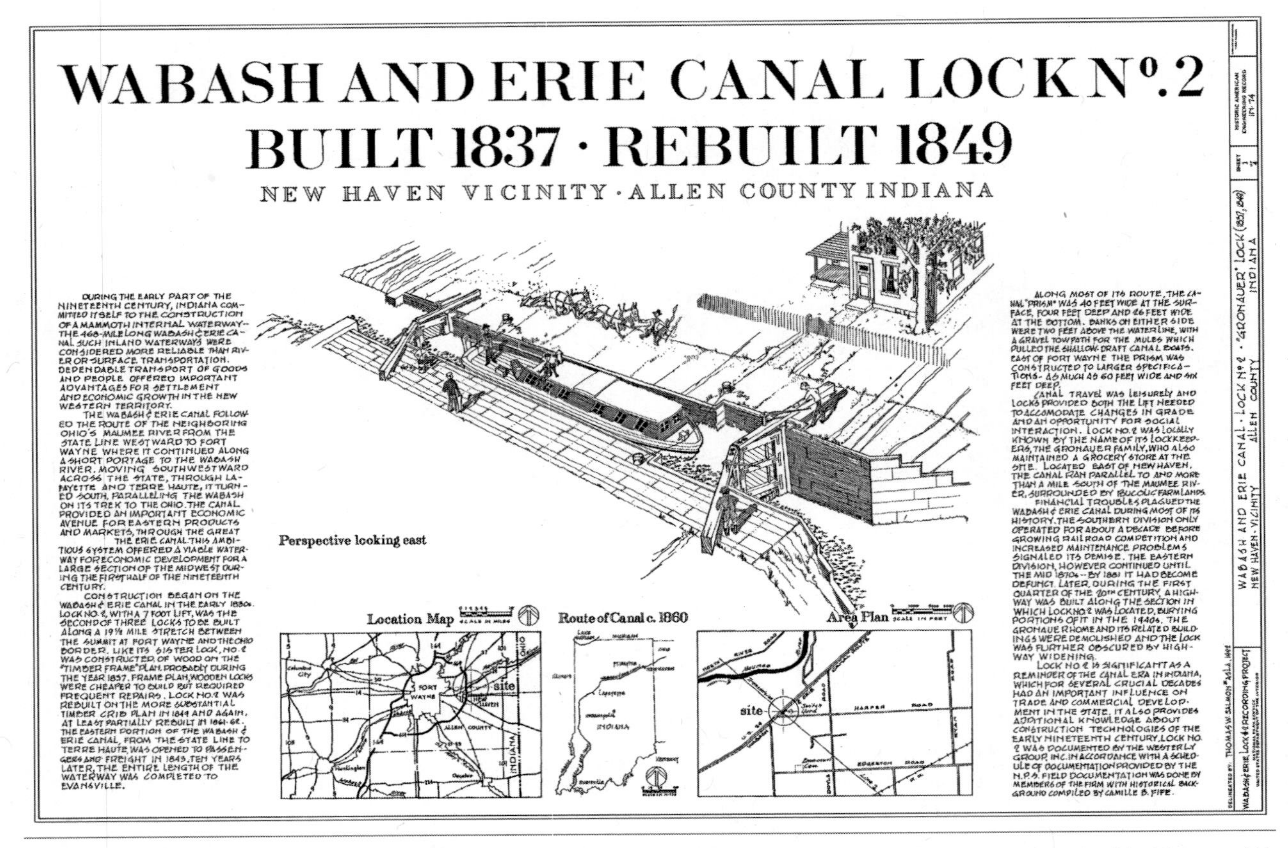

Figure 5.1. Canal Lock. Wabash and Erie Canal Lock No. 2. HAER IN-74, National Park Service. Delineated by Thomas W. Salmon, ASLA, 1992.

for navigation in January 1985. All locks on the waterway are 110 feet wide and 600 feet long.[4]

Towpath

Towpaths were provided for the use of animals, usually horses or mules, that towed canal boats (Figure 5.1). If the canal was close to a river, the towpath usually lay on the side closer to the river. Normally constructed of packed earth, flagstones were sometimes used when the path might be exposed to floodwater. In his book, *An Elementary Course of Civil Engineering,* Mahan described canal towpaths:

> The tow-paths should be from ten to twelve feet wide to allow the horses to pass each other with ease; and the foot-path at least six feet wide. The height of the surface of these paths, above the water surface, should not be less than two feet, to avoid the wash of the ripple; nor greater than four and a half, for the facility of the draft horses in towing. The surface of the tow-path should incline slightly outward, both to convey off the surface water in wet weather, and to give a firming footing to the horses, which naturally draw from the canal.[5]

Berm

The berm is the bank of the canal opposite the towpath and is sometimes termed the heelpath. The bank sometimes follows the natural contour of the land where the expense of construction would not be significantly increased.[6]

Prism

The prism is the trapezoidal cross section of the canal's channel. Typically prism dimensions range from 40 to 50 feet at the water line, 26 to 30 feet wide at the bottom, and about 4 feet deep. The side slope is typically 1 foot vertical to 2 feet horizontal.[7]

Lock

A canal lock is divided into three parts: a chamber, the body of the lock between the lock gates at either end of the lock and bordered by the lock wall; the fore (head) bay, the area upstream of the upstream lock gates; and the aft (tail) bay, the part downstream of the downstream lock gates (Figure 5.2).

Most historic American canals employed miter gates, a pair at either end; one leaf of each was placed against the other at an angle. Most gates were constructed of wood and were opened and shut by means of a balance beam, a beam at the top of a gate that served as a lever. These beams are large and long, 24 feet on the Chesapeake and Ohio Canal, mortised into the top of the lock gate heel. Miter gates rest on miter sills—beams, usually of wood, in the shape of a V—upon which the lock gates come to rest to form a watertight seal.[8]

Typically water flows in or out of a lock through sluices, also known as wicket gates, paddles, or gate paddles, in the lock gates. The wicket gates are usually built into the lower part of the lock gate. If a device slides up and down in grooves by means of a rack and pinion, it is called a valve. If it opens and shuts by turning on a horizontal or vertical axis, it is termed a paddle, paddle gate, or butterfly gate.[9]

Dam

Dams were erected in nearby rivers to provide a source of water for the canal. Nineteenth-century canal dams were most often constructed of timber cribs filled with rubble. Engineer John Millington describes the construction and purpose of a canal dam:

> A dam is a permanent obstacle to the passage of water, except over its top surface, and is built transversely across a river for the purpose of keeping back the necessary depth of water above it. The surface of the water which had before one general slope or inclination, is now rendered more level, and at this nearly level line will intersect the general slope of the river at some point behind the dam, another dam will then be necessary for the similar purpose of maintaining depth in the reach behind it. A river is thus converted, as it were, into a kind of staircase or series of nearly level reaches, each of which must contain enough water to float craft within itself.[10]

The dams on Pennsylvania's Schuylkill Navigation, like many canal dams, were constructed of timber frameworks filled with rubble. Each timber crib was formed of logs measuring 18 by 20 inches, dovetailed

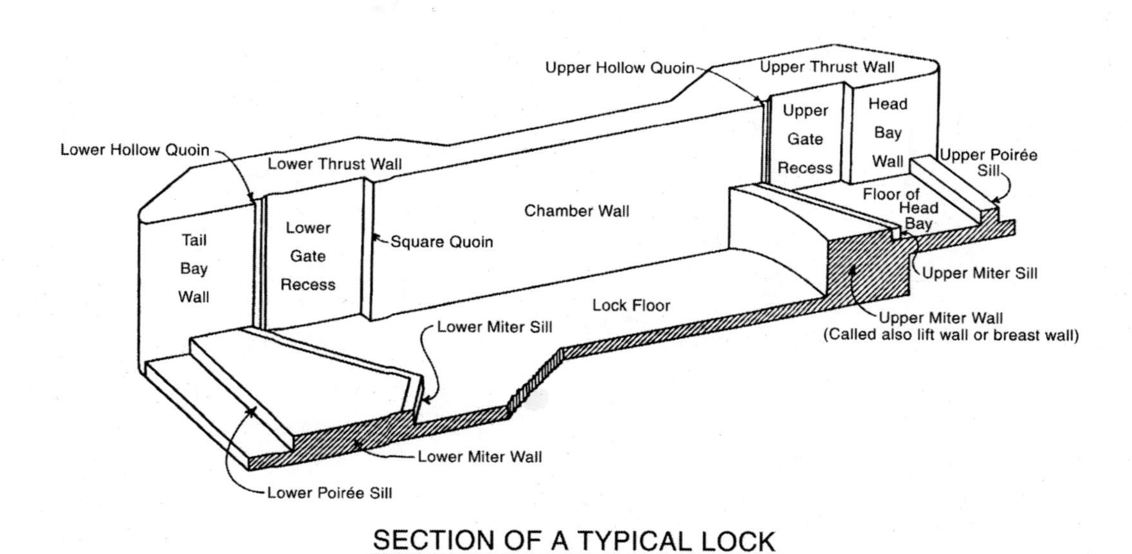

Figure 5.2. Isometric section of a lock. Reprinted from the Engineer School, *Canalization* (Fort Belvoir, VA: The Engineer School, 1940).

together. Each crib measured approximately 20 feet in breadth and was filled with stones and sunk in the line of the dam. The upper portions of the crib were connected together to form a continuous structure, and the whole was backed by a large mass of heavy rubble. The entirety was covered by planking, 6 inches thick.[11]

Thomas and Watt describe later timber crib dams (Figure 5.3) in their book, *The Improvement of Rivers*:

> The crib dam of modern design is usually built of sawn timbers, 10 to 12 inches square, laid crosswise so as to form pens 8 to 12 feet centers, and filled with riprap. The stone is usually in small sizes and irregular shapes called "one-man stone," but sometimes these are taken out in large blocks as blasted in the quarry. . . .
>
> Both the upstream and the downstream sides . . . are carried up vertically, the former being covered with a single or double row of sheet piling, driven as deep as possible. . . . Immediately upstream of this sheeting should be placed an embankment of gravel and clay or of good earth, preferably riprapped on its upper surface for a distance of 10 to 15 feet away from the dam. . . .
>
> The line of the crest is ordinarily placed from 8 to 12 feet from the upper face, and the sloping decking . . . connecting these points should be practically water-tight.[12]

Feeder Canal

To supply sufficient depth of water for navigation, water was carried from a reservoir, river, or other body of water to a canal by means of a feeder canal. Millington describes the feeder canal:

> All canals receive their water by feeders, which are level cuts, sometimes of miles in length; while at others, the canal itself runs into and joins the river, a part of which is frequently used instead of a canal, provided it is free from impediments and not subject to much variation in height of water.[13]

The lock at the head of a feeder is termed a guard lock. It was often designed with higher upper gates and more substantial head walls than usual to withstand floodwaters. It could and sometimes did provide boat access to a river.[14]

Change Bridges

Towpaths did not always remain on one side of a canal. Animals and their drivers changed from one side to the other when necessary to avoid obstructions. When the

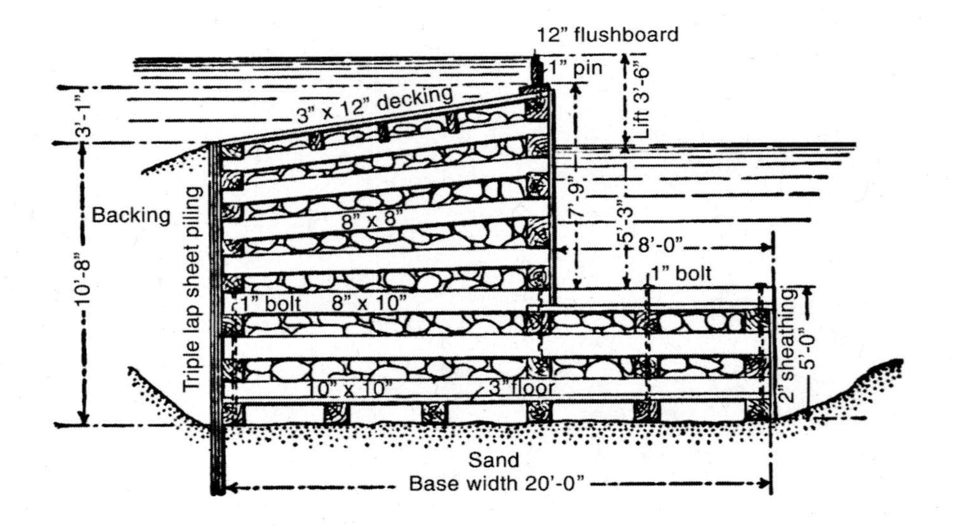

Figure 5.3. Section of timber dam as used on the Fox River, Wisconsin, 1898. Reprinted from B. F. Thomas and D. A. Watt, *The Improvement of Rivers: A Treatise on the Methods Employed for Improving Streams for Open Navigation, and for Navigation by Means of Locks and Dams* (New York: John Wiley and Sons, 1913).

towpath changed sides of the canal, special bridges, called change or crossover bridges, were necessary to permit draft animals to cross the canal without being unhitched from the boat.

Used first in England, change bridges had two important features to permit the towrope to pass from one side to the other. First, the bridge had a low rail with smooth curves to avoid snagging the towrope. Second, the bridge had a ramp system with a circular, cloverleaf-style ramp on one side and a straight ramp on the other. Both ramps came off the bridge in the same direction.

The last remaining change bridge in the United States is the Aldrich Chain Bridge on the Erie Canal at Palmyra-Macedon, New York (Figure 5.4). This bridge, which uses Squire Whipple's patented bridge design, was one of hundreds of Whipple bridges that once crossed the Erie Canal.[15]

Inclined Planes

Inclined planes on waterways were first employed by the Chinese. The first modern use of the inclined plane in a canal system was by William Reynolds, who introduced them on England's Shropshire Canal in 1792. Inclined planes were later extensively used on English waterways. The first systematic use of inclined planes in American navigation occurred on New Jersey's Morris Canal, constructed between 1825 and 1831. Later, steam-powered inclined planes were used on Pennsylvania's Mainline Portage.[16]

Inclined planes were used on the Morris Canal to help surmount the considerable changes in elevation along the canal's route. Had the canal employed only locks, one lock would have been needed every 2 miles and the cost would have been prohibitive. The use of inclined planes was recommended by the eminent engineer

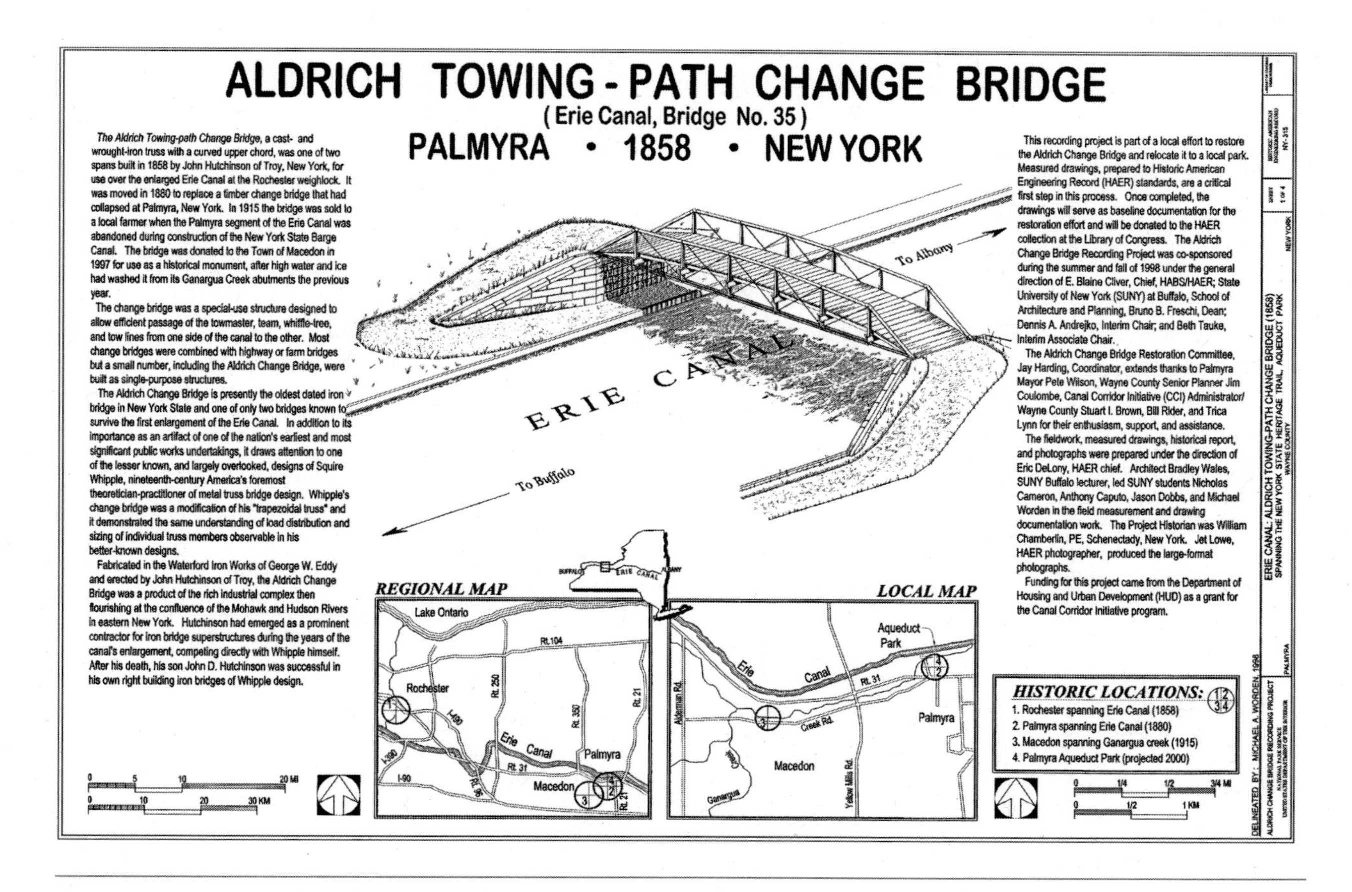

Figure 5.4. Aldrich Towing-Path Change Bridge, Palmyra, New York. HAER NY-315, National Park Service. Delineated by Michael A. Worden, 1998.

James Renwick of Columbia University. The eastern portion of the canal included 12 inclined planes with a total lift of 750 feet, and 16 locks with a combined lift of 164 feet. The western division included 11 inclined planes with a combined lift of 690 feet, and 7 locks with a combined lift of 68 feet. The average grade of the inclines was approximately 1:11.

The inclined planes used on the Morris Canal consisted of a set of tracks running from the canal bed up the incline, over the crest of the hill and into the next stretch of canal bed. The tracks were submerged in both the upper and lower levels of the canal bed and were exposed only along the plane. Morris Canal boats were constructed in two hinged sections, so the trucks used to carry the boats were also divided into two sections, each section having eight wheels with flanges on each side of the rails. The trucks were provided with stanchions, and the boats were fastened with hawsers.

Power for the operation of the planes was provided by a waterwheel. The levers for regulating water supply and control of the breaks were in a high tower with a view of the entire plane. The tower, located about midway between the top and bottom of the plane, also contained the waterwheel. The wire ropes and trucks used for the planes were manufactured by J. A. Roebling & Sons of Trenton, New Jersey.

The wire cables were arranged so that as one wound on the drum, the other unwound. The two ropes passed around submerged horizontal sheaves at the top and bottom of the plane. If the car was to be drawn up, the engineer in the plane house would turn a wheel allowing water into the waterwheel, and the drum wound up the cable at one end and unwound it at the other, drawing the car up. To take an empty boat down the plane, it was hauled out of the upper reach, the water was shut off the wheel, and the car descended by its own weight.[17]

Culverts

In their progress across the landscape, canals frequently crossed streams or creeks, often tributaries of the major river paralleled by the canal. Culverts, short-span structures, were used to carry a stream beneath the canal and towpath. These structures were constructed of a stone barrel arch built on a timber or stone foundation. The structure was covered with earth so that the width of the canal and the towpath was maintained throughout the crossing.[18]

Aqueducts

When the waterway to be crossed was too wide and/or too deep for a culvert, an aqueduct was employed. Typically the trough was anchored to each stream embankment with stone abutments. Spans longer than 30 feet were usually supported upon stone piers. The aqueduct trough, generally wide enough for only one boat, was constructed of wood, iron, steel, concrete, or stone.[19] Civil engineer John Millington describes the construction of an aqueduct:

> The best and most substantial method of building an aqueduct . . . is to form it, piers and all, of masonry or brickwork, elliptic or segmental arches, worked up to a perfectly flat surface on the upper side, upon which the side walls of the channel are afterwards erected. . . . The platform upon the arches must be as much wider than the intended water-course as will leave a towing path on one or both of its sides.[20]

SLACKWATER NAVIGATION SYSTEMS

The most important component of a slackwater navigation system or of a canal is the lock. In contemporary navigation systems, as a tow heads downstream and approaches the lock, the operator pulls levers to open the hydraulic filling valves that allow the water to enter the chamber through outlets. Once water has filled to the upper level, the lock gates open and the tow proceeds into the lock chamber. The tow ties up in the chamber, and the valves open to allow the water to flow from the chamber through culverts and empty downstream. Once the water level in the chamber is equal to the level of the lower pool, the lower lock gates open and the tow proceeds downstream (Figure 5.5).

Historic Overview

One of the earliest river canalization projects in the United States was on the Monongahela River in Pennsylvania and West Virginia. Four locks were built between 1840 and 1844, measuring 158 feet long and 50 feet wide. Lock size on the river increased as coal traffic developed. Locks were rebuilt in the 1903–1916 period and were all 56 feet wide and 182, 360, or 720 feet long.[21]

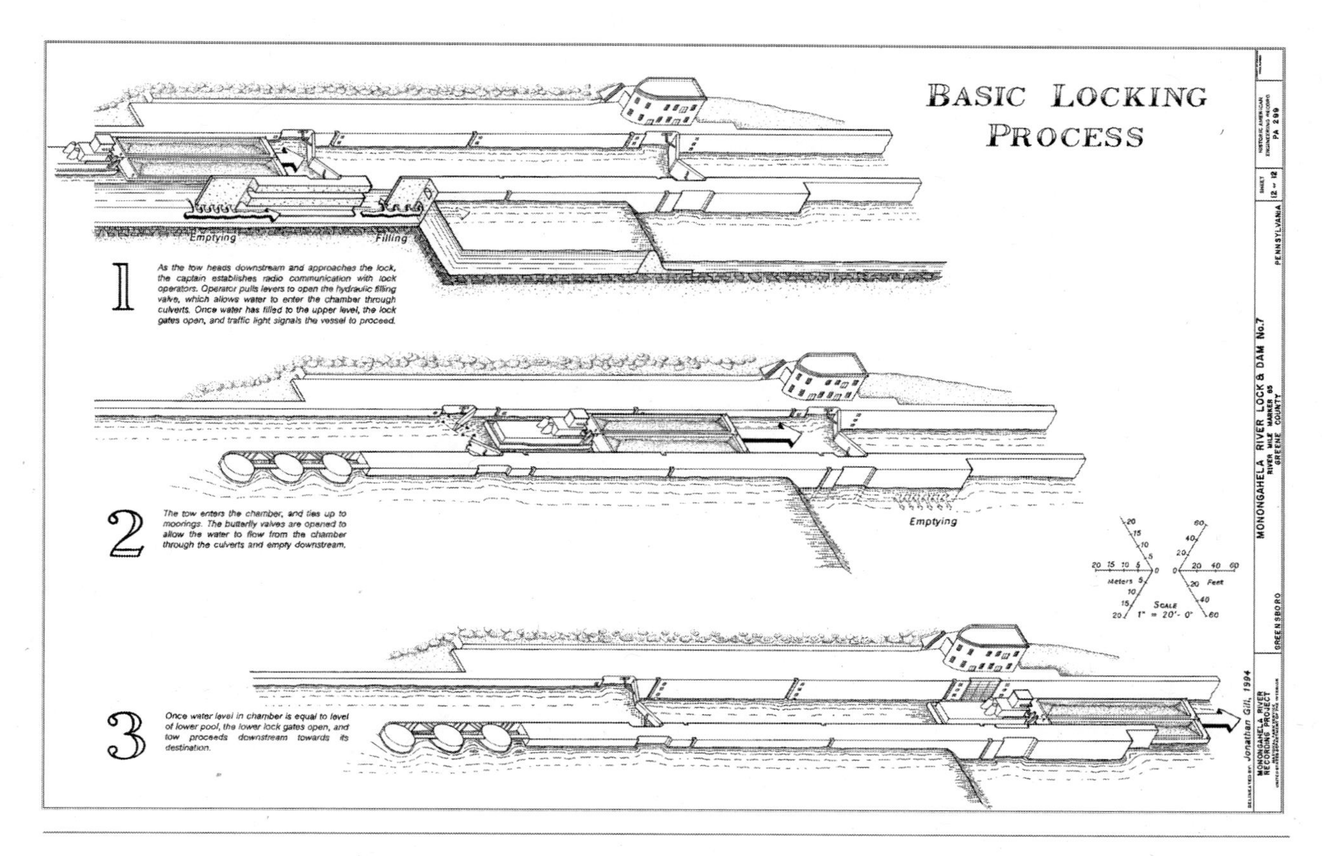

Figure 5.5. Basic locking process. From Monongahela River Lock & Dam No. 7. HAER PA-299, National Park Service. Delineated by Jonathan Gill, 1994.

A canal with locks had been built to bypass the falls of the Ohio in the early 1830s. The first lock and dam leading to canalization of the Ohio River was constructed between 1877 and 1885. Four other similar dams were constructed in the period 1898–1908 on the Upper Ohio. The lock size selected for the initial canalization was based on passing a fleet of ten barges, a fuel flat, and a steamboat. In 1878, a lock size of 110 feet wide and 600 feet was chosen. Beginning in 1954, the deteriorating and undersized locks and old movable dams on the Ohio were replaced by 13 higher-lift, fixed, nonnavigable dams.[22]

Upper Mississippi River canalization did not begin until the twentieth century. In 1930, Congress authorized a 9-foot channel to be constructed by canalization and dredging from Minneapolis to the mouth of the Missouri River just upstream from St. Louis. Most Mississippi locks were 110 feet wide and 600 feet long. During the 1980s and 1990s, the Melvin Price replacement locks and dams were constructed.[23]

Commercial traffic on the Arkansas River began in the 1820s. River improvements began in the 1830s, but dredging, snagging, and other improvements proved insufficient. However, it was not until 1949 that funds were appropriated by Congress for canalization and hydroelectric power generation. A series of delays followed, but all the navigation locks and dams along the river in Arkansas and Oklahoma were under construction by 1968.[24]

Locks

A lock is a structure designed to pass vessels from one water level either up or down to another water level. It is an open chamber with gates at both ends—at the upper pool and the lower pool.

One of the first issues considered in the design of a lock is the type of filling and emptying system to be used. The types of service gates, sills, walls, floors, and approaches often hinge on this decision (Figure 5.6). Filling and emptying systems used in the lock chamber include wall ports, laterals, bottom longitudinals, and multiple wall ports. Filling and emptying of very low-lift locks have been accomplished by sector gates,

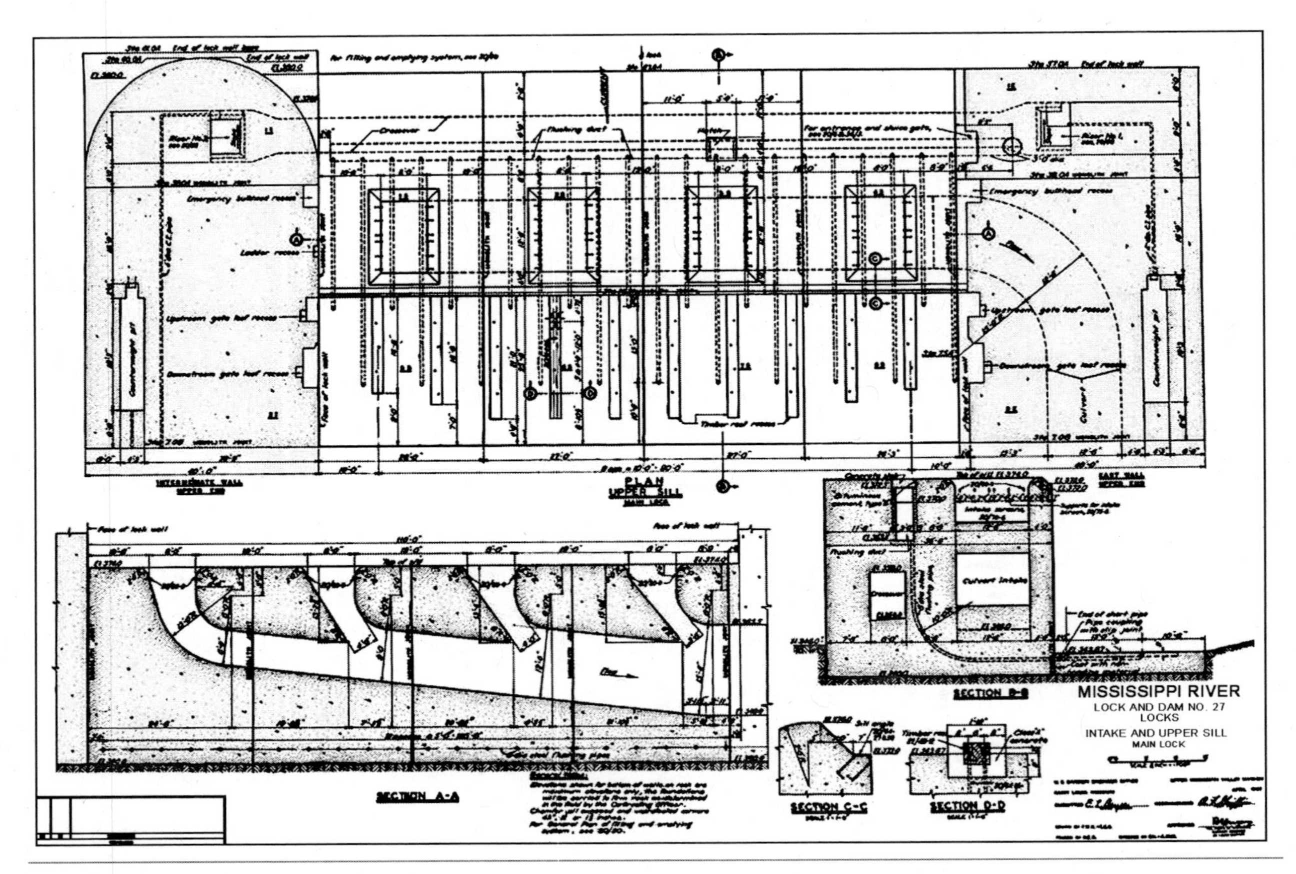

Figure 5.6. Mississippi River Lock and Dam No. 27. Upper sill intake and culverts. U.S. Army Corps of Engineers, as reproduced in HAER IL-33:79.

shutters in lock gates, and longitudinal flumes adjacent to the lock chamber.[25]

Early lock chambers were approximately 50 by 150 feet in size. By the early twentieth century, chamber widths ranged from 27 to 110 feet. The widest locks were on the Ohio River Navigation System. The lock chamber length was usually chosen to accommodate one barge length for the smaller locks and two for the larger locks. With the advent of tows, lock size increased on slackwater navigations. For instance, in the 1920s, lock chambers measuring 56 by 360 feet became the norm. Modern lock chambers measure as much as 110 by 1,200 feet in size.[26]

Approach Walls

The presence of approach walls at each end of a lock reduces hazards and increases the ease of the entrances and departures of tows. Historically, longer approach walls have been located on the landward side of a lock. These walls aided operation close to the riverbank where adverse currents made navigation difficult.

A general rule for the longer approach walls is that their length should equal the usable length of the lock chamber. The longer approach walls are usually straight-line extensions of the lock walls. Approach walls may be built of mass concrete or reinforced concrete, steel sheet pile cells, prefabricated concrete beams for underwater wall sections, cast-in-place concrete for above-water sections, caisson-supported guide walls, floating guide walls, sheet-pile guide walls, and timber guide walls.

Guide walls are long extensions of the lock walls, either in an upstream or downstream direction, that are parallel to the lock wall. These walls serve primarily to guide long tows into the lock and to provide mooring facilities for tows too long to be accommodated in a single lockage. Guide walls can be placed on either the land side, riverside, or both sides of the lock approach, depending on channel conditions.[27]

Guard walls, constructed on the side of the lock closest to the bank, are designed to minimize entrance or exit difficulties caused by currents. They also act as barriers to vessel movement in the direction of navigational hazards and serve as the final means of directing a misaligned front end of a tow, thus avoiding a head-on impact with the end of the lock wall.

Lock Construction

Several types of construction have been employed in modern navigational locks: mass concrete, reinforced concrete, tied-back concrete, steel sheet piling, or earth. Earlier locks were of either stone or wood.

Gravity mass-concrete locks employ soil, rock, or pile foundations and have few limitations as to height or lift. The mass-concrete lock is constructed of monolithic wall sections which resist applied loading by their weight and have floors formed of in-situ bedrock, concrete paving, and/or concrete struts. The floors in mass-concrete locks are usually constructed separately from the walls and are not used as structural members.

Reinforced concrete locks have walls containing steel reinforcement. Approach walls, abutments, areas around culverts, culvert valve recesses, filling and emptying laterals and longitudinals, and other parts of most modern gravity locks are typically constructed of reinforced concrete.

A tied-back concrete wall consists of a reinforced concrete wall that is tied back to sound material. Ties can be steel reinforcement, prestressed threaded bars, or stranded tendons. These reinforced concrete walls are thinner than gravity mass walls, but must be thick enough to contain culverts, galleries, and appurtenances such as floating mooring bitts.

Steel sheet pile lock walls are used only for lock chamber walls and upper and lower approach walls. This type of wall is sometimes employed at locations where traffic is not heavy or where a temporary lock is needed. Sheet pile locks are filled and emptied through sector gates, loop culverts, a combination of sector gates and culverts, valves in miter gates, or special flumes.

In a few tidal locks on the intracoastal waterways, earth levees have been used as lock chamber walls with concrete gate bay monoliths. Riprap protection is provided on the levee slopes and on the floor of the lock chamber to prevent scour by towboat propeller action. A continuous timber wall lines the sides of the lock chamber.[28]

Gate Bay Monoliths

The gate bay monoliths are the portion of the lock that houses the gate recesses, gate anchorages, gate machinery, and, sometimes, culvert valves and culvert bulkheads. The top width of the gate bay walls must be wide enough to house the operating mechanism, provide space for gate anchorages, enclose the valves, allow the gates to be recessed flush with the face of the wall for miter and sector gates, and provide sufficient concrete between the culverts and the gate recess for stability.[29]

Lock Sills

Lock sills provide the damming surface beneath the service gates or temporary closures. Lock sills are sometimes constructed integrally with lock walls and sometimes are not. Typically the lock gate sill elevation is higher than the adjacent lock floor. It is set low enough to provide a cushion between the bottom of the vessel and the top of the sill. Sometimes the gate sill masonry can be used to form intake ports for culvert filling and emptying systems and for crossovers containing the various utilities. The elevation of sill tops in relation to the water surfaces of the upper and lower pools dictates the draft of vessels that can use the lock.[30]

In the early twentieth century, miter sills were typically made of 10 by 10 inch or 12 by 12 inch timbers with tops a foot to 18 inches above the floor. In modern locks, gate sill depths are typically made as great as possible to lessen entry and exit times and chamber surges. Modern gate sills are usually 2 to 3 feet in height above the lock floor.[31]

Lock Gates

The primary purpose of lock gates is to hold the water in the lock chamber. Seven types of gate structures have been used over time in navigation locks in the United States: miter gates, submergible vertical-lift gates, overhead vertical-lift gates, submergible Tainter gates, vertical-axis sector gates, rolling gates, and tumbler gates. The horizontally framed miter gate is the preferred gate type due to operational efficiency and lower maintenance requirements.

Miter Gates

Miter gates are the only type of service gate used in a lock that cannot be operated with a differential head upstream and downstream (Figure 5.7a). A miter gate has two leaves that provide a closure at one end of the lock. It derives its name from the fact that the two leaves meet at an angle pointing upstream, resembling a miter joint. The steel gates have a girder framework covered by a

a

b

HALF PLAN
DOWNSTREAM LEAF
DOWNSTREAM LEAF
UPSTREAM LEAF
UPSTREAM LEAF
SECTION C-C
HALF DOWNSTREAM ELEVATION
SECTION A-A
END VIEW B-B

MISSISSIPPI RIVER
LOCK AND DAM NO. 27
LOCKS
LIFT GATE - BOTH LEAVES
GENERAL ASSEMBLY

Figure 5.7. ***a,*** Miter gates. Starved Rock Lock and Dam, Peru vicinity, Illinois. HAER IL-127, National Park Service. Photograph by Jerry Mathiason, 1994. ***b,*** Vertical-lift gates. Missssissippi River Lock No. 27 (Upper Mississippi Valley Division, U.S. Army Corps of Engineers, 1947).

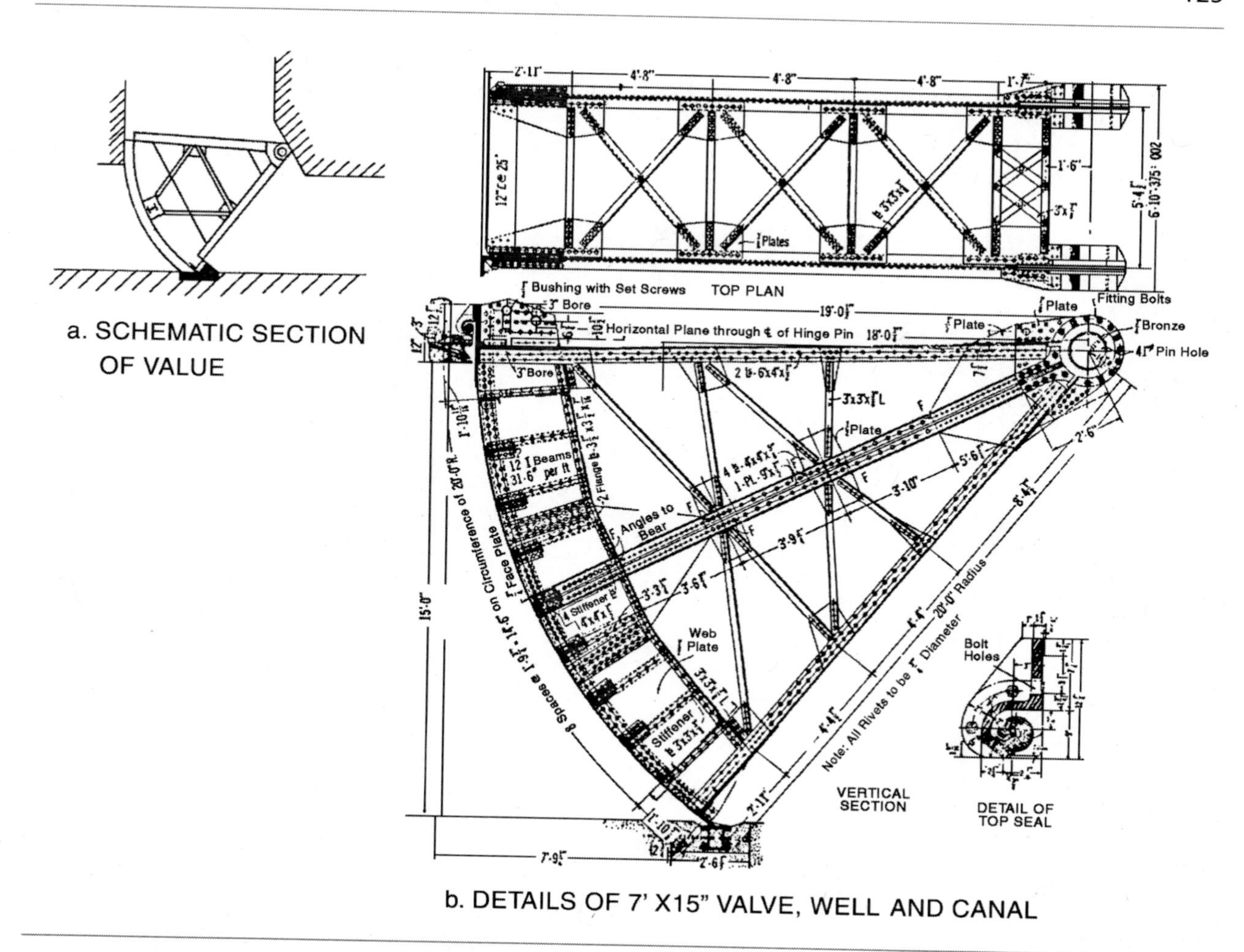

Figure 5.8. Tainter valve (Engineer School, 1940).

the lock, two or three feet below the lowest opening of the culvert.[44]

Where the foundation of a modern lock is of erodible material, such as sand or gravel, the lock floor is constructed of concrete. If the lock is excavated in rock, a concrete floor may be unnecessary.

Floating Mooring Bitts

When tows have entered the lock chamber, barges must be kept under control and stationary while the lock is being filled and emptied. Floating mooring bitts attached by lines to vessels serve this function. These bitts consist of a watertight floating tank that rises and falls as the lock chamber water level changes. The floating tank is mounted with wheels that ride inside steel guides in the mooring bitt recesses. Four to eight mooring bitts are usually provided in each chamber wall.[45]

Navigation Dams

Navigation dams impound water in a pool above a dam to provide adequate depth for navigation throughout the reach to the next dam upstream. A navigation dam is composed of one or more types of structures that operate together to limit passage of a pool of water. Navigation dams may be either fixed or movable. Fixed dams provide a permanent barrier to passage, with boats diverted through the associated lock(s). Movable dams provide passage in the main river channel in some circumstances.

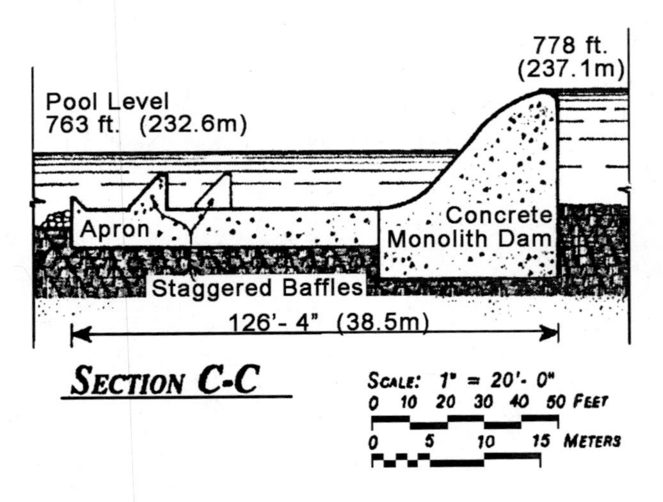

Figure 5.9. Fixed dam, Monongahela River. From U.S. Army Corps of Engineers 1897, in Douglas C. McVarish, *Monongahela River Navigation System: Historical Engineering Evaluation*, 1999, Figure 93.

Fixed Dams

A fixed dam permanently blocks the channel to some height above the riverbed (Figure 5.9). Water passes over the top, either over an ungated spillway or through gated or ungated tunnels in the dam itself. Fixed dams range in construction from embankments of earth or rock to massive concrete structures.

A fixed dam may be a low-overflow weir with a crest lower than the adjacent banks, designed to be overtopped at all stream flows. It also may be designed with a fixed crest at an elevation lower than normal pool, with conduits through the dam to pass small flows and flashboards on the spillway crest that can be removed manually or are designed to break loose at a particular stage.[46]

Movable Dams

In a movable dam, all or part of the damming surface can be raised out of the water or lowered to the streambed to pass flood flows. Dams with gates that lift out of the water normally require piers on the spillway crest, due to limitations on the length of the gate and the need for an operating bridge. This precludes the use of the channel for navigation at any flow; all traffic must flow through the lock.

A movable dam that lowers to the riverbed without obstructive piers or bridges may permit traffic over the dam at higher flows when the dams are down. Such a dam is termed navigable.[47]

Navigable Movable Dams

Navigable movable dams consist of a concrete sill at bed level supporting movable gates or wickets that form the damming surface in the raised position but can be lowered to the streambed to pass flood flows and permit traffic to pass over the dam at higher flows. Structural members supporting the damming surface are hinged to lie flat on the sill when the dam is open. Primary types of movable dams include the Chanoine wicket, the Chanoine-Pascaud wicket, the Bebout wicket, the Chittenden drum, and the bear-trap.

The Chanoine wicket is a wood panel supported in an angled position by a prop and attached to the sill by a metal frame (horse) hinged to the sill and to the downstream face of the wicket (Figure 5.10). Wickets are raised and lowered by a workboat operating upstream of the dam. This wicket type was used on the original canalization program on the Ohio River and was designed with a standard width of 3.75 feet set 4 feet on centers.

The Chanoine-Pascaud wicket is a modified Chanoine wicket constructed of steel with higher and wider panels for more rapid and flexible operations. The wickets can be positioned at two angles, permitting flow to pass between the wickets. Tests on the Ohio River indicated that a wicket with a length of up to 35 feet and a width of 6 feet could be satisfactory.

The Bebout wicket is similar to the Chanoine wicket. It is supported by a prop hinged to the sill and to the wicket and is designed to trip and lower automatically when overtopped by a sufficient depth of flow to raise the resultant water pressure above the fulcrum. The wicket is raised by a hoisting line attached to the top.[48]

Developed by Col. H. M. Chittenden of the U.S. Army Corps of Engineers as a modification of the French Desfontaines wicket, the Chittenden drum consists of a sector of a circle, airtight and working in a closed chamber. The water is admitted under head into this chamber and pushes up the wicket. When the head is shut off and the chamber opened to the lower pool, the wicket is pushed down into the recess. This type of wicket was used at Dam No. 2 on the Monongahela River.[49]

The bear-trap dam, the first movable dam used in the United States, was first employed in 1818 on the Lehigh River of Pennsylvania. This type of dam consists of two leaves, both hinged to the sill. The upstream leaf is solid, and the lower leaf, which serves as the prop, is buoyant. The chamber under the two leaves

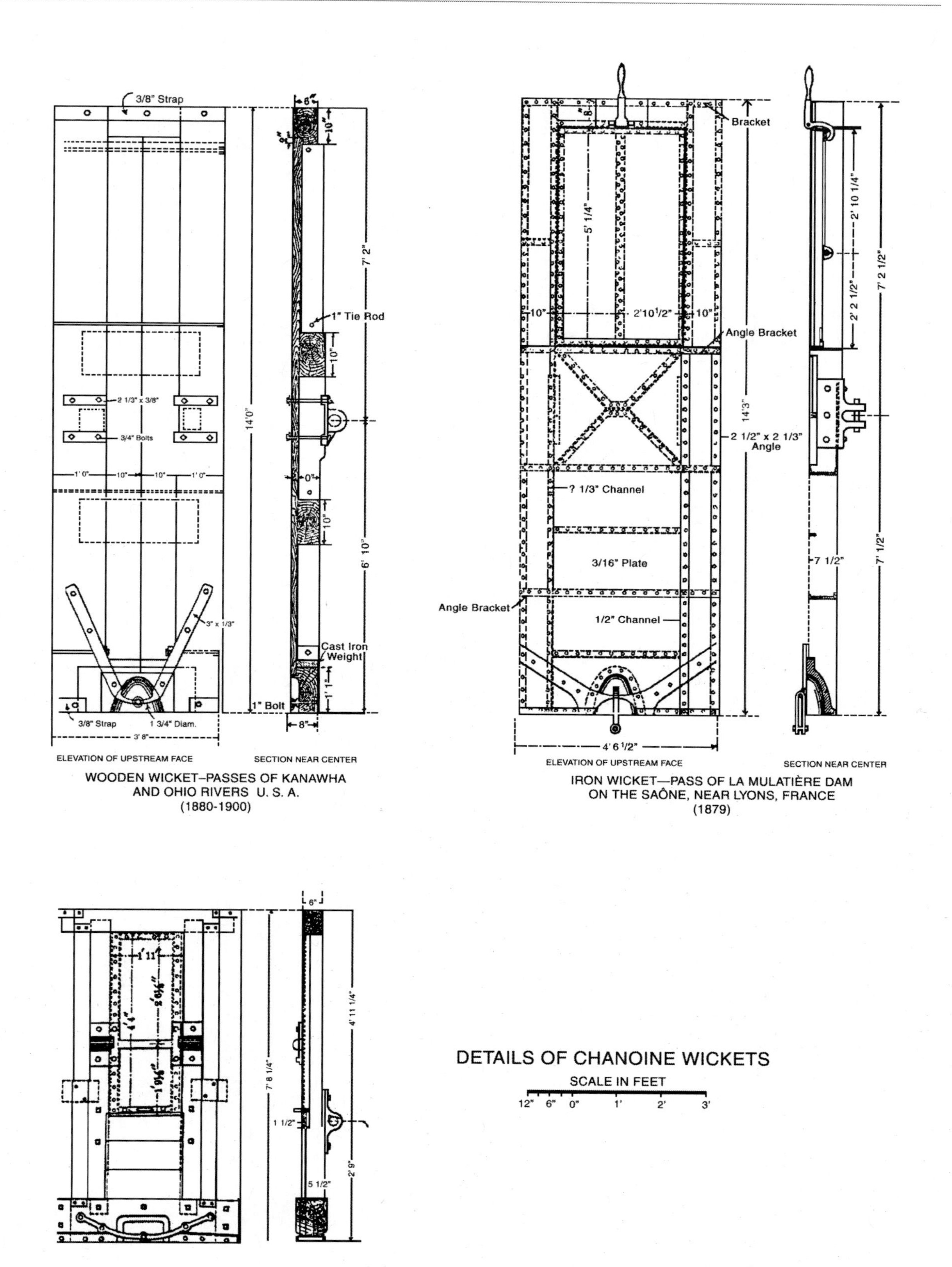

Figure 5.10. Chanoine wicket dam (Thomas and Watt 1913: Figure 60).

is nearly watertight with sliding seals between the sides of the leaves and the adjoining piers. Conduits in the piers connect this chamber with the upper and lower pools. Valves are operated to admit water to the chamber from the upper pool, thus raising the dam into position. When water is released from the chamber to the lower pool, the dam collapses.

Nonnavigable Movable Dams

Nonnavigable movable dams typically consist of a concrete sill at about streambed elevation, piers on the crest, and movable gates. The type of gate used also controls the dam sill and associated piers. Spillway gates include Tainter gates, radial gates, hinged-crest gates (bascule, Pelican, and flap), vertical-lift gates, and roller gates. The most commonly used gate on navigable projects is the Tainter. A bridge mounted on the piers and extending across the spillway serves as an operating bridge for mobile hoisting machinery and access to the gate-operating machinery in the piers.

Tainter gates are a segment of a cylinder mounted on radial arms that rotate on trunnions embedded in piers on the spillway crest (Figure 5.11a). Because of their simple design, relatively light weight, and low

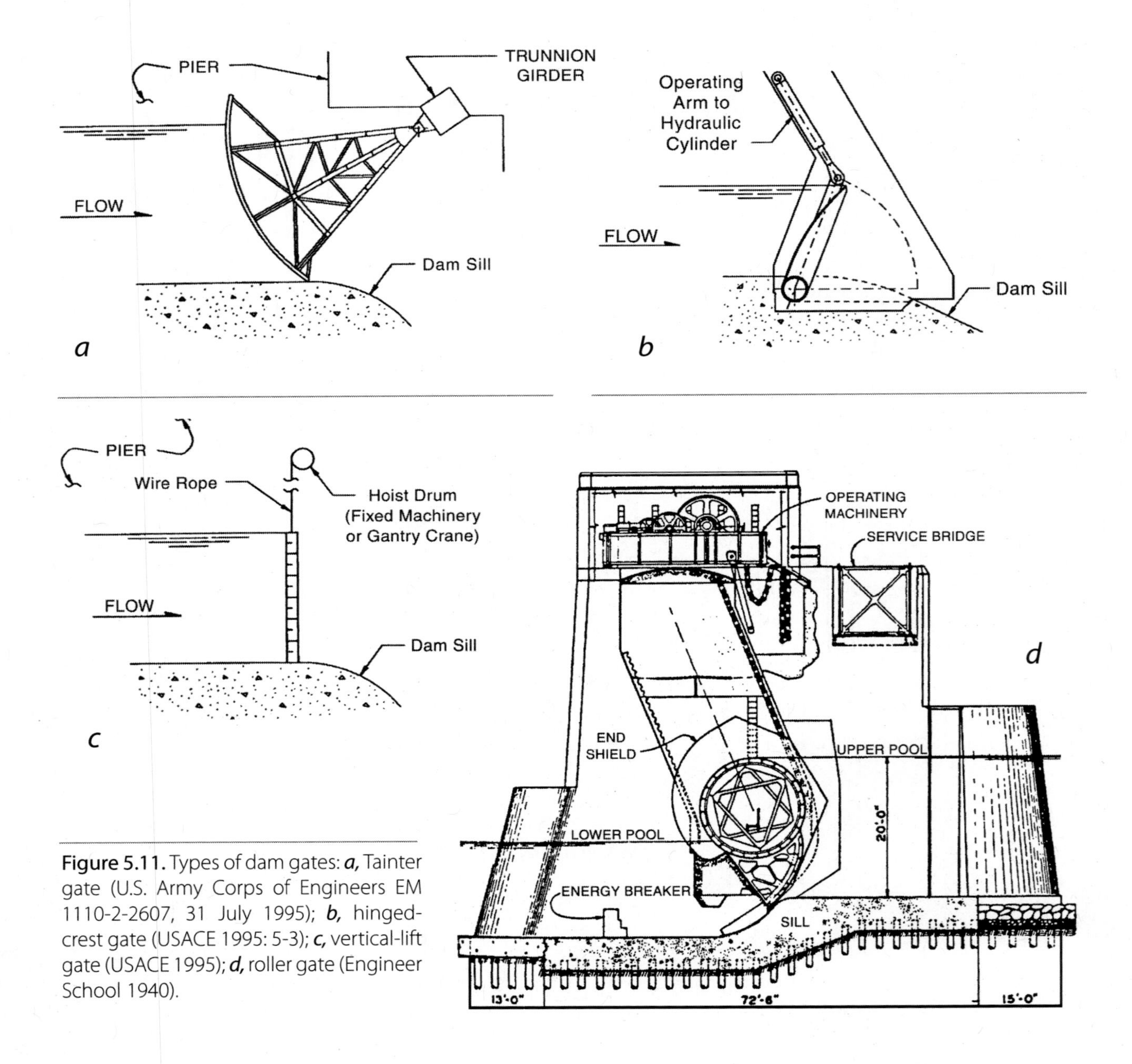

Figure 5.11. Types of dam gates: ***a,*** Tainter gate (U.S. Army Corps of Engineers EM 1110-2-2607, 31 July 1995); ***b,*** hinged-crest gate (USACE 1995: 5-3); ***c,*** vertical-lift gate (USACE 1995); ***d,*** roller gate (Engineer School 1940).

hoist-capacity requirements, Tainter gates are considered one of the most economical and most suitable gates for controlled spillways. Tainter gates are raised and lowered by wire rope or cables anchored at each end. Gates are usually operated by individual hoists, one at each end of the gate. Tainter gates built to heights of 75 feet and lengths of 110 feet have been used for navigation dams. Tainter gate piers are concrete with a precast/prestressed concrete or steel trunnion girder anchored into the pier. The thickness varies with height but is usually 10 to 15 feet.

Submergible Tainter gates were developed to allow passage of ice without having to use large gate openings. Two types have developed, one in which the top of the gate can be lowered below the normal upper pool elevation, and the piggyback gate, in which a shaped lip on the top of the gate or a double skin plate can be used to keep the flow off the back of the gate.

Radial gates are similar in configuration to Tainter gates but are raised or lowered with hydraulic cylinders instead of cables. In Europe, these gates are normally used instead of cable-hoisted gates.

Hinged-crest gates are known by a variety of names, including bascule, Pelican, and flap gates (Figure 5.11b). These gates are hinged at the base to a dam sill and are raised to retain pool levels and lowered to pass flows. They can be straight or curved to fit the dam sill crest when in the lowered position. The simplest form of hinged-crest gate is a flat, stiffened plate hinged at the bottom and operated by a screw stem or hydraulic cylinder connected to the top of the gate at one end. This type of hinged-crest gate is limited to approximately 35 feet of length and 8 feet of height. Automated operation of these gates sometimes occurs, as in the hydraulically operated bascule gate on the Jonesville Dam on the Ouachita-Black Rivers Navigation System.[50]

Vertical-lift gates consist of a skin plate over horizontal girders that transmit the water load to the piers (Figure 5.11c). The gate is stiffened vertically by several full-depth diaphragms. High piers are required for the gates in the fully raised position above high water level. The gate is mounted on wheels or rollers to permit movement under water load and is raised by chains at both ends. Vertical-lift gates have been designed for spans in excess of 100 feet and heights of 60 feet.

When very high gates are required, the vertical-lift gate may be designed in two or more horizontal sections (leaves) to reduce hoist capacity or reduce pier height. Multiple-leaf gates are more complex to operate than single-leaf gates because the top leaf must be removed from the guide slot and stored before the lower leaf can be raised. Double-leaf gates are used at McNary Lock and Dam on the Columbia River.[51]

Roller gates are metal cylinders with ring gears at each end that travel on inclined metal racks on the piers (Figure 5.11d). The roller gate is raised and lowered by a chain around one end of the gate operated by a hoist mounted in the pier. Water can be admitted to or released from the interior of the gate to change the gate's buoyancy. Roller gates have been designed with heights of up to 30 feet and lengths of up to 150 feet. Roller gates are employed on the Kanawha and on the Upper Mississippi River.[52]

GLOSSARY

Most of the following definitions are paraphrased from either *Canalization*, published by the Engineer School, Fort Belvoir (1940), Thomas Swiftwater Hahn and Emory L. Kemp's *Canal Terminology in the United States* (1999), B. F. Thomas and D. A. Watt's *The Improvement of Rivers* (1913), or Terry K. Woods's *The Ohio & Erie Canal: A Glossary of Terms* (1995).

Apron. A structure built downstream from a dam to prevent scouring of the river bottom by falling water. Aprons are generally made heavier near the dam and lighter downstream and are constructed of concrete, heavy riprap, or timber cribs filled with stone.

Aqueduct. A structure for carrying a canal channel and towpath across a waterway or valley too wide or deep for a culvert.

Baffles. A series of rectangular concrete blocks placed in one or more rows across the stilling basin to dissipate the energy of water cascading over a dam.

Balance beam (or *balance lever*). The beam at the top of, and projecting from, a miter lock gate that serves as a lever to swing the gate open and shut and to balance the weight of the gate on its pintle. It consists of a long timber mortised into the top of the gate heel and miter posts that extend over the lock wall.

Basin. A pool connected to a canal or a wide spot in a canal used for docking, loading, unloading, and turning of boats. Canal offices and warehouse were often located adjacent to basins.

Bear-trap dam. The original bear-trap design consisted of two straight leaves, hinged at the bottom, the upstream leaf overlapping the downstream one when lowered; when water was introduced underneath, both leaves were pushed up, the end of the downstream one sliding along and helping to push up the other. Bear-traps were used in movable dams on the Allegheny and Ohio Rivers.

Bear-trap lock. A type of flash lock constructed with hydraulically operated leaves. Frequently used to pass ice and debris on the Ohio River locks and dams. First employed on the Lehigh River.

Berm. The bank of a canal opposite the towpath.

Bollard. A stout wooden or cast-iron post at the side of a lock used to secure lines and to prevent a boat from surging in the lock.

Boule dam. A dam equipped with a number of collapsible trestles which support movable shutters. Each trestle is a structural steel A-frame placed at right angles to the dam. Each frame is hinged to the foundation so that the trestles may be lowered behind the sill when the dam is down. The trestles are spaced at intervals of 5 or 6 feet. The top of each is permanently connected to the top of the next by a chain.

Breast wall. A transverse wall at the upstream end of a lock chamber. This wall steps the canal bottom from the floor of the lock chamber to the bottom of the canal upstream. The lift of the lock can be determined by measuring the distance from the lock floor to the top of the coping of the breast wall.

Butterfly valve. A lock valve installed to swing on either the vertical or horizontal axis that consists of a rectangular or circular blade with a stem or strut. The blade of the valve is attached to the supporting strut or the stem and revolves about it.

Capstan. A manually or powered rotating vertical spindle-mounted drum on which a rope, cable, or chain is wound to pull a heavy object such as the large lock gates of a ship lock.

Chamber. The portion of the lock that extends from the upper to the lower gate sill and is enclosed by the lock gates and lock walls.

Chamber floor. A paved floor constructed for locks founded on sand or gravel. Where the chamber is founded on rock, it is usually left exposed. The chief function of the chamber floor is to prevent erosion from turbulence within the chamber.

Chamber walls. The two sides of the lock chamber. Initially, chamber walls were usually constructed of coursed stone; now walls are generally formed of concrete. Concrete chamber walls usually contain the culverts and valves for filling and emptying the chamber, galleries to carry electric wiring, and piping used for hydraulic systems.

Chanoine wicket. Based on a design developed in 1852 by French engineer Jacques Chanoine and first used on the Seine River near Paris. A narrow wooden leaf which, when raised, is supported in an inclined position by a prop, and when lowered, lies flat on the foundation, immediately downstream from the sill. It is used as part of a movable dam system.[53]

Composite lock. A lock consisting of a rough stone structure lined with wood plank to make it watertight.[54]

Coping stone. A well-dressed stone comprising one of the top courses of a lock or other structure.

Crib dam. A dam constructed across a river or stream in the form of a series of log chambers, each about 6 feet square and filled with stone and gravel.

Culverts. (1) Openings or closed channels located in lock walls that carry water to or from the filling or emptying ports. Culverts are of two general types: longitudinal, which parallel the length of the lock walls; and lateral or cross culverts. (2) (*Singular*): A short-span structure, generally stone, for carrying a stream under a canal and towing path.

Cylindrical valve (or *drum valve*). A valve consisting of a cylindrical ring 4 to 8 feet in diameter and 4 to 6 feet high, sliding up and down inside a covered steel drum, open at the bottom and directly above, and completely covering a vertical section of culvert. The cylinder, sliding up and down in the drum, opens and closes the ports between the pedestals, thus permitting and stopping flow from the horizontal to the vertical culverts or vice versa.

Dam piers. Concrete monoliths erected to support the gates of movable, nonnavigable dams. These piers are generally surmounted by a service bridge.

Double lock. (1) Two locks in series with a common pair of gates between. (2) A pair of locks side by side.

Drop gate. An upstream lock gate, pivoted at the bottom, so that when opened, it lies flat on the bottom of the lock.

Drum valve. An improvement on the ordinary cylindrical valve, with the closed cylinder on the inside and the open sliding cylinder on the outside, making the former stationary.[55]

Entrance lock. On a ship canal, the lock accommodating vessels traveling from the larger body of water into the canal.

Feeder canal. A channel carrying water from a lake, reservoir, river, etc. to a canal.

Feeder dam. A dam built across a stream to create a reservoir that supplies water to a canal through a feeder canal.

Feeder lock (or *outlet dock, guard lock*). A lock used to pass boats from an impounded river to a canal.

Flash lock. A single gate built into a weir or at one end of a low dam to enable small boats or log rafts to pass downstream over rapids and low falls on a shallow stream.

Flashboards. Boards installed atop dams to temporarily raise their heights. These boards are secured by iron pins inserted in bored holes at the top of the dam.

Flight of locks. A series of locks constructed close together.

Floodgate. The gate of a waste weir used to drain excess canal water.

Forebay. The portion of a lock immediately upstream of the upstream lock gate(s).

Gate iron. An iron strap attached to the coping stones that hold the gate in place.

Gate pocket (or *gate recess*). The recessed part of a lock wall that holds the lock gate in place when it is open.

Gated dam. A dam constructed to permit increased control over the water levels in the navigation pool upstream of the dam. Machinery mounted on tall concrete piers moves large chains or cables which lift gates that are hinged into the body of the piers or can move vertically. As the gates are raised or lowered to control the amount of water flowing under or over them, the upstream pool is maintained at a relatively constant level.

Guard lock. The lock at the head of a canal or a feeder or dam, providing access to an impoundment of water. It was often designed with higher upper gates and more substantial head walls than usual to withstand floodwaters.

Guard wall. An extension of a lock wall that protects craft from the draw of current toward the overflow sections of the dam when the current is likely to be strong enough to suck vessels from the lock approach and sweep them over the dam.

Gudgeon pin. A portion of the top hinge of a miter gate leaf. A fixed part of the leaf, it seats in a gudgeon or eye connected to the lock wall by adjustable anchorage bars. The gate is suspended in position by the gudgeon and the pintle, the two points about which it pivots when in movement.

Guide wall. An extension of the wall of the lock nearest the land and generally parallel to the bank. The guide wall's primary purpose is to guide and align vessels and tows entering or leaving the lock. Its secondary purpose is often to retard erosion of the riverbank.

Head bay (or *forebay*). The area of the lock immediately upstream from the lock chamber.

Head gate. A gate in the bulkhead wall of a reservoir which regulates the flow of water from the reservoir of a canal.

Hollow quoin. A recessed piece of masonry generally cut in an arc of the same radius as the heel of the gate and about two-thirds of a semicircle in length. This shape provides a wide bearing for the gate thrust, affords support when the gate is struck from above by a boat, and provides excellent protection against debris.

Inclined plane. An inclined railway used where the change in elevation in the line of the canal would have required an excessive number of locks. Power was provided by stationary steam engine or by hydraulic turbine.

Land chamber. In a two-lock complex, the lock that is located closest to the riverbank.

Lift gate. A lock gate that resembles the vertical-lift gate used as a movable element of a dam. This gate is opened either by raising or lowering it between two piers or uprights carrying vertical supporting tracks. These gates were employed at the Little Falls Lock of the New York State Barge Canal.[56]

Lock key. A metal device used to operate the wicket gate mechanism.

Middle wall. In a two-lock complex, the wall that separates the two lock chambers.

Miter gate sill or *miter sill.* A raised portion of the floor, in plan like a flattened V with the point upstream, against which the lower edges of the closed gate rest.

Miter gates. Lock gates, originally wood, now generally steel, consisting of a pair of symmetrical leaves, movable about a vertical axis, shutting against each other at one end and against the miter sills at the bottom, and abutting against the hollow quoins at the other end.

Movable dams. Dams in which the framework could be lowered to the sill of the dam, permitting vessels open navigation on canalized rivers. Types of moveable dams included the bear-trap, needle, wicket, curtain, and drum types.

Outlet lock. A lock used to pass canal boats from a canal to river.

Paddle (or *wicket* or *sluice*). The valving in a lock gate that allowed water to enter or leave a lock chamber. Such paddles were usually operated from the top of the gate by a crew using a wrench-like device on the paddle stem.[57]

Paddle stem. A long iron rod extending from the pivot point of a lock gate paddle through the top of the gate and the balance beam.[58]

Piers. Concrete monoliths erected to support the gates of movable, nonnavigable dams.

Pintle. A carefully finished nickel-steel, hemispheric member, fixed to a base on the shelf in the lock wall formed by the bottom of the miter gate recess. The gate is suspended in position by the gudgeon and the pintle, the two points about which it pivots when in movement.

Pintle shoe. A socket at the bottom of the gate leaf that bears upon and semi-encloses the pintle.

Prism. The trapezoidal cross-sectional shape of the canal's channel.

River chamber. In a two-lock complex, that lock located adjacent to the dam.

Roller gate. A dam gate consisting of a massive steel cylinder extending horizontally from one pier to the next. The cylinder is designed with circumferential teeth at each end which engage in a tooth track slanting upward and downstream in a recess in the pier.

Rolling gate. A lock gate forming a massive steel frame, sheathed on one side and running on flanged wheels over a track embedded in the floor. The gate somewhat resembles a large boxcar. When open, the gate is withdrawn into a recess, usually through the land wall and into the bank. The American-type rolling gate was devised for use on the Ohio River, where a number were originally installed before the application of vertical framing adapted the miter gate to the conditions of low lift and wide span typical of the river.

Service bridge. In a nonnavigable gated dam, the tops of the piers are commonly joined by a service bridge. This bridge allows the movement of operators from pier to pier and ordinarily supports a traveling repair crane, well as an emergency bulkhead hoist or crane.

Sheet piling. A row of driven, long, slender planks, frequently used to prevent water from flowing under a lock.

Ship canal. A canal capable of carrying large, ocean-going vessels.

Ship lock. A lock capable of handling large vessels.

Sidney gate. A special type of Tainter gate with movable or sliding trunnion pins. The gate was devised in an effort to overcome the disadvantage of limited maximum clearance of other Tainter gates that results from a fixed trunnion. The pins of the Sidney gate are seated in slots on the pier face, which causes the gate to be opened by rotation about the trunnions until the center of gravity of the gate is vertically below the point of lift. From this point, the entire gate is lifted to full clearance, the pins rising vertically along the pier tracks.

Slackwater. Water impounded behind a dam. The slackwater allows the river upstream of the dam to be used for navigation.

Snubbing post. A post in a lock to which a boat was tied up while in the lock so that the forward motion of the boat could be stopped as it entered the lock. Generally made of cast iron, they were usually not less 8 eight inches in diameter and 12 to 18 inches high.

Stilling basin. A structure with a masonry bottom constructed adjacent to the downstream side of the dam to dissipate the energy of the falling water. A stilling basin is an apron with pedestals or baffle piers to reduce the energy of the falling water.

Stoney valve (or *rising stem valve*). A valve resembling a thin, flat box which is raised and lowered vertically to and from its sealed, closed position. The downstream side rests on a roller train running on a metal frame set in the masonry, or fixed rollers on the gate rest directly on the supporting track frame. The valve moves in a recess only slightly inset into a side wall culvert. This valve was employed in Lock No. 9 of the Monongahela River Navigation System.

Stop gate. (1) A gate, normally kept open, for the purpose of guarding against flooding or to isolate a level of a canal containing a break. (2) Normally closed gates behind a dam. The gates could be opened to flood the nearby countryside, lessening the effect of a flood.

Strut pin. A steel pin that connects the strut to the sector arm and the gate connection in Ohio-type gate-operating machinery.

Strut tube (or *strut*). A steel tube, part of Ohio-type gate-operating machinery, situated between the sector arm and the gate connector and gate bracket, and linked to these adjacent components by steel struts pins.

Tail bay. The part of a lock downstream of the downstream lock gate.

Tainter gate. An internally framed sector of a cylinder, with a system of radial bracing (trunnion arm) at each end which carries the gate load to a trunnion pin. The trunnion pin is anchored to the pier, and the arm and gate rotate about the pin. The Tainter gate is usually raised and lowered by chain or steel cables, one near each end, which are carried upward to hoists on the piers or the service bridge.

Tainter valve. A lock valve consisting of a steel-plate sector of a cylinder, supported by ribs which are in turn supported by beams which carry the load to radial struts at either end. The end struts bear upon and revolve about trunnions which are at the axis of the skin-plate cylinder.

Tide lock. An outlet lock of a canal or canalized river that terminates in a body of water subject to tidal fluctuations.

Venturi-shaped. A tubular shape with a constricted middle and flare at either end.

Vertical- lift gate. A dam gate, in principle the basic form of the ordinary Stoney gate, its distinguishing features in canalization dams being its large size and the types of rollers used. Vertical-lift gates for long spans are usually framed with trusses or built-up girders. The water loads are transmitted to steel girders set in recesses in the pier faces. The actual contact points of horizontal support are made through fixed rollers rather than the roller trains of the Stoney types. Vertical-lift gates were used at the Emsworth and Montgomery Island Dams on the Ohio River.

Waste weir. A stone, concrete, or wooden structure built in the towpath bank of the canal with gates or stop planks, the lifting of which enabled the draining of a canal for repairs, cleaning, or protection from ice in the winter.

Wasteway. A cut-down section of a towpath, sometimes more than 100 feet long, lined with riprap or concrete that allowed excess water to drain from the canal.[59]

Weigh lock. A lock fitted with scales on which a boat comes to rest when the lock gates are closed and the water is emptied from the lock chamber.

Wicket gate. A small gate built into the lower part of a lock gate to let water in and out of the lock chamber. Built to slide up and down in grooves with the aid of a rack and pinion, it was commonly termed a valve. If it opened and shut by pivoting on a horizontal or vertical axis, it was termed a paddle, a paddle gate, a pivot valve, or a butterfly gate.

Wing wall. The wall that angles away from a lock at its extreme ends.

BIBLIOGRAPHY

American Canal Society. *The American Canal Guide: A Bicentennial Inventory of America's Historic Canal Resources.* Freemansburg, PA: American Canal Society, 1988.

American Society of Civil Engineers, Waterways Division. *Manual on Lock Valves.* Manual of Engineering Practice No. 3. New York: Committee on Lock Valves, Waterways Division, American Society of Civil Engineers, 1930.

Casto, James E. *Towboats on the Ohio.* Lexington: The University Press of Kentucky, 1995.

Clement, Dan. Morris Canal Inclined Plane 10 West. Historic American Engineering Record, NJ-30. 1983.

The Engineer School. *Canalization.* 2 volumes. Fort Belvoir, VA: The Engineer School, 1940.

Hahn, Thomas F., T. Gibson Hobbs, Jr., and Robert S. Mayo. *Towpaths to Tugboats: A History of American Canal Engineering.* York, PA: The American Canal and Transportation Center, 1992.

Hahn, Thomas Swiftwater, and Emory L. Kemp. *Canal Terminology of the United States.* Morgantown, WV: West Virginia Institute for the History of Technology and Industrial Archeology, 1999.

Hullfish, Dr. William. "The Change Bridge." *American Canals* 33 (Summer 2004): 3–6.

Johnson, Leland R. *The Davis Island Lock and Dam, 1870–1922*. Pittsburgh: U.S. Army Engineer District, 1985.

Kemp, Emory. *The Great Kanawha Navigation*. Morgantown, WV: Institute for the History of Technology, 2000.

Mahan, Dennis Hart. *An Elementary Course of Civil Engineering for the Use of the Cadets of the United States' Military Academy*. New York: Wiley and Putnam, 1838.

McVarish, Douglas C. *Monongahela River Navigation System: Historic Engineering Engineering Evaluation*. Prepared by John Milner Associates, Inc. for the Pittsburgh District, U.S. Army Corps of Engineers, 1999.

Millington, John. *Elements of Civil Engineering*. Philadelphia: J. Dobson, Richmond, Smith & Palmer, 1839.

O'Brien, William Patrick, Mary Yeater Rathbun, and Patrick O'Bannon. *Gateways to Commerce*. Denver: National Park Service, 1992.

Oxx, Francis H. "The Ohio River Movable Dams." *The Military Engineer* XXVII, No. 151 (January-February 1935): 49–58.

Peterson, Margaret S. *River Engineering*. Englewood Cliffs, NJ: Prentice-Hall, 1986.

Sadowski, Frank E., Jr. The Erie Canal. www.eriecanal.org.

Shank, William H. *Towpaths to Tugboats: A History of American Canal Engineering*. York, PA: The American Canal and Transportation Center, 1995.

Stevenson, David. *The Principles and Practice of Canal and River Engineering*. 2nd edition. Edinburgh: Adam and Charles Black, 1872.

Thomas, Benjamin F., and D. A. Watt. *The Improvement of Rivers*. New York: John Wiley and Sons, 1913.

U.S. Army Corps of Engineers, Engineering and Design. Lock Gates and Operating Equipment. Engineer Manual EM 1110-2-2703, 30 June 1994.

—. *Planning and Design of Navigation Dams*. Engineer Manual EM 1110-2-2607, 31 July 1995.

—. *Planning and Design of Navigation Locks*. Engineer Manual EM 1110-2-2602, 30 September 1995.

Wegmann, Edward. *The Design and Construction of Dams*. New York: John Wiley and Sons, 1900.

Whitford, Noble E. *History of the Canal System of the State of New York Together with Brief Histories of the Canals of the United States and Canada*. Volume I. Albany: Brandow Printing Company, 1905.

Wilson, Herbert M. The Morris Canal and Its Inclined Planes. Scientific American Supplement, February 24, 1883. Available at www.catskillarchive.com/rrextra/abnjmc.html.

Woods, Terry K., compiler. *The Ohio & Erie: A Glossary of Terms*. Kent, OH: Kent State University Press, 1995.

NOTES

1 Thomas F. Hahn, T. Gibson Hobbs, Jr., and Robert S. Mayo. *Towpaths to Tugboats: A History of American Canal Engineering* (York, PA: The American Canal and Transportation Center, 1995), 4–6.

2 Hahn, Hobbs, and Mayo, 14–15; Margaret S. Petersen, *River Engineering* (Englewood Cliffs, NJ: Prentice-Hall, 1986), 279.

3 For brief discussions of canals in numerous states, see Hahn, Hobbs and Mayo 1995.

4 Petersen, 283–285.

5 Dennis Hart Mahan, *An Elementary Course of Civil Engineering for the Use of the Cadets of the United States' Military Academy* (New York: Wiley, 1861), 314.

6 Thomas Swiftwater Hahn and Emory L. Kemp, *Canal Terminology of the United States* (Morgantown: West Virginia Institute for the History of Technology and Industrial Archeology, 1999), 9.

7 Hahn and Kemp, 96.

8 Hahn and Kemp, 6.

9 Hahn and Kemp, 132.

10 John Millington, *Elements of Civil Engineering...* (Philadelphia: J. Dobson; Richmond, Smith & Palmer, 1839), 692.

11 David Stevenson, *The Principles and Practices of Canal and River Engineering* (Edinburgh: Adam and Charles Black, 1872), 146.

12 B. F. Thomas and D. A. Watt, *The Improvement of Rivers* (New York: John Wiley & Sons, Inc., 1913), 507–508.

13 Millington, 695.

14 Hahn and Kemp, 43, 51.

15 Dr. William Hullfish, "The Change Bridge," *American Canals* 33 (Summer 2004): 3–6.

16 James E. Held, "The Canal Age," *Archeology* magazine online, July 1, 1998. www.archeology.org/online/features/canal.

17 Herbert M. Wilson, "The Morris Canal and Its Inclined Planes," *Scientific American Supplement*, February 24, 1883 (www.catskillarchive.com/rrextra/abnjmc.html); Dan Clement, Morris Canal Inclined Plane 10 West, Historic American Engineering Record, NJ-30.

18 Hahn and Kemp, 31.

19 Hahn and Kemp, 3.

20 John Millington, 713–714.

21 Douglas C. McVarish, *Monongahela River Navigation System: Historic Engineering Evaluation* (West Chester, PA: John Milner Associates, Inc., 1999).

22 Petersen, 269–270.

23 Petersen, 272.

24 Petersen, 276–278.

25 A detailed discussion of filling and emptying systems is found in Petersen, 357–381.

26 Thomas and Watt, 341–342; U.S. Army Corps of Engineers, *Planning and Design of Navigation Locks*, EM 1110-2-2602, 30 September 1995, 5-1.

27 Petersen, 356–357; U.S. Army Corps of Engineers, 4-4.

28 U.S. Army Corps of Engineers, 4-1–4-4.

29 U.S. Army Corps of Engineers, 4-4.

[30] U.S. Army Corps of Engineers,
[31] Woods, 24; U.S. Army Corps of Engineers, 5-2.
[32] Petersen, 350-351.
[33] U.S. Army Corps of Engineers, 7-1.
[34] U.S. Army Corps of Engineers, 7-1; Petersen, 351, 354, 359, 361.
[35] U.S. Army Corps of Engineers, 7-2; Petersen, 361–362.
[36] Petersen, 354–355.
[37] Petersen, 364.
[38] Petersen, 377–378.
[39] Petersen, 373.
[40] Petersen, 368–369.
[41] American Society of Civil Engineers, *Manual on Lock Valves* (New York: ASCE, 1930).
[42] The Engineer School, *Canalization* (Fort Belvoir, VA: The Engineer School, 1940), II, 368.
[43] The Engineer School, II, 368.
[44] Description of a Lock Floor, 1837; RG 79, Entry 2227, Drawings and other records concerning construction, National Archives, College Park, Maryland, as cited in Hahn and Kemp 1999, 72.
[45] U.S. Army Corps of Engineers, *Planning and Design of Navigation Locks*, 10-1.
[46] Petersen, 329–330.
[47] Petersen, 330.
[48] Francis H. Oxx, "The Ohio River Movable Dams," *The Military Engineer* XXVII, No. 151 (January–February 1935): 54.
[49] Thomas and Watt, 649.
[50] U.S. Army Corps of Engineers, *Planning and Design of Navigation Dams*, EM 1110-2-2607, 31, July 1995, 5-3.
[51] Petersen, 340–341.
[52] Petersen, 339.
[53] James E. Casto, *Towboat on the Ohio* (Lexington: University Press of Kentucky, 1995), 46.
[54] Terry K. Woods, compiler, *The Ohio & Erie Canal: A Glossary of Terms* (Kent, Ohio: Kent State University Press, 1995), 8.
[55] Thomas and Watt, 497–498.
[56] Thomas and Watt, 453.
[57] Woods, 16.
[58] Woods, 16.
[59] Woods, 42.

CHAPTER 6

Shipyards and Marine Structures

Many of the nation's major cities are located on either fresh- or saltwater waterways, and as a result marine structures are components of the industrial landscape of many communities. Military and private shipyards are found in small numbers but are prominent features of the cities that possess them.

The peak of U.S. ship construction was reached during the Second World War. At that time Navy yards in Boston, Charleston (South Carolina), Mare Island (California), Brooklyn, Norfolk (Virginia), Philadelphia, Portsmouth (New Hampshire), and Bremerton (Washington) built combatant ships for the Navy. In addition, large vessels (battleships, aircraft carriers, and cruisers) were built at the Bethlehem Steel Shipbuilding Division in Quincy, Massachusetts; at the Newport News Shipbuilding and Dry Dock Company in Virginia; and at the New York Shipbuilding Company in Camden, New Jersey. Other shipyards located throughout the United States built small naval vessels. The height of wartime ship construction was reached in 1944, when 2,505 vessels with a gross tonnage of 10,142,376 were built.[1]

WHARVES AND PIERS

Carleton Greene defines a wharf as a structure at which vessels may land and load their cargoes and passengers. Wharves can be located either along the shore or projecting into a body of water, but in most localities the term is used only for structures paralleling the shore. A pier is defined as a wharf projecting from the shore.[2]

Wharves

Wharves require a wall to retain the filling. These retaining walls may be divided into three classes: gravity walls, those that depend on the weight of the structure for stability; walls with relieving platforms; and sheet-pile walls, held in place by tie-rods and earth anchors.[3] Wharves have been constructed of timber, concrete, stone masonry, and steel. Historically, timber was, in most areas of the country, the cheapest material but was subject to destruction by decay, marine animals, and fire. By the early twentieth century, concrete piles had become a widely used alternative to timber. Stone masonry was so expensive that it was rarely used except as a facing of monumental structures. Steel for bearing piles and sheet piles had become more frequently used in the early twentieth century.[4]

Piers

Three types of piers are commonly in use: pile platform, block-and-bridge, and solid-filled. Most common is the pile platform. The platform may be constructed of wood, steel, or concrete and may form the deck of the pier itself. The piles may be of ordinary timber or concrete or may take the form of metal or concrete columns of diameters ranging from 2 to 10 feet or more.

The block-and-bridge type consists of blocks usually of timber cribwork, but sometimes of concrete or stone masonry, with bridges extending from block to block. The solid-filled type consists of a retaining wall of some kind along the sides and out end of the pier, with the enclosed area filled with earth or other material usually dredged from the slips alongside the pier.[5]

In a letter written in the early nineteenth century, Fennimore Cooper described New York's piers:

> All the wharves of New York are of very simple construction — a frame-work on hewn logs is filled with loose stone, and covered with a surface of trodden earth. . . . The Americans . . . are daily constructing great ranges of these wooden piers, in order to meet the increasing demands of their trade.[6]

Similar construction was used in Philadelphia's early piers, described as a series of connected wooden cribs infilled with stone, sand, mud, rubble, and/or brick. The exposed vertical surfaces of the cribs, which formed the exterior faces of the piers, were constructed of hand-hewn timbers. The nonexposed surfaces, which formed the interior sides of the crib, were constructed of rough-cut logs, some with bark attached. Hewn timbers of Philadelphia's Pier 35 South measured roughly 9 by 16 inches in section and varied in length from 36 to 38 feet. The timbers were laid in horizontal courses with overlapping ends. Each course was fastened to the course below by a series of 1-inch-diameter wrought-iron rods inserted into holes bored or drilled in the timbers. Transverse timbers or headers extended into the interior of the pier and provided additional support for the face wall. The headers were connected to the face wall by means of half-dovetail notching and were secured in position by wood treenails or iron spikes.

By mid-century, cribwork piers were being replaced by timber pile structures. The Bush Terminal Company pier of 1902 exhibited common construction characteristics for early twentieth-century timber piers (Figure 6.1). Measuring 150 feet wide and 1,315 feet long, it consisted of vertical timber sheet piling with an 80-foot-wide core with solid fill and 35-foot-wide, pile-supported timber decks on either side of the fill. Tie-rods joined the two walls of sheeting. The fill consisted of stratified sand and gravel dredged from the bottom. Transverse pile rows under the deck occurred at about 6-foot intervals. The original deck consisted of two layers of 3-inch-thick timber, laid at 45-degree angles and resting on the fill or on pile-supported underdecks.[7]

In the early twentieth century, many piers were erected with timber foundations and concrete decks. For example, a 1918–1919 pier at the Brooklyn Army Supply Base had piles arrayed in transverse rows 10 feet apart, with the piles within each row driven at 4-foot centers. Bents also included diagonal 6 by 12–inch timber braces. Timber 12 by 12–inch pile caps topped each bent. Longitudinal 12 by 12–inch stringers were placed on the north side of the pier below the caps, providing additional support under sections with railroad tracks. The 18-inch-thick concrete deck had 5/8-inch reinforcing rods, and was poured directly on the pile caps, in 10-foot sections. Timber fender systems consisted of 12 by 12–inch pilings set in 36-inch-wide arrays.[8]

At the same time, concrete piers were also being erected. The Huntington Beach (California) Municipal Pier, erected in 1913–1914, measures 1,314 feet long and 25 feet wide (Figure 6.2). As constructed, it was supported by 203 piles varying from 30 feet to 60 feet in length. All piles were made of concrete, and each was

Figure 6.1. Bush Terminal Company Pier #7. Brooklyn, New York. HAER NY-201A:1, National Park Service. Photograph by Thomas R. Flagg, 1988.

Figure 6.2. Huntington Beach (California) Municipal Pier. HAER CA-80:11, National Park Service. Photograph by William B. Dewey, 1989.

reinforced with eight ¾-inch round steel rods tied together with steel wire hoops and embedded to a depth of 2 inches of clear concrete cover. The remainder of the substructure was also constructed primarily of reinforced concrete. The cross-girders or pile bent caps measured 12 inches by 24 inches, and were reinforced with 1-inch square steel rods. Longitudinal beams located between the center piles of the bents measured 10 by 18 inches and were reinforced with 7/8-inch round rods. Oregon pine joists supported a plank deck.[9]

Sheds

Sheds built on wharves and piers were constructed to shelter freight and passengers from the elements and to prevent theft of merchandise. They were typically used for sorting both inbound and outbound freight and for temporary storage while in transit between the vehicle and the warehouse. The sheds ranged from those of wood-framed construction to massive stone structures.

Pier sheds were typically one or two stories high. The upper story, when present, was often used for passengers. The inshore ends of pier sheds were often joined to bulkhead sheds, extending along the bulkheads on each side of the pier, and to headhouses on the bulkhead, which contained offices and were often two or more stories in height.

Shore facades and outshore ends of piers frequently featured architectural ornamentation. An example of such an ornamented pier shed is that on Manhattan's Pier 95, with its Roman arched primary opening, Roman arched flanking window ensembles, and dentilled cornice (Figure 6.3).[10]

Shed posts, except those located in walls, could obstruct vehicular movement, so the second floor was sometimes suspended from the roof trusses, and sheds of less than 100 feet wide were frequently built in one span. For those of 125 feet, two spans were used, while for 150 feet, three spans provided a good arrangement of posts. Clearance under floor beams in a two-story shed typically varied from 17 to 22 feet. If traveling cranes were used, the clearance was increased by 10 to 15 feet.[11] The materials most frequently used for a shed side covering on pile piers were wood, galvanized sheet iron, and reinforced concrete.[12]

SHIPYARDS

History

Early large-scale shipyards contained extensive manufacturing shops capable of building propelling machinery, boilers, turrets, and substantial portions of the ship's

Figure 6.3. 55th Street Pier Shed, New York. HAER NY-147:5, National Park Service. Photograph by Gerald Weinstein, 1986.

mechanical equipment, as well as producing forgings and steel structures.

As turbines began to supersede reciprocating engines, engine shops were adapted to the construction of the new propelling units. When the watertube boiler replaced the earlier Scotch boiler, boiler shops could be adapted to handle armor or other heavy plate work, and boilers were generally procured from manufacturers specializing in them. Because of their size, they were generally delivered in a partially completed form, and final assembly and tubing were performed in shops specially equipped for this work.[13]

Three of the most prominent privately owned World War II–era shipyards are the Bethlehem Steel Company's Quincy, Massachusetts plant, established in 1899; the Newport News Shipbuilding and Dry Dock Company's plant established in 1886; and the New York Shipbuilding Corporation yard, established in Camden, New Jersey, in 1898 to 1900.[14]

Elements of these shipyards include buildings, berths, outfitting piers, dry docks, shipways, marine railways, cranes, and manufacturing shops. Types of manufacturing shops include anglesmith, pipe, joiner, erecting, heat treating copper, machine, plate and angle, brass foundry, pattern, door, blacksmith, sheet metal, turret and armor, mold loft, and shipwright.

Harry Gard Knox summarizes the principles that governed the location of manufacturing facilities in the "ideal shipyard":

> The ideal shipyard permits the orderly flow of hull materials from their initial unloading, through fabrication, subassembly, and storage to final assembly on the ways. It likewise provides that subassemblies, machinery, and equipment purchased from outside vendors be stored according to plan until requisitioned for installation. . . .
>
> After preliminary processing, plates and shapes are moved to storage areas where they await welding into small subassemblies. In most cases small subassemblies are later joined to form larger ones.[15]

The following sections describe major components of twentieth-century public and private shipyards.

Figure 6.4. Marine railway #3, Thames Tow Boat Company, New London, Connecticut. HAER CT-1:30. Photograph by Jack E. Boucher, 1975, 1978.

Marine Railways

Marine railways have played a critical role in the repairs and construction of boats and ships (Figure 6.4). Smaller facilities were often powered by steam engines or even mules or oxen. Metal cables extended from engine-driven winches to wooden undercarriages that rode on top of the rails. Theodore Wilson describes the marine railway in his book, *An Outline of Ship Building*:

> The marine railway consists of three parallel ways, the top surface of which is level with the deck of the sections. The centre way is intended to sustain the vessel on her keel when undergoing repairs; the other two ways, at equal distance from the centre way, answer the purpose of a launching way, on which the bilge-way and cradle rest during the operation of hauling the vessel on shore. The cradle is constructed in the same manner as for launching from an inclined slip. Temporary ways are laid upon the deck of the dock, being a continuation of the ways on shore. The bilge-ways are got in place, and the cradle placed in position. The hydraulic cylinder is attached to the head of the sliding frame by large wrought iron hauling beams, and the movement commenced, which draws the vessel along eight feet at a time, until the vessel is off the dock, and on the bed-ways in the navy yard.[16]

By 1945, a total of 85 marine railways of 200-foot length or more were in use in the United States. Most of these railways had up to 3,000 tons of capacity. Among shipyards with marine railways were Bethlehem Steel in Quincy, Massachusetts (two railways), Colonna's Shipyards (Norfolk, Virginia) (four railways), Consolidated Shipbuilding Corporation (Morris Heights, New York) (three), Electric Boat Company (Groton, Connecticut) (one), Norfolk Shipbuilding and Dry Dock Company (Virginia) (four), Peterson Boat Works (Sturgeon Bay, Wisconsin) (three), St. Louis Shipbuilding and Steel Company (three), General Engineering and Dry Dock Company (Alameda, California) (two), and Moore Dry Dock Company (Oakland, California) (two). The widespread use of hydraulic boat lifts has largely eliminated marine railways.

Dry Docks

A dry dock, also known as a graving dock, is a critical component of a shipyard, allowing for a ship to be constructed or repaired out of water and floated into a waterway upon completion. The water outside is excluded by gates, and the water inside is removed by pumps.

The 926-foot-long Dry Dock 3 at the Puget Sound Naval Shipyard was constructed between 1917 and 1919. A two-section dry dock constructed of concrete, it is supported directly on soil. The dry-dock facility includes an underslab hydrostatic pressure-relief drainage system, perimeter culverts, and two pumphouses. These components represent a system for flooding, draining, and dewatering the dry dock. Other key components of the dry dock include a removable steel caisson and a divider bulkhead.

A double row of timber sheet-piling cut-off walls, spaced approximately 15 feet on center, encompasses the shoreward perimeter of the facility. Hydrostatic relief is provided by weep holes in the walls and broken stone drains spaced transversely and longitudinally below the entire floor slab.

The entrance caisson is a reversible, hydrometer-type floating steel caisson, of riveted steel construction, with hull plates varying in width from 5/16 to ¾ inch. Watertight vertical bulkheads and trim tanks are provided near each end of the caisson. Within the base of the caisson is fixed concrete ballast. The caisson was originally designed to provide dry-dock flooding through the structure.

A divider bulkhead was constructed in 1945 as part of the subdivision of the dock. The bulkhead consists of a facing of tongue-and-groove planking and timber sheathing that spans horizontally between steel wide-flanged beams. These beams are supported by a concrete sill at the base and a welded steel box beam girder situated near the top of the bulkhead. The box beam is watertight and can be floated into position or lifted into place by a portal crane.[17]

In 1946, dry docks were in use at naval yards in Portsmouth, New Hampshire; Boston; New York; Philadelphia (Figure 6.5); Norfolk; Charleston, South Carolina; San Diego; Roosevelt Base (San Pedro, California); Mare Island (Vallejo, California); Puget Sound (Bremerton, Washington); Pearl Harbor; and Hunters Point (San Francisco). These dry docks are of granite, granite and concrete, concrete and steel, reinforced concrete, and wood and concrete construction. The oldest active dry dock at that time was Dry Dock No. 1 in Boston, a 348-foot granite structure built in 1833 and engineered by Loammi Baldwin. The largest dry docks were over 1,000 feet long and approximately 150 feet wide, the largest being Boston Navy Yard's Dry Dock No. 4 of granite and concrete construction which measured over 1,158 feet long and was built in 1919.[18]

Cranes

Twentieth-century shipyards employed some or all of the following types of cranes: portal gantry cranes, hammerhead revolving cranes, and bridge cranes (Figure 6.6).

Portal Gantry Cranes

The portal gantry crane resembles an old-fashioned steam shovel. In this type of crane, a movable bridge

Figure 6.5. Dry Dock No. 3, Philadelphia Naval Shipyard. HAER PA-381:C:1. Photograph by Jet Lowe, 1995.

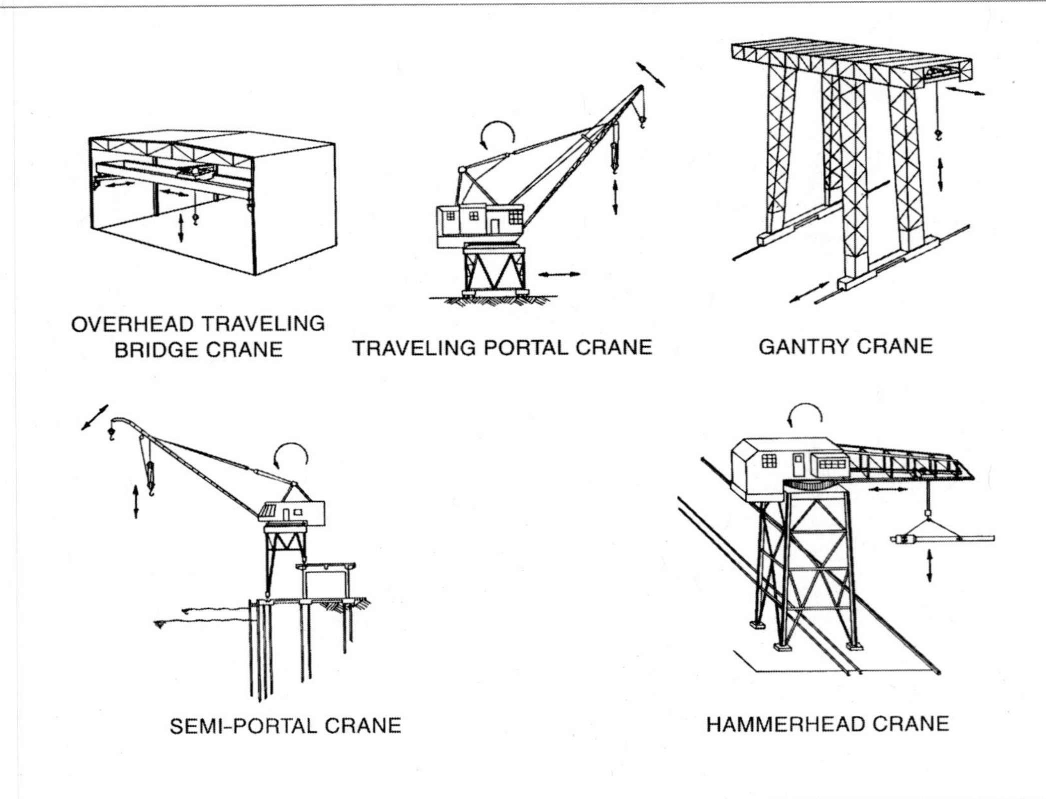

Figure 6.6. Types of shipyard cranes. Reprinted, with permission, from F. G. Fassett, Jr., editor, *The Shipbuilding Business in the United States of America* (New York: The Society of Naval Architects and Marine Engineers, 1948).

spans a roadway or one or more railroad tracks on the side of a wharf or pier. Such cranes were once essential for lifting materials or supplies onto and off ships, at piers and at dry docks. Typically such cranes were constructed by heavy steel platforms supported by four steel legs with steel wheels. The steel wheels enabled the cranes to move along tracks laid on the piers or around the dry dock.

The legs formed a rigid frame, with the platform comprising a base for the crane. This platform was constructed of structural steel plates and plates to form a rigid base for the machinery and stiffened to form the anchors for the A-frame and boom. The platform was often equipped with a geared turntable that permitted the crane to face in any direction. The cranes had a large, steel-clad operator's cab/machine house typically about 10 feet by 30 feet by 50 feet, frequently with a rank of steel-framed windows on one side. A boom was attached to the machinery house, often fabricated with the main chord, consisting of angles of alloy steel of the standard box girder type and braced with angle-iron latticing. The boom was attached to the cab/machinery house. A boom hoist had an average capacity of about 60,000 pounds, while an auxiliary hoist might have a capacity of 18,000 pounds. Two sets of wire ropes passed from drums in the machine house through a series of sheaves to raise and lower the boom and hoist hook. Power was provided by an electric motor that ran on 230-volt DC electric current.[19]

Hammerhead Revolving Cranes

Hammerhead cranes were employed at shipyards to provide greater lifting capacity than portal gantry cranes. A typical example is the 120-ton McMyler crane, originally built in 1916 and last used at the General Dynamics Corporation Shipyard in Massachusetts. Its original portal base had four legs of box-lattice construction curving inward toward the center of the tower and reinforced with V-shaped bracing. The base supported a steel platform, atop which the pintle bearing rested. In operation, the center turned with the pintle column, while the perimeter remained stationary. The operator's cab, which was suspended with and turned with the crane boom, was entered at this level.

The operator's cab was constructed of steel plate pierced with wood-framed sash windows. The windowed front side of the cab angled outward, allowing clear sight lines to the end of the boom, to both sides, and to the ground below. Above the main pintle platform, at the level of the bottom chord of the boom truss, was the machine house platform. At the rear of the platform was the steel-framed machine house, sheathed in steel plate with an enclosed counterweight below. The machine house contained the three motors, transmissions, and drums for the main hoist, auxiliary trolley, and main trolley lines.

The main drive motor for rotating the pintle was located near the center of the machine house platform. A series of bevel gears and shafts extending to either side of the crane carried power out and down to the four sets of paired double flange wheels which rode on a circular track below the main pintle platform.

The boom measured 110 feet long and was constructed of steel truss, C-beam chords, and angular lacing. The main and auxiliary hoist trolleys traveled on rails on the lower chord. The main trolley rode on pairs of 24-inch-diameter wheels, while the auxiliary trolley rode on four individual 16-inch wheels. In lifting, wire ropes extended from the drums via a series of sheaves along the top of the boom to the trolleys and hoisting hook mechanisms.[20]

Bridge Cranes

A bridge crane is a structure found in many industrial facilities in which heavy steel parts were stored and moved outdoors. Bridge cranes consist of riveted steel towers, set some distance apart and supporting the lattice track truss.

The bridge crane structure itself was constructed of two parallel sections of lattice truss, and it traveled on steel or cast-iron trucks with cast-iron wheels. The motor for crane movement and hoisting was mounted on steel plates between the two truss sections. Power was transmitted to the drive wheel by shafts and bevel gears mounted within the truss. A single hoist trolley moved on double flange wheels along the bridge tracks. The hoist hook was capable of being lowered to the ground and raised to the height of the towers by means of wire ropes.

The operator's cab was mounted at the bottom of the crane truss and was generally constructed of steel plate, with viewing windows on one side. Access to the cab was by means of a ladder mounted on the tower, another on the crane truss, a service catwalk on top of the truss, and a ladder connecting the catwalk and the cab floor.[21]

Shipways

A shipway is the cradle or platform on which the hull of the vessel is laid down and from which the completed ship is launched into the water. Shipways are used for final assembly and launching of ships. They generally consist of a wood deck of planking, carried on piling for the length of the ship where the ground is soft, or for the inboard end under the bow where the ground is firm. A shipwright first lays down keel blocks on the middle of the ways. Personnel later place side blocks under portions of the ship where extra weight is to be sustained and carry stages up the sides as work progresses on the hull of the ship. As a rule, the entire work of erecting a ship hull is done outdoors. In the case of Camden's New York Shipbuilding Company, however, the shipways were roofed over.[22]

Shipyard Buildings

In laying out a shipyard, attention is paid to the various processes in the building of a ship, from raw material to finished vessel. The layout of storehouses and shops is arranged to require the minimum amount of transport and handling of raw, fabricated, and welded material. To transport heavy materials and parts, locomotive and freight cars on standard-gauge tracks have often been used.[23]

Ropewalks

Among the surviving historic shipyard buildings at the Charlestown (Massachusetts) Navy Yard is the ropewalk. This building, designed by prominent Boston architect Alexander Parris, is ¼ mile long and was constructed between 1834 and 1837. After construction, most of the cordage used by the Navy was made here using steam-powered machinery. This building is the last complete ropewalk in the United States.[24]

Ropewalks were equipped with machines to convert hemp to finished rope. The first preparation machine was the "spreader" which combed and straightened out fibers. From this machine the fiber passed to a smaller and finer one, where the fibers in the sliver were further straightened and the sliver evened and drawn down. Then it passed to a drawing frame, where the sliver was reduced in volume, straightened, and evened more thoroughly. From the preparation machines, the hemp passed to the spinner, where it was spun into yarn and wound on a bobbin.

The next step was twisting strands of yarn to form rope. In the process, the strands were drawn down the ropewalk by steam power. The final step, called closing, put the strands into rope. This was accomplished by two machines, one at either end of the ropewalk. The one at the lower end was called a layer, as it lay up or closed the rope. The upper machine was stationary, and was used to keep the proper twist in the strand while laying. The Charlestown ropewalk was capable of making rope of up to 170 fathoms (1,020 feet) in length.[25]

Sail Loft

A sail loft is a space used to fabricate sails for wind-powered ships. Few lofts remain standing at shipyards of the United States, though. Among those remaining is Building No. 55 at California's Mare Island Shipyard, constructed in 1854–1855. The building is two stories high, 400 feet long, and 60 feet wide and is constructed with solid timber trusses and brick side walls. The sail loft occupied the second floor of the building.[26]

Storehouse

The variety and quantity of materials used in ship construction made storehouses a necessity at a shipyard. The configurations of storehouses have depended on the materials stored there. Storehouses used for heavy construction materials such as steel plate were generally one story high, while general ship outfitting storehouses could be multiple stories in height and served by freight elevators, such as the one at the Philadelphia Naval Shipyard shown in Figure 6.7.

Mold or Pattern Loft

Blueprints were taken from the drafting room to the mold or pattern loft. The mold loft was the space in which ship plans, in blueprint form, were laid out in the actual size of the ship to be built. The loft usually occupied the top floor of the largest building in a yard and was often 100 feet or more wide and as much as 600 feet long. Some yards had small mold lofts as well as a

Figure 6.7. Supply Department Storehouse, Philadelphia Naval Shipyard, Philadelphia, Pennsylvania. HAER PA-387-I, National Park Service. Photograph by Jet Lowe, 1995.

large one, so that plans for parts of ships could be handled at the same time work was proceeding in the main loft. The loft had a specially prepared, smooth floor of wood, upon which patterns were expanded and marked out. The work of the loft was to make molds or "templates" of heavy paper and thin boards, from drawings or blueprints of the ships.[27] With improved lofting processes developed in the 1960s, 1/10-scale drawings were able to be used. CADD (computer-aided design and drafting) has now replaced the lofting process.[28]

Shipfitters' Shop

Shipfitting consists of the marking of steel plates, channels, and other parts before their fabrication in the steel mill and installation on a ship. The shipfitter had charge of assembling and erecting all metal parts upon the hull of the steel ship. He measured up pieces of work, made templates, and fabricated certain parts and bolted them into place.[29]

Frequently, as at the Norfolk Navy Yard, a shipfitters' shop was located in the same building as the assembly shop.

Anglesmith Shop

Workers in the anglesmith shop were variously known as anglesmiths, frame benders, and furnacemen. The anglesmith bent and welded small lengths and light angular shapes. Among the jobs done were bending and welding tank frames, door frames, frame staples, and bulkhead staples using molds furnished by the mold loft. These actions were performed with red-hot material, similar to blacksmithing.[30]

Plate Shop

The plate shop was where plates and frames were cut to size, punched for rivets, bent into shape, and prepared for the hull.[31] It included the milling, shaping, and assembling of parts in sections for erection upon the ship.

Typically, the plate shop or steel mill was a large, long building open at both ends with a roof high enough for the use of cranes (Figure 6.8). It consisted of two sections or sides with car tracks running lengthwise to divide the two sides. All frame parts or "shapes" were handled on one side, and all plates on the other.

Figure 6.8. Plate shop. Interior, first floor. Bethlehem Steel Shipyard, Hoboken, New Jersey. HAER NJ 95:7. Photograph by Rob Tucher, 1994.

The parts for some portions of the ship were assembled in shops within the mill.[32]

Sheet Metal Shop

Historically, the sheet metal shop manufactured a variety of articles from sheet metal and tin. Early twentieth-century products of the shop included tanks, ventilating systems, lining boxes, lockers, wire work, boiler and engine lagging, ship insulation, corrugated bulkheads, exhaust and intake headers, fuel oil tanks, and ventilating cowls. Typical equipment in a sheet metal shop consisted of brakes, rolls, shears, punches, beading machine, flanging machine, combination angle shears, rotary cutter, hydraulic press, electric wire crimper, wiring machine, drill presses, lathes, and shapers.[33]

An example of a shipyard sheet metal shop was Building 9 at the General Dynamics Shipyard in Quincy, Massachusetts. The building was a rectangular, 72 by 643 foot, brick and timber-framed building with a low-pitched gable roof. Exterior walls were pierced by large, multi-light, steel framed windows. Both floors were single, huge open spaces with machinery and work areas in the two exterior bays. The first floor was used as a sheet metal shop, while the second floor was used as joiner shops. Sheet metal shop machinery included forming rolls, a squaring sheer, drill presses, a loft press brake, and a turret punch.[34]

Blacksmith and Forge Shop

The blacksmith and forge shop was where laborers worked iron and steel used for ship repair and construction (Figure 6.9). Some such shops manufactured anchor chain. The typical equipment in a blacksmith shop of an early twentieth-century shipyard consisted of steam hammers, pneumatic hammers, oil furnaces, large coal furnaces, small coal furnaces, shears, anvils, tongs, dies, sledgehammers, hammers and other small tools, and cranes for manipulating large work.[35]

Foundry

Depending on the shipyard, copper or brass, iron, or steel foundries may have been present. Two massive foundry buildings, one at Norfolk and one at Philadelphia (Figure 6.10), were erected in the World War I and postwar era. The standard foundry plan is described as follows:

> The typical foundry building comprises a high center aisle, 80 feet wide; two lower side aisles, one of 55 feet width with a mezzanine floor,

Figure 6.9. Forge. Mare Island Naval Shipyard. Alameda, California. HABS CA-1547, National Park Service.

and one of 45 feet width; and a 100-foot material and flask yard adjacent to the latter.

The material yard is served by an overhead traveling crane of 10 tons capacity and 40 feet lift. The adjacent side aisle, into which materials are moved from the open yard, or from bins directly into the foundry, is of one story for the greatest part of its length, 32 feet high to the bottom chords of roof trusses. This aisle contains the cupolas and the various converters, furnaces, etc. At the cupolas a second floor is provided for charging, with an intermediate floor to house the blowers for the cupolas. The single-story portion of the side aisle is provided with 2-ton and 5-ton traveling cranes. The center (main) aisle, in which the large castings are molded, poured and handled, is 75 feet high to the bottom of roof trusses, and is provided with three tiers of cranes. . . . The 55-foot side aisle with a gallery floor 22 feet above the main floor houses molding machines, crucibles, cleaning and grinding apparatus, etc., and is served generally by monorail cranes.[36]

Machine Shop

Early twentieth-century machine shops built reciprocating and turbine engines, propeller shafts and propellers, and fittings for boilers, motors, and pumps. A

Figure 6.10. Propeller foundry, interior. Philadelphia Naval Shipyard. HAER PA-387, National Park Service. Photograph by Jet Lowe, 1995.

Figure 6.11. Structural assembly shop. Philadelphia Naval Shipyard. First floor interior. HAER PA 387S:21, National Park Service. Photograph by Jet Lowe, 1995.

great variety of the hull and superstructure elements were drilled, machined to size, and fitted, such as manholes, scuttles, mast fittings, deck hoists, windlasses, turrets, ammunition hoists, air ports, and gratings.[37]

One such machine shop stood at the Bethlehem Steel Company's Hoboken (New Jersey) Shipyard, a late nineteenth-century building, expanded in the twentieth century. The interior was divided into a ground floor and timber-framed second-story galleries lining the side walls. Later concrete galleries replaced the timber-framed galleries. It included machine tools of various types and sizes such as lathes, drill presses, and stamping machines. The machine shop was open its entire length and capped with a monitor roof.[38]

Structural Assembly Shop

In a large naval yard such as the Philadelphia Naval Shipyard, the structural assembly shop played an essential role in ship construction and repair (Figure 6.11). Manufacture of vessel parts for new construction and repair began in the forge shop, where blacksmiths produced forging and metal shapes. Steel components were then machined and assembled in a number of different shops. The inside machine shop was responsible for planing and drilling parts. From there, components were forwarded to the structural shop, where shipfitters, riveters, and craft workers from other trades installed large structures such as stern sections and turrets. A series of cranes and railroad tracks enabled components to be moved into the building for assembly and for the assembled components to be moved to nearby shipways or dry docks for installation.[39]

Joiner Shop

At the turn of the twentieth century, joiners built and installed furniture for staterooms and storerooms, completed the inside finishing of these rooms and the pilot house, constructed decks, fitted up the ship carpenter's workshop, and built accommodation ladders, tables, desks, and lockers. Sections of the joiner shop included the drafting room, mill work, cabinet, bench, and ship work, and the finishing room.[40]

The 1899 joiner shop in the Percy and Small shipyard in Bath, Maine, had a first floor with a clear span of 40 by 75 feet supported by post and beam construction. The space contained a circular sawmill with a 42-inch blade and a 43-foot carriage, a large bevel-cutting jigsaw for cutting changing bevels into the huge curved frame pieces, and a huge planer with a 93-foot bed and 43-foot carriage.[41]

Later, some shipyards, such as the General Dynamics Corporation's Quincy, Massachusetts shipyard combined the functions of the pattern and joiner shops. This facility featured large, open workspaces for the fabrication and construction of wooden patterns and models. While the mold loft fabricated patterns for massive plate and structural elements, the pattern and joiner shops built models for parts such as boilers, rigging parts, interior machinery, and piping systems.[42]

Pattern Shop

The pattern shop produced the wooden patterns from which castings of iron, brass, and other metals were made in the factory. Patterns were made for ship and deck fittings such as hawse pipes, chocks, and bitts. As deck fittings were standardized, the patterns for them could be used many times, but hawse pipes had to be made to suit the shape of the bow of the ship. Each pattern was a full-size wooden model of the piece to be cast. Tools used were essentially the same as those used by other woodworkers.[43]

As at Quincy, Massachusetts's Fore River Shipyard, the pattern shop often shared quarters with the mold loft, occupying the floor below.[44]

Shipwright Shop

The shipwright did all woodwork about a ship except cabinet work and finishing. He installed heavy wood decks and foundations for machinery, cut and fastened cargo battens, and made and installed wooden masts, spars, and booms.[45] Other work included making keel blocks, wedges, shores, staging, launching ways, scaffolding, and ribbands.[46]

Carpenter Shop

One example of a carpenter shop was a three-story brick building measuring 209 feet by 50 feet in plan that stood at Bethlehem Steel's New Jersey shipyard. The interior of the building consisted of three largely open working floors that housed woodworking and sheet-metal working activities. At the time of its recordation for HAER, the only remaining significant piece of machinery was a 24-inch American swing saw.[47]

Electrical Shop

The electrical shop was responsible for installing the electrical wiring and electrical equipment on ships. Modern electrical shops, such as the one at the Norfolk Naval Shipyard, repair, test, and install a great variety of electrical and electronic equipment. The Norfolk shop is also equipped to rewind motors and generators.[48]

Pipe Shop

The pipe shop provided all the piping for engine and boiler rooms, bilge and ballast systems, freshwater and sanitary piping throughout the ship. Its work included all kinds of pipe fitting, flanged and threaded, of brass, copper, and steel, and in all sizes.[49]

Paint Shop

At early shipyards, the paint shop was often a wood-framed building in which the paint used for wood schooners was stored. Modern paint shops employ air-powered equipment to do such tasks as abrasively blasting a ship's hull and undertake a variety of painting on board ships. Typically, workers in the paint shop also apply deck coverings.[50]

Plate Yard

The plate yard was an outdoor space used to store steel plate used in ship construction. Typically, a bridge crane traveled across the yard to retrieve plate for use. The plate yard at the Fore River Ship and Engine Company in Quincy, Massachusetts, is described in a 1902 booklet on the yard:

> Large enough to encamp a regiment comfortably, or contain the Madison Square Garden . . . with 200 feet to spare, the plate yard now contains over 15,000 tons . . . of steel. From end to end travels an electric crane that spans the width of the yard (150 feet) —a great lattice-like structure that seems gossamer-like in lightness, and yet is capable of picking up an inch-thick plate, 15 by 20 feet, and carrying it at a speed of 500 feet a minute. A track is laid through the center of the plate yard, over

which the ubiquitous locomotive draws the company's own flatcars.[51]

Outfitting Pier

An outfitting pier was where ship outfitting took place, including installation of boilers and engines, running pipelines for water and steam, electric wiring, fitting of masts, booms, and smokestacks, and the joinering and carpentery on decks, pilot houses, and basins (Figure 6.12).[52] In other piers, the quality of previous work was checked, decks were painted, and propulsion machinery was tested prior to sea trials. To facilitate outfitting, related facilities were located close to the pier, including pipe shops and storehouses, sheet metal shop and stores, electrical shops, and the other various components.[53] Outfitting piers were equipped with cranes and were usually of steel-bearing pile and bent or masonry construction designed to handle heavy loads.[54]

Figure 6.12. Outfitting Pier #2. General Dynamics Corporation Shipyard, Quincy, Massachusetts. HAER MA-26C:2, National Park Service. Photograph by Robert Brewster, 1989.

GLOSSARY[55]

Anglesmith. A worker who has acquired special skill in bending, welding, and forming metals into angular shapes.

Annealing. To soften metal by heating and cooling it.

Bending floor. The floor in the plate and angle shop on which plates and frames are bent.

Bending rolls. A large machine used to give curvature to plates.

Bending slab. Heavy cast-iron blocks arranged to form a large solid floor on which angles are bent.

Forging. A mass of steel worked to a special shape by hammering when red hot.

Ground ways. Timbers fixed to the ground, under the hull on each side of the keel, on which the ship is launched.

Hawse pipe. Casting extending through the deck and side of the ship for passage of the anchor chain.

Keel blocks. Heavy blocks on which a ship rests during construction.

Launching. The operation of placing a hull in the water by having it slide down the launching ways.

Lay out. To transfer marks from a template to a plat or shape.

Lay-out table. A steel floor on a concrete foundation upon which articles are measured and marked in the machine shop. The surface is covered with vertical and horizontal lines.

Laying off. Marking plates, shapes, and the like for shearing and punching from a template.

Lines. The plans of a ship that show its form. From the lines, which are drawn full size on the mold loft floor, templates are made of the various parts of the hull.

Mold. A light pattern of a part of a ship, generally made of thin wood or paper.

Mold loft. A large room with a smooth finished floor on which the decks, frames, floors, and other parts of a ship are drawn full size.

Mold or *template*. A pattern of a part to be constructed.

Plates. Flat metal sheets of various thicknesses, used for covering the outside of vessels, for decks, and for partitions.

Plate and shape (or *angle*) *shop*. Mill where plates and angles are fabricated.

Rigging. Manila and wire ropes, lashings, and the like used to support masts, spars, and booms.

Superstructure. Deck houses and the like located above the shelter deck of a ship.

Ways. Where hulls of ships are constructed and made ready for launching.

REFERENCES

Bethlehem Steel Company, Shipbuilding Division. *An Introduction to Shipbuilding*. New York: Bethlehem Steel Company, 1942.

—. *Shipbuilding, Ship Repairing, Ship Conversion*. New York: Bethlehem Steel Company, 1946.

—. *Ten Repair Yards on Atlantic and Pacific Coasts*. New York: Bethlehem Steel Company, 1947.

Bureau of Yards and Docks. *Building the Navy's Bases in World War II*. Volume I. Washington, DC: Government Printing Office, 1947.

Carmichael, A. C. *Practical Ship Production*. New York: McGraw-Hill Book Company, 1941.

Dickinson, Rogers. *The Plant of the Fore River Ship & Engine Company, Quincy, Massachusetts*. Boston: H. B. Humphrey Company, 1902.

Fassett, F. G., Jr., editor. *The Shipbuilding Business in the United States of America*. New York: The Society of Naval Architects and Marine Engineers, 1948.

Greene, Carleton. *Wharves and Piers: Their Design, Construction, and Equipment*. New York: McGraw-Hill, 1917.

Hepburn, Richard D. *History of American Naval Dry Docks: A Key Ingredient to Maritime Power*. Arlington, VA: Noesis, Inc., 2003.

Kelly, Roy Willmarth, and Frederick J. Allen. *The Shipbuilding Industry*. Boston: Houghton Mifflin Company, 1918.

MacBride, J. D. *A Handbook of Practical Shipbuilding with a Glossary of Terms*. New York: D. Van Nostrand, 1918.

Raber, Michael S. *Marine Railways of Southeast Connecticut: Historical Survey and Inventory*. South Glastonbury, CT: Raber Associates, 2006.

Sun Shipbuilding and Dry Dock Company. *Products and Service: Sun Shipbuilding and Dry Dock Company*. Chester, PA: The Company, 1946.

Swanson, William Elmer. *Modern Shipfitter's Handbook*. New York: Cornell Maritime Press,1941.

Wilson, Theodore D. *An Outline of Ship Building, Theoretical and Practical*. New York: John Wiley & Son, 1873.

NOTES

1 F. G. Fassett, Jr., editor, *The Shipbuilding Business in the United States of America* (New York: The Society of Naval Architects and Marine Engineers, 1948), 68, 134. A comprehensive list of World War II–era shipbuilders and their production is contained in Chapter 3.

2 Carleton Greene, *Wharves and Piers: Their Design, Construction and Equipment* (New York: McGraw-Hill, 1917), 1.

3 Greene, 3.

4 Greene, 4, 13, 14.

5 Greene, 3–4.

6 As cited in Celia Orgel, South Street Seaport, Piers 17 and 18, HAER No. NY-156, 1995, 3.

7 Michael S. Raber, Bush Terminal Company, Pier 5, HAER No. NY-201B, 2-3.

8 Michael S. Raber, Brooklyn Army Supply Base: Pier 1, HAER No. NY 202-A, 1988, 1–4.

9 Rebecca Coward, Huntington Beach Municipal Pier, HAER No. CA-80, 1990, 3–4.

10 Greene, 159–161.

11 Greene, 166.

12 Greene, 166.

1 Fassett, 201.

14 Fassett, 203, 207.

15 Henry Gard Knox, "Multiple Yards," in *The Shipbuilding Business in the United States of America*, edited by F. G. Fassett (New York: The Society of Naval Architects and Marine Engineers, 1948), 211.

16 Theodore D. Wilson, *An Outline of Ship Building, Theoretical and Practical* (New York: John Wiley & Son, 1873), 310–311.

17 Katheryn H. Krafft, Puget Sound Naval Shipyard, Dry Dock No. 3 (Facility 703), HAER No. WA-116-E, 1991, 1–3.

18 Fassett, 179–180.

19 David W. Harvey, Puget Sound Naval Shipyard, Portal Gantry Crane No. 51, HAER No. WA-116-C, 1994.

20 Virginia A. Fitch, General Dynamics Corporation Shipyard, McMyler Crane (Structure 33S), HAER No. MA-26-G, 1989, 1-4.

21 Virginia A. Fitch, General Dynamics Corporation Shipyard, XYZ Crane and Towers (Structure 21S), HAER No. MA-26-E, 1989, 3–4.

22 Roy Willmarth Kelly and Frederick J. Allen, *The Shipbuilding Industry* (Boston: Houghton Mifflin Company, 1918), 138; J. D. MacBride, *A Handbook of Practical Shipbuilding with a Glossary of Terms* (New York: D. Van Nostrand, 1918), 81.

23 A. C. Carmichael, *Practical Ship Production* (New York: McGraw-Hill Book Company, 1941), 169–170.

24 Charlestown Navy Yard: Architectural and Technological Highlights, http://www.nps.gov/bost/planyourvisit/upload/NY%20Arch.pdf.

25 George Ripley and Charles A. Dana, editors, *The American Cyclopaedia* (New York: D. Appleton Company, 1873), 424–425.

26 Mare Island Naval Shipyard, Sail Loft (Building No. 55), HAER CA-3-C.

27 Kelly and Allen, 116–117; MacBride, 4.

28 Robert C. Stewart, Naval Base Philadelphia-Philadelphia Naval Shipyard, Structural Assembly Shop, HAER No. PA-387-S, 1.

29 Kelly and Allen, 128; Carmichael, 182.

30 Kelly and Allen, 126.

31 Kelly and Allen, opposite 82.

32 Kelly and Allen, 124.

33 Kelly and Allen, 172.

34 Virginia A. Fitch, General Dynamics Corporation Shipyard, Joiner and Sheet Metal Shops (Building 9), HAER MA-26-A, 1989.

35 Kelly and Allen, 165.

36 Activities of the Board of Yard of Docks in the World War, 1917–1978, 129, as cited in Naval Base Philadelphia–Philadelphia Naval Shipyard, Foundry/Propeller Shop, HAER PA-387-O:4-5.

37 Kelly and Allen, 166.

38 Bethlehem Steel Company Shipyard, Machine Shop, HAER NJ-95A:2–3.

39 Naval Base Philadelphia-Philadelphia Naval Shipyard, Structural Assembly Shop, HAER PA-387-S:9.

40 Kelly and Allen, 187–189.

41 Maine Maritime Museum website: www.bathmaine.com/.

42 General Dynamics Corporation Shipyard, Quincy, Massachusetts. Joiner and Sheet Metal Shops. HAER MA-26-A: 5.

43 Kelly and Allen, 191; MacBride, 7.

44 Rogers Dickinson, *The Plant of the Fore River Ship & Engine Company, Quincy, Massachusetts* (Boston: The H. B. Humphrey Company, 1902), 14.

45 Kelly and Allen, 136.

46 Carmichael, 187.

47 Richard L. Porter, Meredith Arms Bzdak, and Rob Tucher, Bethlehem Steel Company Shipyard, Carpenter Shop, HAER NJ-95-D, 1994.

48 Norfolk Naval Shipyard, Major Production Shops, www.nnsy1.navy.mil/shops/shops.htm.

49 MacBride, 7.

50 Norfolk Naval Shipyard website: http://www.nnsy1.navy.mil/shops/shops.htm.

51 Dickinson, 14.

52 Kelly and Allen, 103.

53 Fassett, 212.

54 General Dynamics Corporation Shipyard, Outfitting Pier 2, HAER MA-26-C: 3–4.

55 Most definitions are taken from the glossary in Kelly and Allen's *The Shipbuilding Industry* (1918).

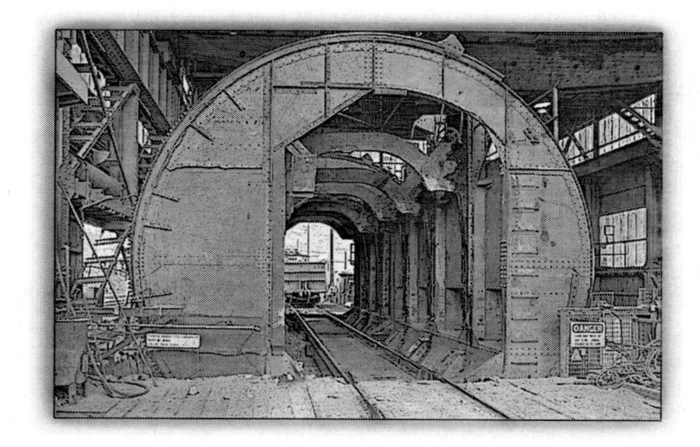

CHAPTER 7

Power Generation

In January 2007, 33 percent of the energy generated in the United States was produced by coal, 26 percent from natural gas, 15 percent from crude oil, 13 percent from nuclear energy, 2.6 percent from natural gas liquids, 4.3 percent from hydroelectric power, 4.2 percent from biomass, 0.5 percent from geothermal sources, and 0.4 percent from wind energy.[1]

This chapter considers the major types of structures related to the country's historic power generation system. The chapter is divided into three major sections: hydroelectric power, fossil fuel–powered generation plants, and nuclear power plants. Due to the newness of large-scale use of the technology, the physical components of renewable energy generation complexes, including wind turbines and biomass and geothermal facilities, are not considered in this chapter.

HYDROELECTRIC POWER

In 2005, the total hydroelectric generating capacity in the United States was 77,541 megawatts. Hydroelectric facilities are found in every state except Delaware and Mississippi. The state with the largest hydroelectric generating capacity is Washington, whose 57 plants have a capacity of 21,778 megawatts.[2] The nation's largest hydropower plant is the 6,180-megawatt Grand Coulee power station located on the Columbia River in Washington. The largest privately owned hydropower project in the United States is the 512-megawatt Conowingo Project on the Susquehanna River in Maryland.

Hydroelectric facilities are composed of a few basic elements: the headwater, the dam, the penstock, the turbine, the generator, the tailwater, and the afterbay. The dam creates a "head" or height from which water flows. A pipe (the penstock) carries the water from the reservoir to the turbine. The rapidly moving water pushes the turbine blades, and the force of the water on the turbine blades turns the rotor, the moving part of an electric generator. When coils of wire on the rotor sweep past the generator's stationary coil or stator, electricity is produced.[3]

Plant Types

Hydroelectric plants may be divided into four basic types: run-of-river plants without pondage; run-of-river plants with pondage; reservoir or impoundment plants; and pumped storage plants (Figure 7.1).[4]

Run-of-River Plants

Run-of-river plants, sometimes called a diversion, channel a portion of a river through a short canal or penstock, using the natural flow of the river to power the turbines. They may not require the use of a dam. A typical reason for the lack of pondage is that the dam is constructed to maintain a given water level, often for navigation, and the power plant is only incidental. The heads of such plants are often quite low, and, at times of flooding, tailwater rises to such an extent that the plants are inoperative. The natural flow of the river remains largely unaltered. The powerhouse is often an integral part of the dam.[5]

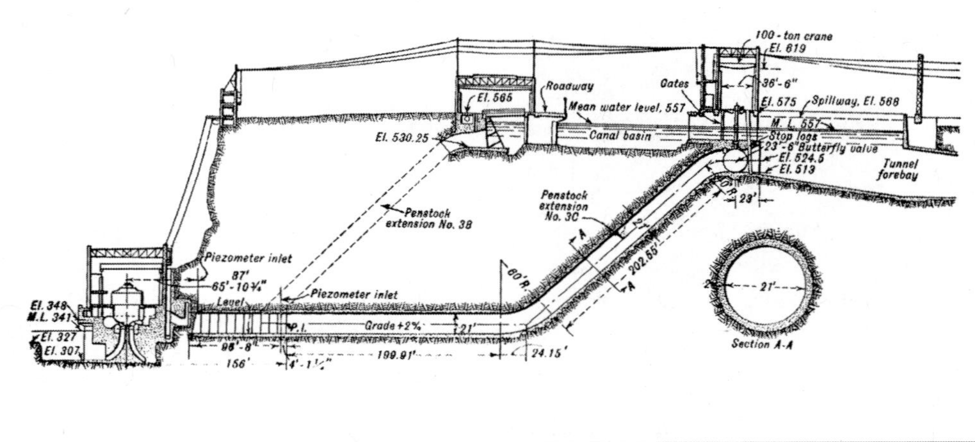
100 - ton crane
El. 619
Roadway
Gates
36'-6"
El. 566
Mean water level, 557
El. 575
Spillway, El. 566
El. 530.25
Canal basin
Stop logs
23'-8" Butterfly valve
El. 524.5
El. 513
Tunnel forebay
Penstock extension No. 3C
Penstock extension No. 3B
23'
Piezometer inlet
87'
65'-10¼"
Level
Piezometer inlet
60' R
202.65'
21'
Section A-A
El. 348
M.L. 341
El. 327
El. 307
P.I.
Grade +2%
21'
98'-8"
156'
4'-1½"
199.91'
24.15'

a

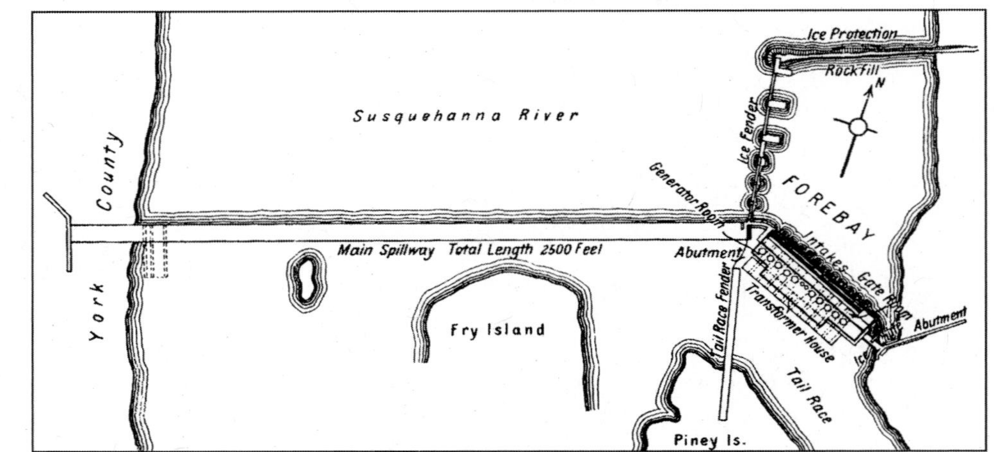
Ice Protection
Rock Fill
N
Susquehanna River
Ice Fender
York County
Generator Room
FOREBAY
Main Spillway Total Length 2500 Feet
Abutment
Intakes
Gate
Tail Race Fender
Transformer House
Abutment
Fry Island
Tail Race
Piney Is.

b

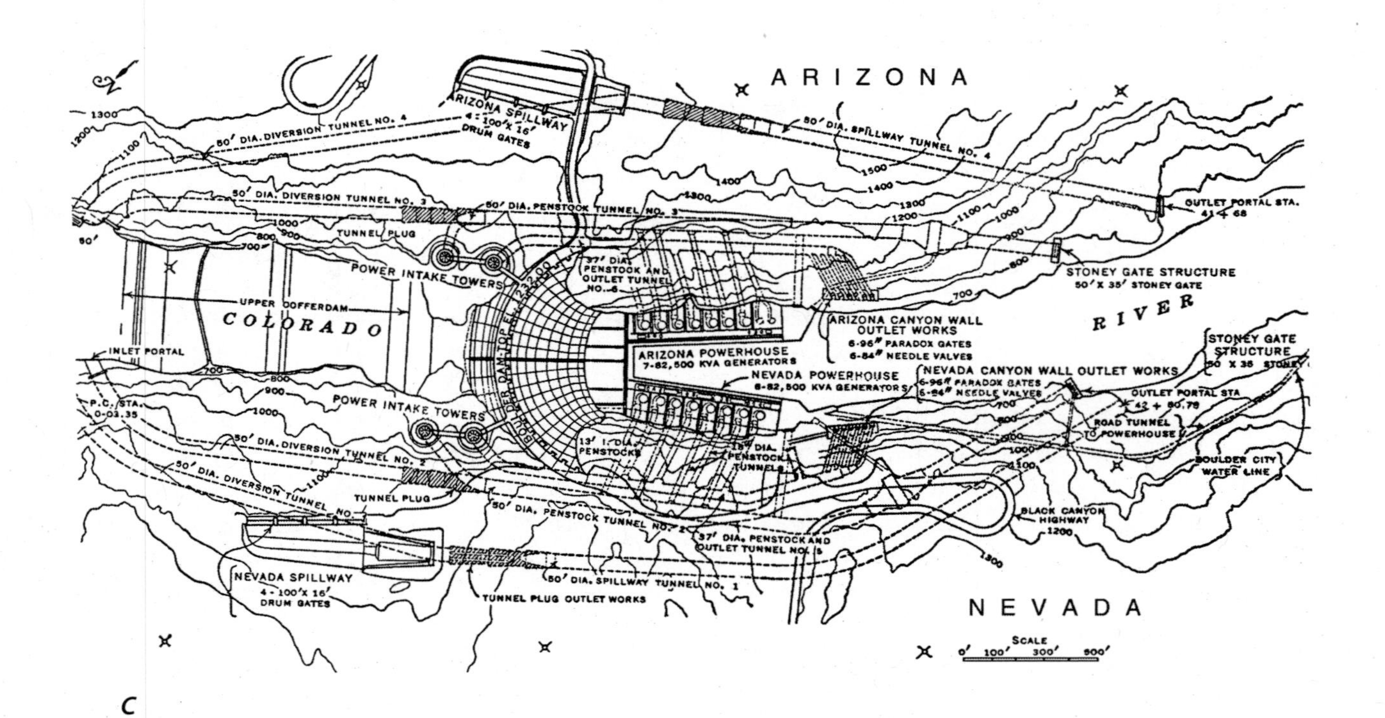
ARIZONA
ARIZONA SPILLWAY
4 - 100' x 16' DRUM GATES
50' DIA. DIVERSION TUNNEL NO. 4
50' DIA. SPILLWAY TUNNEL NO. 4
50' DIA. DIVERSION TUNNEL NO. 3
50' DIA. PENSTOCK TUNNEL NO. 3
OUTLET PORTAL STA. 41 + 68
TUNNEL PLUG
POWER INTAKE TOWERS
37' DIA. PENSTOCK AND OUTLET TUNNEL NO. 6
STONEY GATE STRUCTURE
50' x 35' STONEY GATE
UPPER COFFERDAM
COLORADO
RIVER
ARIZONA CANYON WALL OUTLET WORKS
6-96" PARADOX GATES
6-84" NEEDLE VALVES
INLET PORTAL
ARIZONA POWERHOUSE
7-82,500 KVA GENERATORS
NEVADA POWERHOUSE
8-82,500 KVA GENERATORS
NEVADA CANYON WALL OUTLET WORKS
6-96" PARADOX GATES
6-84" NEEDLE VALVES
STONEY GATE STRUCTURE
OUTLET PORTAL STA 42 + 50.79
P.C. STA. 0-03.35
POWER INTAKE TOWERS
ROAD TUNNEL TO POWERHOUSE
50' DIA. DIVERSION TUNNEL NO. 2
50' DIA. DIVERSION TUNNEL NO. 1
13' DIA. PENSTOCKS
37' DIA. PENSTOCK TUNNELS
BOULDER CITY WATER LINE
TUNNEL PLUG
50' DIA. PENSTOCK TUNNEL NO. 2
BLACK CANYON HIGHWAY
37' DIA. PENSTOCK AND OUTLET TUNNEL NO. 5
NEVADA SPILLWAY
4 - 100' x 16' DRUM GATES
50' DIA. SPILLWAY TUNNEL NO. 1
TUNNEL PLUG OUTLET WORKS
NEVADA
SCALE
0' 100' 300' 500'

c

Among examples of this type of plant are those built by Louisville Gas and Electric Company on the Ohio River and the Winfield, Marmet, and London plants on West Virginia's Kanawha River.[6]

Run-of-River with Pondage

The usefulness of a plant is increased if it has pondage and if tailwater conditions are such that floods will not drown out the plant. The term pondage means that there is sufficient storage at the plant to take care of diurnal variations in power demand. Examples of this type of plant are Conowingo and Safe Harbor on the Susquehanna River in Maryland and Pennsylvania, respectively.[7]

Reservoir Plants or Impoundment Plants

Hydroelectric plants that take their flow from large storage reservoirs are suitable for development as peak-load plants. Water released from the reservoir flows through a turbine, spinning it, which in turn activates a generator to produce electricity. Notable examples include Hoover Dam and the Grand Coulee Dam.[8]

Pumped Storage Plants

Pumped storage plants are peak-load plants that pump all or a portion of their own water supply. They generally consist of a tailwater pond, which may be replaced by a river or natural lake, and a headwater pond. During times of peak load, water is drawn from the headwater pond through the penstocks to operate hydroelectric generating units. During off-peak hours, pumps are operated to shunt the water back from the tailwater pond to the headwater pond. Pumped storage plants are unique among hydroelectric plants in that practically no water supply is necessary.

Pumped storage plants can store power. Power is sent from a power grid into the electric generators. The generators then spin the turbines backward, which causes the turbines to pump water from a river or lower reservoir to an upper reservoir, where the power is stored. To use the power, the water is released from the upper reservoir back down into the river or lower reservoir. This spins the turbines forward, activating the generators to produce electricity.[9]

The largest operating, privately owned pumped storage project is jointly owned by the Virginia Electric and Power Company and Allegheny Generating Company in Bath County, Virginia. The largest federally owned pumped storage project is the Tennessee Valley Authority's 1,530-megawatt Raccoon Mountain project on the Tennessee River in Tennessee.

The typical components of a hydropower facility are hydraulic storage and conveyance facilities, including the reservoir, dam, headrace, headworks, penstock, gates, valves, and tailrace; the powerhouse structure and foundation; the turbine-generator unit; station

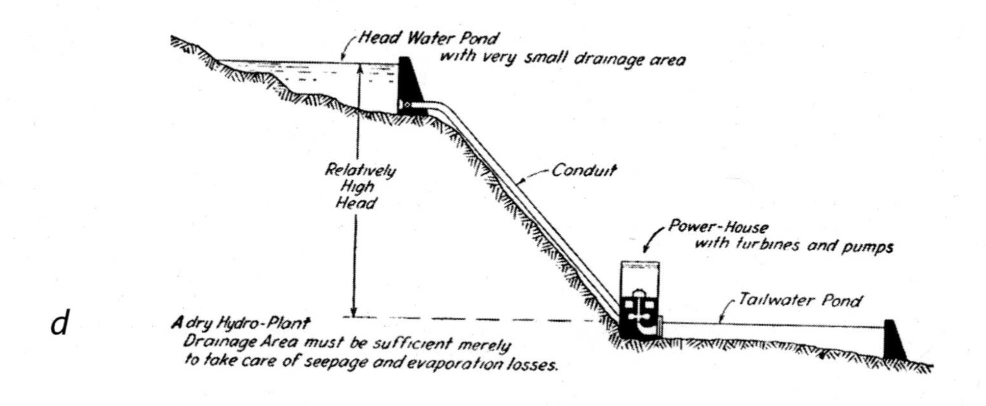

(FACING PAGES): **Figure 7.1.** Types of hydroelectric plants. ***a,*** Run-of-river without pondage. Schoellkoff Plant, Niagara Falls, New York. Reprinted from William P. Creager and Joel D. Justin, *Hydroelectric Handbook*, 2nd edition (New York: John Wiley and Sons, 1950), 192. ***b,*** Run-of-river with pondage. McCall Ferry Development, Pennsylvania. Reprinted from Lamar Lyndon, *Hydro-electric Power,* volume I (New York: McGraw-Hill Book Company, 1916), 350. ***c,*** Reservoir. Hoover Dam development, Colorado River, Nevada-Arizona (Creager and Justin 1950:198). ***d,*** Pumped storage (Creager and Justin 1950: 203).

electrical equipment, including transformer, switch gear, automatic controls, conduit and grounding and lightning systems; and transmission lines.[10]

The reservoir for a hydroelectric facility may be a natural lake but is more typically a dammed stream or river. The largest reservoirs in the United States are 34.8 billion cubic meter Lake Mead associated with Hoover Dam and the 33.3 billion cubic meter Lake Powell associated with the Glen Canyon Dam.[11]

Dams

Several types of dams have typically been used to provide the head of water required for hydroelectric power generation.

Solid gravity concrete dams (Figure 7.2) take the form of vertical upper faces topping a battered downstream face. The lower portion of each dam face is battered. This dam is adaptable to all localities, but its height is limited by the strength of the foundations. As the name implies, it resists the action of water pressure by the gravity action of its mass.[12]

Arch dams are those that reach from one side of the stream or river to the opposite one in the arc of a circle rather than straight across from bank to bank (Figure 7.3). The convex side is placed upstream. The pressure of the water is resisted by the arch action of the structure. When a considerable length is to be spanned, the length can be divided into several short portions, each with a short radius arch.[13] Arch dams may be divided into three types: single arch, multiple arch, and arched gravity dams. Multiple-arch dams consist of a series of arches supported by buttresses, while arched gravity dams are basically gravity dams curved in plan.[14]

The usual type of buttressed concrete dam consists of a series of parallel, equidistant concrete buttresses covered by a watertight sloping face and a downstream face and bucket to support the sheet of spilling water. Because of the considerably sloping upstream face, the vertical component of water pressure is available to assist stability. As a result, less weight of concrete is required than for solid dams. Buttressed concrete dam types include flat-slab, multiple-arch, and round-head buttress types.[15]

Earth dams (Figure 7.4) are typically constructed in several different methods: rolled layers, the hydraulic fill method, and the semi-hydraulic fill method. In the first method, material is excavated by power shovel, scrapers, dragline, or elevating graders, hauled onto the dam, deposited, spread, moistened, and rolled. In typical hydraulic fill dams, water under heavy pressure is delivered to a large nozzle. The stream of water is direct-

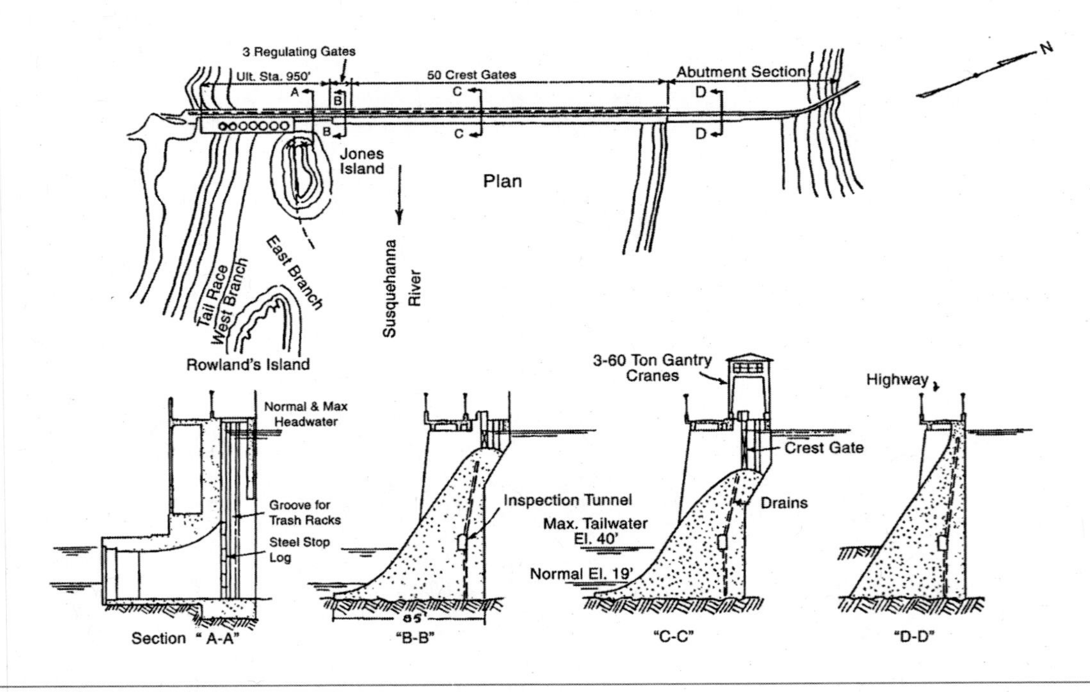

Figure 7.2. Solid gravity concrete dam. Plan and sections, Conowingo Dam, Maryland. Reprinted, with permission, from Conowingo Hydro-electric Development, *Power Plant Engineering* (Anonymous 1928: 635).

Figure 7.3. Arch dam. Owyhee Dam, Nyassa vicinity, Oregon. HAER OR-17, National Park Service. Photograph by Clayton B. Fraser, 1989.

ed against the bank to be excavated. The upstream slope of an earth dam always requires protective cover against erosion due to wave action. Protective covers include dumped stone riprap, hand-placed riprap, grouted riprap, concrete slabs and blocks, bituminous paving, and plantings.[16]

Rock-fill dams have three fundamental parts: the dumped rock fill; an upstream rubble cushion of laid-up stone bonding into the dumped rock; and an upstream impervious facing resting on the rubble cushion.[17]

Movable Crest Gates

Because of variations in stream flow, the elevation of the lake formed by a dam changes from periods of high to low water. Since an important element in considering the value of a hydroelectric facility is the amount of power that may be continually generated during periods of low water, any means by which the lake level may be kept higher than the dam crest will increase the value of the investment.[18]

The principal type of crest gates are flashboards, Stickney gates (Figure 7.5a), automatic crest gates with

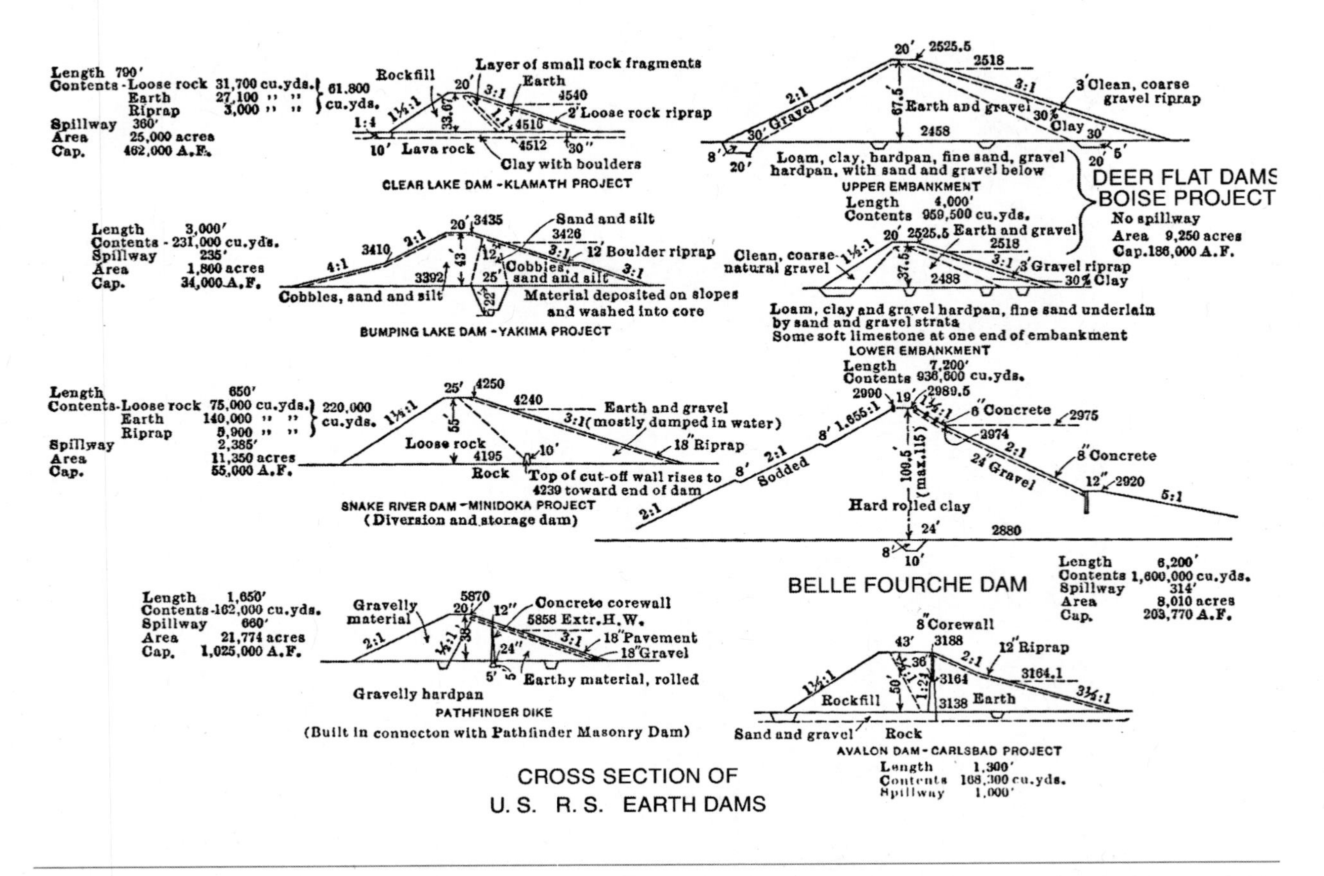

Figure 7.4. Cross sections of earth dams. Reprinted from Lamar Lyndon, *Hydro-electric Power*, volume I (New York: McGraw-Hill Book Company, 1916), 280.

rolling counterweights (Figure 7.5b), Stoney gates (Figure 7.5c), Tainter gates, and rolling gates.[19] Some devices used to raise dam crests are manually operated, while others are either automatic or operated with electrical power.

Flashboards are basically a series of planks set vertically on top of the crest, forming a wooden barrier and held in place by vertical iron rods with their lower ends sunk into the masonry of the dam. Flashboards range in height from 18 inches to 4 feet and are designed to bend over during flood periods. Flashboards can be either temporarily or permanently installed.[20]

Stickney gates consist of two leaves, set at approximately right angles to each other, hinged at the point where they join to the crest of the dam. The dam is provided with a long recess into which one of the leaves may sink, taking a vertical position, the other then taking a practically horizontal position. A counterweight is attached to one of the leaves and produces a moment tending to turn the gate to the open position.[21]

Automatic gates with rolling counterweights turn about a knife edge at the lower end and are held against the water thrust by flat iron links attached to their upper end. These links are, in turn, fastened to chains or wire ropes that wind around a heavy steel drum. The drum rolls on an upwardly inclined path, and as the gate tends to fall on account of the increase in pressure due to the rising water level, the chain or cable must unwind from the drum, thus causing the drum to roll along its inclined path. Opening the gate causes an elevation in the position of the drum. The roller is filled with concrete to add to its weight.[22]

(PAGES 157–158): Figure 7.5. Examples of movable gates. ***a,*** Stickney automatic crest gate (Lyndon 1916: 295). ***b,*** Automatic gates with rolling counterweight (Lyndon 1916: 297). ***c,*** Stoney gates. Wilson Dam and Hydroelectric Plant, Muscle Shoals, Alabama. AL-47. Delineated by Agatha Lesage, 1994.

STONEY SPILLWAY GATES 1925

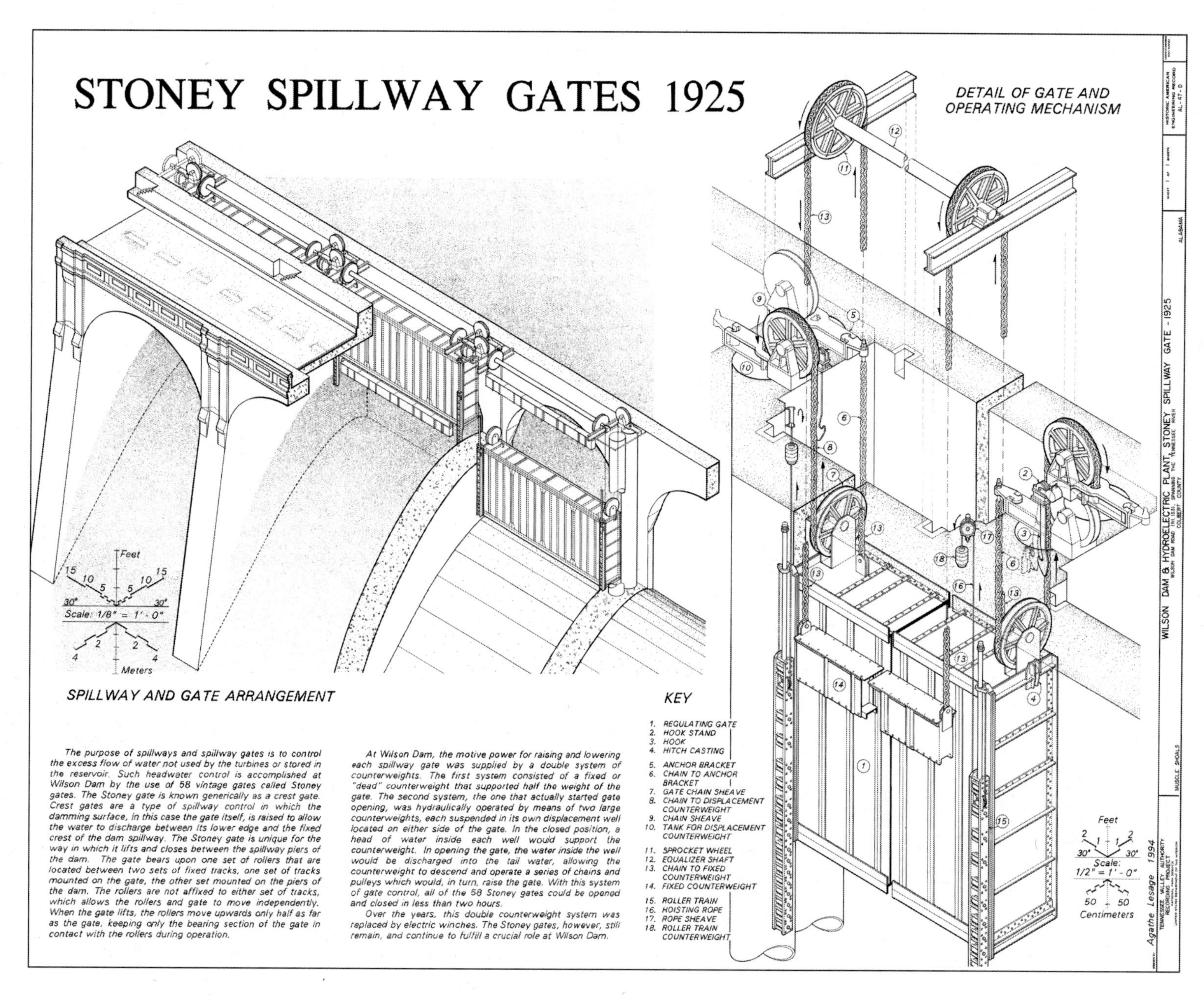

SPILLWAY AND GATE ARRANGEMENT

The purpose of spillways and spillway gates is to control the excess flow of water not used by the turbines or stored in the reservoir. Such headwater control is accomplished at Wilson Dam by the use of 58 vintage gates called Stoney gates. The Stoney gate is known generically as a crest gate. Crest gates are a type of spillway control in which the damming surface, in this case the gate itself, is raised to allow the water to discharge between its lower edge and the fixed crest of the dam spillway. The Stoney gate is unique for the way in which it lifts and closes between the spillway piers of the dam. The gate bears upon one set of rollers that are located between two sets of fixed tracks, one set of tracks mounted on the gate, the other set mounted on the piers of the dam. The rollers are not affixed to either set of tracks, which allows the rollers and gate to move independently. When the gate lifts, the rollers move upwards only half as far as the gate, keeping only the bearing section of the gate in contact with the rollers during operation.

At Wilson Dam, the motive power for raising and lowering each spillway gate was supplied by a double system of counterweights. The first system consisted of a fixed or "dead" counterweight that supported half the weight of the gate. The second system, the one that actually started gate opening, was hydraulically operated by means of two large counterweights, each suspended in its own displacement well located on either side of the gate. In the closed position, a head of water inside each well would support the counterweight. In opening the gate, the water inside the well would be discharged into the tail water, allowing the counterweight to descend and operate a series of chains and pulleys which would, in turn, raise the gate. With this system of gate control, all of the 58 Stoney gates could be opened and closed in less than two hours.

Over the years, this double counterweight system was replaced by electric winches. The Stoney gates, however, still remain, and continue to fulfill a crucial role at Wilson Dam.

KEY

1. REGULATING GATE
2. HOOK STAND
3. HOOK
4. HITCH CASTING
5. ANCHOR BRACKET
6. CHAIN TO ANCHOR BRACKET
7. GATE CHAIN SHEAVE
8. CHAIN TO DISPLACEMENT COUNTERWEIGHT
9. CHAIN SHEAVE
10. TANK FOR DISPLACEMENT COUNTERWEIGHT
11. SPROCKET WHEEL
12. EQUALIZER SHAFT
13. CHAIN TO FIXED COUNTERWEIGHT
14. FIXED COUNTERWEIGHT
15. ROLLER TRAIN
16. HOISTING ROPE
17. ROPE SHEAVE
18. ROLLER TRAIN COUNTERWEIGHT

a

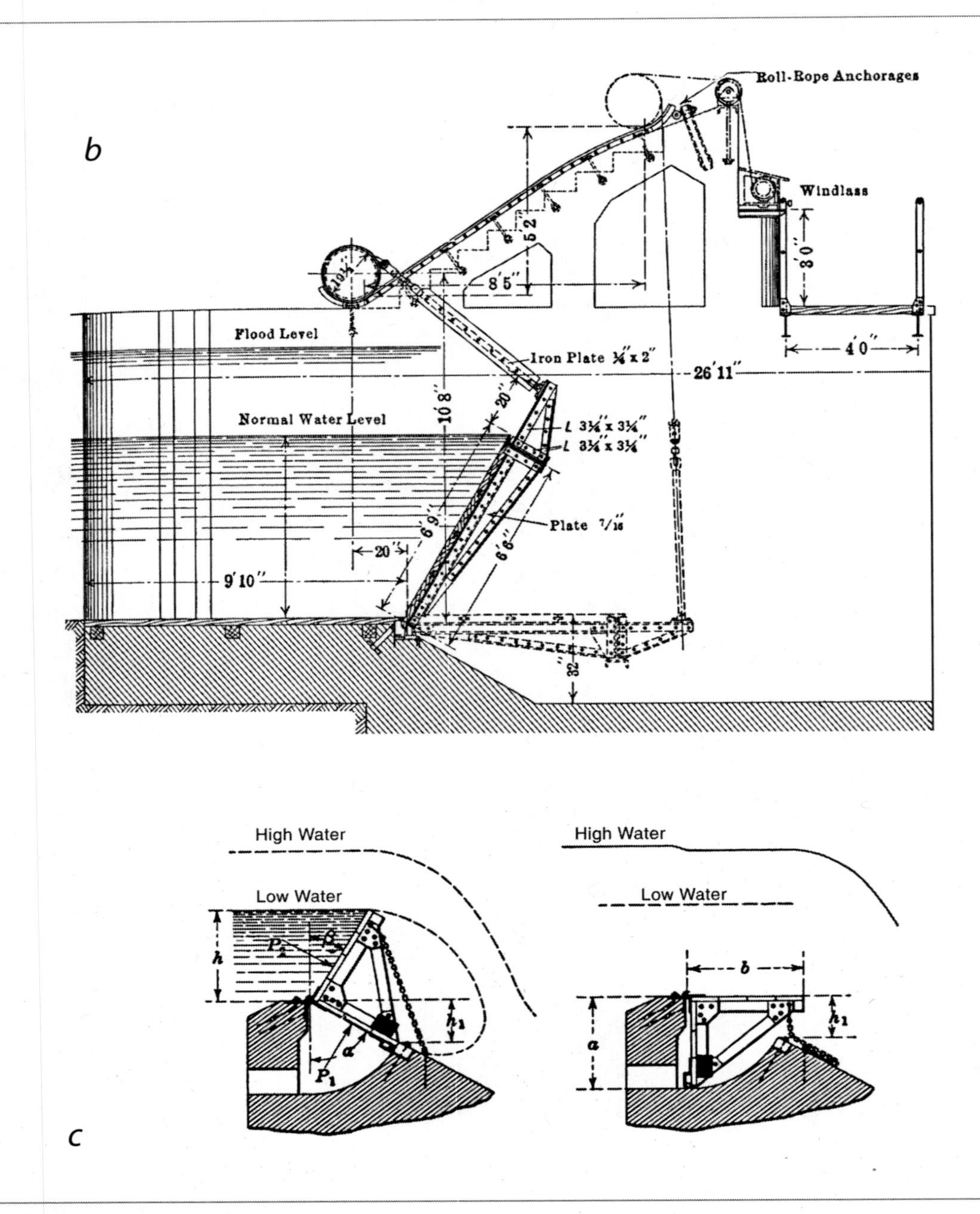

Figure 7.5 *(CONTINUED)*: Examples of movable gates. ***b,*** Automatic gates with rolling counterweight (Lyndon 1916: 297). ***c,*** Stoney gates. Wilson Dam and Hydroelectric Plant, Muscle Shoals, Alabama. AL-47. Delineated by Agatha Lesage, 1994.

Several types of water barriers placed on the crests of dams move themselves automatically when the water reaches a certain level and then restore themselves to normal position when the lake has fallen a certain distance below the maximum allowable level. The simplest, lowest-cost device of this type is a tilting crest gate. Typically such gates consist of a flat or curved slab which revolves about a rocker or bascule.[23]

Stoney gates consist of regular sliding sluices, comprising a panel or slab, which move vertically upward to open and downward to close. A series of rollers, which have their centers along the same vertical line, are imposed between the gate and its supporting pier. When the gate moves, the rollers travel in the same direction as the gate does, but at half the speed of the movement of the gate. The hoisting apparatus consists of a winding drum.[24]

Historically, Tainter gates have been among those most commonly used at hydroelectric dams. These gates are circular arcs in cross section, supported on a framework of steel or wood, convex surfaces being turned upstream, the concave downstream. They are pivoted by trunnions set on the downstream side of the gate. Means of rotation include a rack-and-pinion device or a chain or cable with winding hoist.[25]

Rolling gates were introduced in Germany. By the second decade of the twentieth century, a few installations had been made in the United States. In simple form, they consist of a cylinder placed between two piers, the diameter of the cylinder being equal to the height to which it is desired to elevate the water. The cylinder, or drum, is built up of boiler plate and braced to withstand stresses. Inclined paths are provided, one at each end of the drum, up which the drum may be rolled by hand or machine power. On each end there is a gear that engages with a rack laid on the inclined abutments. By means of a sprocket chain wrapped around one end of the cylinder connecting with the operating mechanism, the dam is rolled up and down the abutments.[26]

Forebay

The forebay is the common term given to the portion of the lake or reservoir immediately upstream above the headworks. It is generally a small bay in the lake in which the water is comparatively still, except for motion produced by the inflow passing through the head gates. A boom is usually placed across the forebay, making an angle with the direction of flow. The boom is designed to prevent floating materials and debris from entering the forebay.[27]

Spillways

Power dams are equipped with spillways to discharge the excess flow not used by the turbines or stored in the reservoir. Spillways may be divided into the following types: spillway dams, side-channel spillways, chute spillways, saddle spillways, shaft spillways, siphon spillways, and emergency spillways.[28]

A spillway dam (Figure 7.6a) may use one of the types of construction cited above. A side-channel spillway, as the name implies, consists of a spillway parallel to the crest of the weir over which the water enters the channel (Figure 7.6b). A chute spillway is isolated from the dam and employs an excavated trench paved with concrete in whole or in part (Figure 7.6c). A saddle spillway takes advantage of natural depressions or saddles in the rim of the reservoir basin away from the stream channel to carry away excess flow. A shaft spillway consists of a vertical, flaring funnel connecting with an L-shaped outlet conduit extending through or around the dam (Figure 7.6d). A siphon spillway discharges water through a closed conduit based on the principle of the siphon (Figure 7.6e).[29]

Intakes

The function of the intake in a hydroelectric development is to let water into the penstock or conduit under controlled conditions. An intake contains racks to prevent potentially destructive debris and ice from entering into the water passages of the plant and injuring the equipment and gates that control the flow of water through the intake (Figure 7.7). Intakes can be characterized as high or low pressure. High-pressure intakes are used where the drawdown is considerable—when, for instance, a reservoir serves both for water storage and as the headwater of a hydro development. Low-pressure intakes are used for smaller drawdowns.[30]

Trash Racks

Trash racks, placed within the intake, are made up of rods or bars, spaced at varying distances apart, set at an angle to the vertical, and placed in front of the head gates. Generally, flat, rolled steel bars are spaced 1 to 2 inches apart.[31]

At some plants, a mechanical rack-raking machine is installed to keep the racks free of debris. Some racks are also heated to prevent ice accumulation in colder weather.[32]

Intake Gates or Valves

Intake gates may be divided into several types: sliding gates, generally constructed of wood or structural steel, which slide directly on their seats without interposition of rollers; wheeled or tractor gates, in which the pressure is taken by wheels attached to the gate; Stoney gates, structural steel lift gates in which the pressure from the gates is taken by trains of rollers attached neither to the gates nor to guides; caterpillar gates, in which the pressure is taken by an endless chain of

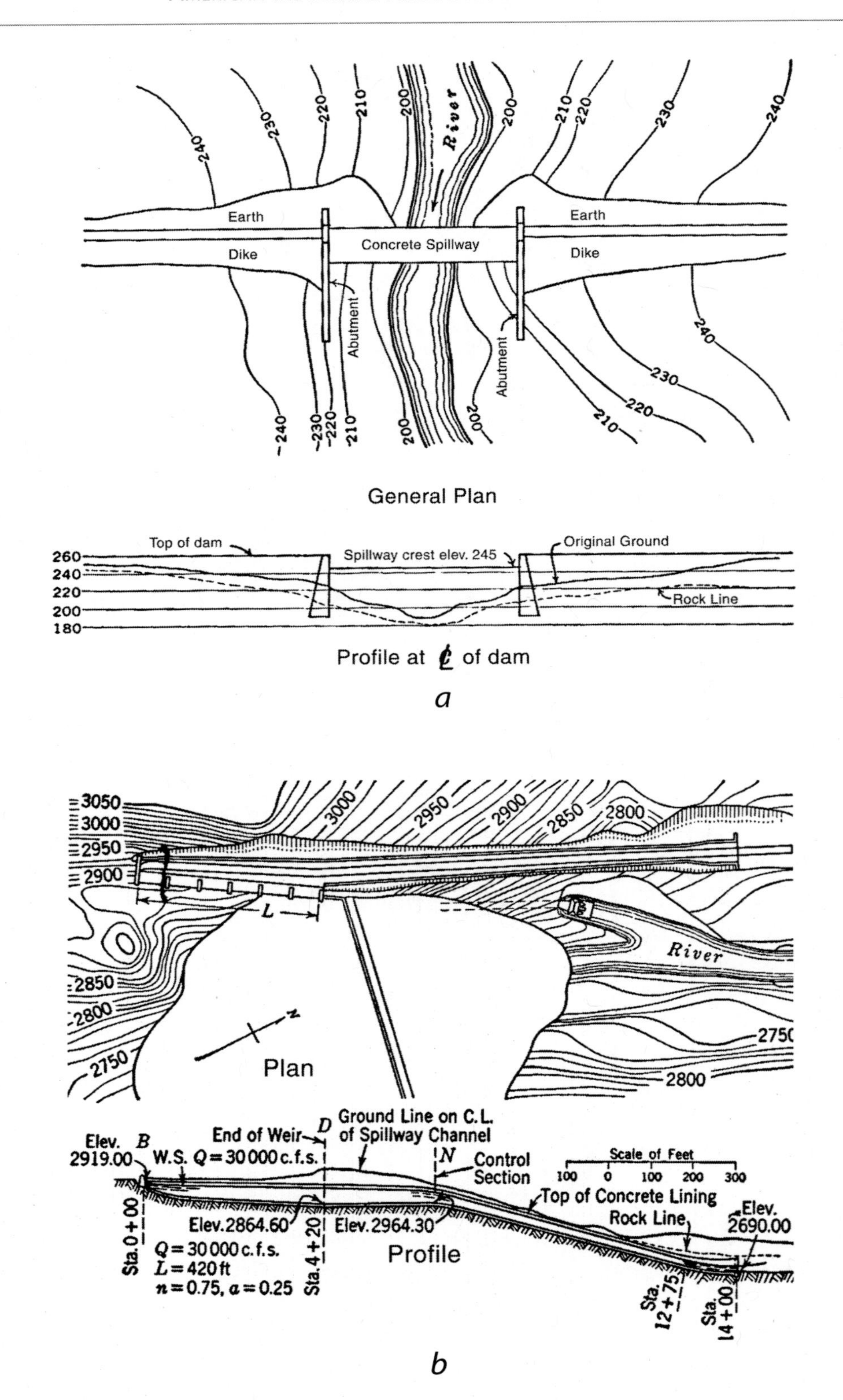

(FACING PAGES): **Figure 7.6.** Types of spillways. ***a,*** Spillway dam. Reprinted from William P. Creager and Joel D. Justin, *Hydroelectric Handbook*, 2nd edition (New York: John Wiley & Sons, Inc., 1950), 495. ***b,*** Side-channel spillway (Creager and Justin 1950: 497).

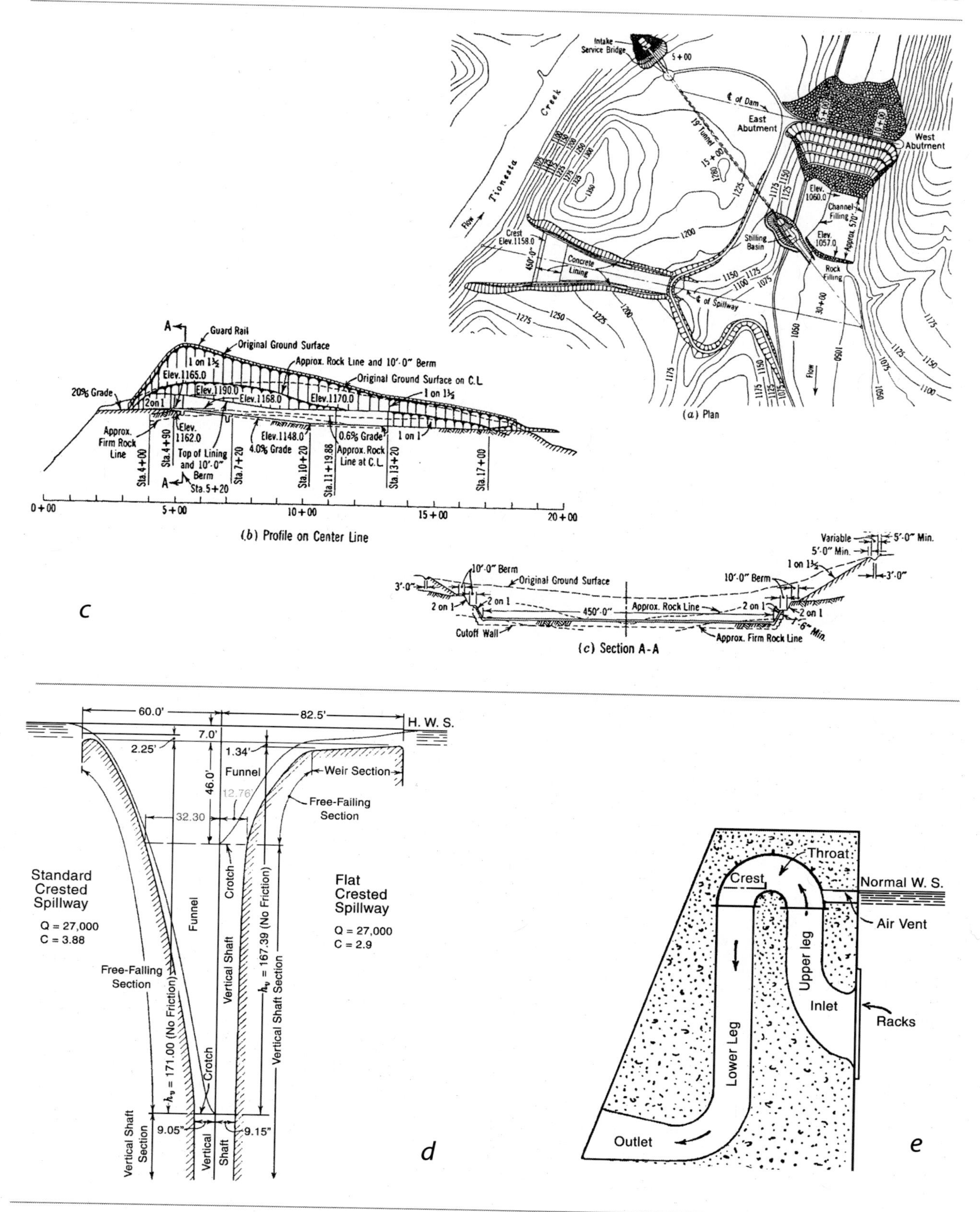

Figure 7.6 (CONTINUED). Types of spillways. *c,* Chute spillway (Creager and Justin 1950: 499). *d,* Shaft spillway (Creager and Justin 1950: 500), *e,* Siphon spillway (Creager and Justin 1950: 502).

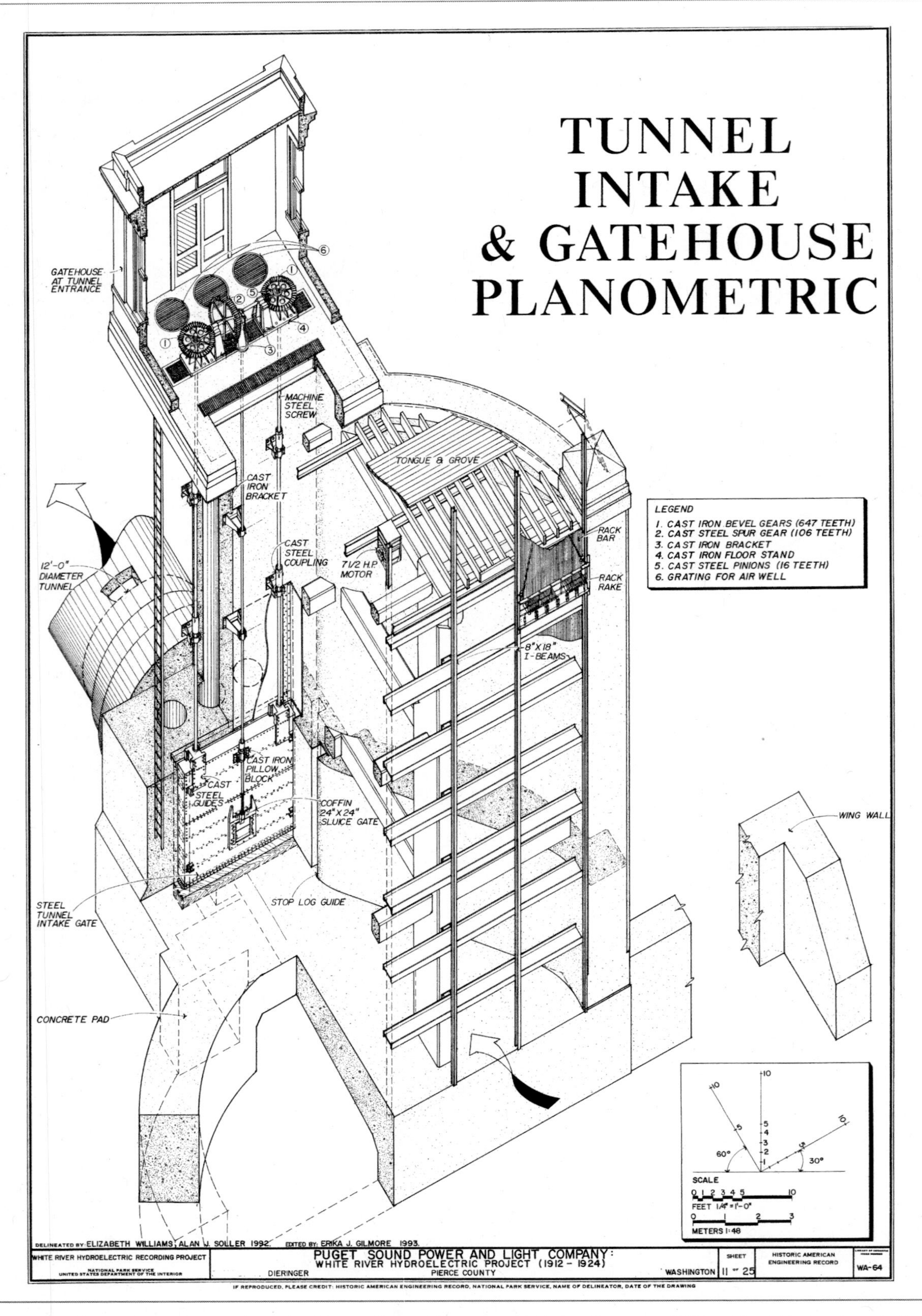

Figure 7.7. Planimetric of tunnel intake. Puget Sound Power and Light Company White River Hydroelectric Project, Dieringer, Washington. HAER WA-64, National Park Service. Delineated by Elizabeth Williams and Alan J. Soller, 1992.

rollers located on each side of the gate and running in a vertical slot in the intake piers; Tainter gates, steel sector gates; butterfly valves, circular vertical or horizontal pivoting valves; and cylinder gates, a steel cylinder open at the top and bottom and having balanced water pressure on the inside and outside surfaces.[33]

The type of gate used depends primarily on the size of the opening, the head on the gate, and the operating conditions. Operating conditions are either of the following: the gate must open and close against a full operating head with free discharge through the intake, or the gate is required to operate only after the conduit is filled through a small auxiliary filler gate and the head on the gate is reduced to a small part of the total head. For low-pressure intakes, plain slide gates, wheeled gates, caterpillar gates, Stoney gates, and Tainter gates are generally used; for high-pressure intakes, caterpillar gates, butterfly valves, and cylinder gates are generally used.[34]

Stoplogs

Stoplogs are used to unwater the intake gates for inspection and repair. If the span is not too great, stoplogs consist of timbers guided into place by means of a vertical slot at each side of the opening above the gates. If the opening to be closed is wide and deep, steel I-beams, with or without bolted timbers, can be used.[35]

Gate Hoists

In some large hydroelectric plants, one or two traveling gantry cranes operate all the head gates. These gantry cranes generally use drum hoists. In other installations, each of the gates is operated by one of several types of mechanisms. These include rack-and-pinion hoists, screw-stem hoists, hydraulic hoists, and drum hoists.

Of these, drum hoists are used exclusively for those gates that close by their own weight, such as the Tainter, caterpillar, and Stoney. For gates that require a positive thrust to close them, other types are used. Gate hoists can be opened by hand, by electric motor, or by hydraulic pressure.[36]

Air Inlets

Air inlets in back of head gates at the entrance to a penstock or pipeline are critical. If they are omitted and the pipeline is suddenly drained when the head gate is closed, the pipe is likely to collapse.

Penstocks

The turbines in a hydroelectric installation must work in concert with the passages and gates designed to control the water as it flows to, through, and away from the turbines. In most hydroelectric installations, a penstock, a conduit used to carry water from the supply source to the turbine, is used (Figure 7.8).[37]

After the water has passed the intake at the dam, it enters a conduit that carries it from there to the turbines of the powerhouse. The general types of conduits include canals, flumes, steel pipe, wood-stave pipe, concrete pipe, and tunnels. A conduit system may be divided into two parts: the "high-line" or non-pressure conduit and the "penstock" or pressure conduit.

The pressure penstock requires that the water discharging to the turbine always be at a pressure above atmospheric pressure. The siphon penstock is constructed so that at points in the penstock, the pressure may be less than atmospheric pressure and sections of the conduit act as a siphon. Valves to stop the flow of water are installed in the upper end of penstocks. Types of valves used include gate valves, butterfly valves, and needle valves.[38]

A high-line conduit may be a canal, flume, tunnel, wood-stave pipe, concrete pipe, or steel pipe. A penstock is usually a steel pipe but may also be a tunnel.[39]

Surge Tanks

Water-hammer pressure is created in long, closed conduits by sudden closure of the turbine gates. For very long conduits, the water hammer in normal operation of the turbine may be substantial and may require extraordinary conduit strength to withstand it. Violent fluctuations in pressure in the conduit may interfere with proper turbine regulation.

The simplest means of eliminating positive water-hammer pressure is to provide a bypass to take the rejected flow. This may be accomplished by installing a relief valve at the turbine or a surge tank at the lower end of the conduit.

Ordinarily such tanks are designed so that water will not spill over the top under the most drastic condition of load rejection. The surge tank is always located as close as possible to the powerhouse, in order to reduce the length of penstock to the minimum, and is

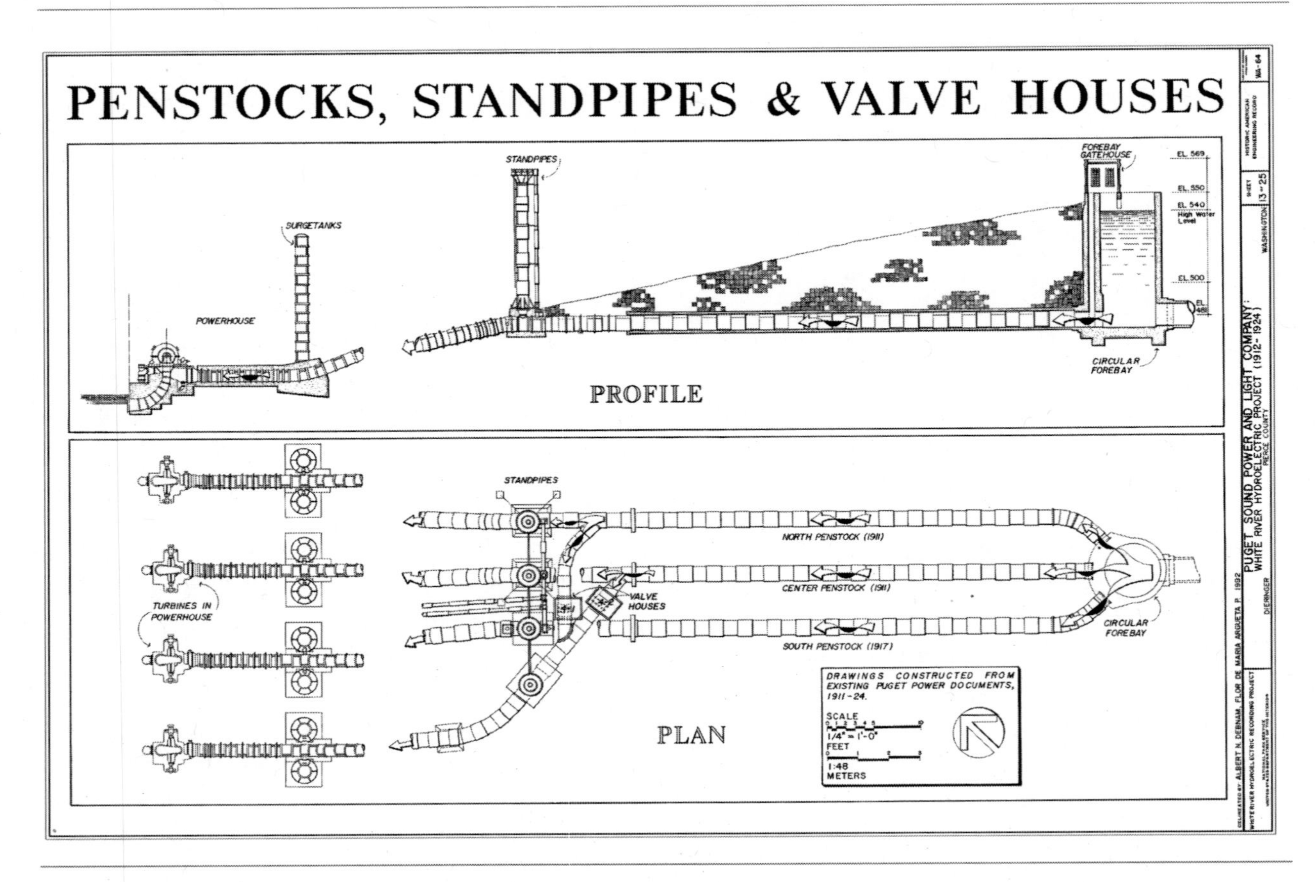

Figure 7.8. Penstocks. Puget Sound Power and Light Company White River Hydroelectric Project, Dieringer, Washington. HAER WA-64, National Park Service. Delineated by Albert N. Debnam and Flor de Maria Argueta P., 1992.

preferably situated on high ground, to reduce the height requirement of the tower.[40]

Powerhouse

Powerhouses support and house the hydraulic and electrical equipment (Figure 7.9). The powerhouse is divided into two major parts: the substructure and the superstructure. The substructure or foundation consists of the steel and concrete components necessary for the draft tube, to support the stay ring and generator, and encase the spiral case. The superstructure provides protective housing for the generator(s) and control equipment, as well as structural support for the cranes.

Powerhouses differ in the ways they are oriented or connected with the dam. Some are structurally connected with the dam. Others are located at a distance from the dam, and a penstock carries the water from the intakes to the powerhouse.

The substructure is divided into a series of bays: one bay for each main unit; one for the exciter units; and an entrance or working bay. The working or entrance bay is usually at one end of the building and provides space where the equipment may be unloaded.

Powerhouse substructures are typically built of concrete or reinforced concrete and may be classified according to the type of turbine setting used: open flumes, vertical concrete spiral casings, vertical metal spiral casings, horizontal metal spiral casings, and impulse turbines. By far the most common type of setting for mid-twentieth-century hydroelectric facilities is the scroll or spiral casing.[41]

Distributor assemblies consisting of spiral cases, head cover, bottom ring, and wicket gates are used to control the amount and direction of water entering into the rotating turbine. For low heads (below 20 feet), an open-flume or pressure case setting may be used. Register and cylindrical gates were used in early hydropower installations, basically comprised of a cylinder with

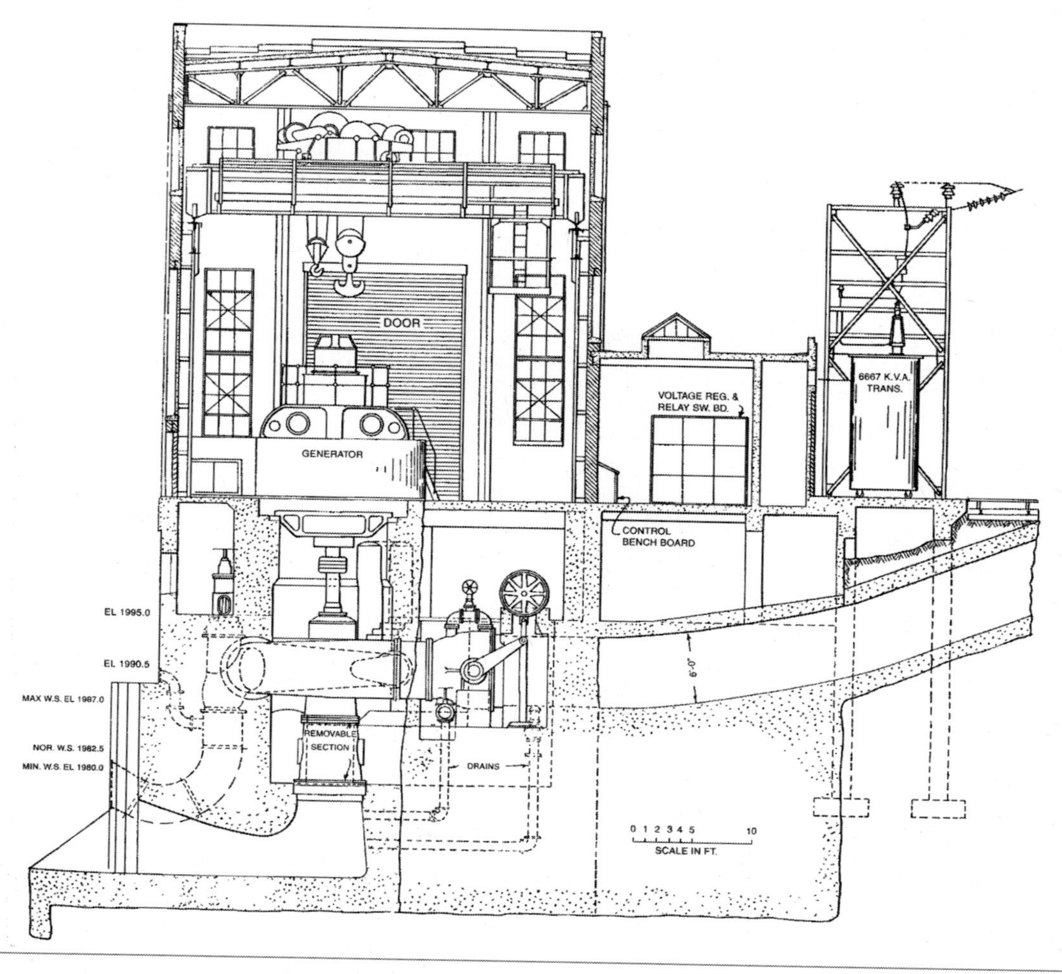

Figure 7.9. Cross section of Prospect No. 2 Hydro Plant. Reprinted, with permission, from Prospect No. 2 Hydro Plant, *Power Plant Engineering* (April 1, 1928): 389.

rectangular openings that could be moved circumferentially to block or permit the flow into a turbine.

Conveying the water from the penstock and directing the proper amount of it correctly against the runner of a turbine requires a scroll case, a speed ring, and turbine gates. The scroll case for medium and high head developments is circular in form. In plan, it extends from the penstock and wraps, in spiral form, around the speed ring. The scroll case is either formed in concrete, made of plate steel, or made of cast iron or steel. A low-head plant with large and short passages from intake to speed ring usually has water passages cast in the substructure concrete. This passage is often rectangular in cross section.[42]

The speed ring is the part of the turbine that joins the discharge ring with the turbine cover and pit liner. Inside the speed ring and bolted to it is the inlet gate mechanism. The mechanism is operated by a governor that maintains speed control under variable load by opening or closing the gates.[43]

Hydraulic Turbines

As water passes through a hydropower plant, its energy is converted into electrical energy by a prime mover known as a hydraulic turbine. The turbine has vanes, blades or buckets that rotate about an axis by the action of the water. The rotating portion of the turbine is often referred to as the runner. The rotation of the turbine in turns drives an electrical generator.

Hydraulic turbines may be divided into two basic types: impulse and reaction. The impulse type, frequently called a Pelton wheel after an early developer, Lester Pelton, uses the kinetic energy of a high-velocity jet that strikes the single or double bowl-shaped buckets of the impulse runner (Figure 7.10). The water

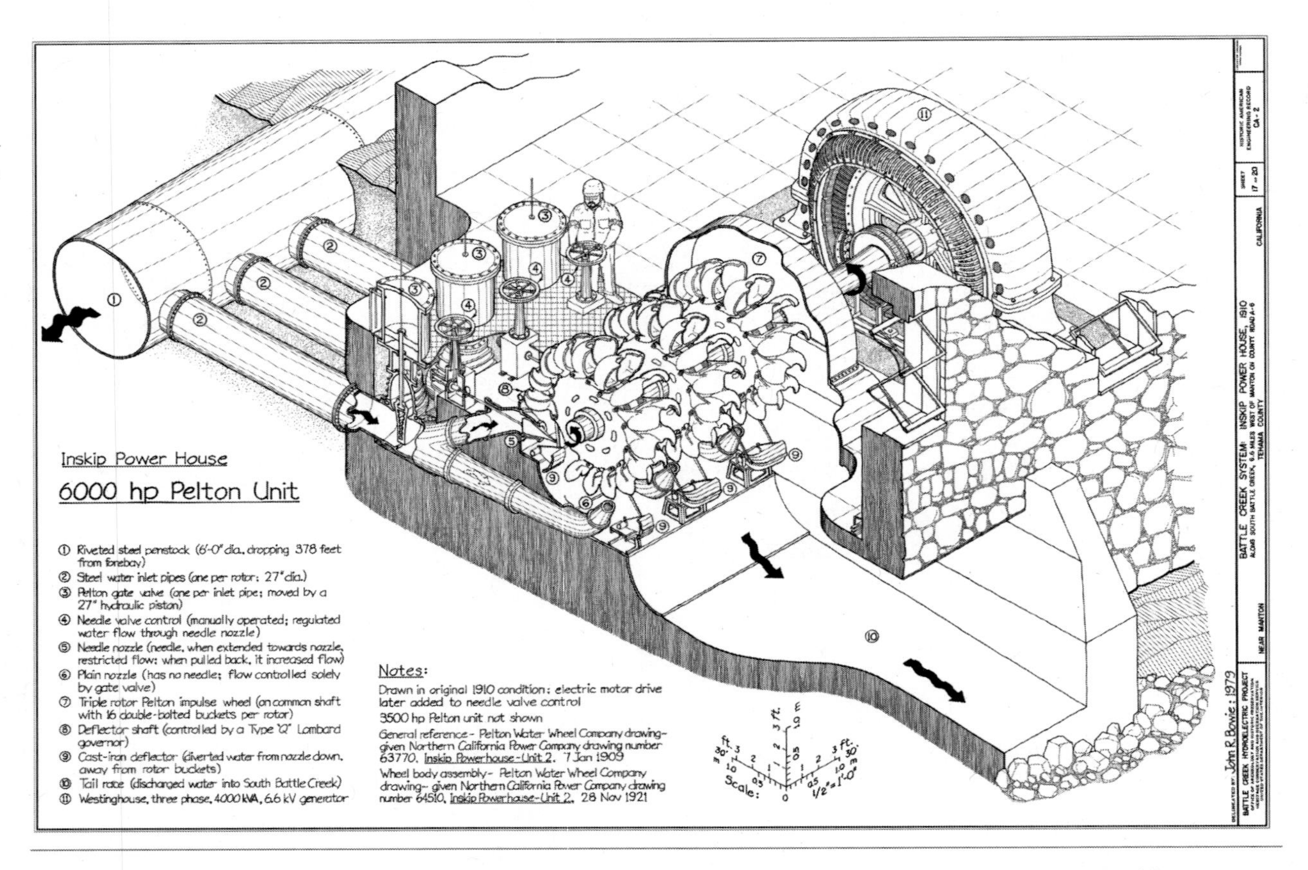

Figure 7.10. Impulse (Pelton) turbine. Battle Creek System. Inskip Power House, 1910, near Manton, California. HAER CA-2, National Park Service. Delineated by John R. Bowie, 1979.

striking the buckets of the runner is regulated through the use of a bulb-shaped needle in a nozzle. The position of the needle determines the quantity of water striking the runner. The impulse wheel is used, in general, for high-head units of 800 to 1,000 feet or more. They are also used for lower heads for smaller capacity units.[44]

In reaction turbines (Figure 7.11), the runner chamber is completely filled with water, and a draft tube is used to recover as much of the hydraulic head as possible. If the flow is perpendicular to the axis of rotation, the runner is called a radial-flow turbine. If the water flow is partially radial and partially axial, it is called a mixed-flow turbine. The most common mixed-flow turbine design was developed by James B. Francis and bears his name. Francis turbines have a crown and band enclosing the upper and lower portions of the bucket. The standard runner of the Francis turbine consists of two crowns between which buckets or blades are placed (Figure 7.12).[45]

When large volumes of water are available at such low heads that the Francis-type runner is either too slow or too bulky, a propeller-type runner is used. The design of the runner of the Nayler turbine resembles a ship's screw propeller. Water enters the scroll case axially as in the Francis turbine and then flows into a transition space in which the radial is changed to axial flow.[46]

The third type of reaction turbine is an axial-flow turbine in which the water flows parallel to the axis of rotation. The most common axial-flow turbine is the propeller turbine. Propeller turbines either may have blades of the runner rigidly attached to the hub (fixed-blade runners), or the blades can be made adjustable so that the turbine can operate at a wide variety of flow conditions with better efficiency. One design of an adjustable blade turbine with both adjustable blades and gates is the Kaplan turbine.

Principal components of the turbine include speed rings, guide vanes or wicket gates, the main shaft, guide

Figure 7.11. Pelton Reaction Turbine, Wise Power House, Pacific Gas & Electric Company, California. Reprinted from Pelton Water Wheel Company, *Pelton Impuse and Reaction Turbine Installations* (San Francisco: Pelton Water Wheel Company, 1920).

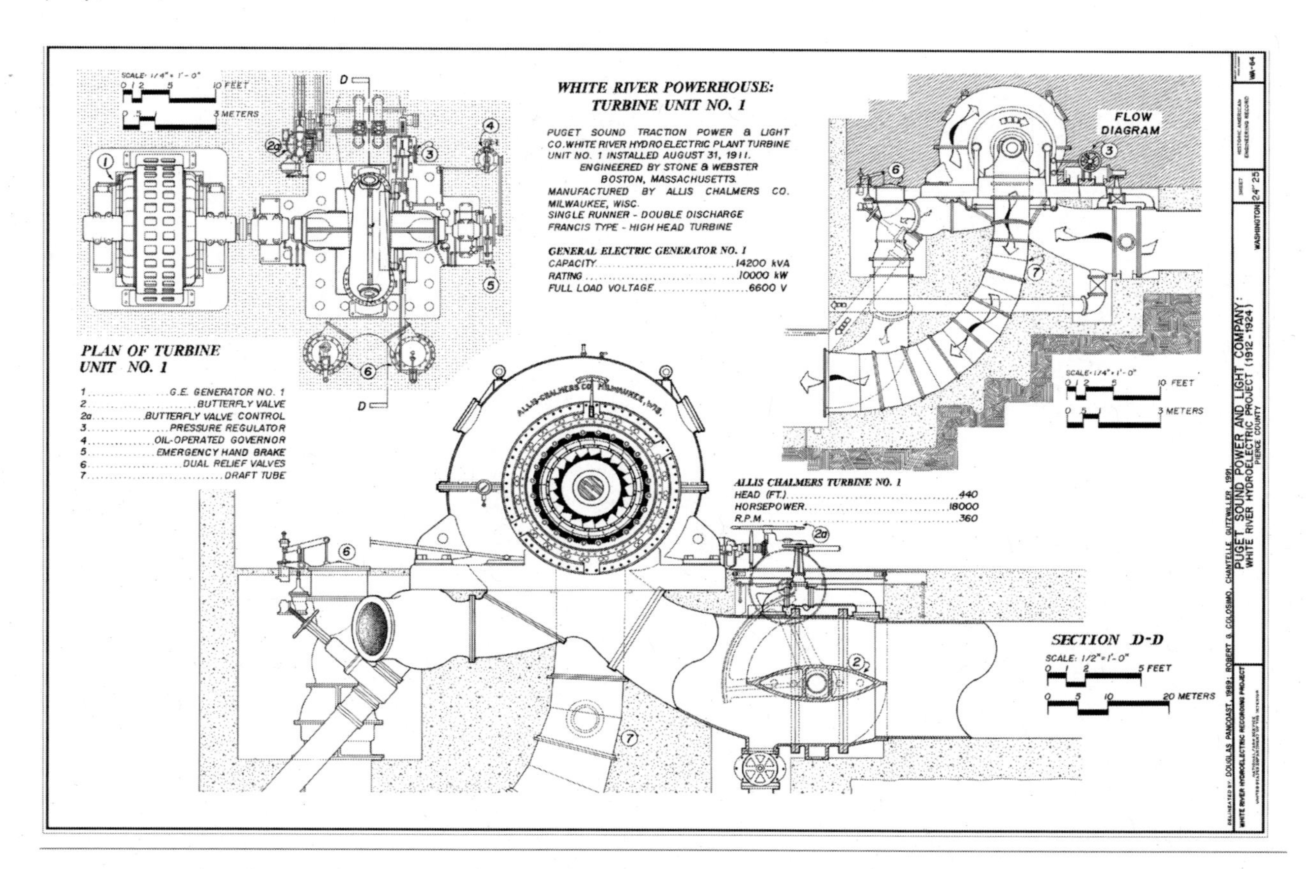

Figure 7.12. Francis turbine. White River Powerhouse Turbine No. 1. Puget Sound Power and Light Company White River Hydroelectric Project, Dieringer, Washington. HAER WA-64, National Park Service. Delineated by Douglas Pancoast, 1989, and Robert G. Colosimo and Chantelle Gutzwiller, 1991.

bearings, thrust bearings, governors, and pressure regulators. The speed ring is that part of the turbine that connects the cover plate and the discharge ring. The movable guide vanes, sometimes called wicket gates, control the flow into the runner. The guide-vane assembly, together with the speed ring, is sometimes called the distributor. The turbine shaft must transmit the full torque of the runner under the maximum head conditions at which the turbine will operate. Turbine guide bearings for vertical units are of two principal types: those for water lubrication, usually of a wood or fibrous material; and those for oil lubrication, which are of the babbitted type. Thrust bearings, usually mounted on the upper or lower bridge of the generator, support a thrust load amounting to several hundred tons. Governors are designed to regulate the speed of the unit by increasing or decreasing the amount of water supplied to the turbine. Pressure regulators are used with turbines with long penstocks to prevent rapid velocity changes in the pipeline in that they open when the turbine gates close.

Pipe Galleries and Electrical Conduits

The space adjacent to the turbines and immediately below the main generator floor is usually occupied by a continuous gallery from end to end of the plant. This gallery contains all the piping for plant operation, including that needed for the governor system, lubricating oil, and so on. The electrical cable is often carried in the same gallery but is sometimes laid through a separate gallery or tunnel or on a rack in which the cables are laid in troughs.[47]

Superstructure

The superstructure framework (Figure 7.13) is usually made of structural steel. Interior walls are constructed of brick, brick veneer backed with another material, reinforced concrete, concrete blocks, stone backed with brick, hollow tile, or gunite on steel lathing. The height of the superstructure is usually determined by the clearance the traveling crane requires to handle the largest pieces of equipment and carry them over the other machinery in operation. The typical equipment in the superstructure includes the following: main generating machinery; turbine governors, pumps, and tanks; motor-generator sets and exciters; compressed-air equipment; water supply pumps; switchboard; high-tension transformers and switches (often outside the building); storage batteries; transil-oil tanks with centrifuge, filters, and pumps; lubricating-oil tanks with filters and pumps; and elevators and hoists. Other interior spaces include locker rooms and washrooms, operating and engineering offices, machine and carpenter shops, storeroom, boiler room, and test room.[48]

Generators

A generator is defined as a machine that transforms mechanical energy into electrical energy. Its primary features include a field or assembly of magnets arranged to produce a magnetic flux and an armature or assembly of electrical conductors arranged across the path of the magnetic flux.

Generators are characterized as one-, two- or three-phase, depending on the number of armature windings. The standard type of generator in a hydroelectric plant is the revolving-field type. Major parts include the stator, the rotor, the shaft, the coupling, bearings, bearing bracket, and brakes. A stator consists of a cast-iron or fabricated steel frame, a laminated magnetic core, and armature windings. The shaft, usually constructed of forged steel, is flanged at the bottom in a vertical generator to connect to the turbine shaft.[49]

The excitation of a generator is the power input required by the field winding to maintain the necessary intensity of magnetic flux. Much greater field current is required with a lagging power factor and heavy load than with a leading power factor and light load. Excitation systems consist either of a centralized system involving a bus fed by one or more exciters and feeding all the generator fields, or an individual excitation system, involving an individual exciter associated with each generator. An exciter is a direct-current generator designed to provide excitation to synchronous machines.[50]

Generator voltage regulators are used to automate control of the generator field excitation as required for voltage control, frequency control, load control, power-factor control, and automatic synchronizing.[51]

(FACING PAGE): **Figure 7.13.** Hydroelectric plant superstructure. Transverse section. White River Hydroelectric Power Plant (1912–1924). HAER WA-64, National Park Service. Delineated by Dennis McGrath, 1989; Robert G. Colosimo and Chantelle Gutzwiller, 1991.

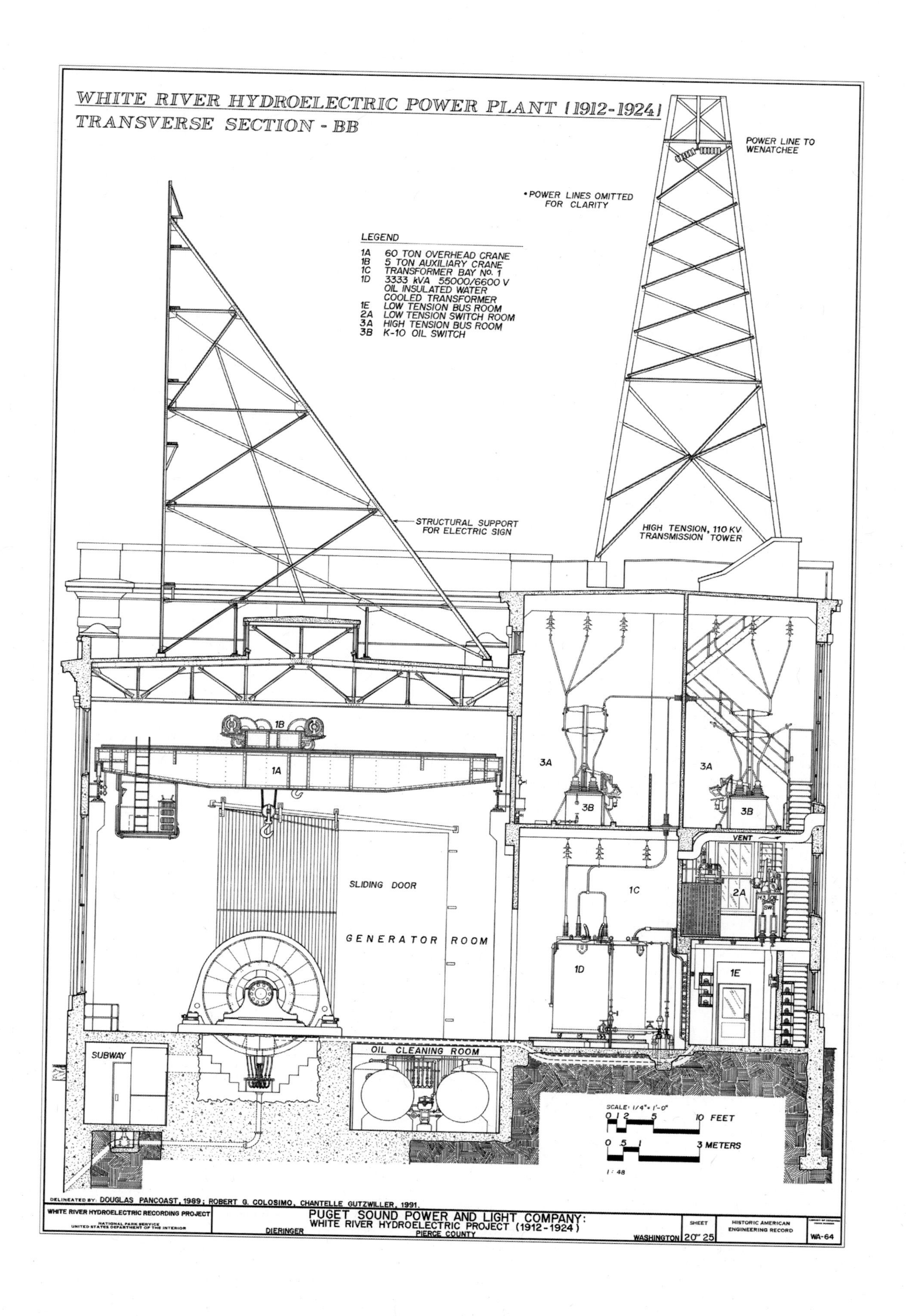

WHITE RIVER HYDROELECTRIC POWER PLANT (1912-1924)
TRANSVERSE SECTION - BB
POWER LINE TO WENATCHEE
• POWER LINES OMITTED FOR CLARITY
LEGEND
1A 60 TON OVERHEAD CRANE
1B 5 TON AUXILIARY CRANE
1C TRANSFORMER BAY No. 1
1D 3333 kVA 55000/6600 V OIL INSULATED WATER COOLED TRANSFORMER
1E LOW TENSION BUS ROOM
2A LOW TENSION SWITCH ROOM
3A HIGH TENSION BUS ROOM
3B K-10 OIL SWITCH
STRUCTURAL SUPPORT FOR ELECTRIC SIGN
HIGH TENSION, 110 KV TRANSMISSION TOWER
1B
1A
3A
3B
3A
3B
VENT
SLIDING DOOR
1C
2A
GENERATOR ROOM
1D
1E
SUBWAY
OIL CLEANING ROOM
SCALE: 1/4" = 1'-0"
0 1 2 5 10 FEET
0 .5 1 3 METERS
1 : 48
DELINEATED BY: DOUGLAS PANCOAST, 1989; ROBERT G. COLOSIMO, CHANTELLE GUTZWILLER, 1991.
WHITE RIVER HYDROELECTRIC RECORDING PROJECT
NATIONAL PARK SERVICE
UNITED STATES DEPARTMENT OF THE INTERIOR
PUGET SOUND POWER AND LIGHT COMPANY:
WHITE RIVER HYDROELECTRIC PROJECT (1912-1924)
DIERINGER
PIERCE COUNTY
WASHINGTON
SHEET 20 OF 25
HISTORIC AMERICAN ENGINEERING RECORD
WA-64

Transformers

A transformer is a device for transferring electrical energy from one alternating-current (AC) circuit to another AC circuit, usually of different voltage. The essential features include a primary winding which receives energy from the supply circuit; a secondary winding which delivers energy to the receiving circuit; and an iron core common to both windings.[52]

Switching Equipment

Among the most important considerations in the design of a hydroelectric plant is its connection to the transmission system. Because of its location, a hydroelectric station is usually associated with long lines and high transmission voltages.

The scheme of bussing and switching on both the low-voltage and high-voltage side of the transformers depends on the number of outgoing lines, the number of transformers and generators, and the desired operation dependability and flexibility. Switching equipment consists of switches, breakers, and other devices used by station operators for opening and closing electrical circuits and connecting or disconnecting generators, transformers, or other equipment.

The switchboards of a hydroelectric plant constitute the centers of control, metering, and relaying of the main power circuits; control, protection, and distribution of energy from the control batteries and chargers; control of the excitation circuits; and control and protection of the auxiliary power circuits.[53]

Outdoor Substations

The substation of a power plant is the place where generator voltage is transformed to transmission line voltage and high-tension switching is performed. Typically outdoor substations include the following elements: transformers, conductors, insulators, lightning arresters, choke coils, disconnect switches, air break switches, current transformers, potential transformers, protective oil circuit breakers, and high-voltage fuses. The choke coil is used to force a lightning surge to take the path provided for it.[54]

Transmission Lines

The hydroelectric plant is usually located a considerable distance from the distribution system into which it feeds. Unlike the steam plant which radiates energy over a number of comparatively small feeders, the hydroelectric plant usually delivers its entire output over two or three heavy transmission lines.[55]

FOSSIL FUEL–POWERED GENERATING PLANTS

These plants convert energy from fossil fuel—coal, petroleum, or natural gas— into steam, usable for powering turbines. The turbines drive electric generators to produce electricity.

By the 1930s, these plants ranged from small plants with a capacity of only a few hundred kilowatts to very large plants, then termed "super-power stations," having nearly one million kilowatts of installed generating capacity.[56]

Gas-Fired Power Plants

Early gas-fired power plants were located in portions of the country such as Texas, where a supply of natural gas was readily available. For example, a 5,000-kilowatt power plant opened in Amarillo, Texas, in 1927, employing a 6-inch natural gas line to furnish gas at 45 pounds pressure for the generation of steam (Figure 7.14). Each boiler was supplied with seven gas burners which were installed in the front wall of the furnace. Steam pressure was maintained by a hand-operated gas supply valve. To guard against shutdown due to interruption of the gas supply, the gas burners were of a combination type suitable for either gas or oil, and an oil storage tank was provided.[57]

Other power plants, such as the 21,000-kilowatt Neches Station in Beaumont, Texas, operated on a combination of natural gas and fuel oil. In this plant natural gas was taken from the main pipeline of the Magnolia Pipeline Company, and the pressure was stepped down before delivery to the burners. The fuel oil was pumped to tanks immediately outside of the plant and was later pumped to fuel oil heaters by steam-driven and motor-driven pumps and delivered to the burners when it reached proper viscosity.[58] Both the Amarillo and Beaumont plants and others like them used gas and/or oil to heat steam to propel turbines.

This most basic process has a relatively low energy efficiency. Typically, only about one-third of the thermal energy used to generate steam is converted to electrical energy.[59] A major evolution occurred with the de-

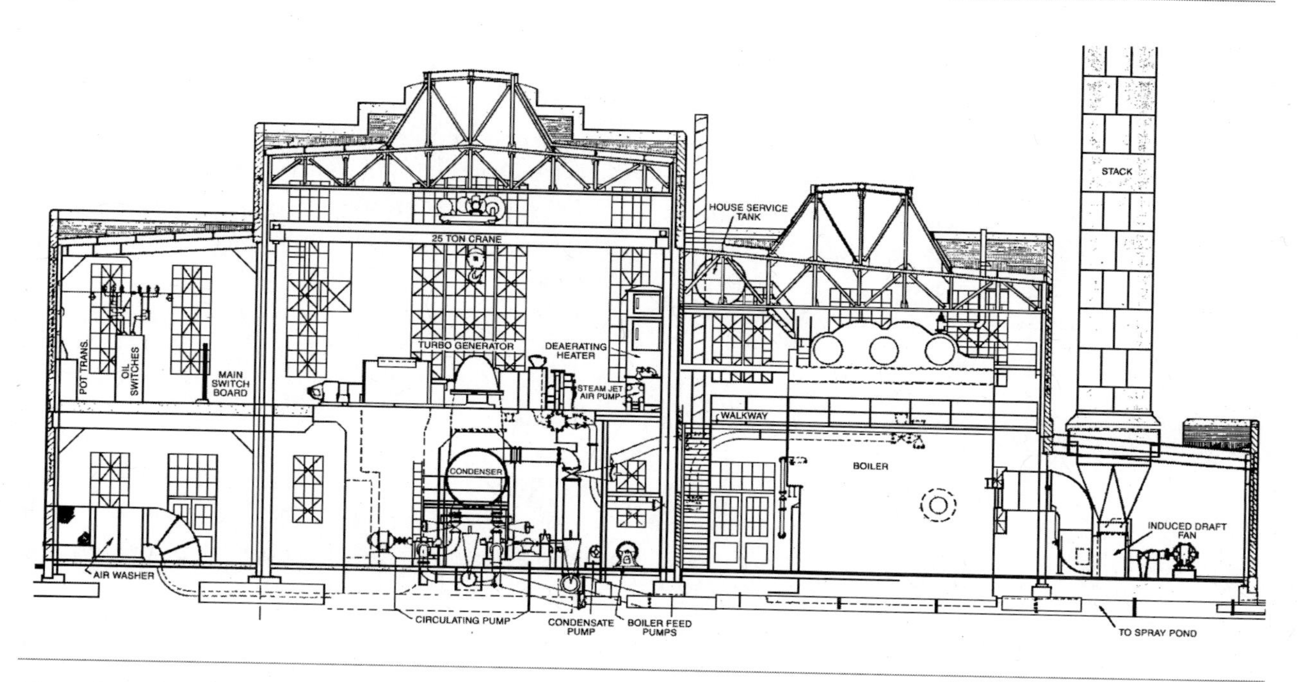

Figure 7.14. Cross section of Amarillo station. Reprinted, with permission, from Alfred Iddles, Amarillo Station Uses Natural Gas as Fuel, *Power Plant Engineering* (November 15, 1927): 1177.

velopment of gas turbines. The predecessor of today's gas turbine was introduced as early as 1791. The modern gas turbine can be traced back to 1906 when H. Holzwarth developed the constant-volume gas turbine. The first gas turbine used to generate electricity was installed in Neuchatel, Switzerland, in 1939 and was capable of producing 4 megawatts of electricity.[60] Instead of using steam to turn a turbine, hot gases from burning natural gas are used to turn the turbine and generate electricity.

The principal components of the modern gas turbine are the compressor, the combustor, and the turbine itself. Compressors draw in air from the environment and by adding mechanical energy, bring the system to the required overpressure. Large modern gas turbines are equipped with multistage axial compressors. The air that leaves the compressor is raised to the preset turbine inlet temperature by burning fuel in a combustor. The fuel and a portion of the air coming through the compressor are brought into a combustion zone through the swirl basket and/or the fuel nozzle. The swirl stabilizes the flow, and the fuel continually being supplied is ignited by gases circulating in the zone's center. The hot gases coming from the combustor expand in the turbine, thus driving the rotors and the blades.

Many recently constructed natural gas–fired plants are combined-cycle units. These plants have both a gas turbine and a steam unit. In combined-cycle plants, the waste heat from the gas-turbine process is used to generate steam, which is then used to generate electricity.

By 2000, 95 percent of the new electric generating capacity added in the United States was represented by gas-fired plants.[61]

Coal-Fired Power Plants

Coal-fired power generation is a simple process. In most coal-fired plants, chunks of coal are crushed into fine powder and are fed into a combustion unit to be burned. Heat from the burning coal is used to generate steam which in turns spins one or more turbines to generate electricity.

Coal has played an important role in electrical production since the first power plants were built in the United States in the 1880s. The earliest electric power plants used hand-fed wood or coal to heat a boiler and produce steam. The steam was initially used in reciprocating steam engines that turned generators to produce electricity. In 1884, the more efficient high-speed steam turbine was developed by British engineer Charles A.

Parsons and quickly supplanted the steam engine in power generation.

In the 1920s, pulverized coal firing was developed. This process had the advantages of a higher combustion temperature, improved thermal efficiency, and a lower requirement for excess air for combustion. In the 1940s, the cyclone furnace was developed. The furnace permitted the use of poorer grades of coal with less ash production and greater overall efficiency.

Coal Handling

Coal arrives at steam-generating plants by rail, barge, or truck. After arrival, the coal is unloaded, prepared, and transferred to either storage or the plant for firing. The sequence of handling varies depending on plant location and means of transport.

For example, at New York's East River Station (Figure 7.15), coal was delivered directly to the station by ocean-going colliers. Coal was conveyed into the plant by two electrically operated coal towers, each capable of handling 350 tons of coal per hour. A typical tower consisted of a steel structure housing machinery and a trolley boom for swinging the bucket over the barge. A ring crusher on each tower broke the coal into smaller sizes.[62]

A similar coal tower system was used in the Chester (Pennsylvania) Waterside Station (Figure 7.16). In Chester, the towers served as points of entry for the gravity-fed distribution system. After passing through sorters and crushers, the coal was guided by hopper into cable-driven cars that followed a set of tracks to one of two main brokers. In the East River Station, the coal was fed by conveyors to either the storage yard or storage bunkers.[63]

In the case of truck transport, all that may be necessary is for a truck to dump to a small outdoor storage pile or into a basement bin. If coal is received by rail, some means of bringing the hopper cars to the appropriate spot for rapid unloading is probably needed. Frequently, a track scale is used to weigh detached cars that are again weighed when empty.

Railroad hopper cars come in three basic types: the conventional, manually locked-door, sawtooth hopper; a bottom drop type; and a top-dump car. The latter is

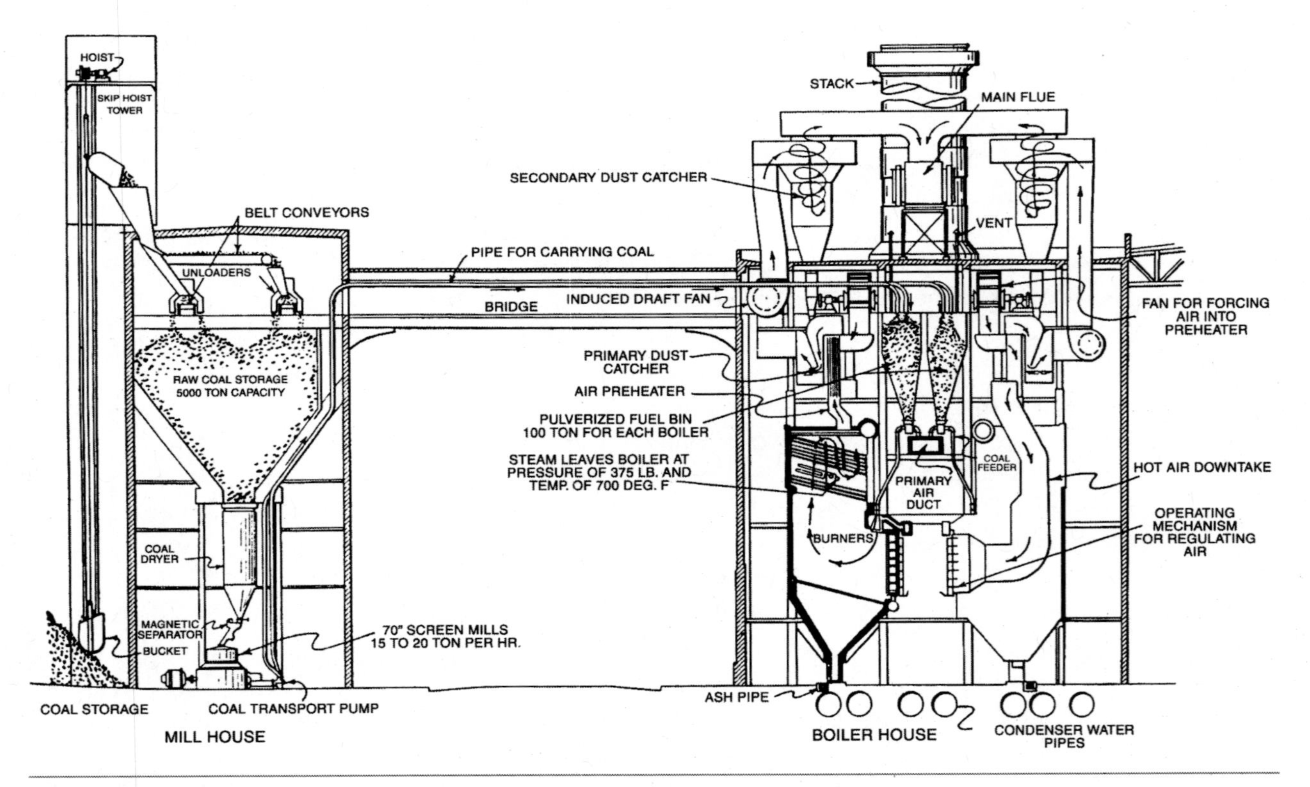

Figure 7.15. East River Station coal delivery system. Reprinted, with permission, from East River Station, *Power Plant Engineering* (March 15, 1927): 41.

unloaded by rotary car dumpers, machinery that can empty a 50-ton of coal in less than three minutes (Figure 7.17). Sawtooth hopper cars generally require some sort of shaker or vibrator to quickly unload the car, especially when the coal is wet or frozen. Modern bottom dumping cars can generally be emptied without the aid of shakers. In each case, the coal is unloaded to an undertrack hopper. The hopper is ordinarily of sufficient size to accommodate two or three carloads of coal.[64]

A typical hopper is equipped with a bottom feeder that places the coal on a belt that conveys it over a magnet pulley where impurities such as scrap iron, loose bolts, spikes, and other debris are separated from the coal by magnetic attraction. The coal is then ready for filling bunkers in the case of a stoker plant or for passing into crushing mills in the case of a pulverized fuel plant, or for passing out into the yard where it is stored for future use.[65]

Coal Storage

The active coal stockpile of a plant is almost always stored in a covered structure, such as a bin, silo, or bunker. Silos typically are built of concrete and hollow tile and are cylindrical in form. Bunkers are usually suspended from beams and girders high above the firing floor. The two basic designs are parabolic and cylindrical in cross section.[66]

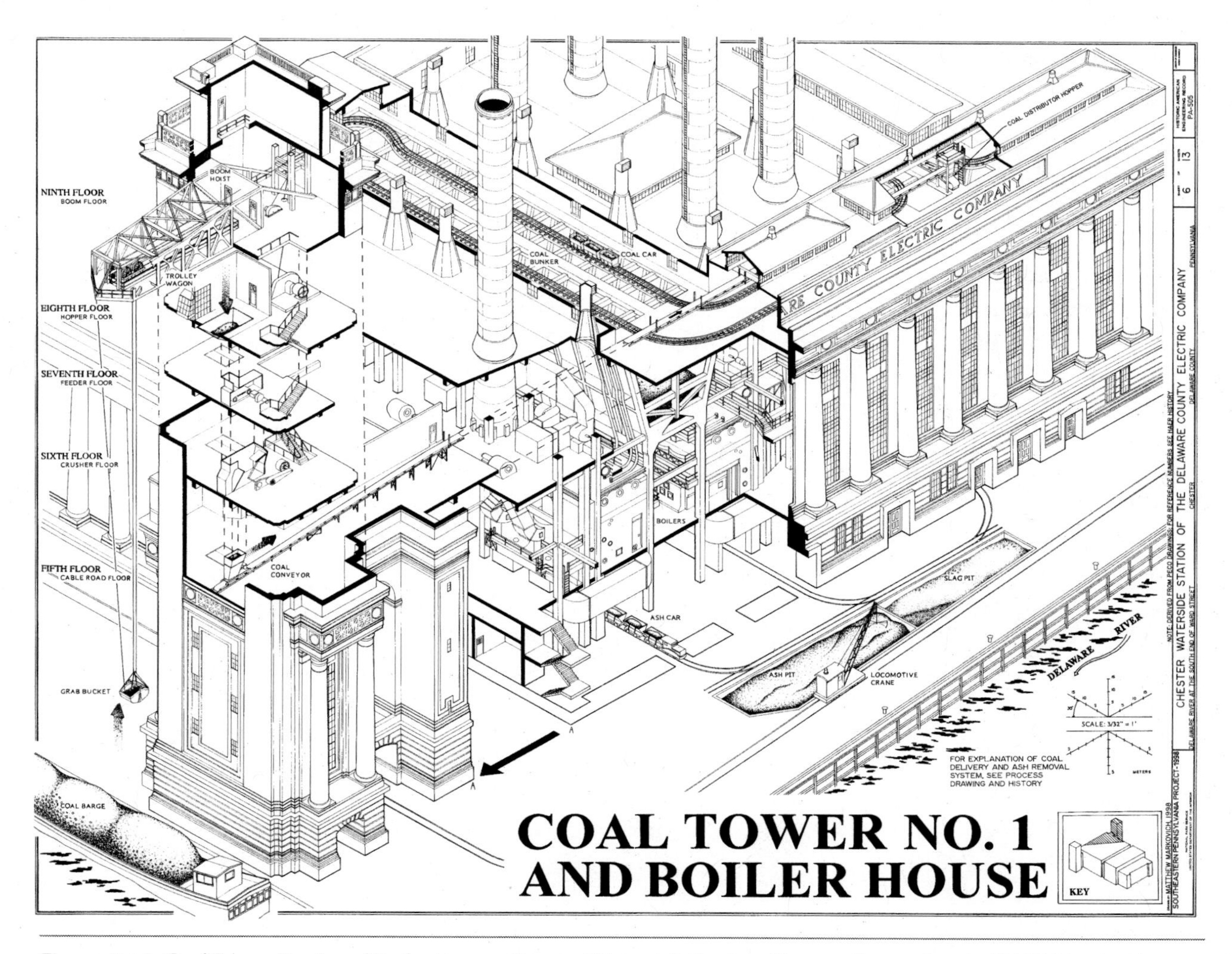

Figure 7.16. Coal Tower No. 1 and Boiler House, Chester Waterside Station, Chester, Pennsylvania. HAER PA-505, National Park Service. Delineated by Matthew Markovich, 1998.

Figure 7.17. Example of rotary car dumper. Utah Copper Company Magna Concentrator, Magna vicinity, Utah. HAER UT-24:4, National Park Service. Photograph by Jack E. Boucher, 1972.

Delivering Coal

Devices called feeders deliver coal at a controlled rate from a storage area to a point where a conveyor can handle it conveniently for transport. Several types of feeders are typically employed: apron feeders, bar-flight feeders, reciprocating feeders, and vibrating feeders. In small-volume plants, screw and belt conveyors generally deliver the coal.[67]

Apron feeders consist of overlapping steel pans mounted on double rolls of steel rolling chain. Bar-flight feeders consist of bars or flights attached to two strands of chain so that the bars slide along the bottom of a trough, dragging the coal. Reciprocating feeders agitate the incoming coal, assisting in feeding it into motor-driven pans. In one type of vibrating feeder, a feeder pan is attached to vibrator arms or bars. The vibration causes the coal to move up- or downhill in a smooth stream, and the rate of movement is changed by variations in the intensity of vibration.[68]

An important step in coal handling is weighing the coal upon delivery and again just before it is burned. The first measurement ensures that only the energy received is paid for; the second provides input for performance calculations.

Moving Coal into Plants

Transferring coal from the unloading point or storage point to the bunkers usually involves one of several types of mechanical conveyance. These conveyances typically include bucket elevators, used widely for vertical lifts; continuous-flow conveyors; and belt conveyors. A continuous-flow conveyor is essentially a duct within which closely spaced skeleton flights act as the conveying element.[69]

Pulverized Coal Systems

Pulverized coal burns as a gas, and its fires are easily lighted and controlled. Coal pulverization is a multistep process. First, the raw coal is crushed to a uniform size by the coal crusher. Several types have been employed, most commonly the swing-hammer crusher, the roll crusher, and the Bradford breaker. The latter, the most satisfactory for large capacities, consists of a large-diameter, slowly revolving cylinder of perforated steel plates.

A coal feeder supplies the pulverizer with an uninterrupted flow of raw coal. The primary types are the belt feeder and the overshot roll feeder. The former uses an endless belt running on two separate rollers receiving coal at one end and discharging it at the other. An overshot roll feeder has a multi-blade rotor that turns about a fixed, hollow, cylindrical core.[70]

To achieve the particle size reduction needed for proper combustion, pulverizers or mills are used to grind the fuel. The four most common types are the ball-tube or ball (Figure 7.18), the ring-roll or ball-race, the impact or hammer mill, and the attrition type.

A ball-tube mill is basically a hollow horizontal cylinder, rotated on its axis. Coal is fed to the cylinder or the ball-tube mill through hollow trunnions and intermingles with the ball charge. Pulverization is accomplished through continual cascading of the mixture. Hot airflow is passed through the mill to dry the coal and remove the fines from the pulverizing zone.

Impact mills consist primarily of a series of hinged or fixed hammers revolving in an enclosed chamber lined with cast wear-resistant plates. Grinding results from a combination of hammer impact on the larger particles and attrition of the smaller particles on each other.

A high-speed mill that relies on considerable attrition grinding along with impact grinding is used for direct firing. In this type of mill, the grinding elements consist of pegs and lugs mounted on a disk rotating in a chamber.

Ring-roll and ball-race mills comprise the largest number of pulverizers in service. The grinding action takes place between two surfaces, one rolling over the other. The rolling element may be either a ball or a roll, while it may roll over either a race or a ring.[71]

Early coal-pulverizing installations received undried coal and relied on ambient air in the mill system. Because no heat was added to the system, maximum pulverizing capabilities were not realized. Present systems use mill drying. Three methods of supplying and firing pulverized coal have been developed: the storage or bin and feeder system, the direct-fired system, and the semi-direct system.

In a storage system, the coal is pulverized and conveyed by air or gas to a collector where the carrying medium is separated from the coal, which is then transferred to a storage bin. The device used to separate the coal, which usually operates on centrifugal force, is the cyclone separator.[72] From the storage bin, the pulverized coal is fed to the furnace.

In a direct-fired system, coal is pulverized and transported with air, or slightly diluted with gas, directly to the furnace where the fuel is consumed. In a semi-direct system, a cyclone collector between the pulverizer and furnace separates the conveying medium from the coal. The coal is fed directly from the cyclone to the furnace in a primary airstream.

Stokers

When pulverized coal was not used for fuel, automatic stokers were employed (Figure 7.19). These were inherently less efficient than pulverized coal firing because of the smaller amount of coal surface area exposed to the combustion air.[73] Stokers were initially used to burn coal in the 1700s. Stokers may be divided into three general groups depending on how the coal reaches the grate of the stoker for burning: underfeed, overfeed, and spreader.[74]

In underfeed stokers, the coal comes from below the air-admitting surface of the grate and is fed by a screw or a ram into a retort. In overfeed stokers the coal is fed from a hopper into the burning zone of the furnace by a moving grate on which the coal is burned. With spreader stokers, the coal is fed into the furnace by a mechanical or pneumatic device located above the grate surface. Spreader stokers are further divided into several types depending on the type of grate: stationary grate, dumping grate, reciprocating grate, vibrating grate, and traveling grate.[75]

Cyclone Furnaces

The crushed coal feed in pulverized coal plants is either stored temporarily in bins or transported directly to the cyclone furnace. The furnace is basically a large cylinder jacketed with water pipes that absorb some of the heat to make steam and protect the burner itself from melting down. A high-powered fan blows the heated air and chunks of coal into one end of the cylinder. At the same time, additional heated combustion air is injected along the curved surface of the cylinder, causing the coal and air mixture to swirl in a centrifugal "cyclone" motion.

The hot combustion gases leave the other end of the cylinder and enter the boiler to heat the water-filled pipes and produce steam. The coal used must have a relatively low sulfur content in order for most of the ash to melt for collection. In addition, high-powered fans are required to move the larger coal pieces and air forcefully through the furnace. This process produces more nitrogen oxide pollutants than does pulverized coal combustion.

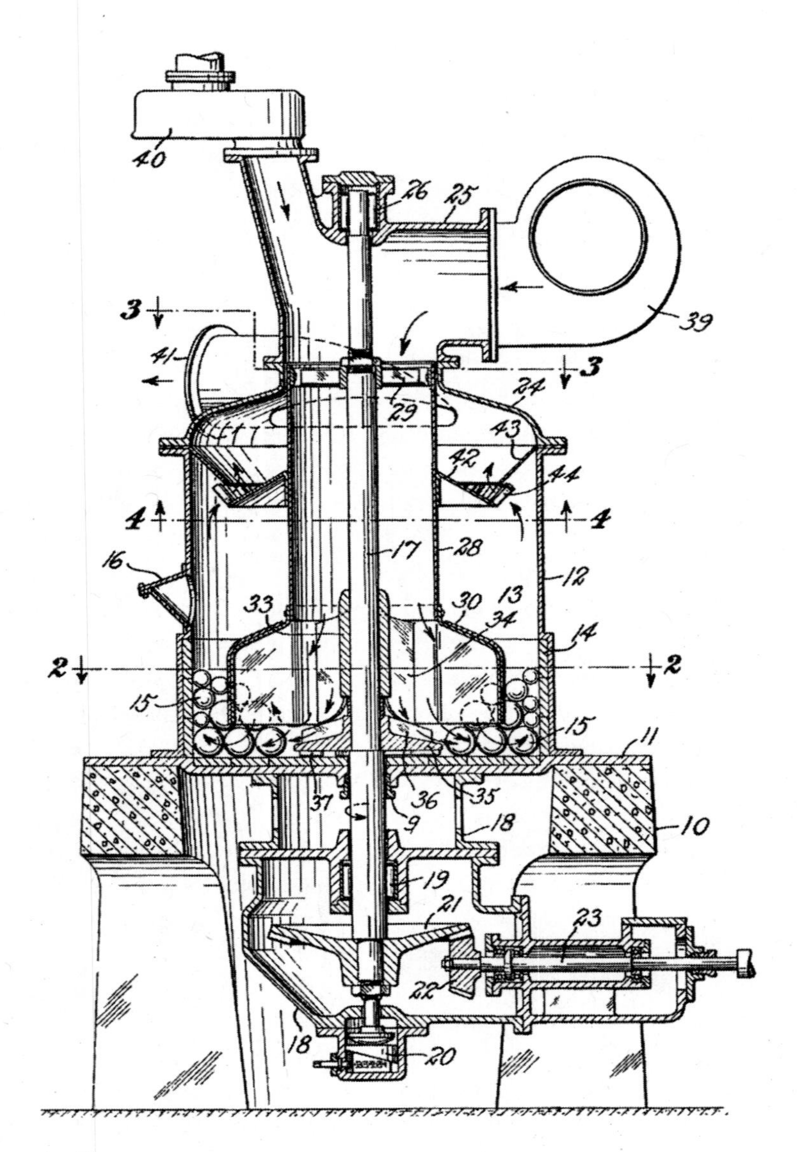

a

Figure 7.18. Ball mills. *a,* Vertical ball mill. M. Frisch. Ball Mill Pulverizer. October 20, 1931. Patent No. 2,041,287. *b,* Horizontal ball mill. P. L. Crowe. Ball Mill. September 30, 1925. Patent No. 1,627,488.

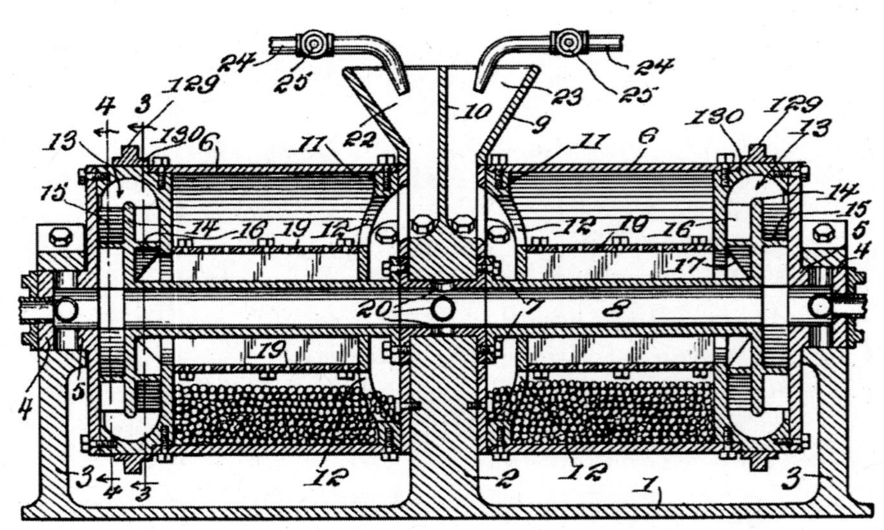

b

Figure 7.19. Example of coal stokers in a power plant. Commonwealth Edison Company: Crawford Electric Generating Station. Chicago, Illinois. HAER IL-114:12, National Park Service. Photograph by Jet Lowe, 1987.

Steam Generator

A steam generator is critical to the modern power plant. The steam generator converts the energy from fossil fuels to steam used to power the turbines to drive the electric generators. In basic terms, the steam generator has two separate but related loops: a combustion loop and the steam-water loop. Two common designs are used: the watertube boiler and the firetube boiler. The latter type is a water-filled cylinder with tubes running through it. Fuel is burned at one end of the cylinder, and the hot gas of combustion passes through the tubes to the other end. In passing through the tubes, the hot gases heat the water to produce steam. Once quite popular, the firetube boiler is rarely used in large utility applications.[76]

The watertube boiler (Figure 7.20) is a box whose walls are made up of tubes through which water flows. The fuel is burned in the box or furnace, as it is usually called. Heat is transferred to the water that flows through the tubes, thus generating steam. The flow of water not only generates steam but keeps the tubes cool enough that they do not fail.

Flow through the waterwall tube is provided by the steam-water loop. In most boilers, water circulates from a drum at the boiler top, through pipes called downcomers to the boiler bottom, and then up through the waterwall tubes to the drum again.[77]

On the furnace side of the boiler, two types of heat transfer predominate: radiation and convection. Almost all of the heat transferred to the waterwalls of the boiler is by radiation. After the fuel burns, the flue gas that remains is very hot. Heat is transferred from the flue gas by convection to other tubes in the boiler.

In some small or old boilers, flue gas is removed by natural convection, also called natural draft. The hot flue gas rises through the stack and draws in cool air for combustion. As boilers became larger, natural draft became inadequate, and a fan was added to blow sufficient air for combustion into the furnace. If the fan is big enough, it will pressurize the boiler furnace and aid in the removal of flue gas. Such a fan is called a forced-draft fan.

Another feature seen in combustion air and flue gas systems of most large modern utility boilers is the combustion air heater. Because the gas that exits from the steam generator can be as hot as 600 degrees and represents a major loss of heat and source of inefficiency, a heat exchanger is often included in the system so that the flue gas can heat the incoming combustion air.

Primary boiler subcomponents include burners, ignitors, powerplant fans, air heaters, superheaters and reheaters, and economizers.

Burners provide controlled, efficient conversion of the chemical energy of fuel to heat energy which is transferred to the heat-absorbing surfaces of the steam generator. To do this, burners introduce the fuel and air for combustion, mix these reactants, ignite the combustible mixture, and distribute the flame envelope and the products of combustion.[78]

The function of a fuel-burning system is to supply an uninterrupted flammable furnace input and to ignite that input continuously as fast as it is introduced. Ignition takes place when the flammable furnace input is heated above the ignition temperature. Ignitors use an electric spark or arc.[79]

All power-plant boilers use mechanical draft fans. They supply the primary air for pulverization and transport of coal to the furnace. They also supply the secondary and tertiary air to the windboxes for completion of combustion.[80]

Justified by the increased efficiency resulting from lower exit-gas temperatures and the higher flame temperature in the boiler furnace, air heaters make pulverized coal firing practical by providing the drying and transporting medium. Two types of air heaters are used: the tubular recuperative design and the regenerative design.[81]

The function of a superheater is to raise the boiler steam temperature above the saturated temperature level. As steam enters the superheater in an essentially dry condition, further absorption of heat increases the steam temperature. The reheater receives superheated steam that has partially expanded through the turbine. Its role is to superheat this steam to a desired temperature.[82]

Economizers help improve boiler efficiency by extracting heat from flue gases discharged from the final superheater section of a radiant-reheat unit. In the economizer, heat is transferred to the feedwater.[83]

Steam Turbines and Turbo-Generators

The idea of the steam turbine can be traced to early Egypt when Hero of Alexandria described what may be considered a crude forerunner of the steam turbine. A practical steam turbine was not developed until the end of the nineteenth century, when De Laval designed a high-speed turbine as the prime mover for a cream separator. Around the same time, the modern steam turbine was developed in designs by Sir Charles A. Parsons and C. G. Curtis. By the mid-twentieth century, steam turbines ranged up to 200,000 kilowatt capacity.[84]

In power plants, steam turbines convert the thermal energy developed by steam generators and boilers into mechanical energy or shaft torque. In the boiler room the heat energy in the coal is transformed to heat energy in steam. The steam is then piped into the turbine room where it is converted immediately into mechanical energy in the turbine and into electrical energy in the generator. The turbine and the generator are

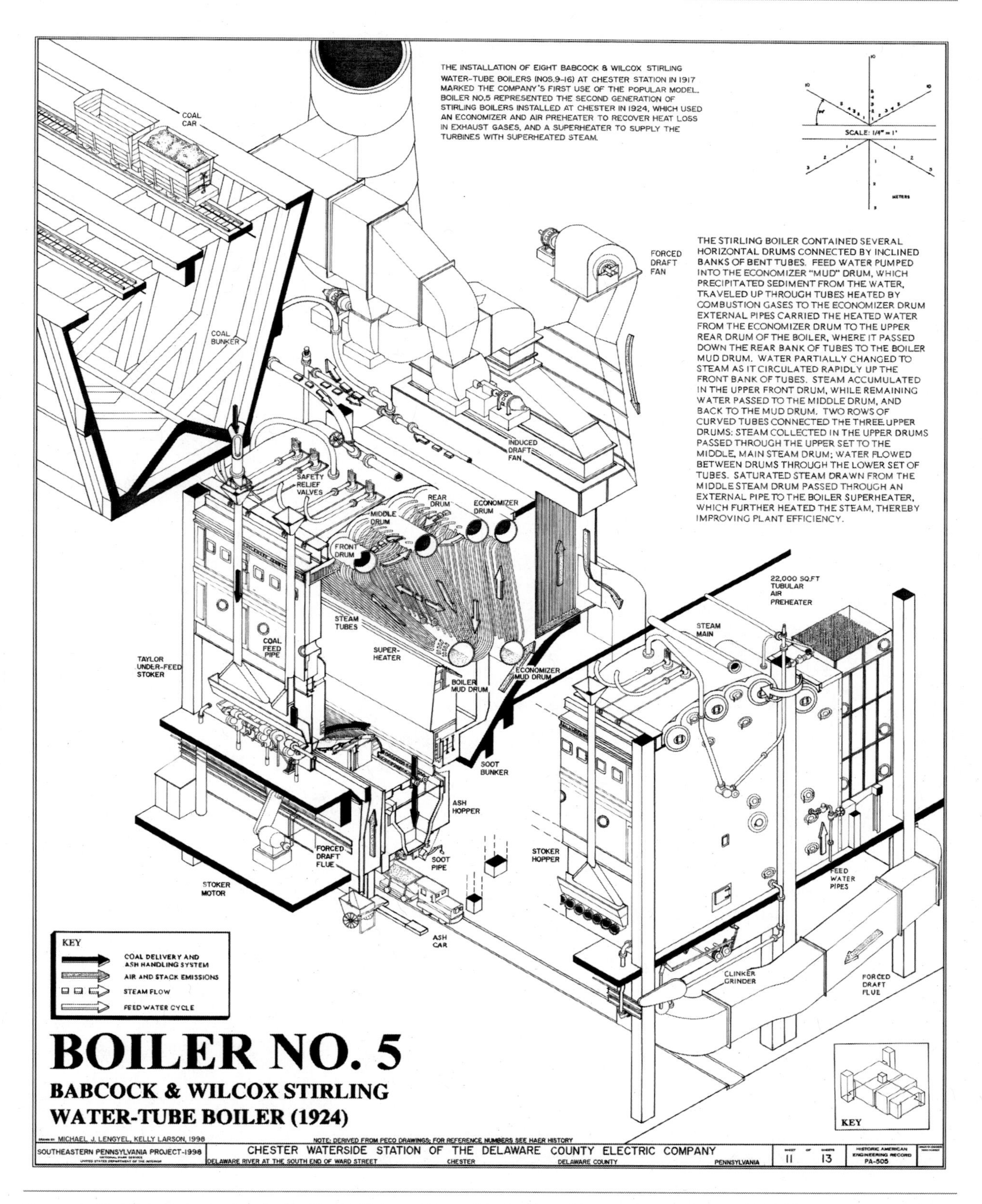

Figure 7.20. Watertube boiler. Boiler No. 5. Chester Waterside Station, Chester, Pennsylvania. HAER PA-505, National Park Service. Delineated by Michael J. Lengyel and Kelly Larson, 1998.

usually solidly coupled together and operate as one unit, called a turbo-generator.[85]

A steam turbine depends for its motive force almost entirely on the dynamic action of the steam. Some of the thermal energy of the steam is converted to kinetic energy by a drop in static pressure in a nozzle. The resultant high-velocity steam is directed by the nozzle into a passage that is integral with the moving rotor of the turbine. In passing through the rotor flow passage, the direction of the steam is changed. A change in momentum results, and a dynamic force is translated into a turning force on the shaft.

If the static pressure drop occurs principally in a stationary nozzle with little or no static pressure drop in the rotor flow passage, the turbine is termed an impulse turbine. If a substantial static-pressure drop occurs in the rotor flow passage, the turbine is called a reaction turbine. The turbine has two important elements: a stationary nozzle and a rotor flow passage. The rotor flow passage consists of blades or buckets.[86]

A modern turbo-generator is typically a compound unit, consisting of two or more units, each with its own generator or with one large generator for several units. This combination typically consists of one high-pressure unit and one or two low-pressure units. When high- and low-pressure units are placed in line with their rotor shafts coupled together, the entire machine is termed a tandem-compound steam turbine. In simple terms, the steam pressure presses against the turbine vanes, causing the shaft to turn. Because the generator shaft is connected with the turbine shaft, it too turns as the result of steam pressure, thus generating electricity. In the 1920s, generator voltage was typically 12,000 to 13,000 volts. In some facilities, the generators were directly connected to the primary windings of step-up transformers.[87]

Condensers

A condenser is a heat exchanger in which the heat remaining in the turbine exhaust steam is transferred to water flowing through tubes. The steam condenses as a result of this cooling. A vacuum is created, which permits more of the available energy to be extracted from the steam while it is still in the turbine. The warm condensing water is then pumped into cooling ponds or spray ponds or permitted to cool in a cooling tower.

ATOMIC POWER–GENERATING PLANTS

Nuclear power was first employed to generate electricity in 1956 at Calder Hall in Cumbria, England. A year later, initial operations began of a 60-megawatt unit at Shippingport, Pennsylvania. By the end of 1968, the combined capacity of U.S. nuclear generating plants in service, under construction, and on order amounted to 64,000 megawatts.[88] In simplest terms, a nuclear reactor uses nuclear fission to generate heat to boil water and power steam turbines.

Nuclear reactors are characterized as light-water or heavy-water units. Two types of light-water reactors are in widespread use: the boiling-water reactor (BWR) and the pressurized-water reactor (PWR). Both use similar fuel, consisting of long bundles of 2 to 4 percent uranium dioxide fuel pellets stacked in zirconium-alloy cladding tubes. The BWR design consists of a single loop in which entering water is turned directly into steam to generate electricity. The PWR is a two-loop system that uses high pressure to maintain an all-liquid-water primary loop. Energy is transferred to the secondary steam loop through two to four steam generators. The Shippingport reactor was a PWR.[89]

Major structures at a nuclear facility include the reactor containment building and the cooling tower. The latter is used to reduce the temperature of waste water used in the generation process. The typical nuclear plant cooling tower takes the hyperbolic form. Hyperbolic towers are also referred to as natural-draft towers. Airflow through these towers is produced by the density differential that exists by the less-dense heated air within the stack and the more-dense cooler ambient air outside the tower. These towers tend to be very large (some handling in excess of 250,000 gallons per minute) and tall (often in excess of 500 feet).[90]

GLOSSARY

Anchor block. A concrete block placed on a hillside to support a portion of a penstock.

Baseload. The minimum constant amount of load connected to the power system over a given time.

Brake jet. The water jet that provides the counterrotational force used to decelerate an impulse runner.

Bulb. A streamlined watertight housing for bulb turbine generators.

Bus. A conductor, or group of conductors, that serves as a common connection for two or more electrical circuits.

Cooling ponds. Shallow ponds into which warm condensing water is pumped and permitted to cool, chiefly by evaporation.

Cooling tower. An enclosure, originally constructed of wood, now generally of concrete, open at the top, in which warmed condensing water trickles down from top to bottom. Air is passed through the housing in the opposite direction.

Cyclone separator. In a coal-fired power plant, a device that separates fine coal dust from carrier air by centrifugal force.

Demand. The amount of electrical power needed or desired.

Design head. The head at which a turbine is designed to operate at maximum efficiency.

Discharge ring. The structural member on a Francis turbine that surrounds the runner band.

Draft tube. The diffuser that regains the residual velocity energy of the water leaving the turbine runner.

Excitation. The power input required by the field winding to maintain the necessary intensity of magnetic flux.

Fixed-blade propeller-type turbine. An axial-flow reaction turbine with blades keyed to the hub.

Forebay. The upstream part of the baylike extension of the river for the location of the powerhouse.

Francis turbine. A radial-inflow reaction turbine, where the flow through the runner is radial to the shaft.

Gate chamber. The part of the power conduit containing the gate.

Gate house. The structure located above the gate chamber that contains the hoist and other equipment for the gate.

Gate valve. A leaflike closing gate, sliding in a plane perpendicular to the penstock axis.

Gross head. The difference between headwater and tailwater elevation of a hydroelectric plant.

Guide vanes. The streamlined movable blades regulating inflow to the turbine runner.

Head. The difference in elevation between the headwater surface above and the tailwater surface below a hydroelectric power plant.

Headrace. That portion of the power canal that extends from the intake works to the powerhouse.

Headwater. The water upstream from the powerhouse.

Impeller. In reversible pump-turbines, the element that converts mechanical energy into hydraulic energy for the pump mode.

Impeller vanes. See *runner buckets.*

Intake. A hydraulic structure built at the upstream end of the diversion canal for controlling the discharge and for preventing silt and debris from entering the diversion.

Kaplan turbine. An axial-flow reaction turbine with adjustable runner blades and adjustable guide blades.

Large hydropower. Large hydropower plants are generally defined as those that have a capacity of more than 30 megawatts.

Load. The amount of electric power delivered or required at any specific point or points on a system.

Low-head hydroelectric plant. A hydroelectric plant with a water drop of less than 65 feet and a generating capacity of less than 15,000 kilowatts.

Main guide bearing. The bearing located closest to the runner.

Micro hydropower. An installation with a capacity of up to 100 kilowatts.

Nozzle. A curved steel pipe supplied with a discharge-regulating device to direct the jet onto the buckets in impulse runners.

Peak load. The greatest amount of power given out or taken by a power distribution system in a given time.

Pelton turbine. A Pelton turbine has one or more jets of water impinging on the buckets of a runner that looks like a waterwheel. Pelton turbines are used for high-head sites of 50 feet to 6,000 feet.

Penstock. A pressurized pipeline conveying the water in a high-head development from the headpond or the surge tank to the powerhouse.

Pondage. The rate of storage in run-of-the-river hydroelectric developments.

Powerhouse. The main structure of a hydroelectric or electric plant housing the generating units and related facilities.

Propeller turbine. The collective term for axial-flow reaction turbines. A propeller has a runner with three to six fixed blades, like a boat propeller. The water passes through the runner and drives the blades. Propeller turbines can operate from 10 feet to 300 feet of head.

Radial-inflow turbine. A term for turbines in which the water enters radially into the runner and leaves it axially. An example is the Francis turbine.

Rated capacity. The capacity that a hydro generator can deliver without exceeding mechanical safety factors or a nominal temperature rise.

Reaction turbine. A term for turbines in which the water jet enters the runner under a pressure exceeding that of the surrounding atmosphere.

Reservoir. An artificial lake into which water flows and is stored for future use.

Reversible pump-turbine. A machine used in pumped-storage developments that operates both as a pump and a turbine.

Runner. The rotating element of the turbine that converts hydraulic energy into mechanical energy.

Runner band. The lower axisymmetric portion of the runner to which the lower or outer ends of the runner buckets attach.

Runner blades. The blades that radiate from the hub, deflect the flowing water, and transfer energy to the runner hub.

Runner buckets. Components of Francis and impulse runners that deflect the flowing water and transfer the energy to the runner crown or disk.

Runner crown. The upper axisymmetric portion (inner shroud) of the runner which provides a mechanical attachment to the main shaft and to which the top ends of the runner buckets attach.

Runner hub. The axisymmetric portion of a propeller runner which provides the attachment to the main shaft and to which the inner ends of the runner blades attach.

Scroll case (spiral case). A spiral-shaped steel intake guiding the flow into the wicket gates of the reaction turbine.

Semi-scroll case. A concrete intake directing flow to the upstream end of the turbine with a spiral case surrounding the downstream portion of the turbine to provide uniform water distribution.

Small hydropower. Hydropower facilities that have a capacity of from 0.1 to 30 megawatts.

Surge tank. A hydraulic structure erected in the power conduit of high-head developments between the pressure tunnel and the penstock to protect the pressure tunnel from water-hammer effects.

Tailrace. The portion of the power canal that extends from the powerhouse to the recipient watercourse.

Tailwater. The water downstream from the powerhouse.

Transformer. A device for transferring electrical energy from one AC circuit to another AC circuit, usually of a different voltage.

Wicket gates. Structures that control the flow of water to the turbine or discharge from the pump.

BIBLIOGRAPHY

American School. *Power Stations and Power Transmission: A Manual of Approved American Practice.* Chicago: American School of Correspondence, 1908.

Anonymous. "East River Station." *Power Plant Engineering* (March 15, 1927): 338–346.

—. "Prospect No. 2 Hydro Plant." *Power Plant Engineering* (April 1, 1928): 388–395.

Carr, T. H. *Electric Power Stations.* 2 volumes. New York: D. Van Nostrand Company, Inc., 1941.

Creager, William P., and Joel D. Justin. *Hydroelectric Handbook.* 2nd edition. New York: John Wiley & Sons, Inc., 1950.

Elliott, Thomas C. *Standard Handbook of Powerplant Engineering.* New York: McGraw-Hill Publishing Company, 1989.

Gulliver, John S., and Robert E. A. Arndt. *Hydropower Engineering Handbook.* New York: McGraw-Hill, Inc., 1991.

Iddles, Alfred. "Amarillo Station Uses Natural Gas as Fuel." *Power Plant Engineering* (November 15, 1927): 1176–1182.

John, Carl F. *Steam-Electric Power Stations.* Scranton: International Textbook Company, 1936.

Lee, John Francis. *Theory and Design of Steam and Gas Turbines.* New York: McGraw-Hill, 1954.

Lyndon, Lamar. *Hydro-Electric Power.* 2 volumes. New York: McGraw-Hill Book Company, Inc., 1916.

Morse, Frederick T. *Power Plant Engineering and Design.* New York: D. Van Nostrand Company, Inc., 1932.

Muller, Richard. *Hydroelectrical Engineering.* New York: G. E. Steichert & Company, 1921.

Pelton Water Wheel Company. *Pelton Impulse and Reaction Turbine Installations.* San Francisco: Pelton Water Wheel Company, 1920.

Stewart, Robert C. "Electricity on the High Iron: Cos Cob Powers the New Haven Railroad." *IA: The Journal of the Society for Industrial Archeology* 23:1 (1997): 25–42.

Struck, Henry W. "Neches Station Increases Texas Power Facilities." *Power Plant Engineering* (April 1, 1927): 392–395.

U.S. Bureau of Reclamation. *Hydroelectric Power.* Denver: Bureau of Reclamation, 2005.

Warnick, C. C. *Hydropower Engineering*. Englewood Cliffs, New Jersey: Prentice-Hall, Inc., 1984.

Weingreen, Joshua. *Electrical Power Plant Engineering*. New York: McGraw-Hill Book Company, 1913.

Wellman-Seaver-Morgan Company. *The Wellman Revolving Car Dumper*. Bulletin No. 63. Cleveland: Wellman-Seaver-Morgan Company, 1921.

Westinghouse Electric Company. *Westinghouse Turbo Generator Sets*. Pittsburgh: Westinghouse Electric Company, 1907.

NOTES

1 United States Energy Information Administration, April 2007 *Monthly Energy Review.*
2 United States Bureau of Reclamation, *Hydroelectric Power* (Denver: Bureau of Reclamation, 2005).
3 United States Bureau of Reclamation.
4 William P. Creager and Joel D. Justin, *Hydroelectric Handbook,* 2nd edition (New York: John Wiley & Sons, Inc., 1950), 191.
5 John S. Gulliver and Roger E. A. Arndt, *Hydropower Engineering Handbook* (New York: McGraw-Hill, 1991), 1.10.
6 Creager and Justin, 191.
7 Gulliver and Arndt, 1.10; Creager and Justin, 191, 195.
8 Creager and Justin, 195, 200.
9 Creager and Justin, 200–202.
10 Gulliver and Arndt, 1.14.
11 *World Almanac and Book of Facts 2006* (New York: World Almanac Books, 2005), 744.
12 Lamar Lyndon, *Hydro-Electric Power*, volume I (New York: McGraw-Hill Book Company, Inc., 1916), 228.
13 Lyndon, 228–230.
14 Creager and Justin, 379.
15 Creager and Justin, 394.
16 Creager and Justin, 405–410.
17 Creager and Justin, 446.
18 Lyndon, 283.
19 Frederick T. Morse, *Power Plant Engineering and Design* (New York: D. Van Nostrand Company, Inc., 1932), 182.
20 Lyndon, 283–284.
21 Lyndon, 295.
22 Lyndon, 297–298.
23 Creager and Justin, 514.
24 Lyndon, 292–293.
25 Lyndon, 299–300.
26 Lyndon, 301–302.
27 Lyndon, 304.
28 Gulliver and Arndt.
29 Creager and Justin, 494–502.
30 Creager and Justin, 533.
31 Lyndon, 307.
32 Creager and Justin, 547.
33 Creager and Justin, 563–565.
34 Creager and Justin, 565.
35 Creager and Justin, 601.
36 Creager and Justin, 593–594.
37 C. C. Warnick, *Hydropower Engineering* (Englewood Cliffs, NJ: Prentice-Hall, Inc., 1984), 122.
38 Morse, 188.
39 Creager and Justin, 606–607.
40 Creager and Justin, 729–730.
41 Creager and Justin, 755.
42 Morse, 207.
43 Morse, 207–208.
44 Creager and Justin, 813–814.
45 Morse, 206.
46 Morse, 206–207.
47 Creager and Justin, 770.
48 Creager and Justin, 734–735.
49 Creager and Justin, 943.
50 Creager and Justin, 967–968, 970.
51 Creager and Justin, 971–972.
52 Creager and Justin, 973.
53 Creager and Justin, 988, 996.
54 Morse, 719.
55 Creager and Justin, 1031.
56 Carl F. John, *Steam-Electric Power Stations* (Scranton, PA: International Textbook Company, 1936), 1–3.
57 Alfred Iddles, "Amarillo Station Uses Natural Gas as Fuel," *Power Plant Engineering* (November 15, 1927): 1176.
58 Henry W. Struck, "Neches Station Increases Texas Power Facilities," *Power Plant Engineering* (April 1, 1927): 390.
59 www.naturalgas.org.
60 Thomas C. Elliott, *Standard Handbook of Powerplant Engineering* (New York: McGraw-Hill Publishing Company, 1989), 2.111–2.112.
61 www.naturalgas.org.
62 Anonymous, "East River Station," *Power Plant Engineering* (March 15, 1927): 340.
63 Matthew Sneddon, Chester Waterside Station of the Delaware Electric Company, HAER No. PA-505, 46; "East River Station," 340.
64 John, 1-31.
65 John, 1-32.
66 Elliott, 3.33–3.36.
67 Elliott, 3.27.
68 Elliott, 3.28–3.29.
69 Elliott, 3.37–3.39.
70 Elliott, 1.170–1.171.
71 Elliott, 1.177–1.180.
72 John, 1-40.
73 John, 1-59.
74 Elliott, 1.183.
75 Elliott, 1.183–1.184.
76 Elliott, 1.59.
77 Elliott, 1.60–1.61.
78 Elliott, 1.88.
79 Elliott, 1.96.
80 Elliott, 1.99–1.100.
81 Elliott, 1.115.

82 Elliott, 1.119.
83 Elliott, 1.119–1.121.
84 John Francis Lee, *Theory and Design of Steam and Gas Turbines* (New York: McGraw-Hill, 1954), 1.
85 John, 2-22.
86 Lee, 2–3.
87 Anonymous, "Modern Power Plant Trends," *Power Plant Engineering* (January 1, 1927): 7.
88 Elliott, 141.
89 Information about Shippingport is contained in its HAER documentation,
90 Elliott, 1–213.

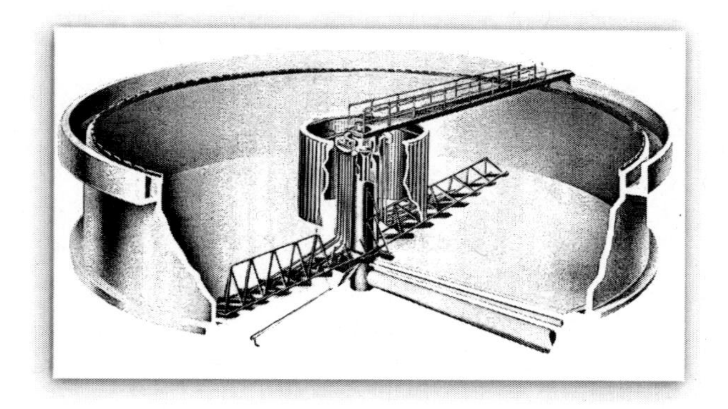

CHAPTER 8

Water Supply and Treatment

WATER SUPPLY

Municipal water systems consist of a few primary components. Water is supplied from a well, a river or lake, or a reservoir. The water is treated in a water treatment plant to remove sediment by filtration and/or settling and bacteria, typically with ozone, ultraviolet light, and chlorine. The output from the water treatment plant is pressurized by a high-lift pump and distributed by the system's primary feeder pipes. The water tower or stand pipe connects with the primary feeder pipes.

Water supply systems may be classified as grid or branching systems or a combination of the two. Engineers generally prefer the former because it can supply water to any point from at least two directions. It also permits any broken pipe sections to be isolated for repair without disrupting service to large areas of a community. A branching system has numerous terminals or dead ends that prevent water from being circulated throughout the system. A combination system is most commonly used, incorporating loop feeders distributing the flow to an area from several directions.

In hilly or mountainous terrain, the distribution system is usually divided into two or more areas or zones. These areas are usually interconnected, and workers may close interconnections using valves during normal operation hours. This prevents the system from having to maintain extremely high pressure in low-lying areas to ensure reasonable pressure at higher elevations.

Among the water sources for municipal supplies are deep wells, shallow wells, rivers, natural lakes, and impounding reservoirs.[1]

Wells

The collection of groundwater is accomplished primarily through the construction of wells. In areas where the primary source of water is an underground aquifer, water is obtained by pumping from wells. In areas where the groundwater table is high and the underground strata are pervious, shallow wells may be drilled at nearly any convenient site.[2]

The well contains an open section through which flow enters and a casing to allow the water to be transported to the ground surface (Figure 8.1). The open section is a perforated casing or a slotted metal screen that permits the flow to enter while at the same time preventing a collapse of the hole. Sometimes gravel is placed at the bottom of the well casing around the screen. The screening is sized to prevent the entrance of fine sand. In addition to a perforated casing, the well also includes a solid casing. Casings are generally constructed of alloyed or unalloyed steel, cast iron, and ingot iron.[3]

When a well is pumped, water is removed from the aquifer immediately adjacent to the screen. Flow then becomes established at locations some distance from the well in order to replenish this withdrawal.

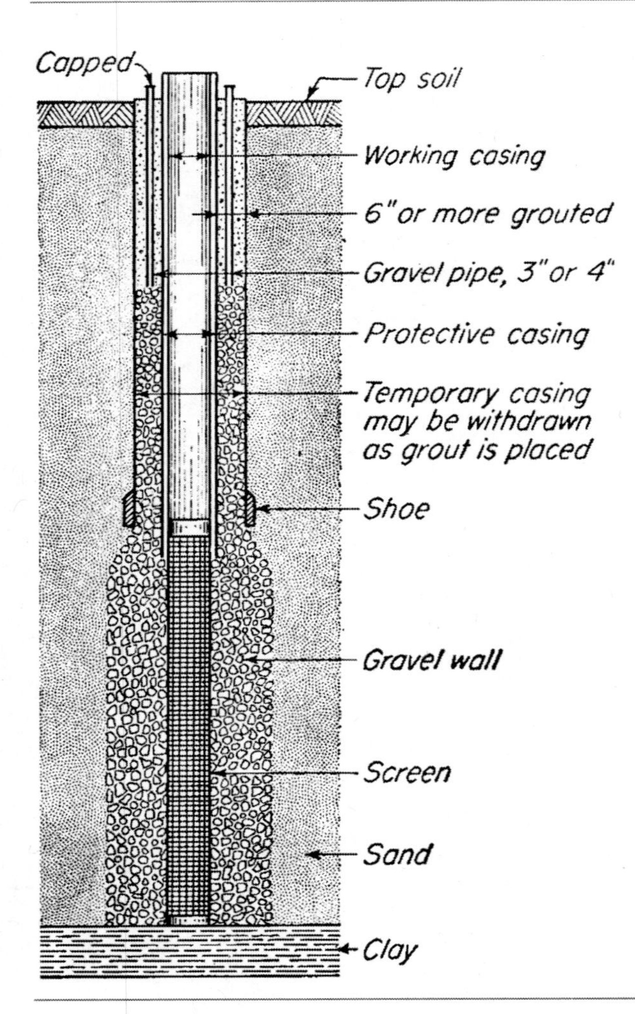

Figure 8.1. Gravity-wall well. Reprinted from Harold E. Babbitt and James J. Doland, *Water Supply Engineering*, 5th edition (New York: McGraw-Hill Book Company, Inc., 1955), 90. Reproduced with permission of The McGraw-Hill Companies.

Impounding Reservoirs

By definition, a reservoir is an artificial impoundment formed by damming of the valley of a stream or river. Many of the largest bodies of water in states are reservoirs, such as Massachusetts's Quabbin Reservoir, Connecticut's Candlewood Lake, and New Jersey's Wanaque Reservoir. Control dams are used to raise the water level of a stream enough to provide gravity transportation of water to pumps.

Dams

Water is impounded behind a dam. Dams may be classified according to either the materials of construction or the basis by which they resist external stresses. Materials of construction are earth, including rolled fill, hydraulic fill, and rock fill; rock-filled timber crib; masonry, including stone, plain concrete, and reinforced concrete; timber; and steel. Types of dams classified by stress resistance are solid gravity, including straight, curved and arched dams; hollow gravity, including slab and buttress, multiple arch, and multiple dome; and arch, including constant angle and constant radius. Further discussion of dam types is found in Chapter 7.

Spillways

Spillways are structures that accommodate the discharge of water from a reservoir and may be divided into three types: uncontrolled, automatically controlled, and manually controlled. Uncontrolled spillways can be divided into overflow, chute, and side-channel.

Manually controlled and automatically operated spillways make it possible to hold the surface of the water in the reservoir at any predetermined elevation for all designed rates of flow over the spillway. Thus, the water surface in the reservoir can be maintained at the same elevation at low rates of stream flow and in floods. Types of automatically controlled spillways include siphon (Figure 8.2a), shaft (Figure 8.2b), flashboards, and float-controlled gates.[4]

Aqueducts

An aqueduct is a conduit that conveys water to a point, usually a distribution reservoir, where distribution begins. An aqueduct may be a canal, a flume, pipe lines, siphons, tunnels or other channel, opened or covered.[5] Perhaps the most famous aqueduct in an American water supply system is the Old Croton Aqueduct, used to carry water from Westchester County's Croton Reservoir to receiving and distributing reservoirs in Manhattan. This water conveyance network was constructed between 1837 and 1842.

Pumps and Pumping Stations

To gain the greatest hydraulic advantage, pumping stations are usually located near the middle of a water distribution system.[6] The pumps used in a water supply system are of several basic types. Low-lift pumps raise surface water and move it to nearby treatment plants. These pumps move large volumes of water at relatively

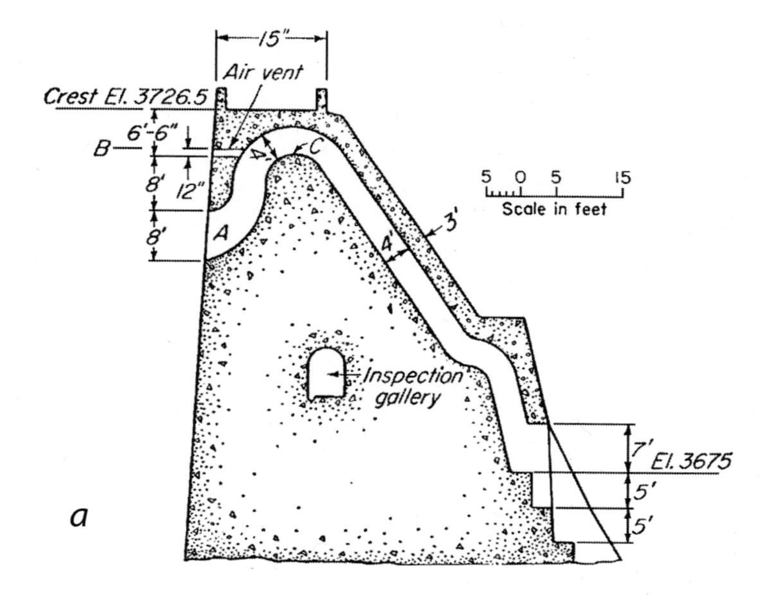

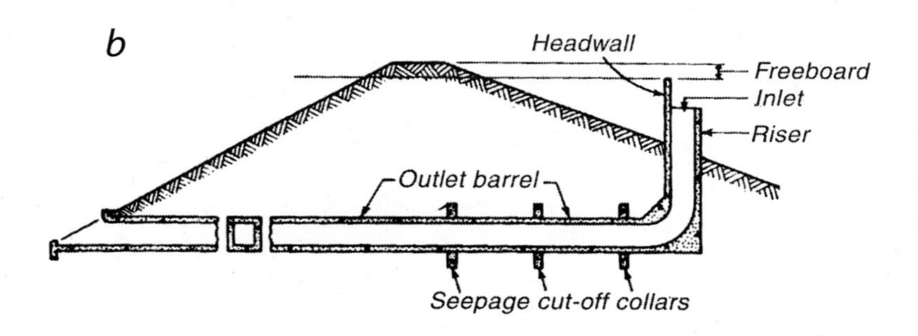

Figure 8.2. *a,* Siphon spillway section (Babbitt and Doland 1955: 133); *b,* section through a shaft spillway (Babbitt and Doland 1955: 134). Reproduced with permission of The McGraw-Hill Companies.

low discharge pressure. High-lift pumps discharge treated water into arterial mains and operate under higher pressure. Pumps that increase the pressure within the distribution system or raise water into elevated storage tanks are called booster pumps. Well pumps lift water from underground and discharge it directly into a distribution system.

The most common pump type in a distribution system is a centrifugal pump (Figure 8.3). These pumps use a rapidly rotating impeller to add energy to the water and raise the pressure inside the pump casing.[7] A second type, the positive-displacement pump, delivers a fixed quantity of water with each piston cycle or rotor. Water is pushed or displaced from the pump casing. Displacement pumps are often of the reciprocating type, where a piston draws water into a closed chamber and then expels it under pressure.[8]

Electrical power is the primary source of energy used for driving pumping equipment, but gasoline, steam, and diesel power are also used.

Water Treatment Plants

The purpose of municipal water supply systems is to provide potable water that is chemically and bacteriologically safe for human consumption and of adequate quality for industrial users. Boiler feed water, water used in food processing, and process water used in the manufacture of textiles and paper have special quality tolerances that may require additional treatment of the municipal water at the industrial site.[9]

Pretreatment processes in municipal water treatment are screening, presedimentation or desilting,

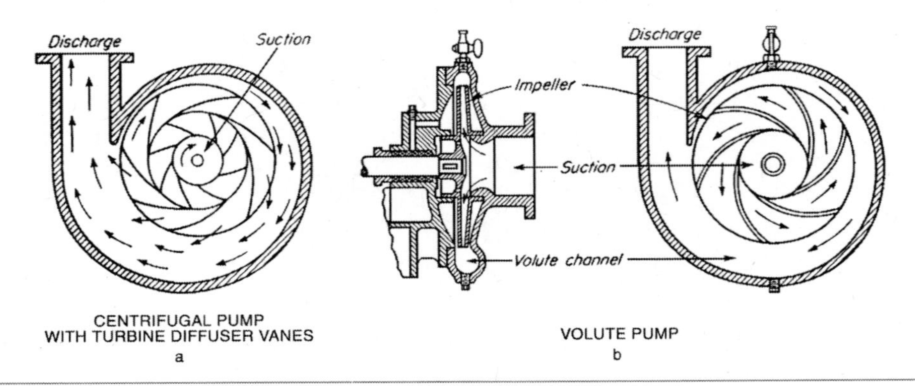

Figure 8.3. Centrifugal pumps (Babbitt and Doland 1955: 206). Reproduced with permission of The McGraw-Hill Companies.

chemical treatment, and aeration. Screening is practiced in pretreating surface waters. Presedimentation is used to remove suspended matter from river water. Chemical treatment prior to in-plant coagulation is most frequently used to improve presedimentation, to pretreat hard-to-remove substances, such as taste and odor compounds and color, and to reduce high bacterial concentrations. Aeration is customarily the first step in treatment for removing iron and manganese from well waters and is a standard way of separating dissolved gases such as hydrogen sulfide and carbon dioxide.[10]

Screening

River water frequently contains large pieces of suspended floating debris varying in size from small rags to logs. Screening is the first step in treating water containing large solids. Coarse screens of vertical steel bars having openings of 1 to 2 inches, or heavy wires forming a mesh with openings about 1 inch square, are employed to exclude large materials (Figure 8.4). These coarse screens are often used with mechanical rakes to clear accumulated material from these screens.

One mechanical device employed is a traveling screen. Trays, sections constructed of wire mesh or slotted metal plates, generally have 3/8-inch openings. As water passes through the upstream side of the screen, the solids are retained and elevated by the trays. As the trays rise into the head enclosure, the solids are removed by water sprays.[11]

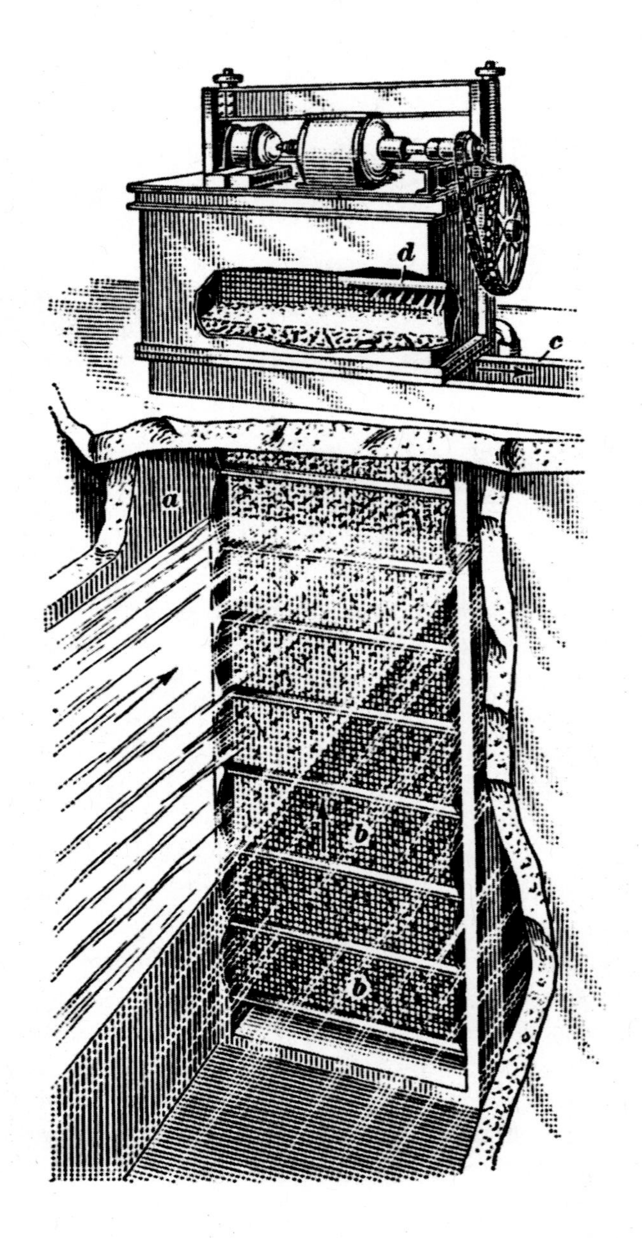

Figure 8.4. Chain-belt fine screen. Reprinted from W. A. Hardenbergh, *Water Supply and Purification*, 3rd edition (Scranton: International Textbook Company, 1952), 329. Reprinted by permission of Thomson Education Direct.

Presedimentation

Surface water of high turbity may require sedimentation prior to chemical treatment (Figure 8.5). Presedimentation basins can have hopper bottoms or be equipped with a continuous sludge removal apparatus. Chemical feeding equipment for prechlorination or partial coagulation is often provided prior to presedimentation for periods when the raw water is too turbid to clarify adequately by plain sedimentation.[12]

Chemical Treatment

Well water supplies are commonly treated to remove dissolved minerals such as iron and manganese. Surface water normally requires chemical coagulation to eliminate turbity, color, and taste- and odor-producing compounds.[13]

Disinfection

The purpose of disinfecting water supplies is to kill pathogenic organisms and prevent the spread of waterborne diseases. This may occur in several ways: physical elimination through coagulation, sedimentation, and filtration; natural die-away of organisms in an unfavorable environment during storage; and inactivation by chemicals introduced for treatment purposes other than disinfection.[14]

Chlorine and chlorine derivatives are conventional disinfectants. Hypochlorous acid is the primary disinfectant. Two primary means are used to dispense chlorine into water for treatment. Direct-feed equipment involves the metering of dry chlorine gas and conducting it under pressure to the water being treated. The solution-feed apparatus meters chlorine gas under vacuum and dissolves it in a small amount of water to form

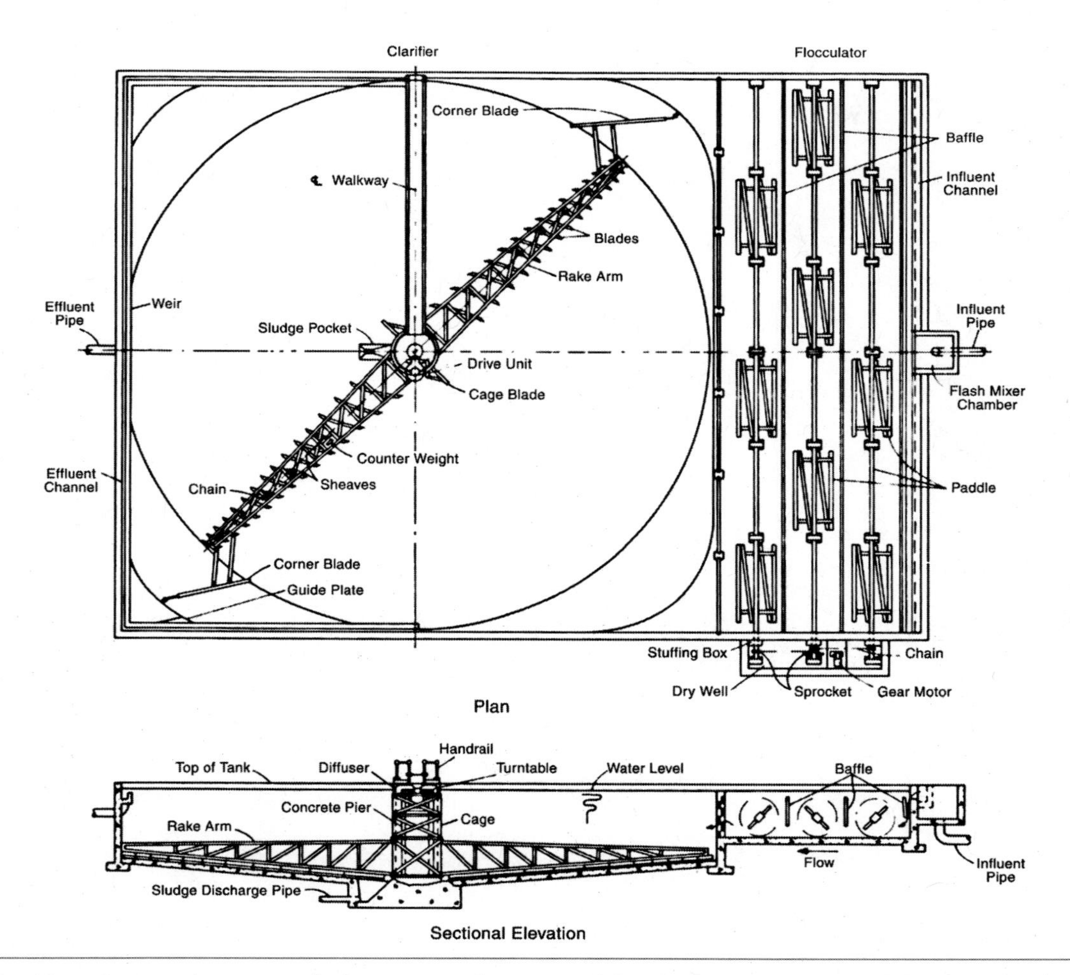

Figure 8.5. Flocculator and square sedimentation tank. Reprinted from John W. Clark, Warren Viessman, Jr., and Mark J. Hammer, *Water Supply and Pollution Control* (Scranton, PA: International Textbook Company, 1971), 343. Reprinted by permission of Thomson Education Direct.

a concentrated solution which is then applied to the water being treated.

Chlorine equipment installations require adequate room for operations and handling chlorine containers. In small plants feeding no more than 200 pounds of chlorine per day, an 8-foot square is adequate. Ventilation is essential for all chlorine equipment rooms.[15]

Iron and Manganese Removal Processes

The simplest form of oxidation uses plain aeration. The aerators most commonly used are of the tray type, where a vertical riser pipe distributes the water on top of a series of trays from which it then drips and spatters down through a stack of three or four of them. In plants using the aeration-sedimentation-filtration process, most of the oxidized iron and manganese is removed by a sand filter.[16]

Chemical Coagulation

Chemical coagulation removes turbidity-producing substances and color in surface water which results from decaying vegetation or industrial wastes. In destabilizing colloids, two basic mechanisms have helped form sufficiently large aggregates to facilitate settling from suspension. Coagulation reduces the net electrical repulsive forces at particle surfaces by electrolytes in solution. The second mechanism, known as flocculation, is aggregation by chemical bridging between particles. The most widely used coagulants for water treatment are aluminum and iron salts.[17]

Taste and Odor Control

The most common control measures in use for taste and odor control are the application of activated carbon, free residual chlorination, combined residual chlorination, chlorine dioxide, ozone, and aeration. Carbon is fed to water either as dry powder or as wet slurry by special equipment.[18]

Aeration

Aeration consists of bringing oxygen into direct contact with nearly all the water being treated. The oxygen tends to oxidize certain undesirable compounds in the water and make the water more palatable. Water can be aerated by exposing it to the atmosphere; flowing over cascades, weirs, steps or troughs; flowing through trickling devices; spraying it through the air; diffusing air through it; aspirating air through the water; and mixing air and water under pressure. The first four methods are most commonly used.[19]

THE FILTRATION PLANT

The first stop in the filtration plant (Figure 8.6) is the mixing tank, where a coagulating chemical is added. The coagulant is stirred into the water generally for about thirty minutes. From the overflow at the top of the mixing basin, the water flows to the settling basin. In that basin insoluble compounds or precipitates resulting from the coagulation treatment settle to the bottom. From the top of the settling basin the partially clarified water flows to the filters, where it passes through sand or other suitable filtration material into a storage basin or reservoir.

Microstraining

Microstraining is a form of filtration whose primary objective is the removal of microorganisms and other suspended solids. A filtering medium consisting of a finely woven stainless steel fabric is normally used. Most microstrainers are the rotating drum type, where the fabric is mounted on the periphery and raw water passes from inside to outside the drum. Microstrainers have been successfully employed in primary clarification of water preceding filtration.[20]

Pressure Filters

Pressure filters employ sand, anthracite, or calcite and underdrains contained in a steel tank. Water is pumped through the filter under pressure, and the media are washed by reversing flow through the bed, flushing out impurities. Pressure filters are not generally used in large treatment works but have been successfully used in small municipal softening and iron removal treatment plants.[21]

Slow Sand Filtration

The use of slow sand filters for water purification began in England in about 1830. The technology was introduced into the United States in the late nineteenth cen-

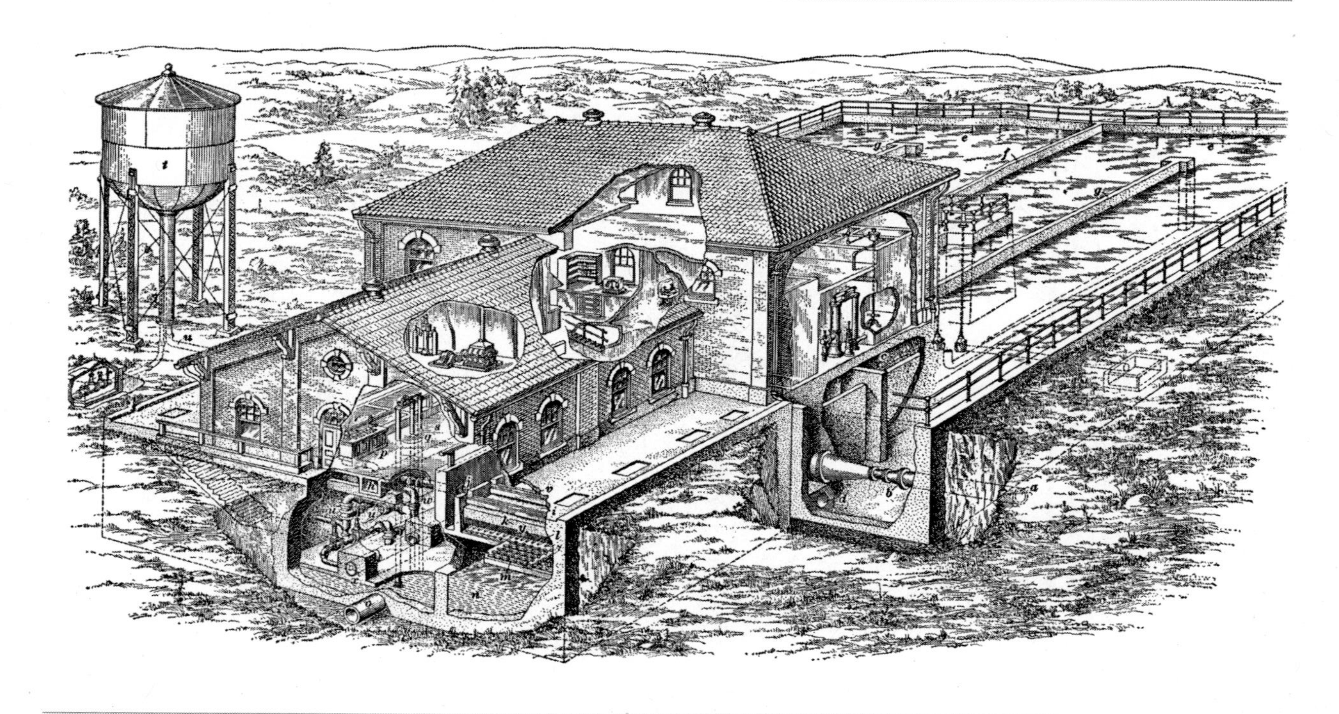

Figure 8.6. Typical water filtration plant with mechanical filters (Hardenbergh 1952: after page 374). Reprinted, by permission of Thomson Education Direct.

tury. The filters are beds of sand 30 to 40 inches deep, each about an acre in extent. The slow sand filter has been adapted to the treatment of water with a turbidity not exceeding 30 ppm and of low color. For high turbidities and high color, coagulation is generally a necessary adjunct. Filtering action is a combination of straining, adsorption, and biological flocculation.[22]

Among the disadvantages of slow sand filters are the large area required for construction; their limitations to clear water with low color; and the relatively high initial cost. Use of slow sand filters has declined because of high construction costs, the large filter area needed, and unsuitability for treating highly turbid and polluted waters requiring chemical coagulation.[23]

Rapid Sand Filtration

To obtain a high rate of filtration through sand filters —approximately 100 to 200 million gallons per acre per day—it is usually necessary to treat the water prior to filtration. Preparatory treatment includes aeration to free the water from dissolved gases and to oxidize iron and organic matter, if present; coagulation; and settling.

The most common type of device for treating municipal water supplies is the rapid sand filter which removes nonsettleable floc and impurities remaining after chemical coagulation and sedimentation of the raw water (Figure 8.7). Water oozes downward through the filter media by a combination of positive head and suction from the bottom.

Filters are generally placed on both sides of a pipe gallery, which contains inlet and outlet piping, washwater inlet lines, and backwash drains. The pipe gallery is decked by an operating floor where control consoles are placed near the filters. A typical filter consists of a 15- to 24-inch-thick layer of gravel, and a 24- to 30-inch-thick layer of sand.[24]

Water Softening

Although not strictly necessary for water purification, some water systems include softening as a part of water treatment. By the mid-1950s, two primary methods of water softening were used: the lime-soda process and base exchange. The lime-soda process includes a thorough mixing of the chemical with the water, followed by agitation for up to an hour to allow the completion of the chemical reaction.[25]

Ion-exchange materials are substances so loosely bound chemically that when placed in a solution of

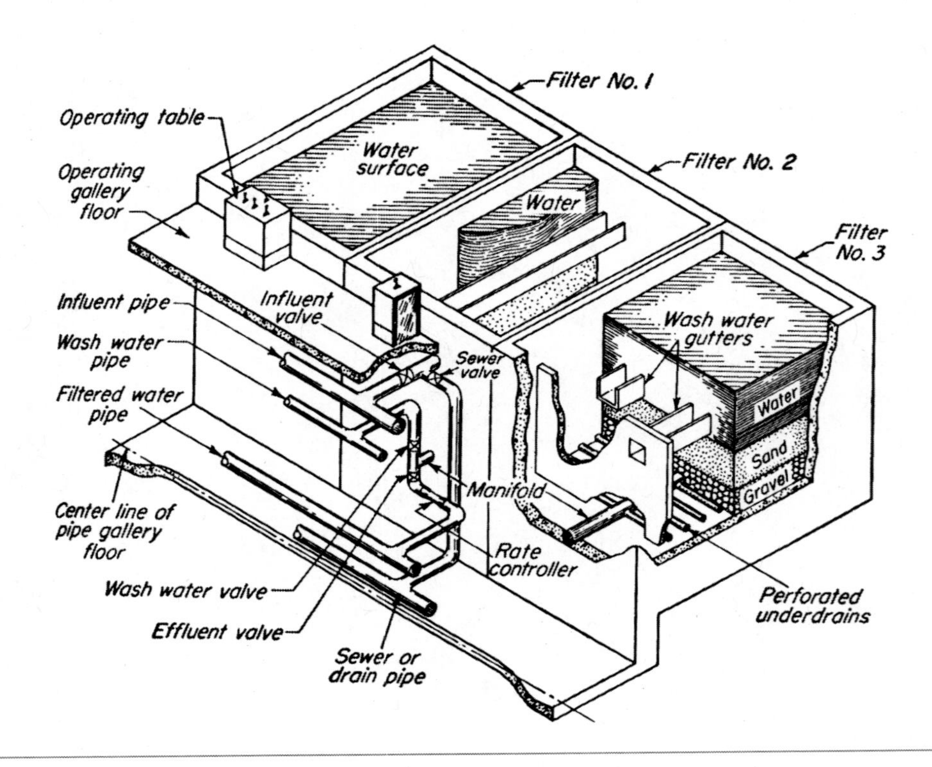

Figure 8.7. Schematic cutaway drawing of a rapid sand filter (Babbitt and Doland 1955: 449). Reproduced with permission of The McGraw-Hill Companies.

greater ionic concentration, cations and anions will be exchanged. Water may be softened by passing it through a layer of sodium base-exchange materials at a temperature of 100 degrees or less. A base-exchange water softener resembles a rapid sand filter of either the gravity or the pressure type.[26]

While water softening generally is beneficial, it increases the tendency of water to deposit calcium carbonate on pipes and tubes. The carbonate balance may be restored by recarbonation of water. This process is accomplished by diffusing carbon dioxide gas through the water, by underwater combustion, by the use of dry ice, or by the use of liquid carbon dioxide.[27]

WATER STORAGE

A critical part of a water distribution system is the storage facility or facilities. Storage tanks are used as a means of providing head for a water distribution system. During periods of low consumption, water is pumped to elevated tanks. During periods of high consumption, the stored water is drawn upon to augment pumped water. Storage tanks also provide excess capacity for fire protection.[28]

Distribution reservoirs are usually strategically located near the center of use. In a large metropolitan area several distribution reservoirs may be located at key points. Reservoirs providing service storage must be high enough to develop adequate pressures in the system they are to serve.[29]

Water Towers

Water towers first became a feature of urban industrial architecture in the nineteenth century. Occupying prominent spots in the landscape, some became civic monuments known for stylistic elaboration. One of the best known is the Chicago Water Tower of 1869 designed by architect William W. Boyington. This Gothic Revival structure was built to conceal the tower's utilitarian function.

Within the decorative outer walls stood a 154-foot tall, 3-foot diameter pipe. The pipe was required to overcome the pulsation of the pumping system. It provided for the expansion of water up into the vertical pipe from the horizontal mains between the strokes of the pump, thus maintaining a constant water pressure. When an improved system of pumping water was developed, the tower and its "standing pipe" were no longer needed. Other nineteenth-century landmark water towers are Milwaukee's North Point Water Tower (1874) and the High Bridge Water Tower in New York City (1872).[30]

Standpipes and Elevated Tanks

As municipal water systems were being developed in the 1880s and early 1890s, some form of elevated storage reservoir was needed. Otherwise, continuous pumping was required except in the case of compressed air systems for very small operations. On flat land, standpipes and elevated tanks were used. Since standpipes were rather inefficient at that time, towers and tanks became the accepted means of water storage.

Commonly referred to as water towers, modern water storage structures are more properly known as standpipes and elevated tanks. Standpipes or elevated tanks are generally needed where construction of a surface reservoir will not supply needed pressure.[31]

In basic terms, a standpipe or elevated tank is simply a large, elevated tank of water. The tank is built of a sufficient height to provide pressure to the water system. Each foot of height provides .43 PSI (pounds/square inch) of pressure. A typical municipal water supply system runs at between 50 and 100 PSI (major appliances generally require a minimum of 20 PSI). The water tank must be sufficiently tall to supply that level of pressure to houses and businesses in the vicinity of the tank, generally from about 115 to 230 feet in height.

To aid in producing the required pressure, elevated tanks are typically located on high ground. In hilly regions, a tower is sometimes replaced by a simple standpipe on the highest hill of the area.

A tank usually holds approximately one million gallons of water, typically about a day's supply for the community it serves. If pumps fail, for example in a power failure, the water tower generally contains enough water to ensure supply for about a day.

Because of the use of towers as reservoirs, a municipality is able to size its pumps for average rather than peak demand. If the pump produces more water than is needed by the system, the excess automatically flows into the tank of the standpipe or elevated tower. If the community demands more water than the pump can supply, the water flows out of the tank to meet demand.

Standpipes

Standpipes are flat-bottomed, cylindrical tanks historically constructed of steel, reinforced concrete, or timber that have a height greater than the diameter. Capacities of standpipes can exceed 2 million gallons.[32] Ground storage tanks are normally sized so that the shell height is a multiple of 8 feet and the diameter a multiple of 10 feet so that the fabricator can use standard 8 by 10 foot plates.[33]

A steel standpipe is constructed by joining together a number of thin steel plates. The horizontal and vertical joins between adjoining plates may be either riveted or welded.

Most concrete standpipes are of prestressed construction. Reinforcement is steel wire with an ultimate strength of about 225,000 pounds per square inch. A thin concrete wall is built. The wire is wound around this wall and stretched, and the wall is completed with a gunite covering. Because of the tension in the wire, the concrete is kept under compression even when the reservoir is full. As a result, leaks do not develop.[34] In the 1950s, concrete standpipes were typically limited to 50-foot heads, because it proved difficult to make them watertight with greater heads.[35]

Elevated Tanks

Elevated tanks may be constructed of steel, reinforced concrete, or timber. In general, the walls of an elevated tank are constructed in the same manner as the walls of a standpipe. The bottom of an elevated tank is often constructed in a bowl shape, and the top is generally made either conical or dome shaped.

The earliest elevated tanks were of wood stave construction, similar to that used on barrels and silos. Metal, usually iron, bands held the tank together. The first known Pittsburgh-Des Moines Corporation (PDM) tank was a small wooden, flat-bottomed tank supported on a wooden tower included as part of the Boone, Iowa, water system in 1893. The life of a wooden tank was estimated at about 15 years, while wood tanks made of pine, fir, and cypress could last from 20 to 25 years if heavy hoopage was employed. Few, if any, wood elevated tanks remain in active use in water supply systems in the United States. The sharp-eyed industrial archaeologist can still find examples of small, elevated wood anks, usually rising from the roofs of abandoned industrial buildings.

Many of the manufacturers of elevated water tanks began working for the railroad industry. They would erect water tanks to furnish water for steam engines. Municipalities borrowed the idea and started using them for city water tanks. Tanks were also built by industrialists to furnish water for manufacturing plants and for fire protection.[36]

The earliest elevated water tanks were usually supported on a heavy grillage of beams strong enough to transfer the considerable weight of the tank and its contents to the columns of the supporting tower. Later tanks either consisted of a flat-bottomed tank resting on steel grillage or had a suspended bottom and no grillage. The latter quickly became more common due to a savings in metal and cost and because the bottom could be inspected and painted more readily and thoroughly. The suspended bottom could be flat, hemispherical, segmental, elliptical, conical, or radial cone-shaped. Flat and conical bottoms accentuate stresses.[37] The hemispherical bottom was most common and was referred to as a saucer plate.

Early steel tank towers consisted of columns commonly formed from two channels laced on two sides or a cover plate laced on one side. For large-capacity tanks, a section built of two web plates, four angles, and a cover plate were common.

By the 1880s, wrought-iron or steel elevated water storage tanks had begun to supplant wood stave tanks. The first such tank with a full hemispherical bottom was constructed in Fort Dodge, Iowa, by the Chicago Bridge and Iron Company (CBI) in 1894.[38] In 1897, the first PDM steel tank and tower were erected in 1897 in Scranton, Iowa. These early tanks were assembled atop the towers and were constructed of steel plates fastened with rivets. In some cases, manufacturers sent their own crews to install the tanks, while in other cases communities hired their own contractors.

Depending on the size of the tanks, teams of six to twelve men would build the tank. A 50,000 gallon tank might take three to four weeks to complete, while the largest tanks might take as long as six months.[39]

In the late 1800s and early 1900s, communities would frequently put fancy handrails on the catwalks of tanks. In 1908, the local water company in Gary, Indiana, constructed an elevated water tank and enclosed it with concrete, giving it the appearance of a castle turret. The enclosed tank is still in use.[40]

By the beginning of the twentieth century, CBI had contracted for 85 elevated tanks in 23 states from New Jersey to Washington. On June 25, 1907, George Horton of CBI was issued his first patent (Patent 857,626) for a hemispherical ellipsoidal bottom water tank supported on a riser (Figure 8.8). By 1915 PDM had built water towers in 43 states, the District of Columbia, eight Canadian provinces, and several foreign countries, in sizes ranging from 2,500 to 2 million gallons.

Specifications produced by one early manufacturer of steel elevated tanks, McClintic-Marshall Company, indicated that tanks were available in sizes ranging from 5,000 gallons to 2 million gallons. Their tanks had ellipsoidal bottoms and, as a result, experienced smaller pressure variations than spherical-bottomed tanks. Ellipsoidal tanks could also be built to a much greater diameter than possible with spherical-bottomed tanks.

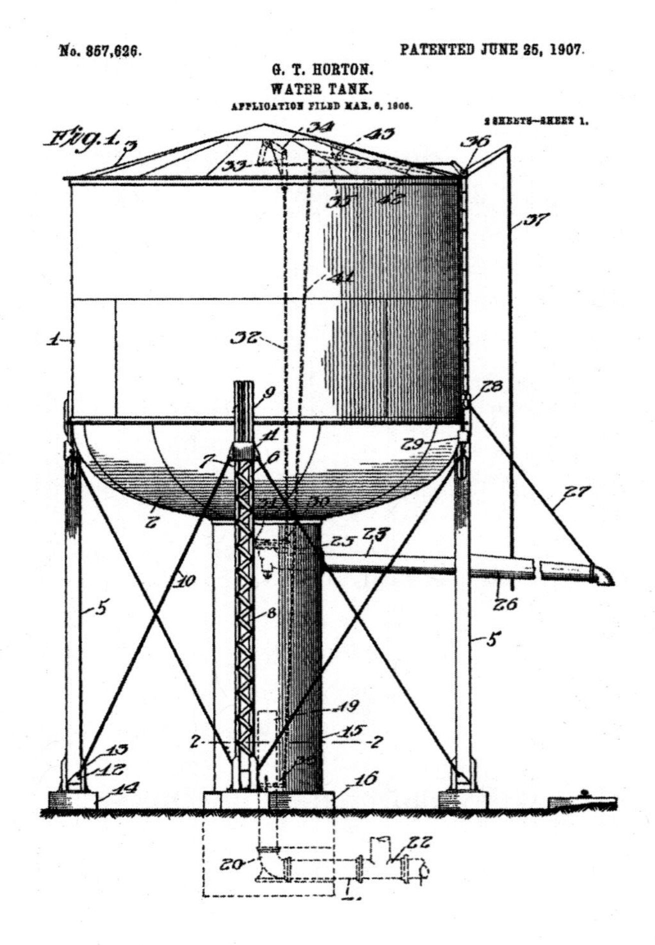

Figure 8.8. Water tank. Patent number 857,626 (Horton 1907).

The smallest available tanks, 5,000 gallons, were constructed with a diameter of 11 feet 5¼ inches and a cylinder depth of 5 feet. These tanks were supported on four posts. A medium-sized 250,000 gallon tank had a tank diameter of 40 feet 7 inches and a cylinder height of 19 feet 3¼ inches. It was supported by six posts. The largest tank, 2 million gallons in capacity, had a diameter of 92 feet, a height of 25 feet 1¾ inches, and was supported by 20 posts.[41]

Elevated tanks are normally sized in increments of 50,000 gallons, up to a 300,000-gallon capacity. For tanks between 300,000 and 500,000, 100,000-gallon increments are used, while for those above 500,000 gallons, 250,000-gallon increments are used.[42]

CBI engineers patented the radial cone bottom in 1929. Between the 1930s and early 1950s, a transition occurred in tank construction, as riveted tanks became increasingly supplanted by welded ones. By the late 1930s, welding had made it possible to introduce tubular support columns, more cost-effective to manufacture and easier to paint and maintain than were lattices. The earlier tanks typically looked like fireworks rockets, cylinders with cone-shaped tops and curved bottoms, sitting atop "fuses" of latticed supports. With welded tanks, new tank shapes came into development.

Manufacturers began making bigger tanks and increasing their diameter. To accommodate this larger diameter, ellipsoidal bottoms were used. For even larger capacities, radial cone bottoms were used, allowing tanks with 1- to 1.5-million-gallon capacity with a reasonable head range.

The next major development in elevated tank construction occurred in 1939 or 1940 with the introduction of the single-pedestal spheroid or pedisphere. There is some question of which company first introduced this design. CBI introduced the first Watersphere® elevated tank in 1939, while Georgia's R. D. Cole introduced its own design around the same time.[43] The first all-welded water sphere tank was built in 1939 by CBI in Longmont, Colorado. In 1954, Chicago Bridge and Iron introduced the first of a larger type of tank, the Waterspheroid®, with a capacity of 500,000 gallons in Northbrook, Illinois.[44]

Waterspheroid elevated tanks range in capacity from 150 thousand gallons to 2 million gallons. The smallest have a spheroid diameter of 35 feet, while the largest have a diameter of 93 feet 8 inches.

Parts of the elevated tank include the riser pipe, the overflow pipe, the access tube, and the outside ladder. If water does not flow in or out of the tank at all times, a heater is used to warm the riser pipe.

The double ellipsoidal tank design dates from about 1950. The torospherical design became popular in the 1950s and remains a common tank shape. It usually employs a multi-legged support structure.

In the 1950s, PDI introduced the fluted tank, a single-pedestal tank surround by a folded-plate tower, known as the Hydropillar™. A design patent for this design was issued in 1963 to Lloyd E. Anderson, Sr., of PDI (Design Patent 197,264). Since that time, more than 900 Hydropillars have been designed and built around the world, more recently by the PDM Division of CBI. The Hydropillar is a steel fluted column with interior structural steel. This construction allows for multiple floors and exterior windows and permits a multipurpose use, such as for a fire station, community center, or water treatment facility.[45]

Hydropillars range in capacity from 1 million to 3 million gallons. The smallest variety has a tank diameter of 66 feet, a head range of 40 feet, and pedestal diameter of 36 feet. The largest have a tank diameter of 118 feet, a head range of 45 feet, and pedestal diameter of 64 feet.

Smaller elevated tanks are often double ellipsoidal, with an ellipsoidal top and bottom and cylindrical sides. Medium-capacity tanks often have an ellipsoidal roof, cylindrical shell, and cone-supported torus bottom.

A less common modern elevated tank form is the composite elevated tank (CET). These structures combine the tensile strength of steel with the compressive feature of concrete. The most recent version of this design, patented in 2001, includes a reservoir supported by a pedestal portion supporting a cylindrical shaft (Patent 6,318,034).

Reservoirs

Reservoirs are flat-bottomed, cylindrical tanks, generally of welded steel construction, that have a height less than or equal to the diameter (Figure 8.9). Reservoirs are usually lined with concrete, gunite, asphalt, or an asphaltic membrane.[46] They are the most economical and effective water storage when located on high ground.

Ground storage tanks are normally sized so that the shell height is a multiple of 8 feet and the diameter a multiple of 10 feet. Reservoir capacities can exceed 2 million gallons.

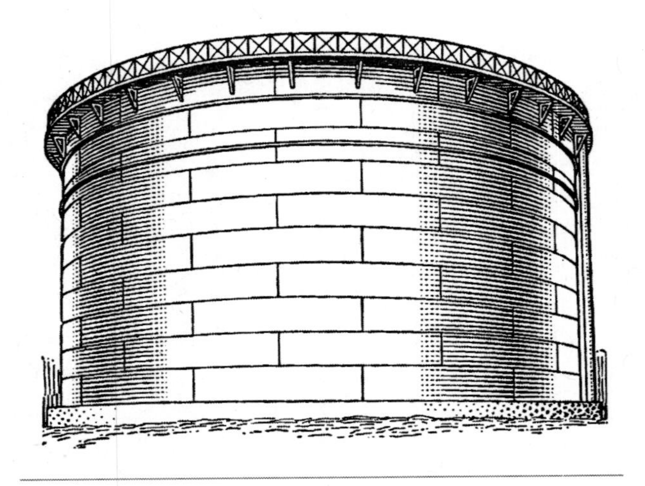

Figure 8.9. Steel reservoir (Hardenbergh 1952: 108). Reprinted, by permission of Thomson Education Direct.

WASTEWATER OR SEWAGE TREATMENT

The purpose of a municipal wastewater treatment system is to prevent pollution of the receiving watercourse. Stream pollution and lake eutrophication resulting from municipal wastes are of particular concern.

Wastewater from households, industries, and combined sewers is collected and transported to the treatment plant, with the effluent commonly disposed of by dilution in rivers, lakes, and estuaries. Other means of disposal include irrigation, infiltration, evaporation from lagoons, and submarine outfalls extending into the ocean.

Because of the changing technology and regulations of wastewater and sewage treatment, few older active facilities retain all their original components. Instead, the typical facility contains some original components still in use and other replacement components. One example is New York's Tallman Island Sewage Treatment Works, shown in Figure 8.10.

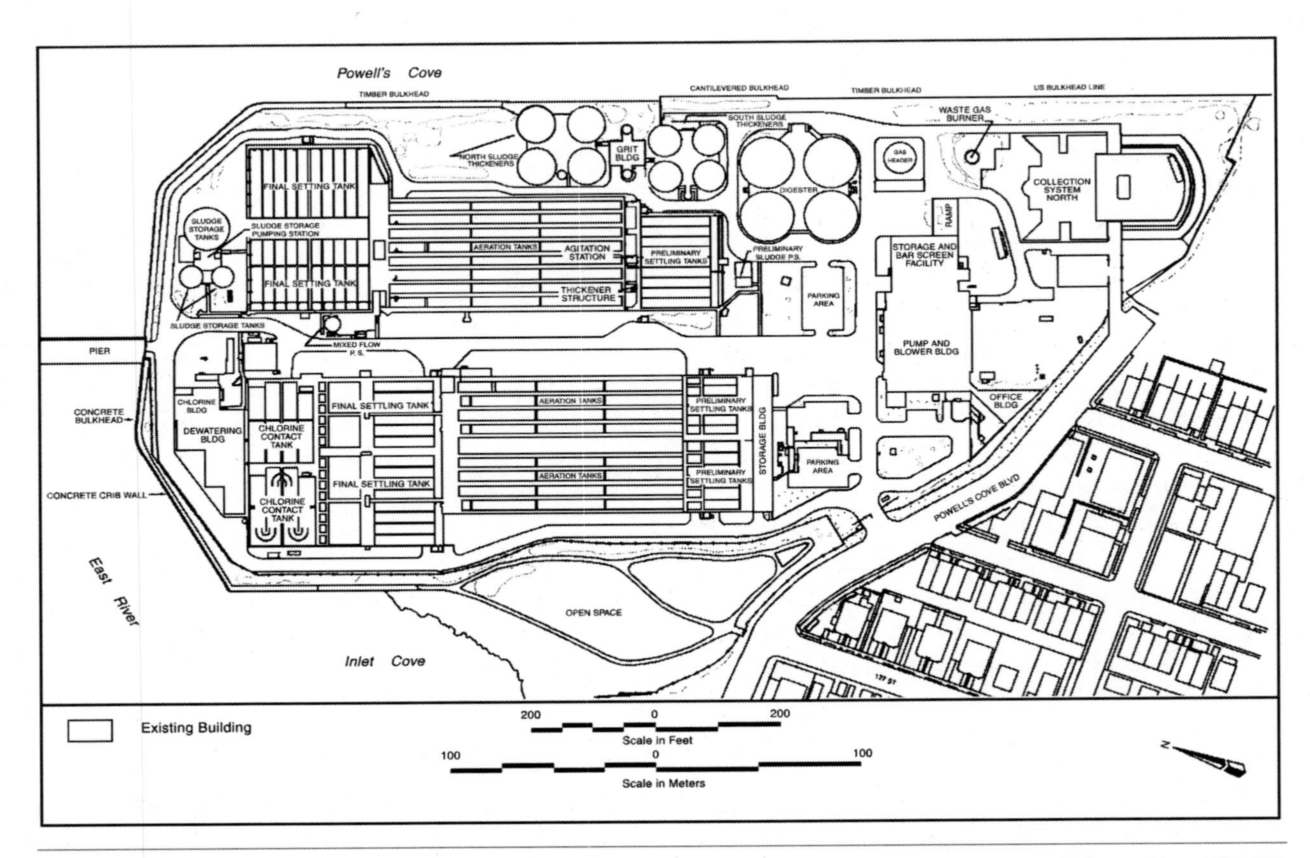

Figure 8.10. Tallman Island Sewage Treatment Works. From *Environmental Assessment Statement: Tallman Island TI-2/TI-3 Water Pollution Control Plant Plant Upgrade*, Blasland, Bouck & Lee, Inc. (Syracuse, NY) and TAMS Consultants, Inc. (New York), February 2006.

Sanitary Sewers

Connections from the main sewer to houses or other buildings were historically made of vitrified clay, concrete, or asbestos cement pipe. Modern connecting pipe is usually made of PVC. Collecting sewers gather flow from individual buildings and transport the material to an interceptor or main sewer. These sewers are usually located beneath the street paving on one side of the centered storm drain. Manholes are generally located at changes of direction, grade, pipe size, or at intersections of collecting sewers.

Intercepting sewers carry the flow from the collecting sewers to the point of treatment or disposal of the wastewater. These sewers normally follow valleys or natural streambeds of the drainage area. Manholes are typically constructed at least every 600 feet.[47]

Wastewater Pumping Stations

In many systems it is necessary to accumulate wastewater at some low point in the collection system and pump it to treatment works or to a continuation of the system at a higher elevation. Pumping stations consist primarily of a wet well, which intercepts incoming flows and permits equalization of pump loadings, and a bank of pumps which lift the wastewater from the wet well. In most cases, centrifugal pumps are used.[48]

Storm Drain Systems

Storm drain systems handle surface water runoff in urban areas. This runoff is collected from the street using one of four basic types of inlets: curb inlet, gutter inlet, combination inlet, and multiple inlets. A curb opening is a vertical opening in the curb through which gutter flow passes. A gutter inlet is a depressed or undepressed opening in the gutter section through which the surface drainage falls, covered by one or more grates. A combination inlet is composed of both curb and gutter openings acting as an integrated unit. (This is the typical storm drain found in many urban and suburban areas.) Multiple inlets are closely spaced interconnected inlets acting as a unit. Inlets are constructed at all street intersections where the quantity of flow is significant.[49]

Storm Drain Systems

Storm drainage systems may be closed-conduit, open-conduit, or a combination of the two. In most urban areas, the smaller drains are frequently closed-conduit, and as the system moves downstream open channels are often used. Because quantities of storm-water runoff are usually large compared with flows in water mains and sanitary sewers, the drainage needed to carry these flows is also quite large. Common practice is to build the storm drains along the center line of the street, offsetting the water main to one side and the sanitary sewer to the other.

Manhole or junction structures are required at changes in grade, pipe size, direction of flow, and quantity of flow. Storm drains are generally built no smaller than 1 foot in diameter. Manhole space for pipes that are 27 inches and under generally does not exceed about 600 feet.[50]

Historic Overview

In antiquity, sewers were constructed primarily as drains to carry storm water and garbage and other waste from urban areas. In the United States, many cities along rivers built combined sewers to convey both storm runoff and sanitary wastewater directly to the river through a series of sewer lines oriented perpendicular to the river. Required treatment of sanitary wastewater resulted in construction of a collector sewer along the river to intercept the dry-weather flow in the combined sewer and to convey this wastewater to a treatment plant.

In the early twentieth century, the object of sewage treatment was the removal of suspended particles and the complete oxidation of the organic matter and ammonia in solution. At that time sewage treatment was broadly divided into two types: processes for preliminary clarification, and methods for the final oxidation of the impurities contained in the clarified liquid.

Preliminary Treatment

Preliminary treatment consists of coarse screening, medium screening, comminution, grit removal, pre-aeration, flotation, flocculation, and chemical treatment.[51]

Coarse Screening

Coarse screening is an essential first step upon the arrival of sewage at the plant. The object of screening is to remove the larger suspended debris such as paper, rags, orange peel, and sticks. Various classes of screens are employed, including hand screens in small works, automatically cleaned screens of a variety of designs, and larger rotary-power-driven screens.

Coarse screens, also known as bar racks (Figure 8.11), are constructed of steel bars with clear openings not exceeding 2½ inches. The bars are usually set in a channel inclined at 30 to 45 degrees with the horizontal to facilitate cleaning. Mechanically cleaned medium screens with clear bar openings of 5/8 to 1¾ inches are commonly used instead of a manually cleaned coarse screen.[52]

Comminutors

Comminutors or shedding devices (Figure 8.12) chop solids passing through bar screens to about ¼ to 3/8 inch in size. They are installed directly in the wastewater flow channel and are provided with a bypass so that the length containing the unit can be isolated and drained for machine maintenance.[53]

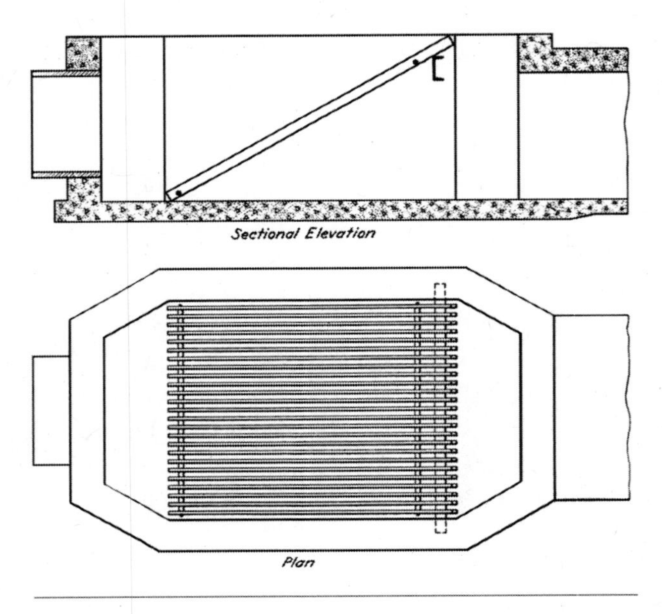

Figure 8.11. Bar screen. Reprinted, with permission, from W. A. Hardenbergh, *Sewerage and Sewage Treatment*, 1st edition (Scranton, PA: International Textbook Company, 1936), 251. Reprinted, by permission of Thomson Education Direct.

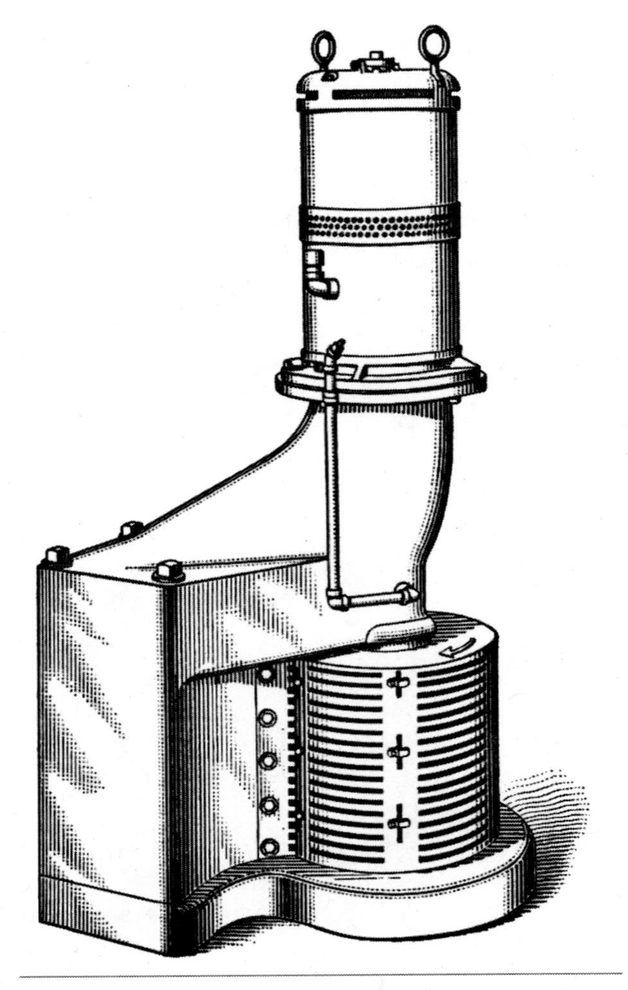

Figure 8.12. Comminutor (Hardenbergh 1936: 254). Reprinted, by permission of Thomson Education Direct.

Grit Removal

Grit includes gravel, sand, and heavy particulate matter such as corn kernels, bone chips, and coffee grounds. Grit removal in waste treatment protects mechanical equipment and pumps from abnormal abrasion and reduces accumulation in settling tanks and digesters.

Several types of grit removal units are employed in wastewater treatment. Standard types are channel-shaped settling tanks, aerated tanks of various shapes, clarifier-type tanks with mechanical scraper arms, and cyclone grit separators with screw-type grit washers (Figure 8.13). The processes of grit removal and pre-aeration can be performed in the same basin: a hopper bottom or clarifier-type tank provided with grit removal and washing equipment.[54]

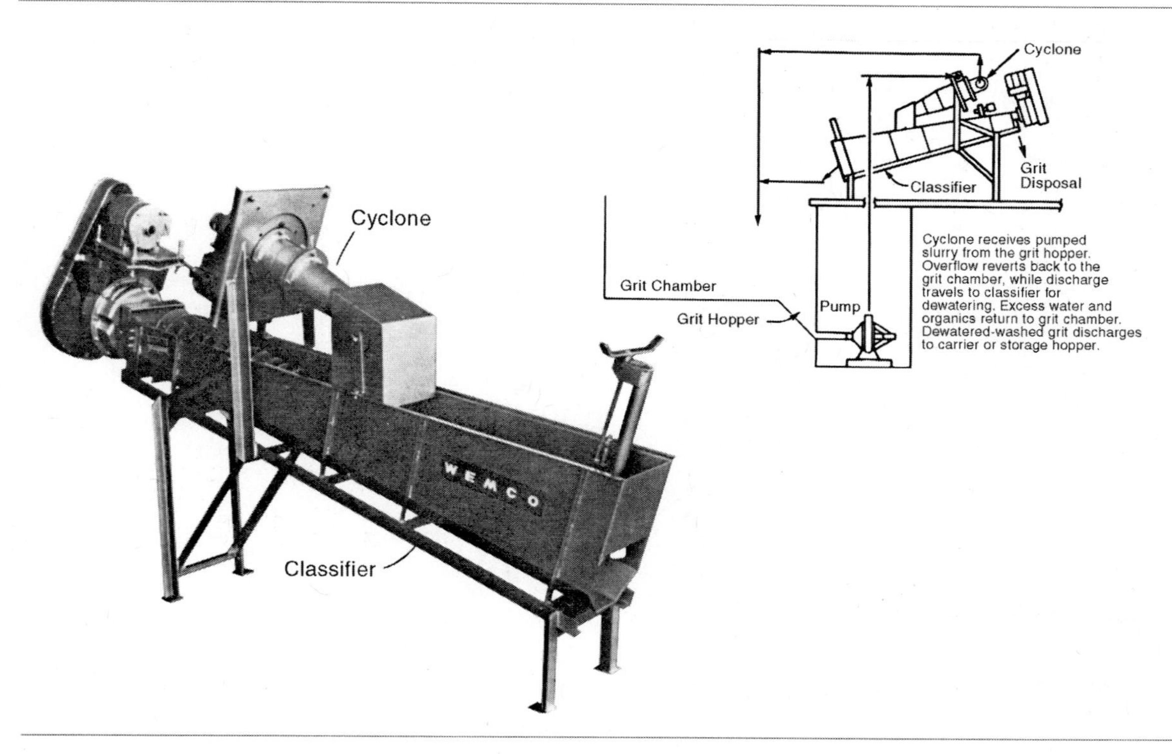

Figure 8.13. Cyclone separator and grit washer (Clark, Viessman, and Hammer 1971 :354). Reprinted, by permission of Thomson Education Direct.

Flotation

Flotation is used to remove fine suspensions, grease, and fats, and is performed either in a separate unit or in a pre-aeration tank used for grease and sometimes grit removal. This machinery is not commonly used in domestic waste treatment but is sometimes dictated by the industrial wastes in municipal wastewater.[55]

Flocculation

Floc is excess sludge. It is produced by chemicals that in solution furnish ionic charges opposite to those of colloidal turbid particles in the water. Coagulants neutralize repelling charges on colloidal particles and produce a jelly-like, spongy mass, the floc.

Flocculation consists of mechanical entrapment of the agglomerated particles by absorption onto the floc. This takes place using coagulant chemicals that cause molecular bridging of the individual molecules of the coagulants.[56]

Various processes have been used for flocculation, among them diffused air, baffles, transverse or parallel shaft mixers, vertical turbine mixers, and walking-beam-type mixers. The most common type of flocculator used in the 1970s was the paddle flocculator (Figure 8.14).[57]

Primary Treatment

Primary treatment is sedimentation. Sedimentation is the removal of solid particles from a suspension through gravity settling.[58]

Detritus and Sedimentation Tanks

A detritus or grit tank is used for intercepting heavy mineral matter in sewage, including grit and sand. The velocity through the detritus tank is sufficient to carry forward the suspended organic matter, but slow enough to allow the grit to settle to the bottom of the tank for daily removal.

Several different types of sedimentation tanks have been employed. In rectangular tanks (Figure 8.15), the length is usually 2½ to 3 times the breadth. Depths generally average about 10 feet. The inlet is configured to diffuse the incoming flow. The outlet consists of a weir across the whole width of the tank. Since light material separates by flotation and forms a scum on the surface, a dip-plate is provided to prevent the scum from flowing over the weir. Since a large proportion of the suspended matter settles near the inlet end, it is from this end that the sludge is removed. The floor is swept

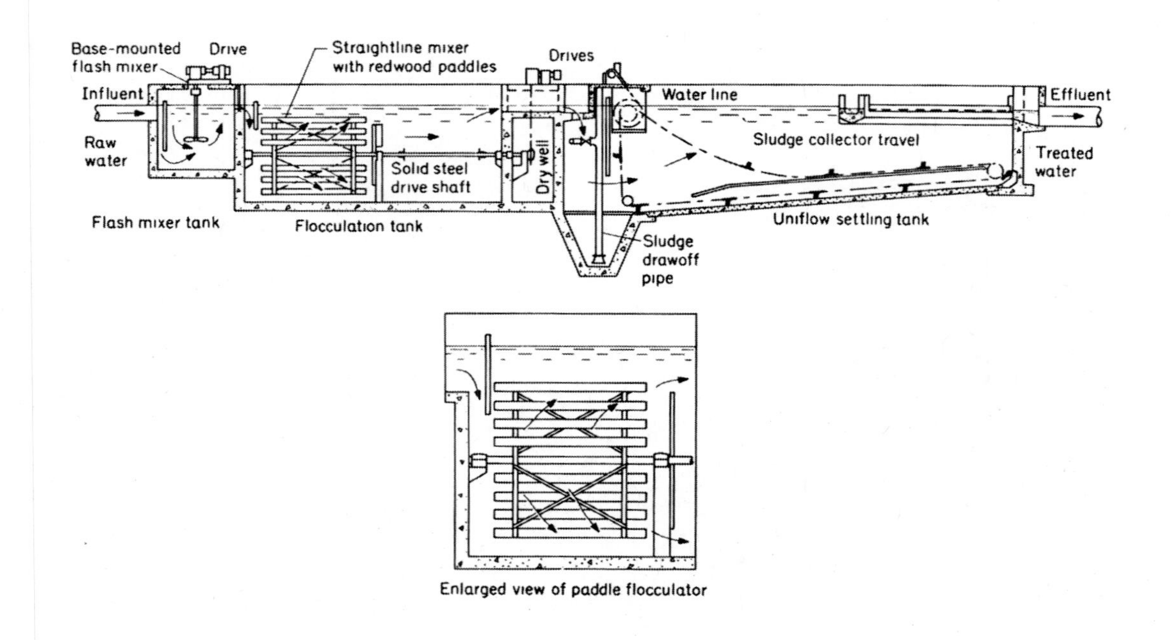

Figure 8.14. Rapid mixing, flocculation, and sedimentation (Clark, Viessman, and Hammer 1971: 321). Reprinted, by permission of Thomson Education Direct.

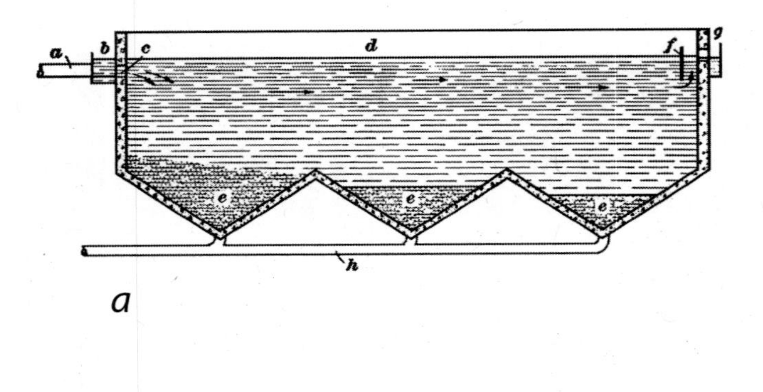

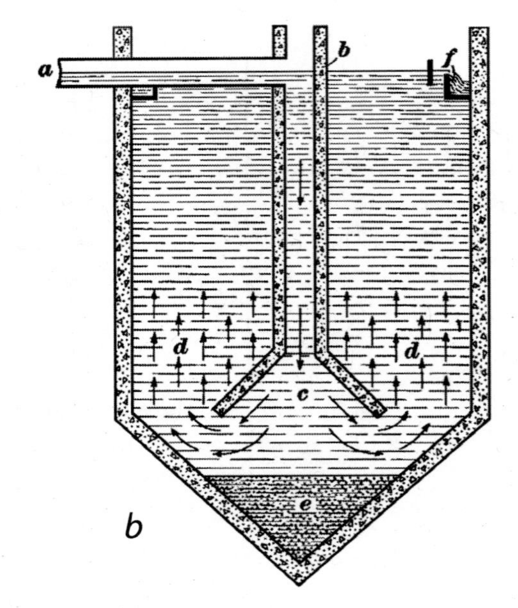

Figure 8.15. Sedimentation tanks: *a*, horizontal-flow tank (Hardenbergh 1936: 262); *b*, vertical-flow tank (Hardenbergh 1936: 263). Reprinted, by permission of Thomson Education Direct.

by scraper blades supported from a moving bridge spanning the breadth of the tank. At the inlet end of the tank there is a pocket in which sludge can be left to consolidate. It is withdrawn under the head of water in the tank by opening the valve in a sludge well.

Circular tanks are frequently used in American treatment plants (Figure 8.16). Diameters are generally in the range of 15 to 30 meters, while depths are generally between 3 and 3.6 meters. The feed pipe passes under the tank and rises vertically in the center, terminating in a bell mouth. A cylindrical baffle plate prevents flows from streaming across the surface. The floors of circular tanks are generally swept by blades supported from a rotating radial bridge. Single scraping blades have been used, but multiple blades set in an echelon at an angle to the direction of motion are more

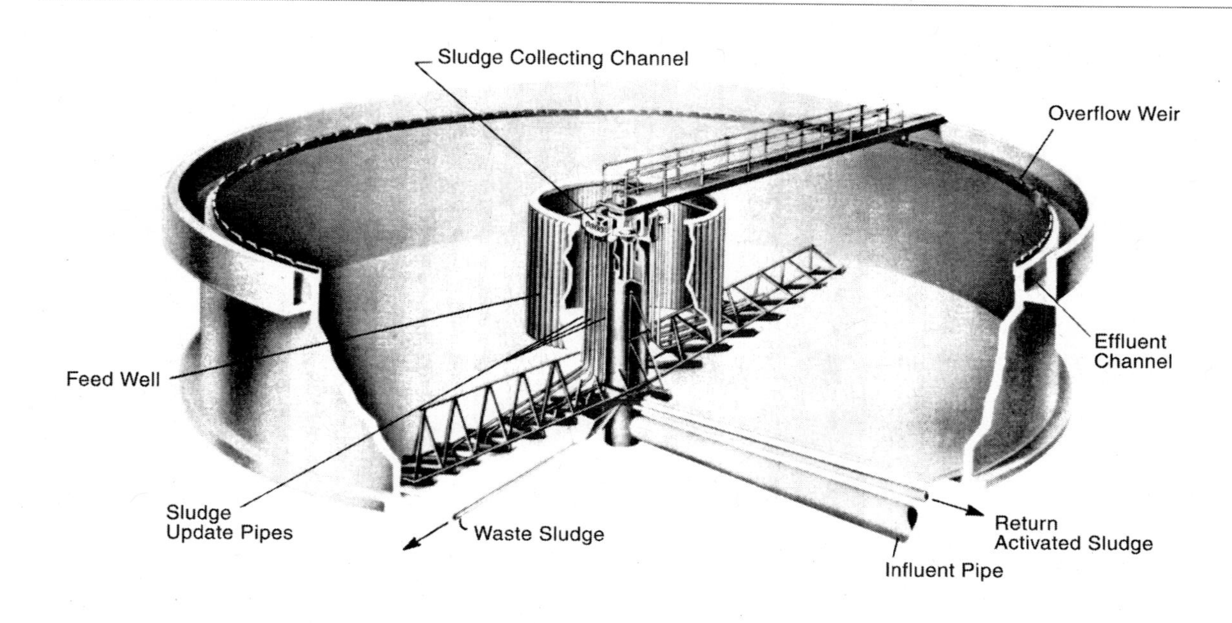

Figure 8.16. Circular clarifier (Clark, Viessman, and Hammer 1971: 349). Reprinted, by permission of Thomson Education Direct.

common. Sludge is drawn off under pressure of the tank's contents.

Upward-flow tanks are most commonly square in plan. The upper part of the tank has vertical walls, while the lower part is an inverted pyramid to avoid adherence of sludge. Flow enters through the feed pipe which terminates in an upward-facing bell mouth. A square or cylindrical baffle prevents flow from passing across the surface. Sludge is drawn off under the pressure of the tank contents. Neither power-driven machinery nor manual labor is required, making this type of tank advantageous for small installations.

The next stage in treating average domestic sewage is often a preliminary preparation tank. This tank removes a large proportion of suspended solids. At the inlet end the heavier portion of the solids remaining are held back in a compartment containing iron baffle plates. Further dissolution and subsidence takes place in storage compartments behind submerged cross walls. In the next stage, sewage undergoes an aerobic treatment in contact with well-vitrified clinker. The partially clarified sewage enters by means of an aerating floor and passes upward and over a sill to the adjoining compartment. The sludge from all the compartments of the tank is removed by opening the outlet valves provided for each separate chamber. These outlets are connected up with a single sludge outlet drain.

Secondary Treatment

Biological treatment is the most important step in processing municipal wastewater.[59] It is performed using activated-sludge processes, trickling filters, or stabilization ponds, and anaerobic digestion.

Activated Sludge

The activated sludge process was developed in England in about 1914.[60] In activated sludge processes, wastewater is fed continuously into an aerated tank where the microorganisms metabolize and biologically flocculate the organics. Microorganisms (activated sludge) are settled from the aerated mixed liquor in the final clarifier and returned to the aeration tank. Several types of activated sludge processes are employed.

Conventional Activated Sludge Process

The conventional activated sludge process is used for secondary treatment of domestic wastewater (Figure 8.17). The aeration basin is a long rectangular tank with air diffusers on one side of the tank bottom to provide aeration and mixing. Settled raw wastewater and return activated sludge enter the head of the tank and flow down its length in a spiral flow pattern. An air supply is tapered along the length of the tank to provide a greater amount of diffused air near the head, where the

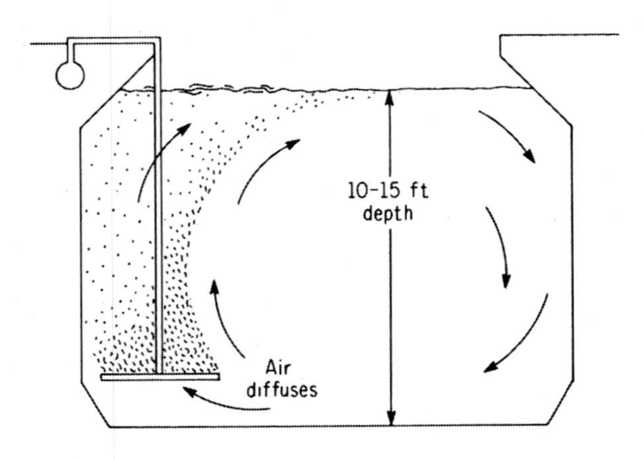

Figure 8.17. Cross section of a typical diffused air, spiral flow, conventional activated-sludge aeration tank (Clark, Viessman, and Hammer 1971: 514). Reprinted, by permission of Thomson Education Direct.

rate of biological metabolism and resultant oxygen demand is the greatest.

The standard activated sludge process uses fine-bubble air diffusers set at a depth of 8 feet or more to provide adequate oxygen transfer and deep mixing. The swing diffuser arm is jointed so the diffusers can be swung out of the tank for cleaning and maintenance.

Step-Aeration Activated Sludge Process

The step-aeration activated sludge process represents a modification of the conventional process and was introduced to the United States at New York's Tallman Island Water Pollution Control Plant by city sanitary engineer Richard H. Gould to overcome some of the problems inherent in conventional activated sludge processes and to conserve aeration tank capacity. His system was based on a system of multi-pass aeration tanks with four channels. The first pass is reserved for re-aeration of the returned sludge to regenerate its absorptive properties. Sewage is then added in incremental steps to the aeration tanks along the course of flow of the returned sludge, to keep oxygen demand at uniform levels. The advantages of step aeration are that it permits more flexibility in operation, produces well-settled sludge, and saves tank volume. All fourteen of New York City's wastewater and sewage treatment plants use the step-aeration process.[61]

Contact Stabilization Activated Sludge Process

The contact stabilization activated sludge process provides for re-aeration of the return activated sludge from the final clarifier, which allows this method to use a smaller aeration tank. Raw wastewater is aerated in a contact tank and then settled. The supernatant from the clarifier is plant effluent, and the subnatant is re-aerated prior to mixing with raw influent in the contact aeration tank.

High-Rate Activated Sludge Process

The high-rate activated sludge process operates with the highest biological oxygen demand (BOD) load per unit volume of aeration tank of any activated sludge system. The BOD loading is approximately three times that of the conventional process, and the aeration period is proportionally shorter. High-rate mixed liquor is in the declining growth stage rather than the endogenous stage. This makes the activated sludge more difficult to settle, and a special clarifier arm with hydraulic pick-up pipes is necessary to ensure adequate suspended solids separation. This process generally uses a combination of compressed-air aeration and mechanical mixing.

Extended Aeration Activated Sludge Process

The extended aeration activated sludge process is often used to treat small wastewater flows from schools, housing developments, trailer parks, institutions, and small communities.[62]

Trickling Filters

In a trickling filter, primary effluent is sprayed on a bed of crushed rock or other media coated with biological films. The surface of the bed may support algal growth, with proper temperature and sunlight conditions. The lower portion of a deep filter frequently supports populations of nitrifying bacteria.[63]

The major components of a trickling filter are the filter media, underdrain system, and the rotary distributors. The filter media, the most common being rock, slag, or fieldstone, provide a surface for bacterial growth, while permitting the passage of liquid and air.

The underdrain system carries away the effluent and permits circulation of air through the bed. The underdrainage system, with provision for flushing, effluent channels, and effluent pipe, is designed to permit the free passage of air. A rotary distributor provides regular soaking of the filter surface. The most common kind is driven by the reaction of the wastewater flowing out of the distributor nozzles.[64]

Stabilization Ponds

Domestic wastewater may be effectively stabilized by natural biological processes that occur in shallow pools. Those suitable for treating raw or partially treated wastewater are referred to as stabilization ponds, lagoons, or oxidation ponds.

A stabilization pond is a flat-bottomed pond enclosed by an earth dike. It can be round, square, or rectangular, with a length no more than three times its width. Operating liquid depth has a range of 2 to 5 feet. A minimum depth of about 2 feet is required to prevent growth of rooted aquatic plants. Depths of greater than 5 feet can create odorous conditions because of anaerobiosis at the bottom.

Influent lines discharge near the center of the pond, and the effluent usually overflows in a corner on the windward side. The overflow is generally a manhole or a box structure with multiple-valved draw-off lines. If the soil is pervious, bottoms and dikes are sealed to prevent groundwater pollution. Clay is commonly used as a sealing agent. Dikes surrounding the pond are typically seeded with grass, graded, and fenced.[65]

Anaerobic Digestion

Anaerobic digestion of sludge consists of two stages that occur simultaneously. The first stage involves hydrolysis of high-molecular-weight organic compounds and their conversion to organic acids by acid-forming bacteria. The second stage is gasification of organic acids to methane and carbon dioxide.[66]

These actions take place in one of two types of digester: single-stage floating-cover digesters or high-rate digesters. In the first type, raw sludge is pumped into the digester through pipes terminating either near the center of the tank or in the gas dome. The biological processes of anaerobic digestion are improved by complete mixing of the digestion sludge. Contents may be mixed either mechanically or by use of compressed digestion gases.

Tertiary or Advanced Waste Treatment

With tightened water pollution standards, tertiary or advanced treatment has been added to many sewage treatment plants. Few such systems are sufficiently old to be included in historic documentation of a plant. Types of advanced treatment include suspended solids removal through microscreening, diatomaceous earth filtration, or chemical clarification; organic removal, through use of activated carbon, oxidation, or foam separation; inorganic removal through distillation, electrodialysis, freezing, ion exchange, or reverse osmosis; and nutrient removal, including phosphate, nitrate, and ammonia removal.[67]

Evolution of a Typical Urban Wastewater Treatment Plant

The Tallman Island Water Pollution Control Plant, located in New York's Queens Borough, was constructed in 1937 to 1939 to treat the sewage flow from the 1939 World's Fair. The initial components of the complex included the pump and blower building, pumping station, three preliminary settling tanks, two aeration tanks, four final settling tanks, four digester tanks, two sludge thickener tanks, two sludge storage tanks, and a grit tank house.

In the original operation, raw sewage would enter the plant at the underground screen chamber in the southwest corner of the pump and blower building. Sewage was then pumped to the preliminary settling tanks by gas-powered engines in the Pump and Blower Building. In the preliminary settling tanks, sewage would separate into solids and grit. The grit was pumped to an underground grit chamber and was removed from the grit tank house silo by truck. Degritted primary sludge was pumped to the plant's sludge handling facilities.

Settled sewage would flow to the aeration tanks where it would be distributed and mixed with return sludge using the step-aeration process. Aeration effluent would be clarified in the final settling tank and discharged into the East River.

Settled sludge would be returned to the aeration tanks or flow by gravity to the sludge thickener tanks. Thickened sludge would be pumped to sludge digester tanks. Digested sludge would be pumped to sludge storage tanks where it was then pumped into a sludge vessel for disposal at sea.

During the 1950s, operations were improved through the addition of additional sludge thickener tanks. In 1959, a chlorine building and contact tank were added to disinfect effluent during bathing season. In 1964, additional improvements, including revising flow patterns and adding tanks, were completed to raise the capacity to 60 million gallons per day (MGD).

In the 1970s, additional upgrades were initiated to raise the capacity to 80 MGD. Additions included an office building, boiler room, and storage and bar screen facility, additions to the pump and blower building, a grit building, new sludge thickeners, a mix flow pumping station, sludge storage tanks, a second chlorine contact tank, two more final settling tanks, one more aeration tank, and two more preliminary settling tanks. New mechanical additions included the bar screen facility, grit building, and the pumping station. The addition of the bar screen facility enabled the plant to remove screenings from the sewage with raked bar screens via a belt conveyor. The grit building received sewage solids from the preliminary settling tanks, from which they were then pumped to cyclone degritters to improve the grit removal process. The mixed-flow pumping settled sludge from the waste sludge sump and enhanced the sludge removal process.

In the mid-1990s, biological nutrient removal facilities were implemented at the plant. Baffles, mixers, and a froth control system were installed in Aeration Tanks 3 and 4.[68]

GLOSSARY

Aeration. The process in which air is brought into intimate contact with water, usually by spraying water through air or bubbling air through water.

Aerobic. A process conducted in the presence of air.

Anaerobic. A process that occurs in the absence of air.

Biochemical oxygen demand (BOD). The measure of the quantity of oxygen needed for unstable material to be oxidized through the action of microorganisms, often measured for 5 days at 20 degrees centigrade.

Coagulation. In water treatment, the process by which alum is mixed into the water to create aluminum hydroxide.

Effluent. The outflow of a water treatment plant.

Filtration. The passing of water through sand and gravel to eliminate impurities.

Flocculation. The process by which aluminum hydroxide adheres to unwanted particles in the water. The resulting clumps of aluminum hydroxide and particles are called floc.

Influent. The untreated water that enters a water treatment plant.

MCL. Maximum contaminant level. The maximum amount of a contaminant allowed in drinking water.

Mechanical filter. A filter designed for the removal of suspended solid particles.

Precipitate. To cause dissolved substances to form solid particles that can be removed by settling or filtering.

Raw water. Untreated water from wells or from surface sources.

Sedimentation. The process that occurs when unwanted material settles to the bottom of the treatment basin and is removed from the water.

Turbidity. The amount of suspended particles in water, measurable by the amount of light that is scattered when light is shone through water.

REFERENCES

Babbitt, Harold E., and James J. Doland. *Water Supply Engineering*. 5th edition. New York: McGraw-Hill Book Company, Inc., 1955.

Blasland, Bouck & Lee, Inc. New York City Landmark and National Register Eligibility Assessment of the Tallman Island Water Control Plant for the Tallman Island Plant Upgrade Project. New York: Blasland, Bouck & Lee, Inc., 2001.

Clark, John W., Warren Viessman, Jr., and Mark J. Hammer. *Water Supply and Pollution Control*. Scranton, PA: International Textbook Company, 1971.

Hammer, Mark J. *Water and Waste-Water Technology*. New York: John Wiley & Sons, Inc., 1975.

Hardenbergh, W. A. *Sewerage and Sewage Treatment*. 1st edition. Scranton, PA: International Textbook Company, 1936.

—. *Water Supply and Purification*. 3rd edition. Scranton, PA: International Textbook Company, 1952.

Hoover, Charles P. *Water Supply and Treatment*. 2nd edition. Washington, DC: National Lime Association, 1936.

Jepperson, Kathy, and Vipin Bhardwaj. Distribution 101: How Does Water Get from the Source to Your Tap? *On Tap: Drinking Water News and Information for America's Small Communities* 1:1 (Spring 2001): 14–17.

Kerns, F. Roger. *Simplicity of Water Purification*. Philadelphia: Dorrance & Company, 1972.

Martin, Francis, and Greg Prelewicz. Storage in Water Distribution Systems. www.cee.vt.edu/ewr/environmental/teach/wtprimer/storage/storage.html.

McClintic-Marshall Corporation. *Elevated Water Tanks and Standpipes*. Bethlehem, PA: McClintic-Marshall Corporation, n.d.

Ramalho, R. S. *Introduction to Wastewater Treatment Processes*. New York: Academic Press, 1977.

Rudolfs, Willem. *Principles of Sewage Treatment*. Washington, DC: National Lime Association, 1955.

Wade, Beth. "Tanks" for the Memories. *American City & County*. November 1, 1998. http://americancityandcounty.com/mag/government_ tanks_memories/.

Woolford, George. *Elevated Tanks for Fire Protection*. Philadelphia: George Woolford, 1900.

NOTES

1 Kathy Jesperson and Vipin Bhardwaj, Distribution 101: How Does Water Get from the Source to Your Tap, *On Tap: Drinking Water News and Information for America's Small Communities* 1:1 (Spring 2001): 14-15.

2 Harold E. Babbitt and James J. Doland, *Water Supply Engineering* (New York: McGraw-Hill Book Company, 1955), 72.

3 Babbitt and Doland, 86–87.

4 Babbitt and Doland, 130–132.

5 Babbitt and Doland, 149.

6 Babbitt and Doland, 178.

7 W. A. Hardenbergh, *Water Supply and Purification* (Scranton, PA: International Textbook Company, 1952), 247.

8 Jesperson and Bhardwaj, 3.

9 John W. Clark, Warren Viessman, Jr., and Mark J. Hammer, *Water Supply and Pollution Control*, 2nd edition (Scranton, PA: International Textbook Company, 1971), 285–287.

10 Clark, Viessman, and Hammer, 302.

11 Clark, Viessman, and Hammer, 310–312; Hardenbergh, 324–330.

12 Clark, Viessman, and Hammer, 342–343.

13 Clark, Viessman, and Hammer, 383.

14 Clark, Viessman, and Hammer, 390.

15 Clark, Viessman, and Hammer, 395–397.

16 Clark, Viessman, and Hammer, 407-408.

17 Clark, Viessman, and Hammer, 409.

18 Clark, Viessman, and Hammer, 423.

19 Hardenbergh, 432; Clark, Viessman, and Hammer, 536– 537.

20 Clark, Viessman, and Hammer, 356.

21 Clark, Viessman, and Hammer, 357.

22 Charles P. Hoover, *Water Supply and Treatment* (Washington, DC: National Limes Association, 1936), 15.

23 Clark, Viessman, and Hammer, 357–358.

24 Clark, Viessman, and Hammer, 358–361.

25 Babbitt and Doland, 489.

26 Babbitt and Doland, 500–502.

27 Babbitt and Doland, 495.

28 Francis Martin and Greg Prelewicz, Storage in Water Distribution Systems, www.cee.vt.edu/ewr/environmental/teach/wtprimer/storage/storage.html.

29 Clark, Viessman, and Hammer, 164–165.

30 The American Water Works Association has designated these three towers and many other water works properties as "Water Landmarks." For a complete list of these landmarks, see www.awwa.org/Membership/Content.cfm?ItemNumber=498&navItemNumber=1458.

31 Babbitt and Doland, 319.

32 Chicago Bridge and Iron Company, Standpipes, www.cbi.com/services/standpipes-and-reservoirs.aspx.

33 Martin and Prelewicz, 4.

34 Hardenbergh, 108–109.

35 Babbitt and Doland, 321.

36 Beth Wade, "Tanks" for the memories, American City and County, November 1, 1998: 1, http://americancityandcounty.com/mag/government_tanks_memories/.

37 Babbitt and Doland, 324–325.

38 Chicago Bridge and Iron Company, Company History, www.cbi.com/about/history.aspx, 1.

39 Wade, 1.

40 Wade, 2. Known formally as the Gary-Hobart Water Tower, it was designated a Water Landmark by the American Water Works Association in 1970.

41 McClintic-Marshall Corporation, *Elevated Water Tanks and Standpipes* (Bethlehem, PA: McClintic-Marshall Corporation, n.d.), 5.

42 Martin and Prelewicz, 4.

43 Wade, 2.

44 Chicago Bridge and Iron Company, 2–3.

45 Chicago Bridge and Iron Company, Hydropillar, www.cbi.com/services/hydropillar-elevated-tanks.aspx.

46 Jesperson and Bhardwaj, 4.

47 Clark, Viessman, and Hammer, 193.

48 Clark, Viessman, and Hammer, 199.

49 Clark, Viessman, and Hammer, 199, 211–214.

50 Clark, Viessman, and Hammer, 211–212, 215.

51 Clark, Viessman, and Hammer, 291.

52 Clark, Viessman, and Hammer, 311.

53 Clark, Viessman, and Hammer, 312–313.

54 Clark, Viessman, and Hammer, 350–351.

55 Clark, Viessman, and Hammer, 292.

56 F. Roger Kerns, *Simplicity of Water Purification* (Philadelphia: Dorrance & Company, 1972), 9.

57 Clark, Viessman, and Hammer, 319.

58 Clark, Viessman, and Hammer, 326.

59 Clark, Viessman, and Hammer, 454.

60 Blasland, Bouck & Lee, Inc., New York City Landmark and National Register Eligibility Assessment of the Tallman Island Water Pollution Control Plant for the Tallman Island Plan Upgrade Project (New York: Blasland, Bouck & Lee, Inc., 2001), D-4.

61 Blasland, Bouck & Lee, Inc., D-4, 8.

62 Clark, Viessman, and Hammer, 509–517.

63 Clark, Viessman, and Hammer, 485.

64 Mark J. Hammer, *Water and Waste-Water Technology* (New York: John Wiley & Sons, Inc., 1975), 366–367.

65 Clark, Viessman, and Hammer, 533–535.

66 Clark, Viessman, and Hammer, 538.

67 Clark, Viessman, and Hammer, 595.

68 Blasland, Bouck & Lee, Inc., D-4–8.

Chapter 9

Manufactured Gas Plants

The last manufactured gas plants in the United States ceased operation in the early 1970s, and remnants of these plants remain throughout the country.[1] Many of these plant sites are still owned by utility companies and are used as electrical substations, storage yards, truck garages, office buildings, and facilities of major generating stations. Many still contain gas regulating facilities, due to their access to the gas distribution system. Some are simply abandoned industrial property, but others have been turned into commercial/retail properties, schools, and even residences. Because of the manufacturing process, most such sites contain significant amounts of hazardous wastes, particularly coal tar, and are the focus of government-mandated clean-up activities.[2]

History

Coal gas was first produced and consumed in Great Britain in 1792. The first commercial natural gas plant in Great Britain was established by the Scotsman, William Murdock, at Smethwick, West Birmingham, by 1805.[3] At the end of the eighteenth century, experiments using manufactured gas as a source of illumination were well underway in the United States. Among these early experiments were those of M. Ambroise & Company of Philadelphia. This company, which specialized in the manufacture of fireworks, used coal gas to illuminate fanciful chandeliers at a gathering at a local amphitheater. In 1802, illuminating gas was used in a sideshow at the Haymarket Gardens in Richmond, Virginia.

In Baltimore, Rembrandt Peale, a member of the nationally prominent family of artists, used coal gas to illuminate an exhibit in his museum and gallery in 1816. Peale's use of gas attracted much attention in the city, and in July of that year, the city passed an ordinance permitting Peale and his associates to manufacture gas, lay distribution pipes, and supply the city with coal gas for street lights. Early the following year, the group chartered the Gas Light Company of Baltimore, the first coal gas company in the United States.

In Newport, Rhode Island, in 1813, Daniel Melville initially used coal gas, generated from a small backyard plant, to light his own home. Soon after, he obtained a patent for his method of gas manufacture and convinced the owners of small textile mills in Massachusetts and Rhode Island to install gaslights for illumination.[4]

Soon, commercial plants sprung up around the country. By 1859, a total of 297 manufactured gas companies served an estimated 4,857,000 customers in the United States. To meet an ever-growing demand, more and bigger manufactured gas plants were constructed. By the mid-1870s, most American communities with a population of 10,000 or more had at least one manufactured gas plant.[5]

New technologies were developed leading to the manufacture of more efficient forms of coal gas, most notably carbureted water gas. This technology was invented by T. S. C. Lowe of Norristown, Pennsylvania, in 1872–1875, and its rights were sold to the United Gas Improvement Company in 1884.[6] This gas had the advantage that it could be economically supplied to small and large markets alike. The peak of manufactured gas production in the United States was reached in the 1920s.

Advances in the manufacture of high-pressure pipes and pumping systems led to a shake-up in distribution technology in the nation's larger communities. The central station distribution system came into use. Utility companies could construct large gas plants and pipe their product under pressure substantial distances to their customers.

In the years between the two world wars, the manufactured gas industry declined in importance. Natural gas, electricity, coal, and petroleum products replaced manufactured gas in the industrial and residential sectors. No longer was coal gas used to illuminate homes and factories; no longer was manufactured gas used to fuel the nation's factories. Demand for the by-products of gas plants lessened as the chemical industry shifted to petroleum-based chemicals. A few manufactured gas plants remained in operation into the second half of the twentieth century, supplying the few industrial operations that had not converted to other energy sources.[7]

TECHNOLOGY

A manufactured gas plant is an industrial facility where gas was produced from coal, oil, and other feedstocks. The gas was stored and then piped to the surrounding area, where it was used for lighting, cooking, and heating homes and businesses.

Two main processes were used to produce manufactured gas. The older and simpler process is coal carbonization. In this process, coal was heated in closed retorts or beehive ovens. Inside these ovens, the coal was kept from burning by limiting its contact with outside air. Volatile constituents of the coal were driven off as a gas, which was collected, cooled, and purified prior to being piped into surrounding areas for use. The solid portion of the coal became a black, granular material called coke. Coke was used as a fuel for many industrial uses and for home heating because it burned hotter and more cleanly than ordinary coal. When coke was the primary product, the facility was called a coke plant.

The second process, carbureted water gas (or blue gas), yielded a mixture that burned hotter and more brightly than the alternative. Introduced in the 1870s, it became the primary method of gas production by the end of the century. In essential terms, it was a combination of the manufacture of water gas, heated fuel with the injection of steam, and oil gas, accomplished simultaneously in connected pieces of apparatus. At the peak of manufactured gas production in the United States in 1926, 252 billion cubic feet of water gas were produced.[8] A variety of water gas processes were developed, all of which involved a first step in which coke or coal was heated in a closed vessel or retort, into which steam was injected. As the result of a chemical reaction, a flammable gas mixture of methane and carbon monoxide was produced. Petroleum products were then sprayed into the hot gas mixture, resulting in another chemical reaction in which petroleum constituents were "cracked" to form methane, which increased the heating and lighting value of the gas.

COAL CARBONIZATION MANUFACTURED GAS PLANTS

Manufactured gas plants were sited in accord with several factors. These include the market for the product, access to water or rail for receipt of raw materials, buffering from nearby residential uses, and the character of the soil and the level of groundwater.[9]

The process of manufacturing gas in the twentieth century is summarized by Jerome Morgan in his 1935 text in the flow chart in Figure 9.1.

The major steps involved in modern coal carbonization gas manufacturing include coal preparation, carbonization, disposal of coke, separation of volatile products, extraction of tar, scrubbing of naphthalene and cyanogen, secondary condensation, and purification of iron oxide.

Coal Preparation

Before charging into the carbonizing apparatus, the coal was weighed and crushed. The degree of crushing depended on the type of carbonizing apparatus. After crushing, the coal was typically stored in elevated bunkers. The coal was charged into the carbonizing apparatus by hand, by gravity, or machine.

The most important machines used for crushing and pulverizing were the Bradford breaker (Figure 9.2), the reciprocating jaw crusher, gyrating or rotating crusher, hammer mill, disk pulverizer, and the roll mill.

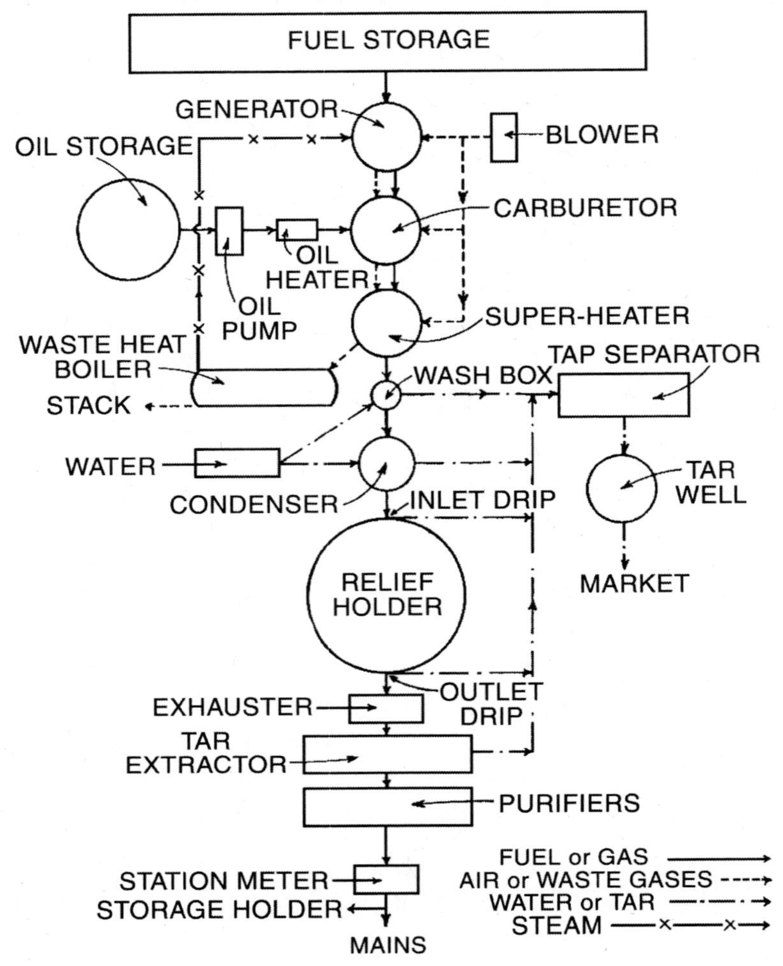

Figure 9.1. Material flow sheet for a carbureted blue gas or water gas plant. Reprinted from Jerome J. Morgan, *A Textbook of American Gas Practice,* volume I (Maplewood, NJ: Jerome J. Morgan, 1931), 454.

Carbonization

Carbonization took place in the retort house of the complex, a brick or concrete structure with a steel frame and a trussed roof (Figure 9.3). The carbonization apparatus could consist of horizontal, inclined, or vertical gas retorts, or of coke ovens. Coke that was destructively distilled forms two types of products: the nonvolatile residue, the coke; and the volatile products, including tar vapors, ammonia, water vapor, cyanogens, naphthalene, light oils, hydrogen sulfide, and the gas itself.

Horizontal retorts were typically made of silica material in a D-shape with the flat side of the D on the bottom. Horizontal retorts were grouped together in a bench. For larger works, benches of eight or nine through retorts were commonly used.

Figure 9.2. Drawing showing construction of a Bradford coal breaker and cleaner (Morgan 1931: 307).

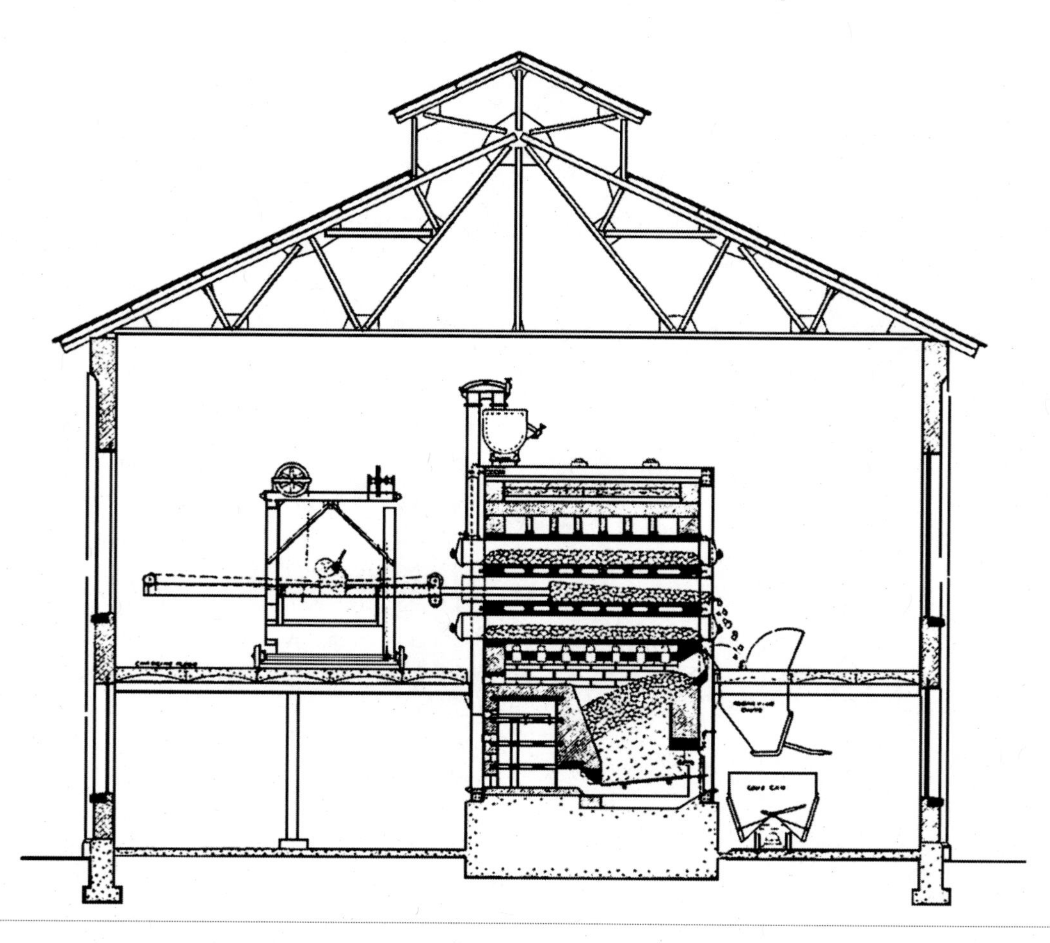

Figure 9.3. Cross section of small retort house (Morgan 1931: 188).

Disposal of Coke

The produced coke was red hot when it emerged from the carbonizing apparatus. A portion was fed directly into the producer, which furnished producer gas for heating the retorts and the ovens. The modern gas producer was a vertical vessel approximately cylindrical in shape. Its body consisted of an iron shell lined with fire brick and mounted on iron supports. The rest of the hot coke was quenched, screened to remove the breeze, and then either used in carbureted blue gas plants or sold as domestic or industrial fuel.[10]

Separation of Volatile Products

Gas was one of the two major products of the destructive distillation of coke. The gas, which carried all other volatile products with it, left the carbonizing apparatus at a high temperature and passed through an off-take pipe to the hydraulic main. The separation of volatile products began in the off-take pipe.[11] An exhauster, a form of rotary gas pump, took the gas away from the hydraulic main under a slight vacuum and furnished the pressure necessary to force it through the rest of the apparatus and into the holder.[12]

Extraction of Tar

Tar extractors collected and removed the remaining tar from the gas which was present as a fog. It did this by passing the gas at high velocity through small holes either against an iron plate or into water.[13]

Scrubbing of Naphthalene and cyanogen

Naphthalene and cyanogen scrubbers were used in larger gas works to further purify the gas. Typically, a horizontal scrubber was used. The type of scrubber used to remove ammonia (discussed below) was also used to remove naphthalene. In fact, a single horizontal rotary scrubber could be used for all three, a single section devoted to each. The scrubbing liquid for absorbing naphthalene was an oil of tar. The scrubbing liquid for removing cyanogen was generally a water solution of ferrous sulfate.[14]

Secondary Condensation

The next impurity to be removed from gas was ammonia. It was absorbed in water, but in order to do this efficiently the gas had to be cooled to about 65 degrees Fahrenheit. If the type of condenser used was one in which the cooling water came into direct contact with the gas, the cooling of the gas and the scrubbing of ammonia could be combined.

The simplest form of ammonia washer was a tower filled with an inert material through which water trickled while the gas passed upward. A more efficient type was a tower where water was sprayed to form a fine rain; the gas met the water as it ascended. A third type was the rotary scrubber, the same type used for naphthalene and cyanogen.[15]

Purification of Iron Oxide

The object of iron oxide purifiers was to remove hydrogen sulfide from the gas. The purifiers were large boxes or tanks containing a mixture of iron oxide and wood shavings supported on wooden grids. The gas was passed either upward or downward through the mixture. To ensure complete removal of hydrogen sulfide, several boxes were typically used in series.[16]

CARBURETED WATER GAS PLANT

The major components of the carbureted water gas or blue gas plant included the generator, the carbureter, the oil sprayer, the superheater, and the relief holder. The relationships of these major pieces of apparatus to one another are shown in Figure 9.4. In the course of producing blue gas, a large amount of the heat generated is absorbed by the gas, and the temperature of the fuel in the generator is lowered. To supply the heat needed for the process, it was often economical to alternate gas-making periods, called runs, during which steam passes through the fuel, with heating periods, called blows, during which air passes through the fuel to bring it up to temperature.[17]

The most common fuel used for the manufacture of blue gas was anthracite. As Jerome Morgan notes in his *Textbook of American Gas Practice*, anthracite is the densest, most concentrated form of commercial carbon, and this gives a greater weight of fuel in the fuel bed and permits more heat to be stored up in the fire during the blow. Other fuels used in blue gas production included coke and bituminous coal.[18]

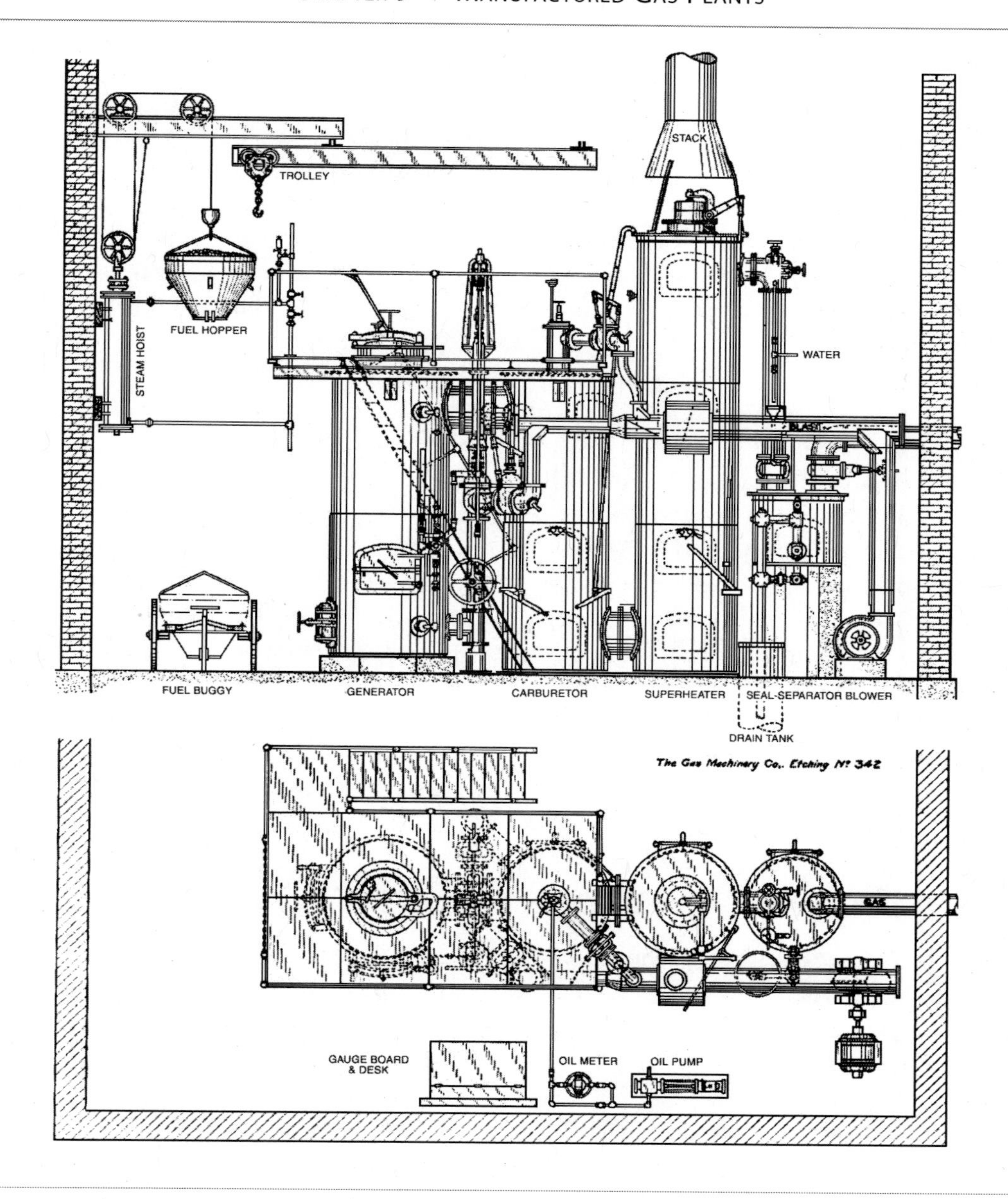

Figure 9.4. Plan and elevation of a small hand-operated carbureted blue gas apparatus (Morgan 1931: 487).

Steam

Steam was used for two primary purposes in manufactured gas plants: for heating around the plant and as a raw material for the manufacture of carbureted gas. The essential elements of steam production in the gas plant include the steam boiler, the feedwater pipe, and the superheater.

Superheater

Superheated steam, a true gas, was produced when all of the water in a vessel was evaporated or when steam was separated from the water and additional heat caused a rise in temperature. If during further application of heat the pressure remained constant, the steam was said to be superheated.[19] In gas manufacture, superheaters could be either integral to the boiler or separately fired.

The superheater and the carbureter were of similar construction. Both were fire-lined steel shells of about the same diameter as the generator. The superheater was about 8 to 10 feet taller than the generator. As with the generator, the lining of the superheater consisted of an inner and outer layer of fire brick, as well as an intermediate insulating layer. The carbureter and the superheater were connected at the bottom by a fire-brick-lined passage or gas-way.[20]

Generator

The generator was the site of the reaction of steam with fossil fuel. It was equipped with a blast connection, bottom and top steam inlets, and bottom and top gas offtakes.[21] The foundation of a flat-bottomed generator consisted of a ring or circle of masonry or concrete. The top of the foundation was at the level of the clinkering floor, while the operating floor was on a level with the top of the generator. The grate was usually at the level of the bottom of the clinkering doors, about 3 feet from the bottom of the generator. Generators were typically lined with fire clay or brick.[22]

Carbureter

The carbureter was the site of injection of light petroleum oils for gas light illumination. It was a fire-lined brick steel shell of about the same diameter and height as the generator. Its foundation was of concrete or masonry. To prevent overheating of the bottom plates, the plate often rested on I-beams spaced a few inches apart. The lining of the carbureter consisted of an inner and outer layer of fire brick, as well as an insulating layer 1 or 2 inches thick. The carbureter and the superheater were connected at the bottom by a fire-brick-lined passage or gas-way.[23]

Oil Sprayer

As Jerome Morgan noted, gas oil represented about half the cost of materials in the manufacture of carbureted water gas, so efficient dissemination of the oil in the carbureter was critical. The oil sprayer was located at the top of the carbureter. Two types of oil sprayers were commonly employed: the solid stream type and the mist type. The former type forced the oil in a number of fine sprays at a pressure of 15 to 30 pounds per square inch, while the latter delivered the spray at higher pressures, up to 250 pounds per square inch.[24]

Relief Holder

Because gas-making is an intermittent process, a small gas holder called a relief holder was provided so that the flow of gas through the main portion of the purifying apparatus could be at a constant rate.[25] The purifying apparatus employed in a blue gas plant was similar to that used in a coal carbonization plant.

Measuring and Storing

From the purifiers, the gas traveled to the station meter where it was measured, calculated at standard conditions of 60 degrees Fahrenheit and 30 inches of atmospheric pressure. Meters could be one of three types: volume or displacement meters, velocity meters, and heat capacity meters.[26]

Following the production of gas, it was generally necessary to store it until needed for use. The vessels used to store gas are known as gas holders and may be classified as low pressure or high pressure and water-sealed or waterless.

GAS HOLDERS

Gas holders were erected to store the gas produced. Most were constructed of iron and were usually double- or single-lift types. A contemporary observer describes how a gas holder looked and worked:

> To the untutored eye they present the appearance, when fully distended, of circular castles or forts, without portholes, embrasures or sally ports, or to the less military mind they might suggest sections of two enormous boilers, one sliding within the other, and set vertically into the ground. This [ground] tank [or pit] contains sufficient water to prevent the gas from escaping under the edge of the holder. When exhausted, the sections slide one within the other, like a telescope when shut up, and the whole affair sits down in the tank so that the top is nearly on a level with the surface of the ground. As the gas is let in and the pressure increases, the huge iron cylinders rise up and the inner one slides up until the holder is fully extended. These are called telescopic holders. Some are made with only a single

section, or "single lift" as it is called (*New York Times*, 7 April 1872).[27]

A large two-lift gas holder had a top section with a diameter of 100 feet and a height of 22 feet, and a lower section diameter of 101 feet 6 inches and a height of 22 feet. Such a gas holder had a capacity of 333,000 cubic feet of gas.

Low-pressure water-sealed holders were of two types: simple or single lift, and multiple lift, telescopic. Low-pressure waterless holders were either tar-sealed or dry sealed. A single-lift holder consisted of an inverted bell which dipped into a tank of water so that it could rise and fall according to the amount of gas confined within it (Figure 9.5). A guide frame prevented the bell from toppling over as it ascended. Pipes drew off the gas above the level of the water.[28]

In the single-lift gas holder, the depth of water in the tank had to equal the height of the bell. Because of the cost of constructing large, deep tanks, the limit of the economical size of this type of holder was soon reached. To increase the capacity of the holder without increasing the size of the tank, the principle of the telescopic or multiple-lift holder was developed (Figure 9.6).[29]

Gas holders were usually set on a foundation consisting of a circular slab of reinforced concrete. In constructing the foundation, the gas inlet and outlet pipe were laid in concrete to the point where they came up through the bottom of the tank, and a valve pit to house the inlet and outlet valves was provided.[30]

Early gas holders were constructed with frames consisting of cast-iron columns and girders. Modern gas holders were constructed of frames consisting of steel structural shapes. To guide the sections of the holder and keep them from tilting as they traveled up and down, carriage bearing guide rollers were fastened to the top curb of each section opposite the uprights of the guide frame.

After the tank was completed, the crown support was erected. This consisted of radiating rafters resting on the columns in the tank. The rafters were connected together to form a skeleton whose top surface had the same shape as the crown of the holder.[31]

Waterless Holders

Low-pressure gas holders not requiring the water-filled tank and water seals were first developed in Europe in the early twentieth century. Two types of waterless

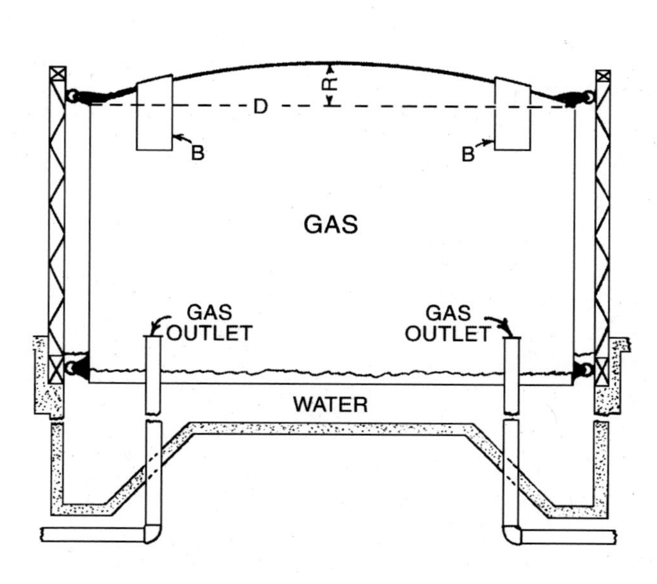

Figure 9.5. Diagrammatic section of a single-lift gas holder. Reprinted from Jerome J. Morgan, *A Textbook of American Gas Practice*, volume II (Maplewood, NJ: Jerome J. Morgan, 1935), 4.

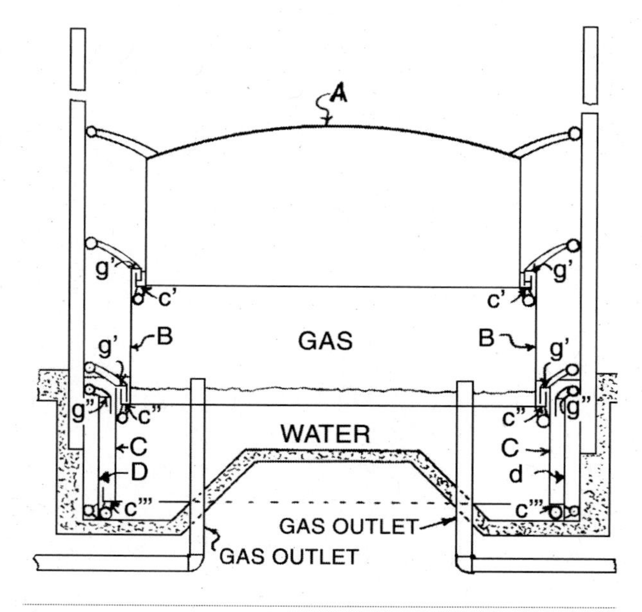

Figure 9.6. Diagrammatic section of a multiple-lift gas holder (Morgan 1935: 5).

holders were developed. In both types the gas was confined in a fixed shell by a piston free to move up and down and was supported by the pressure of the gas beneath it. In the M.A.N.-type of holder, the shell had a polygonal cross section, the undersurface of the piston was a horizontal plane, and the joint between the piston and the shell was made gas-tight by means of a tar seal.[32] The Klonne type of waterless holder had a shell with a circular cross section, a piston with a crowned undersurface, and a dry, friction joint between the piston and the shell. This was called a dry-sealed holder.[33]

Gasholder Houses

Gas holders in colder climates, particularly in New York State and New England, were sometimes constructed with houses, a structure that surrounded the iron gas holder (Figure 9.7). Gasholder houses were constructed for a variety of reasons. The structure protected the iron holder from the elements and enabled it to be built of thinner plates, since the holder itself would not have to withstand wind pressure. The enclosure also prevented the water in the holder pit from freezing; the water formed the seal to prevent loss of gas. The house was also considered an economical measure, as it reduced the condensation of gas in the cold weather, and was also viewed as an attractive architectural element of the gas works complex. In 1971, a total of ten brick, wood, or stone gasholder houses remained in Massachusetts, Rhode Island, New Hampshire, and New York.[34]

COMPRESSORS AND GOVERNORS

In smaller systems distributing manufactured gas, the pressure on the gas in the storage holder was sufficient

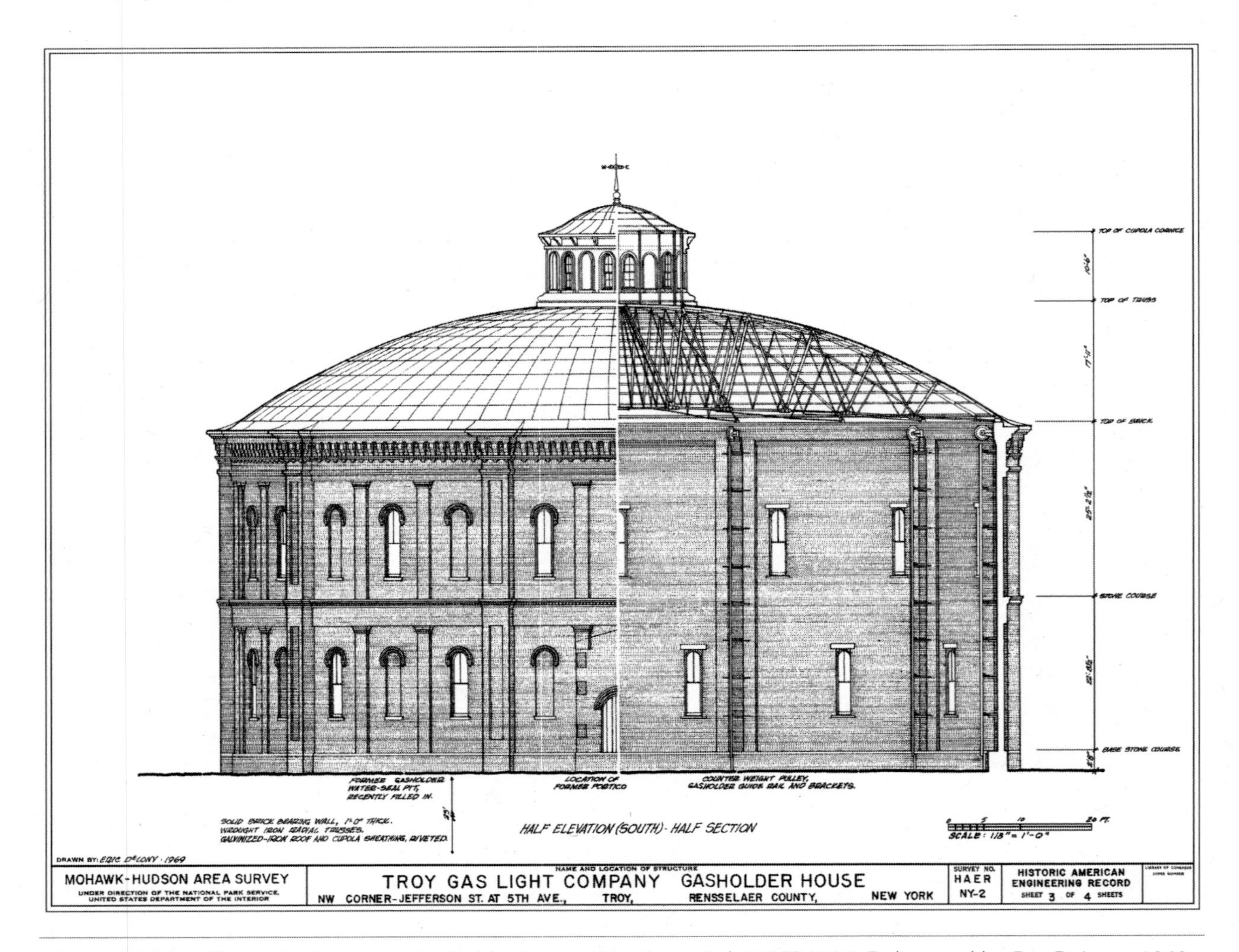

Figure 9.7. Troy Gas Light Company, Gasholder House, Troy, New York. HAER NY-2. Delineated by Eric DeLony, 1969.

to force the gas through the mains. In the modern distribution systems, it was generally necessary to supplement the holder pressure by means of boosters or compressors, and it was almost always necessary to use governors to control gas pressure at various points in the system. In the case of modern natural gas transmission, compressor stations for pumping the gas were located at regular intervals along the line.

REFERENCES

Hatheway, Allen W. Gas Plant Components. Former Manufacturing Gas Plants website, www.hatheway.net/03_gas_plant_components.htm.

Morgan, Jerome J. *A Textbook of American Gas Practice*. Volume I: *Production of Manufactured Gas*. Maplewood, NJ: Jerome J. Morgan, 1931.

—. *A Textbook of American Gas Practice*. Volume II: *Distribution and Utilization of City Gas*. Maplewood, NJ: Jerome J. Morgan, 1935.

Waite, Diana S. Troy Gas Light Company, Gasholder House, Troy, New York. HAER NY-2, 1969.

NOTES

1 According to Allen W. Hatheway, Ph.D., there are between 32,860 and 50,108 individual sites at which substantial amounts of coal tar residuals can be expected to be encountered (Former Manufactured Gas Plants in the United States). See www.hatheway.net/05_fmgp_us.htm.

2 An excellent source of information on manufactured gas plant (MGP) cleanup and former locations of MGP facilities is www.hatheway.net.

3 Allen W. Hatheway, Ph.D., Gas Plant Components, 3. Available at www.hatheway.net/03_gas_plant_components.htm.

4 Jerome J. Morgan, *A Textbook of American Gas Practice*, volume I (Maplewood, NJ: Jerome J. Morgan, 1931), 1–2.

5 Heritage Research Center, Ltd., A Brief History of the Manufactured Gas Industry in the United States, 2005: 1–2, www.heritageresearch.com/manufactured_gas_B.htm.

6 Morgan, 451; Hatheway, Gas Plant Components, 2.

7 Heritage Research Center, Ltd., 2–3.

8 Morgan, 451–452.

9 Morgan, 146-147.

10 Morgan, 149.

11 Morgan, 149.

12 Morgan, 153.

13 Morgan, 155.

14 Morgan, 154.

15 Morgan, 156.

16 Morgan, 156.

17 Morgan, 26.

18 Morgan, 380–384.

19 Morgan, 329–330.

20 Morgan, 485–487.

21 Morgan, 452.

22 Morgan, 401–411.

23 Morgan, 485–486.

24 Morgan, 497–500.

25 Morgan, 452.

26 Morgan, 156–157.

27 Diana S. Waite, Troy Gas Light Company, Gasholder House, HAER NY-2, 1971, 4.

28 Jerome J. Morgan, *A Textbook of American Gas Practice*, Volume II: *Distribution and Utilization of City Gas*, 2nd edition (Maplewood, NJ: Jerome J. Morgan, 1935), 3–4.

29 Morgan 1935, 4.

30 Morgan 1935, 8

31 Morgan 1935, 12–13.

32 Morgan 1935, 32–33.

33 Morgan 1935, 32.

34 Waite, 5.

CHAPTER 10

Cotton Processing and Textile Production

Particularly in the New England states, many former textile mill buildings remain and line the riverfronts of cities. Few active cotton and woolen mills remain in operation in the northeastern states. Many such buildings have been converted to other uses, but occasionally remnants of the former textile production remain.

More recent textile mill buildings remain elsewhere in the country, and some are still operating, although increasing numbers are shuttered as more and more textile production moves overseas. Another textile-related resource primarily found in the southern states is the cotton gin, described below.

COTTON GINNING

After cotton is picked, the first step in its transformation to textiles is ginning. For millennia, separating cotton fiber from seeds proved an almost insurmountable problem. Massachusetts-born Eli Whitney and his invention of the cotton gin solved this problem; he was granted a patent in March 1794. The cotton gin is a machine that separates the seeds, hulls, and foreign material from cotton fibers. While Whitney's original hand-cranked gin was small in size, later gins were large structures capable of processing large loads of raw cotton. In addition to the gin itself, the ginning plant typically contained extractor-feeders, cleaners, hull separators, and driers.[1]

Extractors and hull separators remove hulls or burrs from the seed cotton. The typical method consists of passing the seed cotton over a cylindrical drum with circular saw teeth partially enclosed in a cylindrical screen housing. Cleaners are designed to separate the fine trash and leaf from the seed cotton. Cleaners consist of a series of cylinders with lengthwise bars or rows of pins or fingers. In driers, cotton is either drawn or blown through passages with hot air or drawn over heated coils.[2]

Gins are divided into two general classes: brush gins and air-blast gins. The former uses a large cylindrical brush to remove lint from the saws, while the blast gin uses a blast of air from a specially placed nozzle to remove lint from the saws. A gin consists of series of circular steel saws with specially shaped teeth. The saws have approximately 264 to 282 teeth around their circumference. A series of metal ribs are placed so that the saws project between them with a narrow clearance on each side. The clearance is too small to permit the passage of seeds but large enough to permit cotton fibers to be drawn through.[3]

One cotton gin is now a museum in Burton, Texas, in the Hill Country near Austin (Figure 10.1). This gin, whose design incorporates a technological development resulting from an 1893 patent issued to Robert Munger, used air to bring seed cotton from the wagon to a separator inside the building. The separator cleaned the cotton and separated it from the air flow. Cotton dropped from the separator into a sealed belt-distribution system that carried the cotton to feeders above a row of gin stands. Several gin stands were linked by flues to convey the ginned lint into a battery condenser above the bale press. Seed disposal was also automated,

Figure 10.1. Burton Gin. Burton, Texas. Photograph by the author, July 2005.

using a system of wooden conduits, screw conveyors, and bucket elevators. The machinery was driven by a diesel engine through a line shaft, pulleys, and flat belting.[4]

TEXTILE MILLS

The first textile manufacturing facility in the American colonies was a fulling mill established in Rowley, Massachusetts by Yorkshire immigrants in 1643.[5] The first actual cotton mill in New England and one of the first in America was the Beverly Manufacturing Company, established in the Massachusetts city of that name in 1789. This mill produced 8,000 to 10,000 yards per year.

The Englishman Samuel Slater has been called the father of the American cotton industry. In 1790, he and partners established a mill in Pawtucket, Rhode Island. This mill was the first in North America in which all processes of the Arkwright improved spinning and preparatory methods were placed under one roof.[6] By the end of the first decade of the nineteenth century, a total of 226 cotton mills were in operation in the United States. Pennsylvania and Massachusetts had the largest number with 64 and 54, respectively.[7]

The Embargo Act of 1807, which cut off the flow of English goods, proved an impetus to the young domestic textile industry.[8] Because importation of textile machines, parts, and plans ceased, Americans, including Francis Cabot Lowell and Paul Moody, set about inventing water-powered machinery. Lowell and Moody completed the first plain power loom in 1814. Several years later, the first power weaving in the country occurred in Fall River on a loom invented by Dexter Wheeler. Mule spinning was introduced at the Lyman Cotton Factory in Providence. Beginning in 1838, development of ring spinning and the ring frame started in the United States. In the 1830s, the textile card was introduced in the United States. By 1840, there were a total of 2.3 million spindles in the country, consuming about 297,000 bales of cotton. By 1860, this had increased to 5.2 million spindles using 843,000 bales of cotton.

The most significant technological breakthrough in late nineteenth-century American textile manufacturing was the introduction of the Northrop Loom, developed by J. H. Northrop of George Draper and Sons, Hopedale, Massachusetts (Figure 10.2). The loom, the first automatic power loom used in the United States, was described by its inventors, as a "bobbin changing device, a filling hopper from which bobbins or cops are automatically transferred to the loom shuttle, a peculiar shuttle which can be threaded automatically by the motion of the loom, devices that act to stop the loom if the shuttle is not in position and a warp stop to prevent the making of poor cloth." [9]

In the late nineteenth century, because of the reluctance of some New England cotton manufacturers to embrace changing technology, combined with the presence of lower labor and utility costs in southern states, production began to shift from the north to the south. While in 1880, southern mills included only 522,000 spindles and 12,000 looms, compared with 10.1 million spindles and 212,000 looms in the north, by 1910 the gap began to close, with 11.2 million spindles in the south and 17.4 million spindles in the north. By the mid-twentieth century, 75 percent of the spindles and 70 percent of the looms were located in the southern states.[10]

Cotton Textile Manufacture

Historically, a standard series of steps were followed to transform raw cotton into fabric. These steps include opening and picking, carding and combing, drawing and doubling, spinning, winding, and twisting. In a multi-floor textile mill, the cotton moved upward or downward through the mill during the course of these operations.

Opening and Picking

Cotton arrived at a mill in a compressed bale. Opening and picking the cotton consisted of several steps to transform the cotton from a tightly packed bale containing trash and other foreign matter into a uniform "lap" or sheet in roll form ready for carding. The cotton was separated into particles small enough for cleaning; then it was cleaned and reformed into tufts on a lap or sheet suitable for carding. This operation also typically in-

Figure 10.2. Northrop power looms, manufactured by Draper Northrop, Hopedale, Massachusetts. In Bamberg Textile Mills, Bamberg, South Carolina. HAER No. SC-20, National Park Service. Photograph by Jack E. Boucher, 1986–1987.

cluded blending or mixing the various fiber properties for the final yarn or fabric.[11]

Forming the cotton into an even, flat sheet and rolling it into a lap, as well as the final opening and cleaning, are functions of the picker or scutcher (Figure 10.3). At one time, as many as three different picker operations were needed to perform these functions. In modern cotton mills, these operations are generally combined into a single-process picker assembly.[12]

Carding and Combing

After picking, the cotton was arranged in a series of unopened tufts of fiber varying in size. The card or carding machine separated these fibers into their individual elements, exposing and removing bits of leaf, trash, and other foreign matter enclosed by the unopened fiber aggregates, and formed the cleaned, disentangled fibers into sliver for feeding the next process. Sliver is a continuous, untwisted strand of cotton fibers.[13]

Carding was accomplished by the action of card clothing, a system of inclined wires. Two surfaces covered with cloth usually worked together to achieve the objectives of carding. Two important actions were performed by the surfaces: carding and stripping. Carding action occurred when the wires of the two surfaces were inclined in opposite directions and the direction and rates of motion were such that one surface passed the other, point against point. Stripping action occurred when the wire of the two surfaces pointed in the same direction and the action was point against smooth side. Cards were heavy machines requiring close and accurate adjustment, and needed a firm, rigid frame (Figure 10.4).[14]

The purpose of combing was to separate the short fibers from the longer ones so that the combed fibers would be of a much more uniform length and of a longer average staple. In separating the short from long fibers, the long fibers were straightened to a considerable degree and small dirt particles were removed.[15]

The original comb and its successors were designed to feed the cotton in sheet form. As card sliver fed the comb, it was necessary to arrange the sliver in the form of a narrow lap. The fibers in a card sliver were not straight, and if put through a comb without any straightening, would yield a considerable percentage of long fibers in the comb waste. The less straightening the comb had to do, the fewer long fibers were in the waste.

Combing was quite expensive and added considerably to the cost of finished yarn. Combing added to the value of cotton yarn by improving uniformity and strength, spinning a finer count, and producing a smoother, more lustrous and cleaner yarn. Most yarns finer than 40 gauge were combed.[16]

Figure 10.3. Picker, Aldrich, Greenwood, South Carolina, 1923. Bamberg Cotton Mill, Bamberg, South Carolina. HAER SC-20, National Park Service. Photograph by Jack E. Boucher, 1986–1987.

Figure 10.4. Carding machine, manufactured by Carding Specialists Company, Ltd., Halifax, England. Bamberg Cotton Mill, Bamberg, South Carolina. HAER SC-20. Photograph by Jack E. Boucher, 1986–1987.

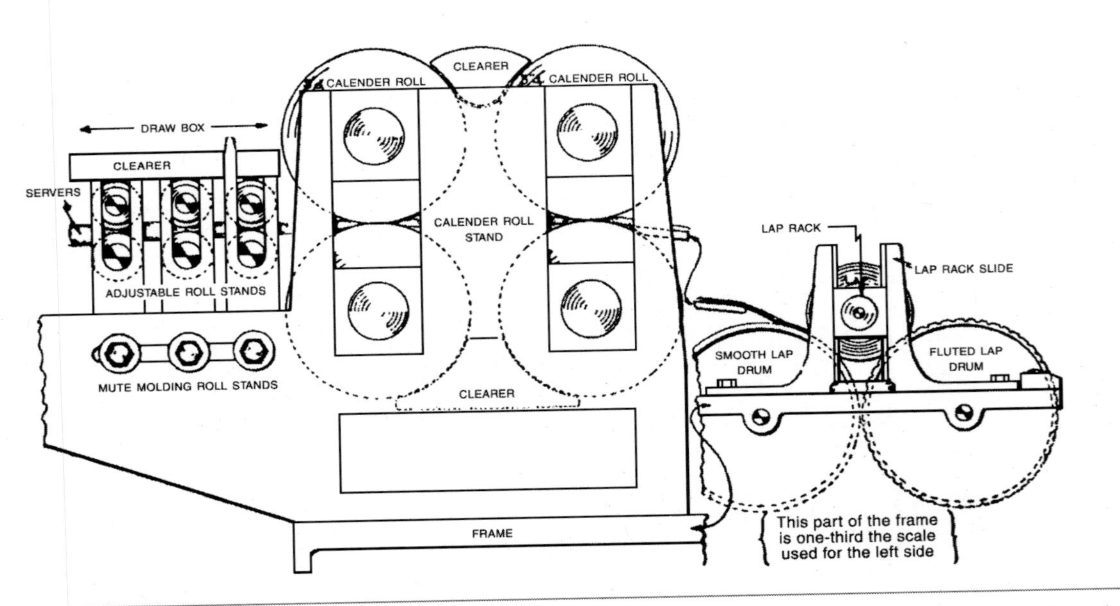

Figure 10.5. Side view of a three-roll sliver lapper. From Gilbert R. Merrill, Alfred R. Macormac and Herbert R. Mauersberger, *American Cotton Handbook* (New York: Textile Book Publishers, 1941), 227.

In the nineteenth and early twentieth centuries, cotton was prepared for the combing process by passing the card slivers through a sliver lapper (Figure 10.5) and then a ribbon lapper. Between 20 and 32 card slivers were fed to the sliver lapper. This machine usually had three pairs of drafting rolls followed by two pairs of calender rolls. A pair of lap drum rolls at the front were used to wind the drafted slivers into a lap varying in weight from 400 to 800 grains per yard in a width of about 9.75 inches and a diameter of 14 inches.

Laps from the sliver lapper were taken to the ribbon lapper. The purpose of this machine was to produce a lap without the lengthwise-ridged effect of the sliver lap and continue the drawing so that the fibers were more thoroughly straightened for feeding at the comber. Most ribbon lappers had four heads (four independent sections), each of which processed a single sliver lap, while some had as many as six heads.[17]

With the development of the lap winder, mills found a more efficient way to prepare materials for combing—namely, by a drawing process after carding, followed by lap winding to package the drawing slivers into a comber lap. The lap winder assembled cotton slivers into a uniform and compact lap and then used it as a supply for the comber.

The original comber, developed by the Frenchman Joshua Heilmann, used stationary nippers, while modern combers used swinging nippers. Combs were built in sections called heads, with the usual arrangement being six or eight heads per comb (Figure 10.6). Products of combs are termed combed slivers. The combing operation was divided into seven basic tasks: feeding the stock from a prepared lap; combing out short fibers, foreign particles, and neps; parallelizing fibers; detaching the combed fibers from the lap; piecing up the fleecy tuft of combed fibers with the fibers in the returned web; condensing the combed web into sliver and doubling the sliver on the table; drafting the doubled sliver through the draw box; and calendering and packaging the combed sliver into a container for further handling and processing.[18]

Drawing and Roving

Drawing is the process of progressively passing or sliding fibers by each other in a way that reduces the size of the strand but does not break its continuity. The action generally required several pairs of rolls running at different speeds. The purpose of all roller drawing was to straighten the fibers being treated and reduce the size of the strand. Straightening was important in arranging fibers more nearly parallel to one another and to the direction of the strand. This helped produce uniform, strong, and smooth yarn.[19]

Drawing frames were usually built in multiple units (Figure 10.7). The main parts of the draw frame are the creel, the drafting system, the calender rolls, coiler, and can-table. Pairs of rolls were the working elements of a drawing frame. They did the drawing, after

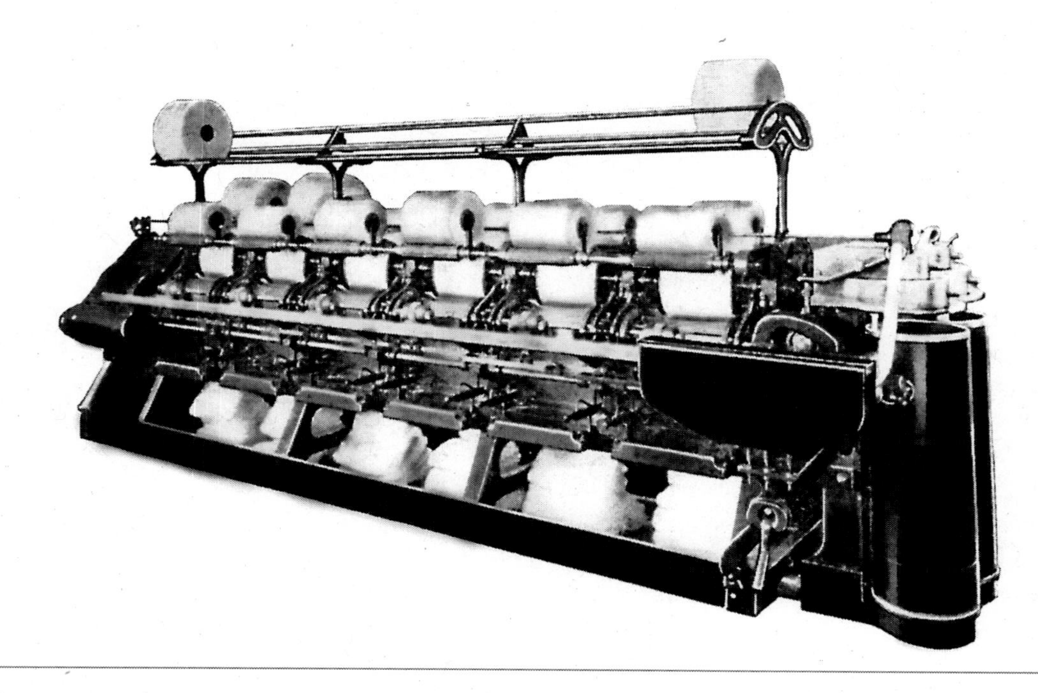

Figure 10.6. Cotton comber. From John Lister, *Cotton Manufacture* (London: Crosby Lockwood and Son, 1894), 27.

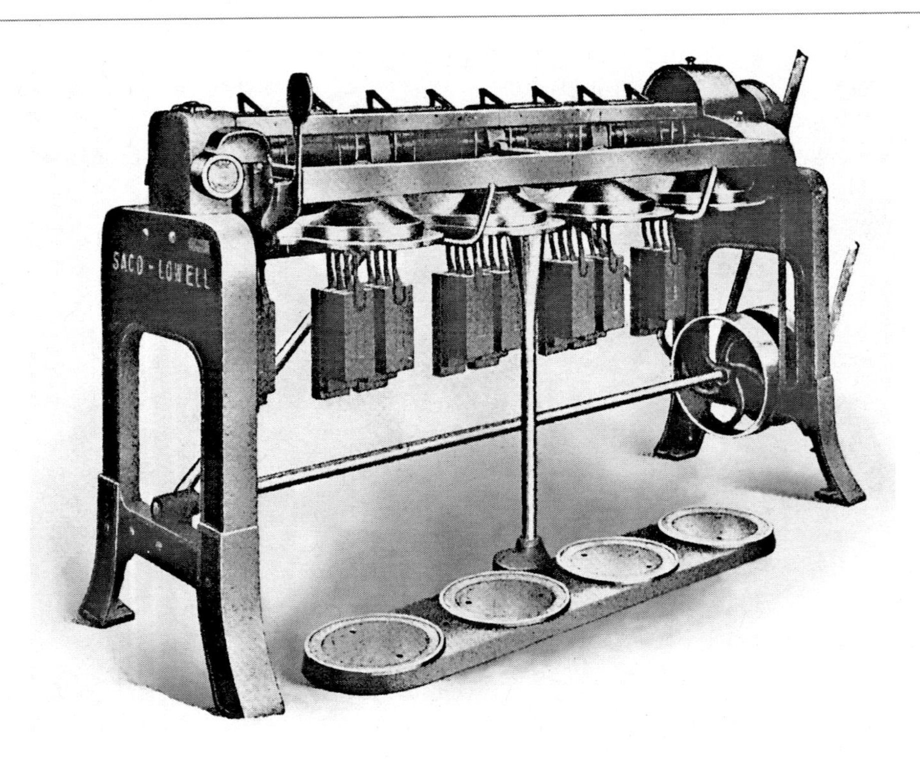

Figure 10.7. Front view of a four-roll drawing frame. Reprinted from Saco-Lowell Company, *Textile Machinery: Cotton Mill Shops* (Boston: Saco-Lowell Company, 1924), 162.

which the calender rolls drew the web produced into a sliver, and coiled it into the can.[20]

Roving is applied to both a process and its product. Historically, most mills had several roving operations, one following another. The number of operations depended on the count and quality of the yarn being spun. A roving, the product of the roving process, is a continuous, slightly twisted strand of cotton fibers that has not received its final reduction by drawing. The machine that produced roving took sliver or other rovings and reduced them by roller drawing, twisted them slightly as needed, and wound the product accurately on a special form of bobbin for further use in other roving frames or in spinning.[21]

Historically, there were four standard roving frames: slubber, intermediate, fine, and jack frames. The simplest and oldest drafting system used on roving frames was the three-roll drafting system in which three lines of roll were running at increasing speeds.

Spinning

The purpose of the spinning frame was to reduce roving to the required size of single yarn and to insert a suitable amount of twist (Figure 10.8). Spinning consisted of three essential steps: reduction of the strand by rolls similar to that used for roving; twisting of the strand after it had been reduced to the required size; and winding of the yarn onto a suitable package for future use. The product of spinning is a single cotton yarn. Spinning frames were usually arranged in long ranks in an open floor.

Winding

Yarn often required rewinding after spinning. Because of the relatively short lengths of yarn on individual spinning bobbins, it was not practical to use them directly in operations such as warping, twisting, and quill winding. In winding, larger packages were formed by combining the lengths from a number of spinning bobbins into a package such as a cone or tube that would allow for more continuous operation in subsequent processes.

In most winding machines, the spinning bobbins were mounted in approximately a vertical position below the winding unit. The yarn was withdrawn over the top end of the spinning bobbin, moved through yarn guides, tensioning devices, and slub catchers before winding onto the core.

Twisting

Twisting was done to produce ply yarn, which comprises two or more yarns twisted in a spiral about each other. If two or more ply yarns are then twisted together, the result is called a cord.

The most common equipment used to twist cotton yarn was the ring twister (Figure 10.9). The basic parts consist of rolls to deliver the yarn at a constant speed, a guide for the yarn, a traveler that could turn freely on a ring, and a rotating spindle which held the package of yarn.

Warping

The object of warping was the arrangement of threads in parallel order in order to compose a sheet of yarn of a specified length and width for the loom. This process was commonly undertaken by what is known as the direct system. In this system, the warp was prepared by placing the individual yarn packages in a large frame called a creel. Each yarn was then threaded through its

Figure 10.8. Spinning frame. Bamberg Cotton Mill, Bamberg, South Carolina. HAER SC-20:24, National Park Service. Photograph by Jack E. Boucher, March 1987.

Figure 10.9. Ring twister. From *Cotton Machinery* (Lowell Machine Shop, 1902).

own tensioning and stop motion device and passed through guides to the front of the creel where they were brought together to form the warp sheet. The yarns were uniformly spaced by placing them into dents of a comb or a reed before they were started around the beam. The reed kept the yarns properly spaced and controlled the width of the warp sheet.

Several machines were developed to simplify the process of warping. One is the warp-balling machine with its reed and curved creel and balling arrangement composed of a steel flyer carried by a series of four grooved rollers. The warper unwound the yarn from a large number of spools and placed it in an even sheet on a beam, a cylinder with heads at each end (Figure 10.10).[22]

The warping mill is a large skeleton drum or reel, with a central vertical shaft from which arms extend to the outside reel, and the whole was moved or rotated by motion from the line shaft. The creel is an upright frame forming the arc of a circle, with divisions from top to bottom wide enough to hold as many as 500 bobbins in a horizontal position.

Normally warps prepared by the direct system could not be used directly for weaving, since the number of ends they contained represented only a part of the total number of ends needed in the loom warp. These warps are termed section warps or back warps, and the beams they were wound on are known as section beams, warper beams, or back beams.

Loom warps were prepared by combining the warps from a number of section beams into one common warp sheet. If the loom warp required 5,000 ends, then a set of ten section beams, each containing 500 ends, could be used to build the loom warps.

The Warper

In one system, the creel was comprised of a series of vertical bars, each carrying nine "cheese holders" or spindles. (These were called cheese holders because the yarn wound on a cylinder resembled a block of cheese.) The holders were built with a spring detent that engaged the groove inside the bakelite sleeve of the cheese and fixed the cheese firmly in position.

The upper and lower ends of the vertical bars, which carried the cheese holders, were connected in endless sprocket chains running lengthwise of the creel. The cheeses were held stationary and the yarn pulled over the end. When the yarn was wound off the cheeses on the outside of the creel, a small motor was started, which caused the vertical bars to move the nearly empty cheeses from the outside to the inside of their respective creel sections.

Dye beams

When warp yarns were to be dyed, they could be wound on specially designed stainless steel beams. These beams had a larger diameter barrel than conventional beams, and the walls of the barrel were filled with holes that provided a means for the liquid to penetrate the warp yarns from inside the package.

Looms

After a warp had been drawn in, it was ready to be delivered to the loom. Because weaving was the interlacing of two or more systems of yarns at essentially right angles to each other, all looms were similar in their essential parts. Figure 10.11 illustrates the essential working parts of a loom.[23]

Woolen and worsted fabric weaving involve a series of processes in sequence: separation of the warp, insertion of the filling; placing picks into the cloth; warp supply; taking away of the woven cloth; filling bobbin/replenishment motion; automatic stop motions for center-filling, warp, and protection stop motions; and box motion, in multiple box looms, where more than one color filling is employed.[24]

Yarn Manufacture

The operational sequence of yarn production was similar to that of general cotton cloth production, with a few significant differences. After picking and opening, the raw cotton was carded. Following carding, the fiber was moved to the drawing frame in order to make the fibers parallel.

The next step in yarn production was the use of the slubbing frame. Slivers produced by the drawing frame were fine, but further attenuation was needed, combined with strengthening by a slight twist, to enable it to hold its shape and make it more convenient to handle. A slubbing frame received the cotton sliver and delivered it as a thinner strand coiled on a bobbin. The next step in yarn manufacture was either ring spinning or mule spinning or both.

The final major step in yarn production was dyeing, a process that usually occurred in a dye house. By the late nineteenth century, hand dyeing had been largely supplanted by machine dyeing. This machine included

Figure 10.10. Barber Coleman Warper. Bamberg Cotton Mill, Bamberg, South Carolina. HAER SC-20. Photograph by Jack Boucher, 1986–1987.

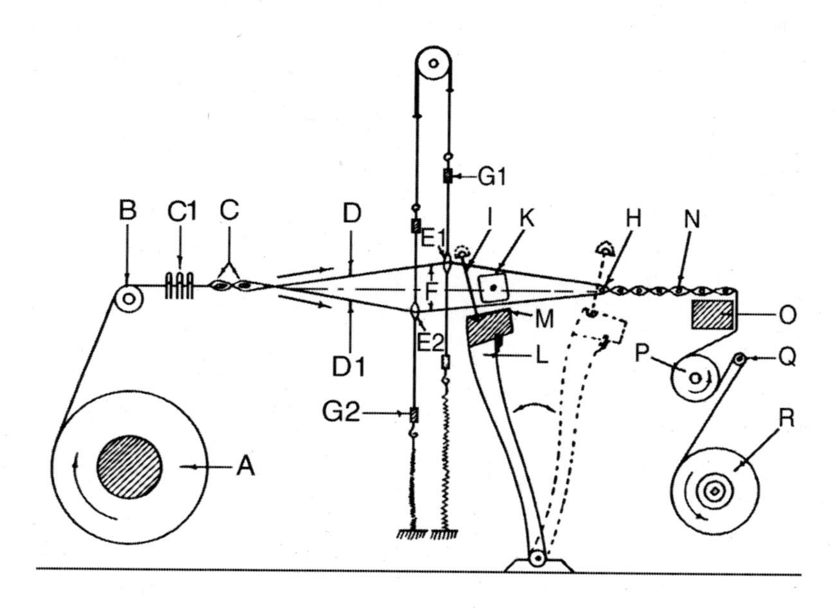

Figure 10.11. Side elevation of the essential working parts of a loom (Merrill et al. 1941: 459). **A,** warp and warp beam; **B,** whip roll; **C,** two lease rods; **C1,** warp stop motion; **D,** one or group of warp yarns in upper shed; **D1,** one or group of warp yarns in lower shed; **E1,** heddle eye in first harness; **E2,** heddle eye in second harness; G1, harness one; **G2,** harness two; **L,** lay or batten; **I,** reed and reed cap; **M,** race plate; **K,** shuttle in shed; **H,** fell of cloth; **N,** already woven cloth; **O,** breast beam; **P,** sand roll or take-up roll; **Q,** guide roll; **R,** cloth roll.

a wooden dye vat. Extending over the vat were reels and bobbins connected with suitable gearing so that they could be revolved. The bobbins were set in revolution and dropped into the dye vat until the yarn acquired the desired shade.

Woolen Manufacture

Initially woolen manufacture was largely home-based, and finishing, consisting of fulling, took place at a local fulling mill. Finer woolens were given additional finishing, including napping. Napping consisted of raising a brushed surface with abrasive implements such as hand cards. Early on, small shops were opened with a water-powered carding machine. By the beginning of the nineteenth century, the cotton carding machine brought from England by the Schofield brothers was widely adapted to wool. Arthur Schofield also invented a wool-picking machine that fluffed up raw wool for carding.

In general, the technology of wool manufacture lagged about 15 years behind cotton. It was not until almost 1830 that most of the machinery to complete the manufacturing process was developed. Napping and shearing were processes used only with wool. A cylindrical power napper was invented around 1797, and a cylindrical shearer became common by 1812.

Wool weaving machinery adopted Lowell's inventions for cotton. Initially, only plain narrow fabrics were woven, but broadlooms developed by the mid-1820s. Small carding shops gradually expanded to include a loom or two. Among the first full-fledged woolen factories was Pontoosuc Woolen in Southbridge, Massachusetts, erected in 1825–1826. By the Civil War period, numerous woolen mills had been erected in Vermont, taking advantage of the large number of sheep raised in the state.[25]

The woolen industry never developed into the gigantic industry that cotton became, and the technology of its machinery advanced at a slower pace. Mechanization generally followed a pattern of adaptation of cotton machines for use with wool.

The Scouring Train

The first major task in wool processing is cleaning. This took place in a machine known variously as a scouring train or a wool washing machine (Figure 10.12). The machine consisted of three or four tubs containing successively weaker solutions of water and sodium carbonate. Greasy wool was introduced at one end of the machine and moved from one tub to the next by means of forks and rollers. After passing through the final tub, the wool was stripped of its greasy coating.[26]

Pickers

The picker was developed for the English woolen industry in the late eighteenth century and was introduced to the United States in the early nineteenth century. The picker eliminated the need to beat the wool with sticks and separate the fibers by hand. The teeth in a rotating cylinder opened the fibers and scoured the wool of impurities, such as burrs and other vegetation. During the later nineteenth century, pickers constructed of cast iron and steel and produced at machine shops replaced the earlier wood machines.

Carding

During the carding process, wool fibers passed through a series of metal teeth that straightened and blended them into slivers. The process also removed residual dirt and other matter left in the fibers. Carded wool intended for worsted yarn was put through gilling and combing, procedures that remove short fibers and place the longer fibers parallel to one another. Then the sleeker slivers were compacted and thinned through drawing. Carded wool to be used for woolen yarn was sent directly for spinning.

Wool carding was typically a multistep process. The first or breaker card did the preliminary rough work on the woolen stock. In a carding machine, wool was carried by the main cylinder until the action of the worker cylinder against the main cylinder carded the wool. The wool that was picked up by the worker cylinder was in turn removed by the action of the stripper cylinder which was stripped of its wool by the main cylinder. Other cylinders transferred the wool from one end of the machine to the other.[27] Final carding was performed by the finisher card, while waste and wool cuttings were carded using a Garnett machine (Figure 10.13).

Spinning

Machine spinning of wool began in the late eighteenth century with the introduction of the spinning jenny in England. The spinning jenny had two major disadvantages. It was hand-powered and thus limited in the number of spindles that could be turned. It was also very difficult to use it to produce yarn of truly uniform size.

Figure 10.12. Improved wool washing machine. Reprinted from James Smith Woolen Machinery Company, *The James Smith Woolen Machinery Company: Manufacturers of Woolen Machinery, Including Machinery for the Preparation of Wool, Also Card Clothing of Every Description* (Philadelphia: James Smith Woolen Machinery Company, 1882), 41.

Figure 10.13. Three-cylinder breast Garnett machine (Smith 1882).

The jenny was superseded by the spinning jack invented in Britain in about 1810, which began to appear in American mills in the 1820s. The jack was powered by water or steam and could be adjusted to place an exact amount of twist and draft in the yarn. In the 1860s, experiments were made to produce a self-operating jack, and the first self-acting machine, termed a mule, was built in 1871.[28]

Woolen spinning involved three principal operations: drafting, or final drawing out; twisting, or insertion of twist; and winding-on, or packaging. The principal motions of a mule were in sequence: drawing-out motion, draft and twist motion, ease-up motion, backing-off motion, winding-on, and drawing-in or re-engaging motion.[29] In the first half of the twentieth century, the ring spinning frame began to replace the mule,

Figure 10.14. *a*, High speed, worsted ring spinning frame; *b*, Cross section of worsted ring spinner. Both from Werner Von Bergen and Herbert R. Mauersberger, *American Wool Handbook: A Practical Text and Reference Book for the American Woolen and Worsted Manufacturer, and Allied Industries* (New York: American Wool Handbook Company, 1938), 442.

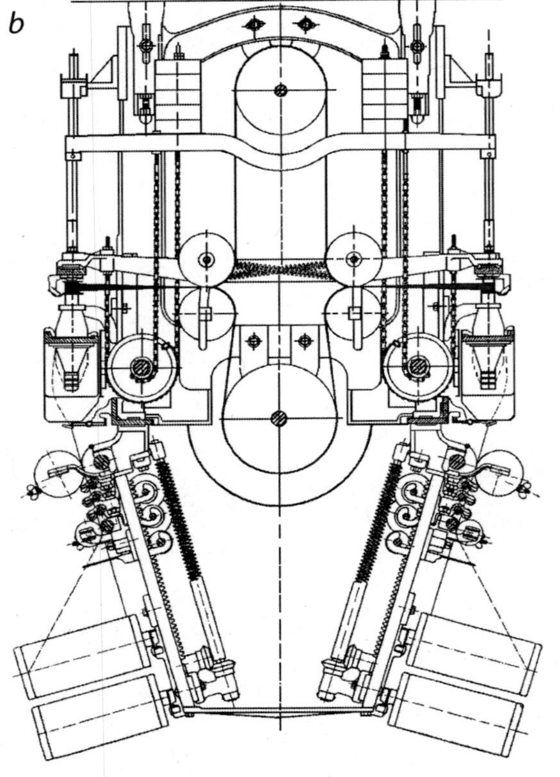

due in part to greater production and reduced floor space (Figure 10.14).

Weaving

Like many textile machines, the power loom was developed by the British textile industry. The first rudimentary power loom was developed by an English clergyman, Edmund Cartwright, in 1785. Power looms first appeared in the United States in 1815, but these early looms were too harsh in operation for delicate woolen yarns. With improved technology, by the 1830s many American woolen mills were equipped with power looms.[30] The first American iron power loom was invented by William Crompton and patented in 1837. In 1857, Crompton introduced his broad fancy loom (Figure 10.15). Among the advantages of this loom were that it could accommodate over a dozen harnesses, the patterns could be easily changed, and the action of the loom was gentle enough to accommodate a warp of woolen yarn.

Fulling Mills

When woolen cloth came off the loom, it had a loose weave, was dirty, and unattractive. Fulling is the process that cleaned, felted, and shrank the cloth. Cloth was placed in a vat of water with a detergent or a caustic substance.

In early fulling mills, a shaft was connected to a waterwheel or horse power at one end and used a set of cams at the other end to raise large wooden hammers (Figure 10.16). These were used in troughs containing cloth solutions. As the hammers hit the cloth, they turned it, changing the area being struck. This process was replaced by the circular fulling machine or mill, perfected by 1830.

The first water-powered fulling mill in the colonies was erected in 1643 at Rowley, Massachusetts. By 1810, there were about 1,682 fulling mills. Soon after, fulling began to be subsumed into the factory process, yet there were still about 1,000 fulling mills remaining in the 1880s. A common practice was to install a fulling ma-

Figure 10.15. Crompton & Knowles Jacquard Loom. Reprinted from Crompton & Knowles Loom Works, *Weaving Machinery for Cotton, Silk, Woolen, Worsted and Special Fabrics* (Worcester, MA: Crompton & Knowles Loom Works, 1920).

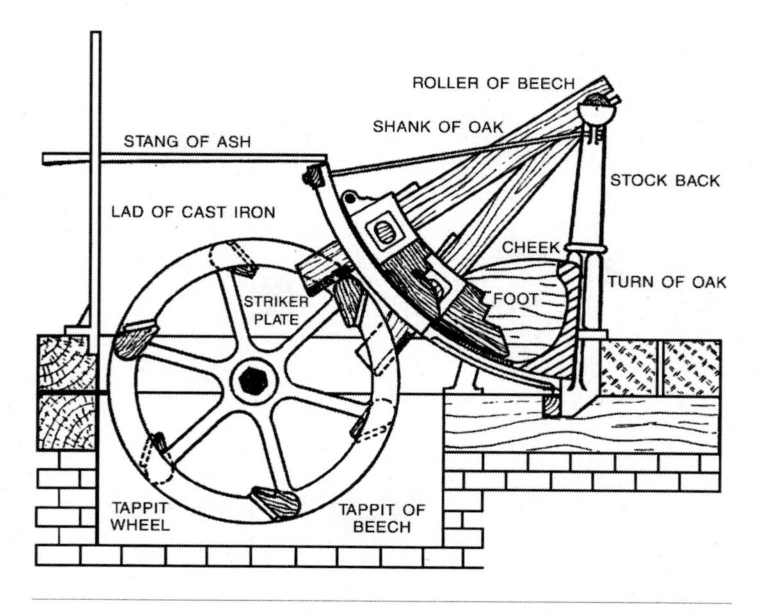

Figure 10.16. Parts of a fulling mill. From E. Kilburn Scott, Early Cloth Fulling and Its Machinery, in *Transactions of the Newcomen Society* (volume 12, 1931).

chine in a gristmill, benefiting from an existing power source.

Textile Mill Complex Components

While early textile mills consisted of a single mill building, by the mid-nineteenth century, cotton mill complexes comprised attached and detached buildings extending considerable distances along riversides or other sites. The principal components of these complexes include picker houses, dye houses, mills, cloth rooms, bleach houses or bleacheries, weave sheds, storehouses, and stock houses.

Picker House

In a typical woolen mill, because of the nature of dried wool, the picker house (Figure 10.17) was made as fireproof as possible. For example, the one-story stone picker house with an interior 9-inch-thick brick wall was typically divided into two sections. One section had two "shoddy" picker machines used for recycling woolen fabric. A shoddy picker would take old and used woolen cloth, rip it apart, and recycle the woolen fibers.

The other section housed a mixing picker, which took the washed new wool. Workers would feed the new wool into a machine equipped with metal pins that would break large chunks and knots of wool into smaller sizes.

Dye House

In the dye house, the textile was washed, dyed, and processed. In early mills, textile and yarn were hand dyed in tubs and vats. By the end of the nineteenth century, hand dyeing was largely supplanted by mechanical dyeing. Twentieth-century dyeing occurred primarily in stainless steel dye kettles.

For example, in North Carolina's Glencoe Cotton Mills, steam from a horizontal tubular boiler ran a variety of dyeing and drying machinery. In the late 1890s, the

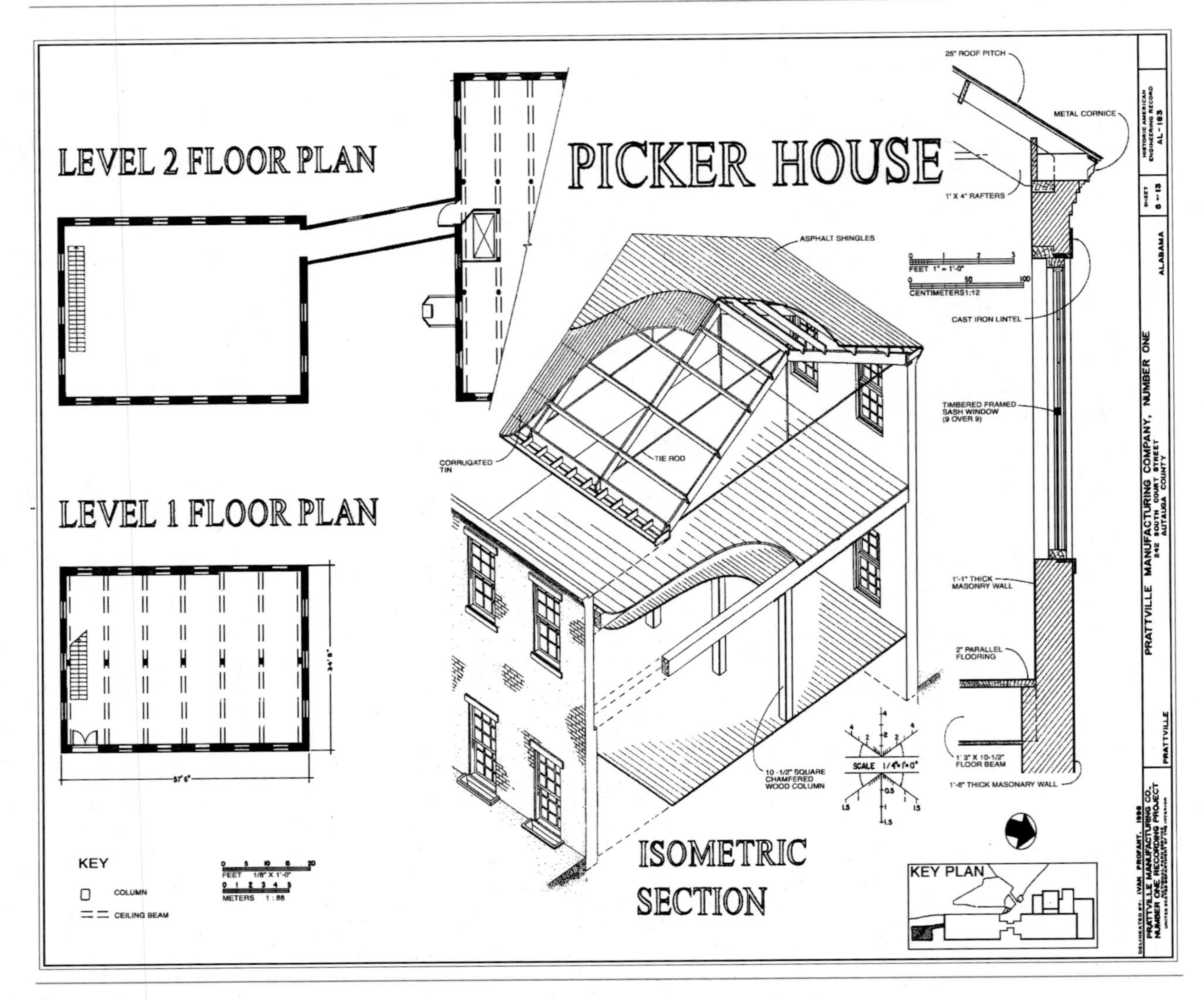

Figure 10.17. Picker house. Prattville Manufacturing Company, Number 1. Prattville, Alabama. HAER AL-183. Delineated by Ivan Profant, 1998.

mill added several hydro extractors, centrifugal drying machines that dried the cotton after it was removed from the dyeing machines. In the 1900s, additional dryers were added, one a nine-cylinder type and the other an eight-warp dryer. At the same time, a 1,000-pound-capacity dyeing machine was used. In the first decade of the twentieth century, 6,000 to 10,000 pounds of cotton were dyed each week.[31]

Mill

A typical New England woolen mill was Massachusetts's Springdale Mill, a three-story building with a basement constructed of locally quarried granite. Burling and fulling of the wool took place in the basement. The basement also included a cold wool dryer and a Cleveland cloth dryer. Weaving and warping took place on the ground floor. Carding took place on the second floor. The third floor was used for spinning.

Cloth Room

After cotton was spun into cloth, the final step was to sew consecutive pieces into a long strip, wind the fabric on a roll, cut the fabric into desire lengths, fold it, and package it for shipment. These activities took place in the cloth room, such as the one in Massachusetts Mills. The interior of the building was primarily composed of open space, articulated only by an internal framework of slow-burning timber and evenly spaced rows of cast-iron columns. Internal stair towers were located in the northwest and southeast corners of the building.

Machinery in a cloth room included a sewing-and-rolling machine which opened small rolls of cloth taken from the loom and rewound them into larger rolls and sewed the ends together. These machines typically had a capacity of between 25,000 and 35,000 yards of cloth per day if operated continuously.[32]

Bleach House or Bleachery

By the mid-nineteenth century, bleacheries were normally one-story buildings with a ceiling about 30 feet high so that a raised platform could be used. The machines were arranged so that normally successive processes would follow one another. Gray goods would be unloaded from a car on a railroad into the gray batching room at one end of the building and would be processed through the operation down the length of the building and back again to the shipping room and storehouse.[33]

Weave Sheds

The weave shed contained row after row of looms (Figure 10.18). In its typical format, it was a single-story, unpartitioned bulding with skylights set into a sawtooth roof aligned to catch the indirect northern light.

Storehouses

Storehouses or storage buildings were built in association with textile mills to store large shipments of cotton or wool. Because of the function of these buildings, windows were generally few in number and small in size. For example, the plan of the Lewiston, Maine, storehouse for the Bates Manufacturing Company consisted of four separate storage rooms, each 62 feet in width, each section defined by masonry fire walls. The framing system was heavy timber. Enclosed elevators served each floor.[34]

Stock Houses

A stock house was a building of a cotton or woolen mill used to store raw materials for production or supplies used for textile production. For example, Massachusetts's Springdale Mill had two stock houses, both of wood-framed construction. The first stored washed and dried wool before it was processed, while the second contained a drug room, where chemicals used in processing wool were kept. Beside dyes, the room stored soaps, an alkali, such as soda ash, and the oils needed to process the wool.[35]

Figure 10.18. Interior. Weave shed, Amoskeag Mill, Manchester, New Hampshire. HABS NH-109-113, National Park Service, ca. 1900. Manchester Historical Association, Manchester, New Hampshire.

GLOSSARY[36]

Automatic feeder. A machine that feeds a steady supply of raw, uncleaned cotton or wool to the carding machine.

Ball warping. The winding of a large number of individual strands of yarn of a specified length onto a beam in the form of a loose, untwisted rope; used chiefly when the yarn is to be dyed.

Beam. A large spool or roll, about 3 feet in diameter, on which warp or cloth is wound.

Beam warping. The transferring of yarn from bobbins or cheeses onto a warp or section beam in the form of a wide sheet. Several such beams were run through the slashing machine to make one loom beam.

Bobbin. A small, wooden or metal core on which yarn is wound.

Breaker picker. The first of two picker machines in which raw cotton was partially cleaned by beating and fluffing and then fed into a finisher picker.

Breast beam. The bar, at the front of the loom, that guides the woven cloth onto the cloth roll.

Calender rolls. A device on the sliver lapper, ribbon lapper, and combing machines that presses the ribbon lap or sliver, as it comes from the drawing rollers, into a loosely matted layer.

Can. A large cylindrical container for receiving and holding lengths of sliver delivered from the front of a carding machine, drawing frame, or combing machine.

Carding. The process whereby the fibers of wool or cotton are combed, straightened, and aligned before being spun into yarn.

Carding drum. The large, rapidly revolving cylinder of the carding machine, covered with several million wire teeth, which pulls out the cotton fibers and, working in conjunction with other rollers, combs the fibers parallel and removes small particles of dirt and knotted fibers.

Cheese. A roll of yarn built up on a paper or wooden tube in a form that resembles a bulk cheese.

Comb. A series of upright metal pegs which separate the individual warp strands and guide them onto a beam in proper order.

Combing. The process of extracting fibers below a predetermined length from cotton sliver and straightening the remaining fibers to make them parallel.

Combing machine. A machine that prepares ribbon lap for spinning into fine yarn by removing short fibers, dirt, and neps and straightening the remaining fibers into parallel alignment.

Creel. The rack for holding packages of roving or yarn on any textile machine.

Drawing frame. A machine in which several strands of sliver are combined into one strand and drawn out so that the combined strands approximate the weight and size of any one of the original strands.

Filling yarn. Also known as woof and welt, the yarn that is interlaced through the warp to produce cloth.

Finisher picker. The second of two older picker machines. This unit received the partially cleaned cotton in the form of lap from a breaker picker and completed the cleaning and fluffing process.

Fly frame. One of several machines that progressively combine two strands of partially processed roving into one, draw out the combined strands until they are of prescribed weight, and twist them loosely in order to give them sufficient strength to withstand subsequent operations.

Fulling. The process in which woolen cloth is cleaned, shrunken, and felted to give it the desired texture and consistency.

Gin. A machine used to remove seeds and to clean dirt from cotton as it comes from the field.

Lap. A general term used to designate wide sheets of loosely matted cotton, formed on machines such as the breaker picker, finisher picker, ribbon lapper, and sliver lapper.

Lint. Long cotton fiber removed by the saws of a gin and delivered to the press box. This is the fiber used in cotton yarn and fabric manufacture.

Linters. Very short cotton fibers remaining on the seed when the ginning is completed.

Loom. A machine for weaving warp and filling yarns to produce cloth.

Mule spinner. A machine that spins many strands of loosely twisted roving into many strands of yarn.

Nap. The woolly or brushed surface of a fabric, typically created by abrasion.

Neps. Short immature fibers, or portions of mature fibers, which are tangled and broken.

Picker machine. A machine that cleans, separates, and fluffs raw cotton, forms the cotton into a uniform layer, and winds it into a roll about a core.

Reeling. Winding yarns from bobbins, onto a revolving reel in the form of a skein or hank.

Ribbon lap. The roll of closely matted cotton fibers, about 10 inches wide, formed on the ribbon lapper from sliver laps.

Ribbon lapper. A machine that draws and combines several rolls of lap from a sliver lapper into one roll of ribbon lap ready for feeding to a combing machine, straightening the fibers slightly and making the lap more uniform in weight and texture.

Ring spinner. A machine that transforms one or more strands of slightly twisted roving into one strand of spun yarn.

Roving. The loosely twisted strand of cotton fibers from the time it leaves the slubber until it goes through the spinner frames and becomes yarn.

Slashing machine. A machine in which warp yarn is arranged in a prescribed sequence, impregnated with sizing to improve its weaving qualities, and wound on a loom beam ready to use.

Sliver. The loose, untwisted strand of cotton fibers produced on the carding machine, drawing frame, and combing machine.

Sliver lapper. A machine that draws and combines several strands of sliver into a sheet of lap and winds it on a spool ready for ribbon lapping or combing.

Slub. A thick place in a strand of yarn caused by improper spinning.

Slubber. A machine that draws out strands of sliver and twists them together loosely in order to give the roving sufficient strength to withstand subsequent operations.

Spinning. The process of making yarn from cotton fibers by drawing out and twisting fibers into a thin strand.

Twisting machine. A machine that twists two or more strands of spun yarn into a heavier, stronger, single strand.

Warp. The set of yarn strands that runs lengthwise in a piece of cloth.

Warp ball. A loose rope of untwisted strands of yarn wound onto a core, usually for dyeing.

Warping. The operation of winding warp yarn onto a beam in suitable arrangement for use as warp in the loom.

Warping machine. A machine that draws yarn from many packages, arranges the strands in parallel in a prescribed sequence, and winds them on beams for use in looms.

Weaving. The interlacing of warp and filling yarn to form a cloth.

Winder. A machine that simultaneously winds yarn from many spinner bobbins onto many cheeses, cones, or filling bobbins.

Yarn. A continuous strand of spun cotton fibers used in weaving or knitting.

BIBLIOGRAPHY

American Society of Mechanical Engineers. *Burton Farmers Gin: A National Mechanical Engineering Landmark*. April 15, 1994. New York: ASME, 1994.

Crompton and Knowles Loom Works. *Weaving Machinery for Cotton, Silk, Woolen, Worsted and Special Fabrics*. Worcester, MA: Crompton & Knowles Loom Works, 1920.

English, Walter. *The Textile Industry: An Account of the Early Invention of Spinning, Weaving and Knitting Machines*. London: Longmans, Green and Company, Ltd., 1969.

International Textbook Company. *Ring Frames, Cotton Mules, Twisters, Spoolers, Beam Warpers, Slashers, Chain Warping*. Scranton, PA: International Textbook Company, 1906.

—. *Yarns, Cloth Rooms, Mill Engineering, Reeling and Baling, Winding*. Scranton, PA: International Textbook Company, 1921.

Lister, John. *Cotton Manufacture*. London: Crosby Lockwood and Son, 1894.

Main, Charles Thomas. *Industrial Plants*. Boston: Caustic-Claflin Company, 1923.

Merrill, Gilbert R., Alfred R. Macormac, and Herbert R. Mauersberger. *American Cotton Handbook*. New York: Textile Book Publishers, Inc., 1949.

Merrimack Valley Textile Museum. *Homespun to Factory Made: Woolen Textiles in American, 1776–1876*. North Andover, MA: Merrimack Valley Textile Museum, 1977.

Rivard, Paul E. *A New Order of Things: How the Textile Industry Transformed New England*. Hanover, NH: University Press of New England, 2002.

Saco-Lowell Company. *Textile Machinery: Cotton Mill Shops*. Boston: Saco-Lowell Company, 1924.

Charles G. Sargent's Sons. *Illustrated and Descriptive Catalogue of Woolen Machinery, Invented and Manufactured by Charles G. Sargent's Sons*. Lowell, MA: Morning Mail Company, 1884.

James Smith Woolen Machinery Company. *The James Smith Woolen Machinery Company: Manufacturers of Woolen*

Machinery; Including Machinery for the Preparation of Wool, Also Card Clothing of Every Description. Philadelphia: James Smith Woolen Machinery Company, 1882.

Von Bergen, Werner, and Herbert R. Mauersberger. *American Wool Handbook: A Practical Text and Reference Book for the Entire Wool Industry*. New York: American Wool Handbook Company, 1938.

Whitin Machine Works. *Cottton Machinery: Cards, Railway Heads, Drawing Frames, Spinning Frames, Spoolers, Wet and Dry Twisters*. East Douglas, MA: Press of C. J. Batcheller, 1896.

Winchester, William E. *The Principles and Processes of Cotton Yarn Manufacture*. In three parts. Philadelphia: Philadelphia Textile School of the Philadelphia Museum and School of Industrial Art, 1902.

Zimiles, Martha and Murray. *Early American Mills*. New York: Clarkson N. Potter, Inc., 1973.

NOTES

1 Gilbert R. Merrill, Alfred R. Macormac and Herbert R. Mauersberger, *American Cotton Handbook* (New York: Textile Book Publishers, Inc., 1949), 15; American Society of Mechanical Engineers, *Burton Farmers* (New York: ASME, 1994), 1. The attribution of the invention of the cotton gin to Eli Whitney may over simplify its origins. See Angela Lakwete, *Inventing the Cotton Gin: Machine and Myth in Antebellum America* (Baltimore: Johns Hopkins University Press, 2005).

2 Merrill, Macormac, and Mauersberger, 153–154.

3 Merrill, Macormac, and Mauersberger, 157–158.

4 ASME, 4–5.

5 From Hayward's New England Gazetteer (1838) as cited on www.newenglandtowns.org.

6 Merrill, Macormac, and Mauersberger, 6–7.

7 Merrill, Macormac, and Mauersberger, 8.

8 Martha & Murray Zimiles, *Early American Mills* (New York: Clarkson N. Potter, Inc., 1973), 112.

9 Merrill, Macormac, and Mauersberger, 15–16.

10 Merrill, Macormac, and Mauersberger, 23, 32.

11 Merrill, Macormac, and Mauersberger, 189.

12 Merrill, Macormac, and Mauersberger, 203–204.

13 Merrill, Macormac, and Mauersberger, 220.

14 Merrill, Macormac, and Mauersberger, 221–225.

15 Merrill, Macormac, and Mauersberger, 243.

16 Merrill, Macormac, and Mauersberger, 244.

17 Merrill, Macormac, and Mauersberger, 245, 248.

18 Merrill, Macormac, and Mauersberger, 251, 253–254.

19 Merrill, Macormac, and Mauersberger, 267.

20 Merrill, Macormac, and Mauersberger, 270–271.

21 Merrill, Macormac, and Mauersberger, 281–282.

22 International Textbook Company, *Ring Frames, Cotton Mules, Twisters, Spoolers, Beam Warpers, Slashers, Chain Warping* (Scranton, PA: International Textbook Company, 1906), 47-1.

23 Merrill, Macormac, and Mauersberger, 634.

24 Werner Von Bergen and Herbert R. Mauersberger, *American Wool Handbook* (New York: Textile Book Publishers, Inc., 1948), 655.

25 Zimiles and Zimiles, 205–210.

26 Merrimack Valley Textile Museum, *Homespun to Factory Made: Woolen Textiles in America, 1776–1876* (North Andover, MA: Merrimack Valley Textile Museum, 1977), 60.

27 Merrimack Valley Textile Museum, 66.

28 Merrimack Valley Textile Museum, 60.

29 The sequence of operations in a modern mule is described in detail in Von Bergen and Mauersberger, 82–485.

30 Merrimack Valley Textile Museum, 82.

31 Brent D. Glass, Glencoe Cotton Mills, Alamance County, North Carolina, HAER NC-6, 1977.

32 International Textbook Company, *Yarns, Cloth Rooms, Mill Engineering, Reeling and Baling, Winding* (Scranton, PA: International Textbook Company, 1921), 68–69; Christopher W. Closs/Valery Mitchell, Massachusetts Mills, Cloth Room/ Section 15, HAER MA-89-A, 1989, 4.

33 Merrill, Macormac, and Mauersberger, 533.

34 Randall Wright, Bates Manufacturing Company, Storehouse, HAER ME-60-A, 4-5.

35 Wachusetts Greenways, Springdale Mill, Holden, Massachusetts, http://www.wachusettsgreenways.org/Mill%20tour.htm.

36 This glossary is based on *Job Descriptions for the Cotton Textile Industry, June 1939* (Washington, DC: U.S. Government Printing Office).

CHAPTER 11

Gristmills, Windmills, and Grain Elevators

The concept of using water to supply power for grinding grain was developed in antiquity. Gristmills were often the first industrial buildings in a North American community, and during the eighteenth century, thousands of mills were operating in the colonies. Few hamlets lacked a mill, and many streams boasted more than one. For example, a small stream near New Preston, Connecticut, supplied the power for nearly 30 mills. Most were gristmills, but others were used to press apples for cider or linseed for oil, to saw wood, or to make plaster, among other tasks.

Many such mill buildings remain standing, though few remain operational. Although both horizontal and vertical waterwheels, as well as turbines, were used as power sources, by far the most common was the vertical wheel, its shaft carrying the power to the one or two pairs of millstones inside.

GRISTMILLS

Although relatively few mill buildings retain their original machinery and even fewer are operable, many remain scattered in the landscape of rural areas, testimony to the former presence of small-scale industry.

Gristmill buildings are typically two- or three-story, stone or timber-framed structures with a full basement and gabled or gambrel roof. Windows are typically constructed with small panes. All such mills are characterized by heavy timber framing, necessary due to both the weight of machinery and the vibrations produced by its operation. Depending on the availability of materials, a mill can be of stone, usually rubble, construction, or its frame can be sheathed in clapboards or wood shingles. Few gristmills are constructed of brick. A wooden cupola sometimes rises from the roof ridge. Floor plans are usually rectangular.[1]

Because gristmills are among the first building types documented by the Historical American Buildings Survey and Historic American Engineering Record, numerous examples of early eastern mills are depicted in measured drawings (Figure 11.1).

Mill Sites

Some early mills were located on natural waterfalls. At most sites, however, it was necessary to dam a river or stream to obtain the required head of water. To withstand water pressure, dams had to be solidly constructed. Some consist of cribbing of oak, pine, or locust logs filled with stones. The cribbing supports a series of strongbacks, spaced about 5 feet apart of both rough and hewn logs. The strongbacks are laid parallel to the stream axis at about a 45-degree angle with a double thickness of 1-inch planking nailed to them. Other dams are built completely of rocks or stone, either laid dry or with mortar, or perhaps bolted with iron bars. Still others are constructed of brush and/or soil.[2]

From the dam to the mill, the water flowed through a headrace, generally 3 to 4 feet wide, sometimes of considerable length. The final channel through which the water flowed before reaching the wheel, the sluice, or

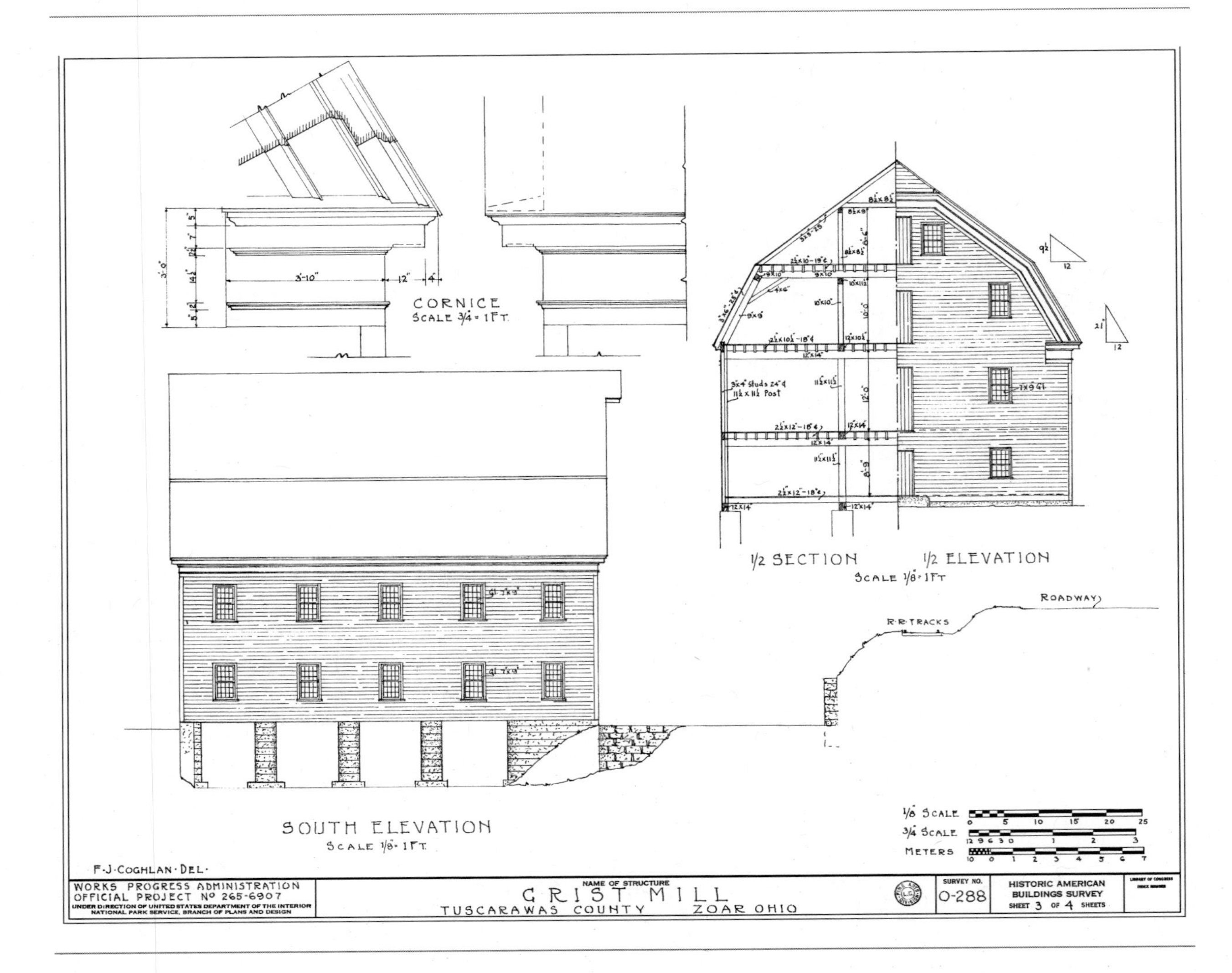

Figure 11.1. Wood-framed gristmill with stone pier foundation. Grist Mill, Zoar, Ohio, HABS O-288, National Park Service. Delineated by F. J. Coghlan, n.d.

spillway, is often of timber construction. Wooden flumes are usually a box section with plank sides and bottom supported on masonry piers, wood cribbing, or piles.[3]

Waterwheels

The waterwheels that supplied the gristmill with power are of four main types: overshot, pitchback, breastshot or breast, and undershot. The names indicate the point on the wheel at which the water was fed to it. In an overshot wheel, water was fed at a (clock) point of about 12:30 or 1 o'clock; in the pitchback mill, at above 11 or 11:30; in the breastshot mill, at between 8 and 10:30; and in the undershot mill, at about 7 o'clock.

During the colonial period, waterwheels were almost completely built of wood and consisted of five main parts: the shaft, the arms, the shrouding or rims, the sole or drum boards, and the partitions forming the buckets or floats. The shafts were almost universally of oak, usually 18 to 24 inches in diameter, dressed in a circular, polygonal, or square form and fitted with iron bands. Gudgeons were inserted in the ends of the shafts so that the protruding ends of the gudgeons ran on the bearings.[4]

Two methods were used to attach the arms of the wheels to the shaft: clasp arms or compass arms. The latter was more common. The arms were inserted in mortises passing through the shaft and were notched and locked together in the center of the shaft.[5]

Overshot wheels were employed at most heads of water greater than 10 feet (Figure 11.2a). The water was conveyed to the top of the wheel by a wooden

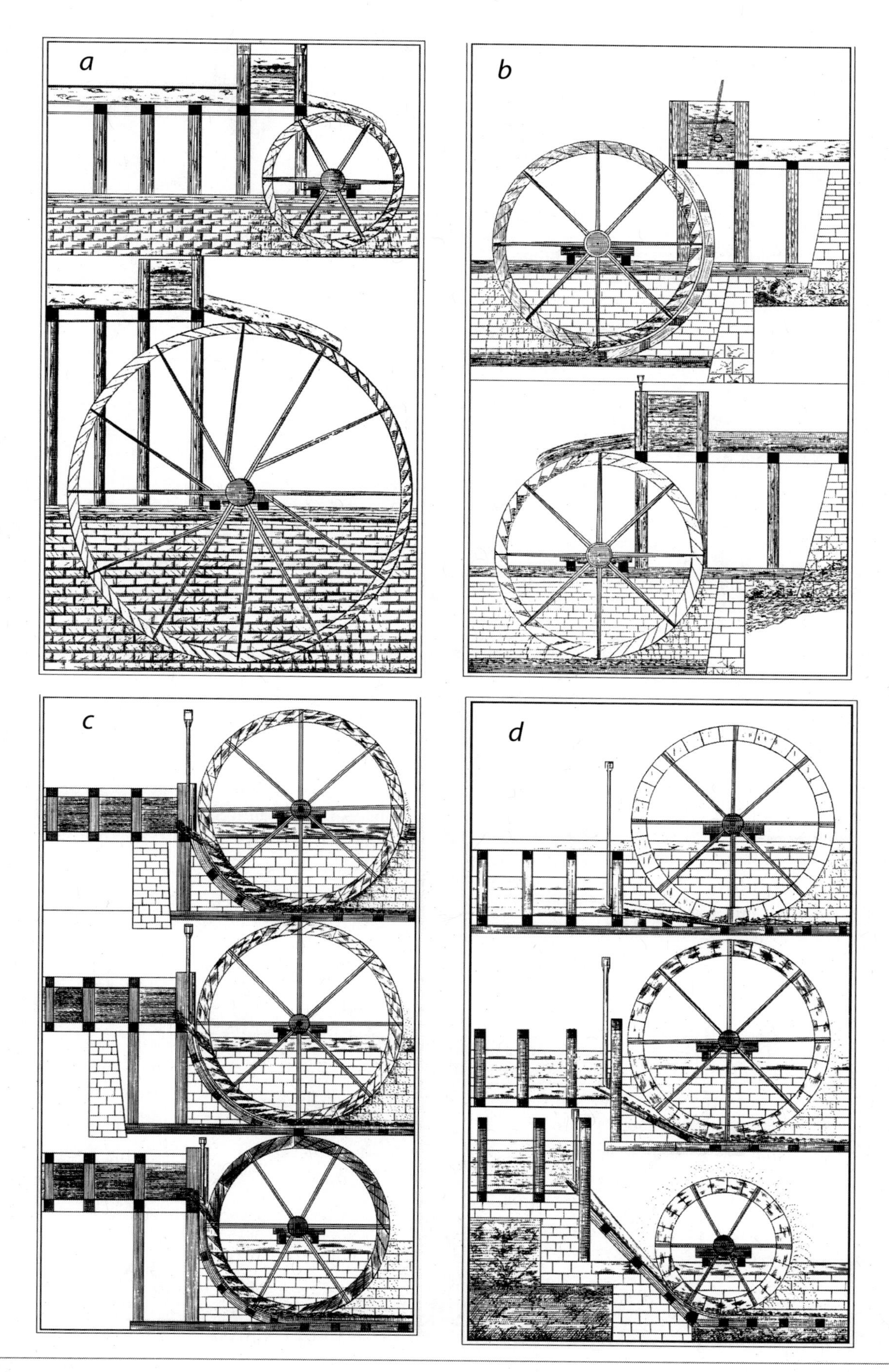

Figure 11.2. Types of mill wheels: ***a,*** overshot; ***b,*** pitchback; ***c,*** breast; ***d,*** undershot. From Oliver Evans, *The Young Mill-Wright & Miller's Guide* (Philadelphia: Oliver Evans, 1795).

trough or flume and fed into the buckets. Power generated by overshot wheels depended almost entirely on the weight of water in the buckets. The forward momentum of the water entering the buckets added a slight increment to their power. On average, these wheels were 70 to 90 percent efficient.[6]

The pitchback wheel was a variation of the overshot wheel in which the water was conveyed to the top or almost to the top of the wheel by a flume (Figure 11.2b). The buckets in the pitchback were set at an angle opposite to those in an overshot. The end of the flume and the control gate were adapted so that the water fed downward into the buckets at the reverse direction to the flow of the stream, causing the wheel to revolve in the opposite direction.[7]

Breastshot wheels, most commonly used for falls between 6 and 10 feet, were constructed in a similar manner as overshot and pitchback wheels (Figure 11.2c). In middle and low breastshot wheels, the buckets were deeper in order to deal with the increased volume of water required for the low head to develop sufficient power.[8]

For low falls of water, undershot wheels were used (Figure 11.2d). These wheels relied entirely on the impulse of the water. As a result, they required much greater quantities of water to produce the same power developed by the other types. On average, these wheels were only 35 percent efficient. In construction, undershot wheels differed little from other types, except that the buckets were replaced by radial floats. Ordinary undershot wheels were built from 10 to 25 feet in diameter with floats from 14 to 16 inches apart.[9]

Tub Wheel

One alternative to the large waterwheel for power was the tub wheel. In its most frequently used form, the tub wheel resembles a wagon wheel, with or without a rim, its spokes replaced by arms with floats or buckets. This wheel with its curved blades was given its name because it ran within a circular enclosure of thick planking assembled in the form of a tub without a bottom. The tub wheel ran upon a vertical shaft. With a head sufficient to impart the necessary velocity, the wheel drove the runner stone without the intervention of gearing. A typical wheel was 7 feet in diameter, with a head of 10 to 12 feet. The floats were placed radially and were fastened to starts inserted in the shrouding pierce by dovetail tenons or keys.[10]

Tub wheels operated much like a child's pinwheel. They were normally less than 30 percent efficient, but if sufficient head of water was present, they could generate appreciable energy. The purpose of the tub was to keep the water from overshooting the wheel and help direct it to the vanes.[11]

Turbines

The turbine was developed in the nineteenth century as an alternative to the waterwheel (Figure 11.3). It operated on a similar principle as the tub wheel. Turbines of French design were introduced into the United States in the 1840s by Ellwood Morris and Emile Geyelin in the Middle Atlantic states and by George Kilburn, Uriah A. Boyden, and James B. Francis in New England.[12] Louis Hunter describes the design of these early turbines:

> The new turbine was a relatively simple mechanism with three principal components: a central fixed disk on which were mounted a number of iron guides that curved downward and outward, forming spiral passages by which the water passed from the penstock to the wheel proper; a horizontal wheel, or runner, mounted on a vertical shaft and having two outer rims, separated by vertical metal strips dividing the space between them into a number of curved passages, or buckets, through which the water received from the fixed guides moved outward; and a gate mechanism by which the admission of water from the penstock to the wheel was regulated.[13]

A turbine develops power by the pressure of water against its blades. This pressure forces the blades to rotate. The turbine quickly became popular as an efficient conductor of energy in gristmills and other small industrial facilities located on rivers and streams with large flows.[14]

Basic Mill Operation

To grind grain to flour, a miller would open a sack of grain and pour it into the millstone hopper or the grain would feed from a bin located in the floor above. To

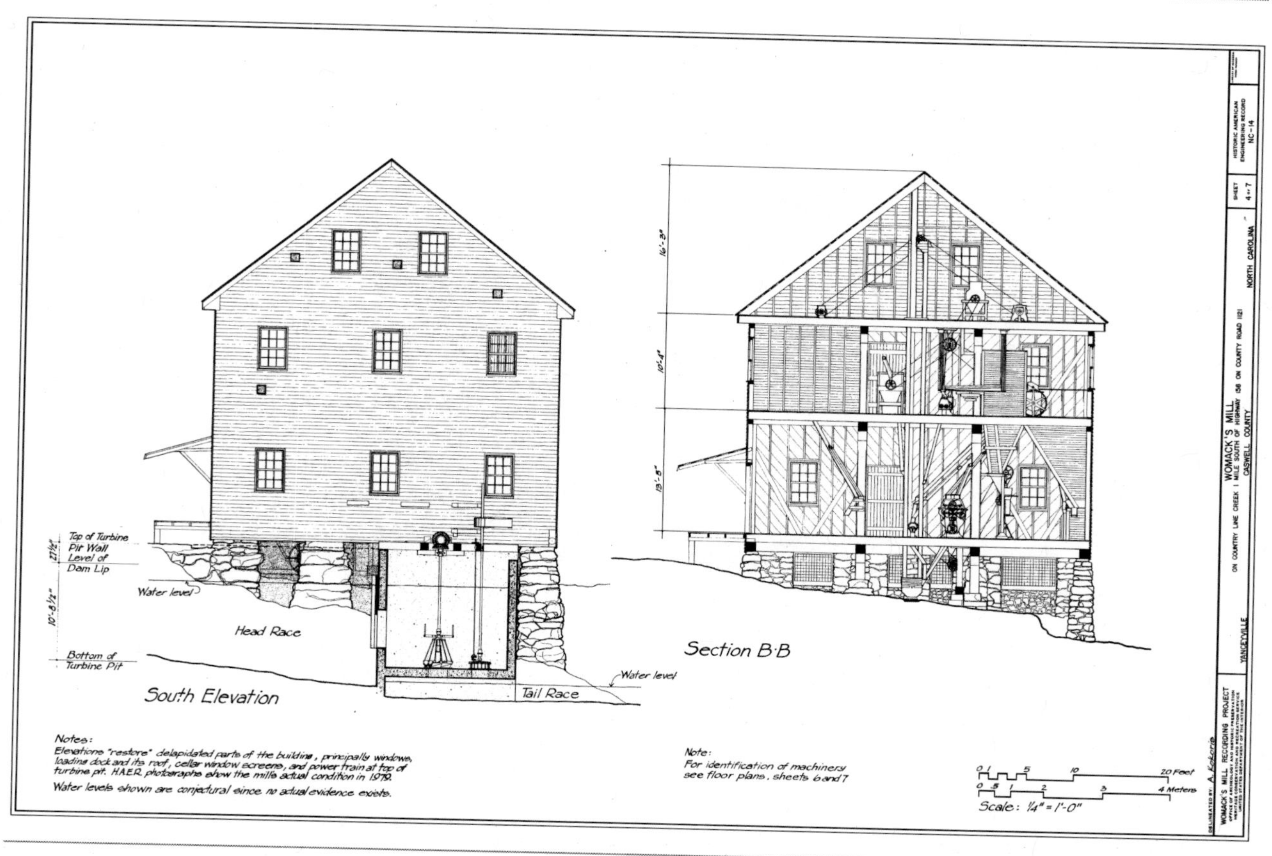

Figure 11.3. Womack's Mill showing turbine in pit. Yanceyville, North Carolina. HAER NC-14, National Park Service. Delineated by A. Kokorie, 1979.

begin the grinding process, the miller would open the control gate on the mill's sluice box.

As the water slowly began to flow over the waterwheel or flow into the turbine, the miller would slowly begin to raise the runner millstone (Figure 11.4). When the waterwheel or turbine began to turn, so would the gears and the millstones. As the grain was slowly fed between the millstones, the miller lowered the runner stone closer to the stationary stone, and the product became finer and finer. Other methods of beginning mill operations are described in Theodore Hazen's *The Art of the Millstones, How They Work.*[15]

In some mills, the volume of grain moving out of the millstone hopper was controlled by a wooden paddle or gate. The grain then fell into a device hung below the hopper. This device with its leather straps is termed the shoe. The shoe could be raised or lowered at one end to allow more or less grain to fall into the millstones. To ensure that the grain flowed at a constant rate, the shoe was vibrated back and forth by a turning damsel. The damsel was mounted atop the balance rynd in the center hole of the upper millstone. The top end of the damsel turned in a hole in the wooden frame that held the shoe and the millstone hopper. Some damsels are adjustable up and down a metal shaft to allow them to be used on different millstones.

The grain fell into the eye of the runner stone, and the flow of grain was controlled by the arrangement of the shoe and the damsel. The millstones worked together in pairs, with a revolving upper millstone called the "runner" stone and a stationary bottom stone, the "bed" stone.

The pattern on both millstones is identical. The furrows in the stones were used to cut the grain like a pair of scissors and to move the grain outward from the center of the millstones to the outer circumference. The furrows in the stone were arranged in groups known as harps or quarters, each group consisting of a "master furrow" running from the eye of the stone to the outer edge. There were a variable number of secondary

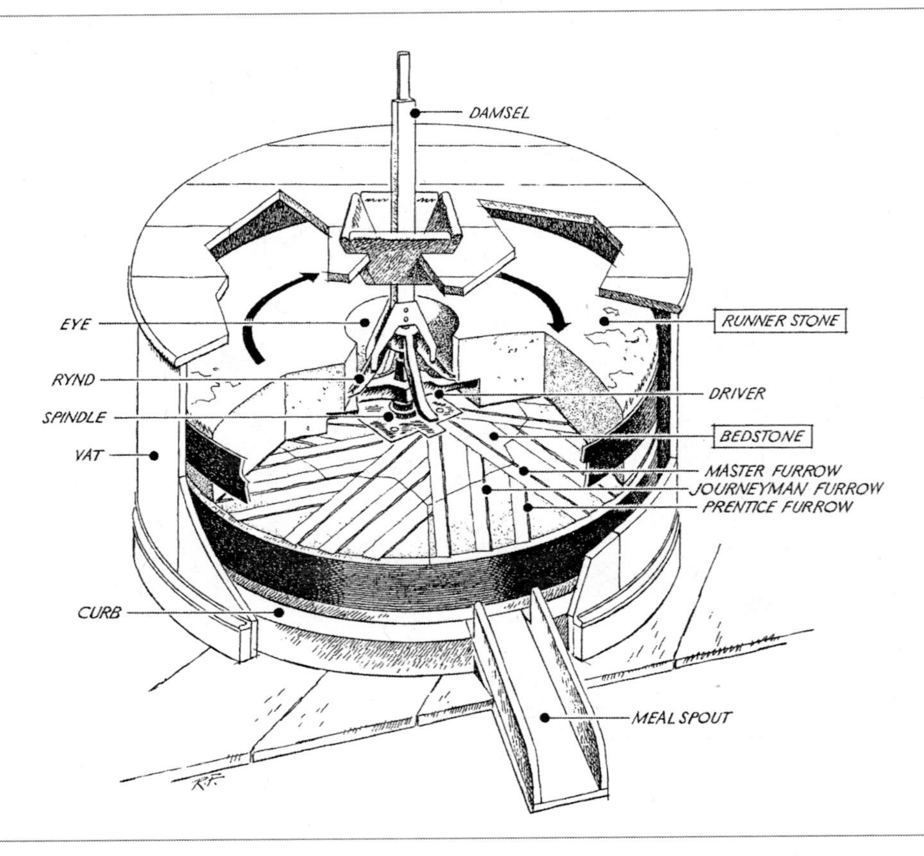

Figure 11.4. Cutaway drawing of millstones in use. Reproduced with permission of Historic Hudson, Inc. from Brooke Hindle, editor, *America's Wooden Age: Aspects of Its Early Technology* (Tarrytown, NY: Sleepy Hollow Restorations, 1975).

furrows. The first secondary furrow was called the journeyman furrow, while the third largest furrow in a quarter was termed the apprentice furrow, and the smallest furrow in a quarter was known as the butterfly furrow. A pair of millstones 48 inches in diameter could grind 400 pounds per hour, while a pair of 56-inch millstones could grind 500 pounds per hour.[16]

Mill Machinery

In colonial gristmills, each waterwheel was geared to drive a single pair of stones by a single-step gearing. To the waterwheel was attached a large face gear that engaged a lantern pinion, often called a wallower. The gears had a dual purpose: to transfer the direction of the drive from the horizontal to the vertical and to increase the speed of the millstone spindle as opposed to the slower motion of the waterwheel shaft. Through gearing it was possible for a runner stone to spin at 100 rpm while a waterwheel spun at only 7 rpm.

Most of the face gear wheels in early mills were constructed with two wooden arms passing through mortises in the waterwheel shaft to form four spokes. The rims, which were pegged or bolted to the arms, were laminated, the two thicknesses being pinned together with bolts. Each ring of the rim was in four segments known as cants. The cogs were driven into mortises in the sides or face of the cants, passed through both thicknesses, and were secured in place by pegs or wedges in the shank ends.[17]

The general layout of a reconstructed late seventeenth-century mill built at Philipsburg Manor, New York, is shown in Figure 11.5. Water to power the mill is conducted from the mill pond to the waterwheel by the flume. The amount of water fed to the wheel is controlled by the flume gate. When the flume gate is raised, water emerges under pressure and strikes the buckets of the waterwheel, causing it to revolve. After powering the wheel, water flows away down the tailrace.

The arms of the waterwheel are mortised into the main shaft, which transmits power into the mill building, where the millstones are located on the stone floor. Attached to the main shaft are face gear wheels, one directly under each pair of stones. The face gear wheels

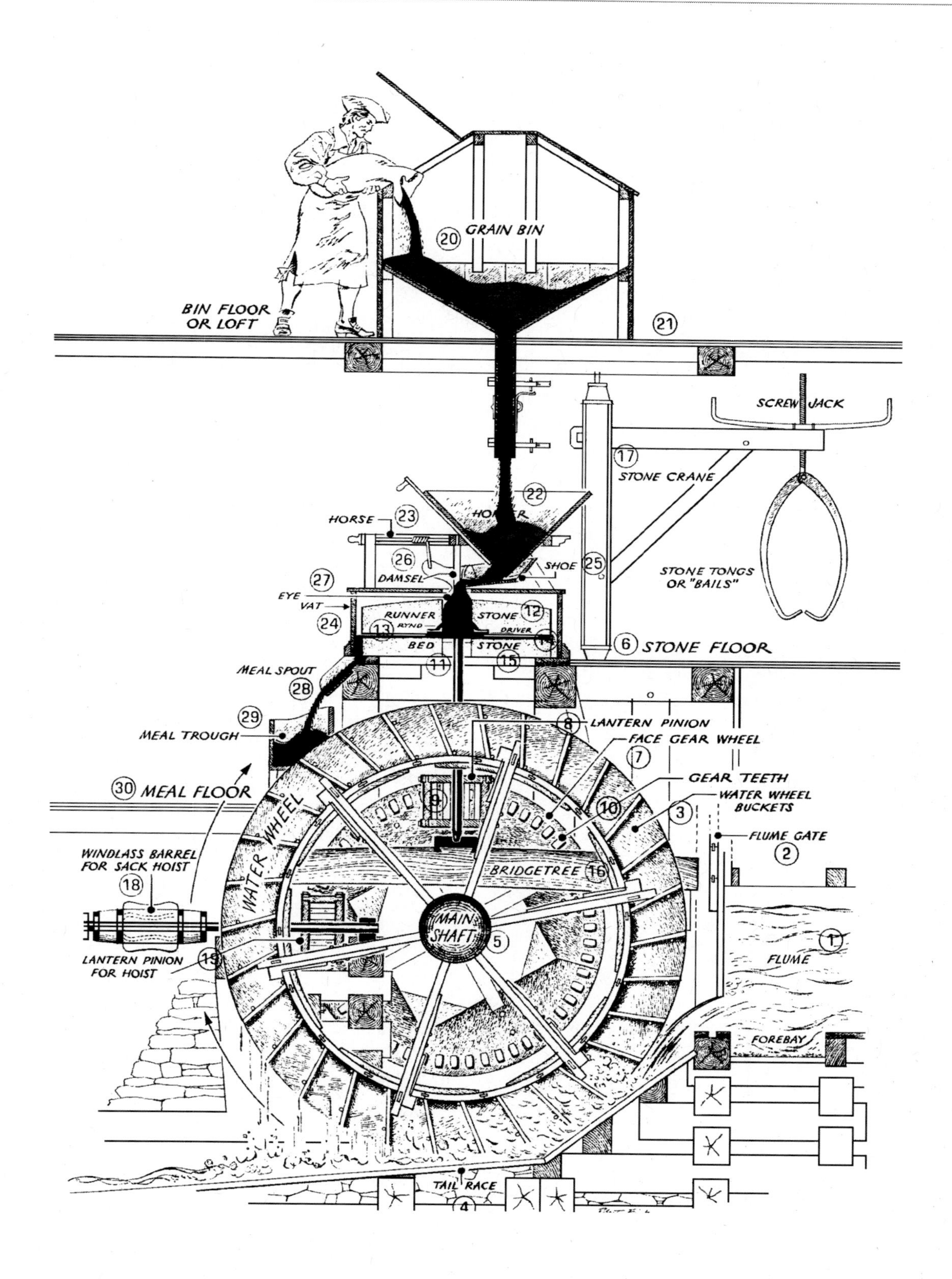

Figure 11.5. Drawing of the Upper Mills, a reconstructed seventeenth-century water mill at Philipsburg Manor, North Tarrytown, New York, showing progress of grain through the mill and source of power (Hindle 1975). Reproduced with permission of Historic Hudson, Inc.

engage into lantern pinions, which are mounted on the millstone spindles, thus transferring the drive from horizontal to vertical and also increasing the shaft speed as the larger number of gear teeth in the face gear wheel engage the fewer staves of the lantern pinion.

The millstone spindles pass through the neck bearing in the center of the bedstone. The runner stone is pivoted atop the spindle by a socket bearing, called the cockeye, in the center of the rynd; the pivot point of the spindle is known as the cock head. Just above the neck bearing is fitted the driver, which engages the runner stone and causes it to revolve while the bedstone remains stationary. The millstone spindles are supported by footstep bearings fitted into bridging boxes mounted on the bridgetree, which can be raised or lowered in a process known as tentering. Next to the millstones is the stone crane, used to lift and invert the runner stone. The windlass barrel for the sack hoist is driven by another lantern pinion from one of the face wheels.

Grain in the grain bins on the grain floor flows by gravity into a spout which delivers it into the hopper, supported by the horse atop the stone case or vat. The base of the hopper feeds grain into the shoe, an inclined, tapering wooden trough. The revolving runner stone turns the damsel, a square shaft which taps against a block of wood in the shoe, causing it to vibrate and thus to feed grain into the eye of the runner stone.

Grain ground between the stones emerges as meal around the periphery of the stones and is trapped in the vat. The runner stone carries meal around to the meal spout, where it is discharged into the meal trough on the meal floor, where it is put into sacks or barrels for delivery or sifting and bolting. Bolting involves the use of a series of cloths of varying knit that allow only certain sizes of materials to flow through them.[18]

The operation of the gristmill was revolutionized by Oliver Evans in his mill built in New Castle County, Delaware, in the 1790s. Mechanical conveyors, driven by the mill's waterwheel, moved wheat, meal, and flour throughout the building, eliminating the need for lifting and carrying sacks of wheat to the top floor and tubs of ground meal to the second floor. A mill that previously required four men to operate could be operated by two. Millers were slow to adopt Evans's improvements, but by the time of Oliver Evans's death in 1819, bucket elevators, screw conveyors, and belt conveyors had become common features of American gristmills.[19]

With increased demand for flour caused by increasing population growth, some entrepreneurs built additional mills, while others enlarged existing ones. Larger and more powerful waterwheels were installed. Even more useful was the adoption of gearing systems that permitted more than a single pair of stones to be driven from a single waterwheel.

Some waterwheel shafts were simply made longer in order to accommodate two face gear wheels, each of which would drive a pair of stones. In other installations, a two-step gear train was developed, usually in one of two patterns.

In one method, a larger and more strongly constructed face gear was mounted on the waterwheel shaft. The big face wheel drove one or two lantern pinions or wallowers on lay-shafts set at right angles. On the wallower shafts were "little face wheels," which meshed into lantern pinions on the millstone spindles. The wallower gudgeon nearest the main shaft rested in a sliding block so that either wallower could be disengaged by the use of a lever. This gear system was referred to as counter gears.

In the other method, a large face gear was mounted on the waterwheel shaft. This gear wheel was usually called the pit wheel because the bottom half ran in a pit. The pit wheel meshed into a lantern wallower attached to a sturdy wooden upright, the upright shaft. Higher up on the upright shaft, frequently directly above the wallower, was mounted a large spur gear wheel, usually called the great spur wheel, in which the cogs or teeth were driven into mortises in the edge of the rim and meshed into lantern or spur pinions attached to the millstone spindles. This arrangement was called spur gear drive. It usually came from below the millstones and was known as underdrift.

In addition to allowing additional millstones to be driven from a single waterwheel, two-step gearing also became necessary when waterwheels were built of a larger diameter so that millstones could be operated at a more efficient speed.[20]

Nineteenth-Century Gristmill

The entire operation of the ca. 1805 Mount Pleasant Grist Mill in Chester County, Pennsylvania, is illustrated in HAER drawings prepared for the Mount Pleasant Grist Mill Recording Project undertaken in 1986–1987 (Figure 11.6). In this mill, feed grinding

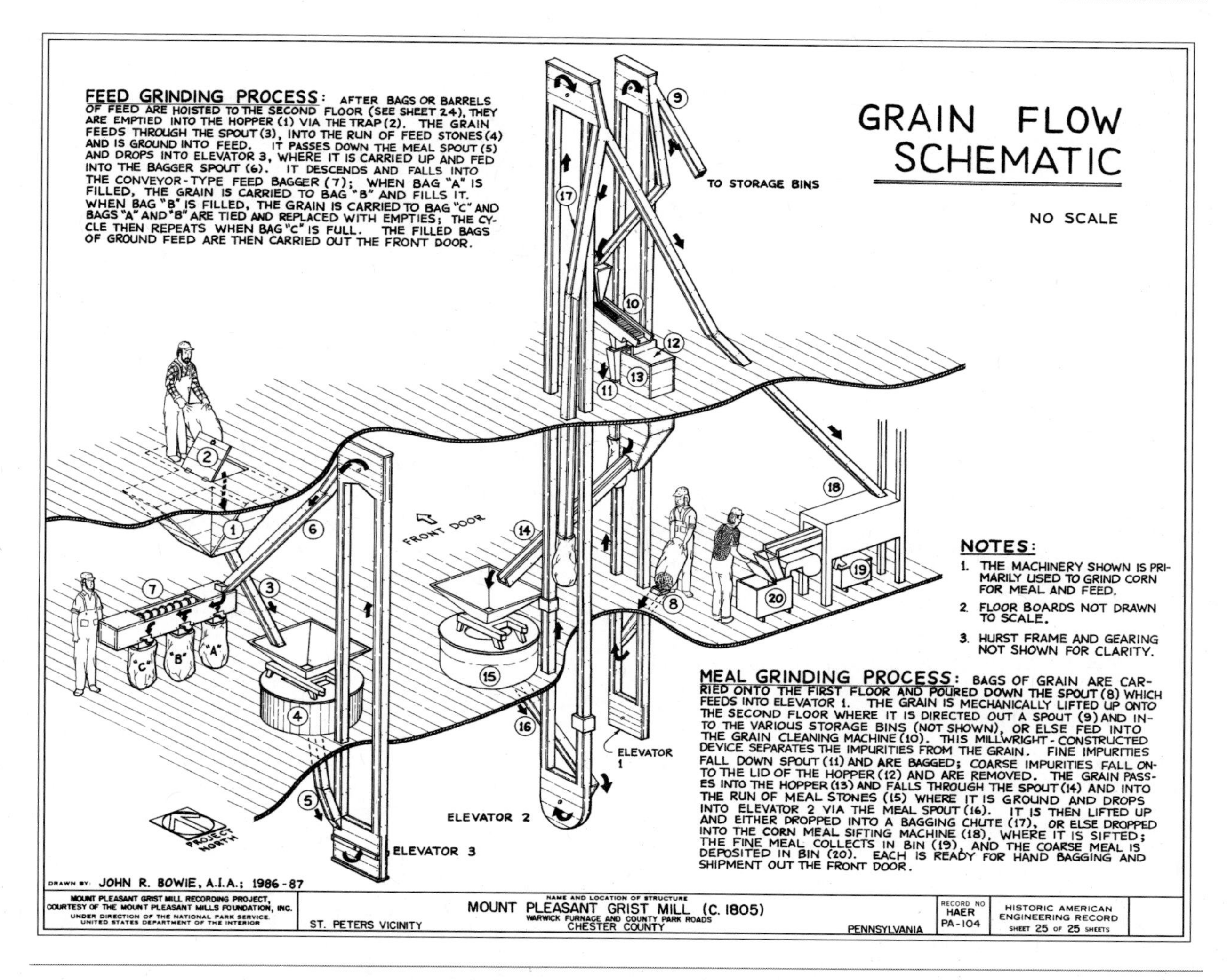

FEED GRINDING PROCESS: AFTER BAGS OR BARRELS OF FEED ARE HOISTED TO THE SECOND FLOOR (SEE SHEET 24), THEY ARE EMPTIED INTO THE HOPPER (1) VIA THE TRAP (2). THE GRAIN FEEDS THROUGH THE SPOUT (3), INTO THE RUN OF FEED STONES (4) AND IS GROUND INTO FEED. IT PASSES DOWN THE MEAL SPOUT (5) AND DROPS INTO ELEVATOR 3, WHERE IT IS CARRIED UP AND FED INTO THE BAGGER SPOUT (6). IT DESCENDS AND FALLS INTO THE CONVEYOR-TYPE FEED BAGGER (7); WHEN BAG "A" IS FILLED, THE GRAIN IS CARRIED TO BAG "B" AND FILLS IT. WHEN BAG "B" IS FILLED, THE GRAIN IS CARRIED TO BAG "C" AND BAGS "A" AND "B" ARE TIED AND REPLACED WITH EMPTIES; THE CYCLE THEN REPEATS WHEN BAG "C" IS FULL. THE FILLED BAGS OF GROUND FEED ARE THEN CARRIED OUT THE FRONT DOOR.
GRAIN FLOW SCHEMATIC
NO SCALE
TO STORAGE BINS
FRONT DOOR
NOTES:
1. THE MACHINERY SHOWN IS PRIMARILY USED TO GRIND CORN FOR MEAL AND FEED.
2. FLOOR BOARDS NOT DRAWN TO SCALE.
3. HURST FRAME AND GEARING NOT SHOWN FOR CLARITY.
MEAL GRINDING PROCESS: BAGS OF GRAIN ARE CARRIED ONTO THE FIRST FLOOR AND POURED DOWN THE SPOUT (8) WHICH FEEDS INTO ELEVATOR 1. THE GRAIN IS MECHANICALLY LIFTED UP ONTO THE SECOND FLOOR WHERE IT IS DIRECTED OUT A SPOUT (9) AND INTO THE VARIOUS STORAGE BINS (NOT SHOWN), OR ELSE FED INTO THE GRAIN CLEANING MACHINE (10). THIS MILLWRIGHT-CONSTRUCTED DEVICE SEPARATES THE IMPURITIES FROM THE GRAIN. FINE IMPURITIES FALL DOWN SPOUT (11) AND ARE BAGGED; COARSE IMPURITIES FALL ONTO THE LID OF THE HOPPER (12) AND ARE REMOVED. THE GRAIN PASSES INTO THE HOPPER (13) AND FALLS THROUGH THE SPOUT (14) AND INTO THE RUN OF MEAL STONES (15) WHERE IT IS GROUND AND DROPS INTO ELEVATOR 2 VIA THE MEAL SPOUT (16). IT IS THEN LIFTED UP AND EITHER DROPPED INTO A BAGGING CHUTE (17), OR ELSE DROPPED INTO THE CORN MEAL SIFTING MACHINE (18), WHERE IT IS SIFTED; THE FINE MEAL COLLECTS IN BIN (19), AND THE COARSE MEAL IS DEPOSITED IN BIN (20). EACH IS READY FOR HAND BAGGING AND SHIPMENT OUT THE FRONT DOOR.
ELEVATOR 1
ELEVATOR 2
ELEVATOR 3
PROJECT NORTH
DRAWN BY: JOHN R. BOWIE, A.I.A.; 1986-87
MOUNT PLEASANT GRIST MILL RECORDING PROJECT, COURTESY OF THE MOUNT PLEASANT MILLS FOUNDATION, INC.
UNDER DIRECTION OF THE NATIONAL PARK SERVICE. UNITED STATES DEPARTMENT OF THE INTERIOR
NAME AND LOCATION OF STRUCTURE
MOUNT PLEASANT GRIST MILL (C. 1805)
WARWICK FURNACE AND COUNTY PARK ROADS
ST. PETERS VICINITY
CHESTER COUNTY
PENNSYLVANIA
RECORD NO HAER PA-104
HISTORIC AMERICAN ENGINEERING RECORD
SHEET 25 OF 25 SHEETS

Figure 11.6. Grain flow schematic. Mount Pleasant Grist Mill (ca. 1805), St. Peters vicinity, Pennsylvania. HAER PA-104, National Park Service. Delineated by John R. Bowie, AIA, 1986–1987.

and meal grinding took place in two different areas of the mill.

To grind feed, bags and barrels of feed were hoisted to the second floor, where they were emptied into a hopper via a trap. The feed then fed through the spout into the run of feed stones and was ground into feed. It then passed down the meal spout and dropped into the elevator. The elevator carried it up and fed it into the bagger spout. It then descended and fell into the conveyor-type feed bagger. When the first bag was filled, the grain was carried to the second bag and so forth. The filled bags of ground feed were then carried out the front door.

In grinding meal, bags of grain were carried onto the first floor and poured down the spout that fed into an elevator. The grain was then mechanically lifted up to the second floor, where it was directed out a spout and into the storage bins or fed into the grain cleaning machine. This mechanism separated impurities from the grain. Fine impurities fell down a spout and were bagged. Coarse impurities fell onto the lid of the hopper and were removed. The grain passed into the hopper and fell through the spout and into the run of meal stones where it was ground and dropped into a second elevator via the meal spout. It was then lifted up and either dropped into a bagging chute or dropped into the corn meal sifting machine, where it was sifted. The fine meal collected in one bin, while the coarse meal collected in another. Both were ready for hand bagging and shipment out the front door.

The HAER documentation for the mill includes comprehensive documentation of the power system for the facility (Figure 11.7). The overshot wheel drove a master gear. The 144-tooth master gear meshed with a cast-iron pinion gear and drove the steel counter shaft. Two mortise gears were attached to the counter shaft and meshed with spindle gears that caused steel stone spindles to rotate the two runner stones. The two bed-stones were stationary.

Roller Mills

The basic difference between older gristmills and roller mills is that instead of grinding grain by means of mill-stones, roller mills grind grain using metal rollers. The roller mill proved a good alternative for those who wanted a product of improved quality and increased quantity. The first use of roller mills was to break up grain before millstones would regrind the particles into flour. Eventually rollers were added to replace the mill-stones. Early roller mills were largely gear-driven, while later mills used belt pulley drives with metal shafting.

Rollers began to be installed widely in mills in the United States in the 1870s. E. P. Allis and Company installed a set of rolls in a mill in Winona, Minnesota, in 1873, while George H. Christian tried a set in the Washburn B mill the same year. The former used marble rolls, while the latter used cast iron. Both proved unsatisfactory because they wore down very rapidly. In 1876, mills began to use the more durable, though expensive, porcelain rolls, and these were, in turn, supplanted by rolls made of chilled iron.

The initial smooth rolls proved unsatisfactory for wheat grinding and had to be used in conjunction with millstones. In 1879, Wisconsin miller John Stevens invented a new roll with corrugations that worked well for

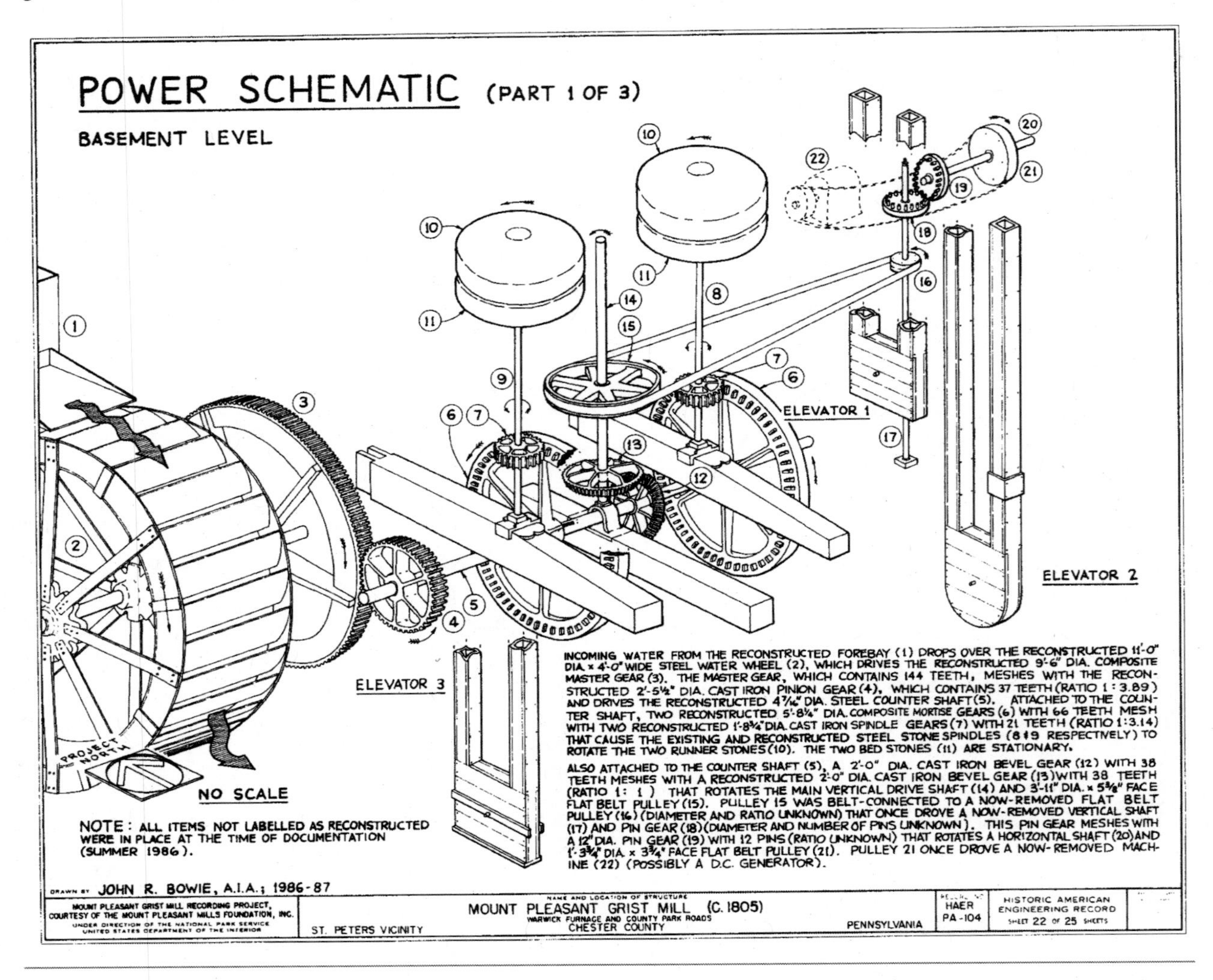

Figure 11.7. Power schematic (basement level). Mount Pleasant Grist Mill (ca. 1805), St. Peters vicinity, Pennsylvania. HAER PA-104, National Park Service. Delineated by John R. Bowie, AIA, 1986–1987.

the preliminary grinding, and Stevens was given a patent for the roller method of flour manufacture. Within a few years, most large flour mills had begun using rollers, and some had converted completely from millstones. Rolls required less space, less power to operate, and less oversight than millstones.[21]

These newer mills featured corrugated cast-iron rollers. The non-touching rollers ran at differing speeds. The slower roller tended to hold the grain, while the faster roller sheared open the kernel of the grain.

As the millers installed roller mills, their mills were no longer referred to as gristmills but were instead known as roller mills. Roller mills became identified with less nutritious, whiter flour, while other mills continued to use millstones to make their stone-ground flour.[22]

The Mascot Roller Mills in Ronks vicinity, Lancaster, Pennsylvania, is an example of a gristmill converted to a roller mill. Built as early as 1740, the millstones were replaced by roll stands in 1906.[23] The machinery of the mill is shown in sectional views in Figure 11.8.

The raw grain was initially weighed by the receiving scale, after which it passed to the dustless double receiving separator. After passing through the separator, dirt was lifted by elevator to the fourth floor, where it fell into a chute to the second floor to be reweighed and given back to the farmer. A second elevator lifted the cleaned wheat up to the turnhead. The turnhead directed the movement of cleaned wheat into various bins on the third floor.

Cleaned wheat dropped through a chute into the first roller mill, with the flow controlled by the feed governor. It was then lifted into the self-balancing sieve bolter. Subsequent stages involved separating the wheat

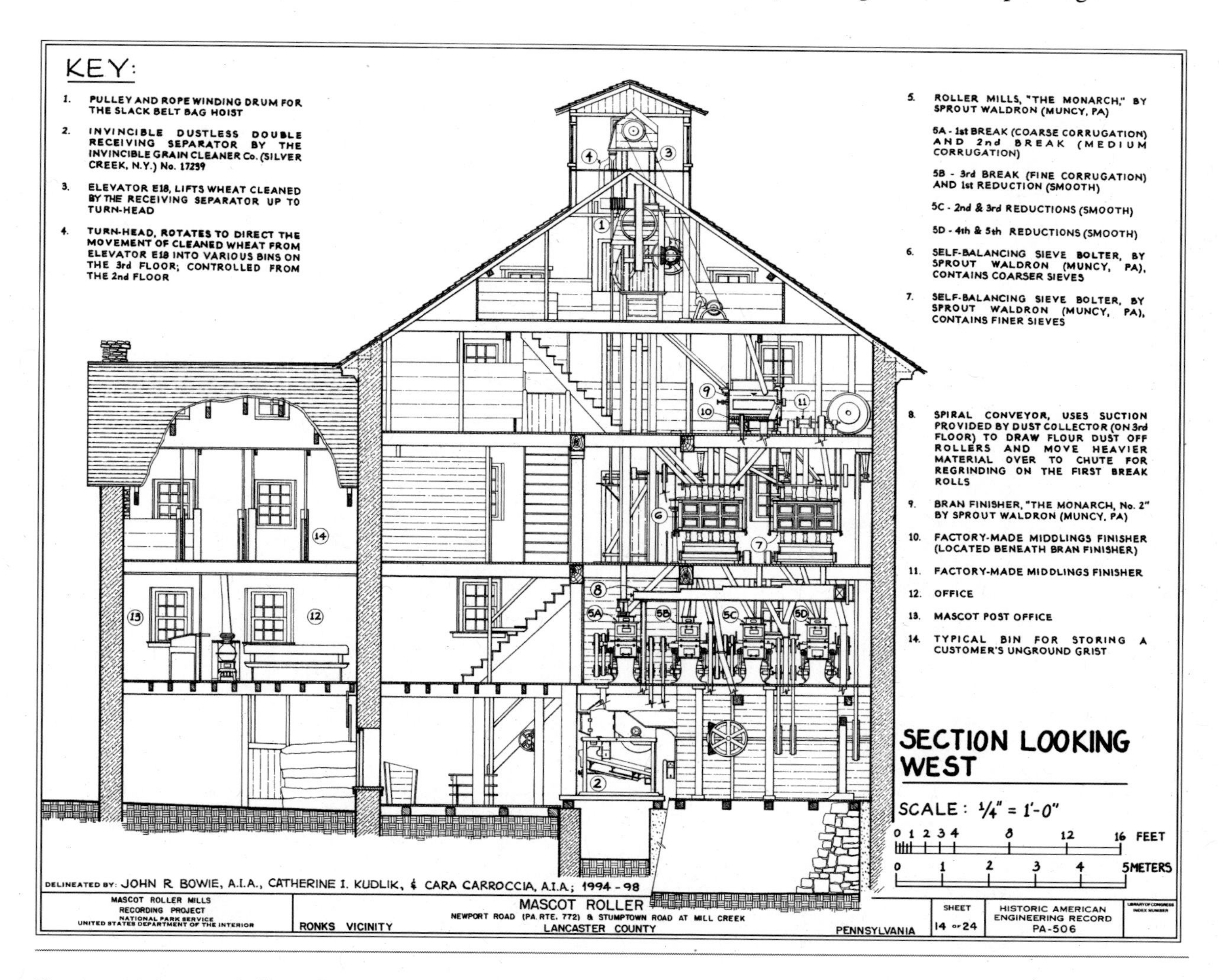

Figure 11.8. Mascot Roller Mills. Ronks vicinity, Pennsylvania. Section looking west. HAER PA-506, National Park Service. Delineated by John R. Bowie, Catherine I. Kudlik, and Cara Carroccia, 1994–1998.

into three products, bran, middlings, and flour, using the various sections of the sieve bolter.

WINDMILLS

According to some sources, the first use of wind power to grind grain occurred in Persia and, according to tradition, was brought to northern Europe during the Crusades. Windmills were erected in many European countries but reached the height of development in England and Holland. The first documentary evidence of English windmills dates from the twelfth century.[24]

Windmills erected to grind grain were concentrated in two areas of the northeastern United States, Long Island[25] and Massachusetts, with others in Rhode Island, Virginia, and Maryland. The surviving wind gristmills in the United States are of two types: the post mill and the smock mill.

Post Mills

In post mills, the mill's primary structure is balanced on a large upright post. Because of this mounting, the mill is able to rotate to face the wind direction. To maintain the upright post, a structure consisting of horizontal crosstrees and angled quarterbars is used. The most common arrangement was two crossbars at right angles to each other under the base of the post, together with four quarterbars. To prevent rotting, crosstrees were frequently placed on brick piers. Post mills were the type built in Long Island during the seventeenth century and most of the eighteenth. No early post mills survive in the United States.[26]

Smock Mills

Smock mills are the more common surviving American mill and represent an improvement over the post mill. Instead of the whole body of the mill rotating to face the wind, this mill has a fixed wooden body containing the milling machinery, together with a rotatable cap that holds the roof, the sails, the windshaft, and the brake wheel (Figure 11.9).

By rotating only the mill cap, the body of the mill could be made much larger than in a post mill and was able to house more pairs of stones and additional ancillary machinery. In addition, the body could be higher, and the extra height allowed longer sails to catch more

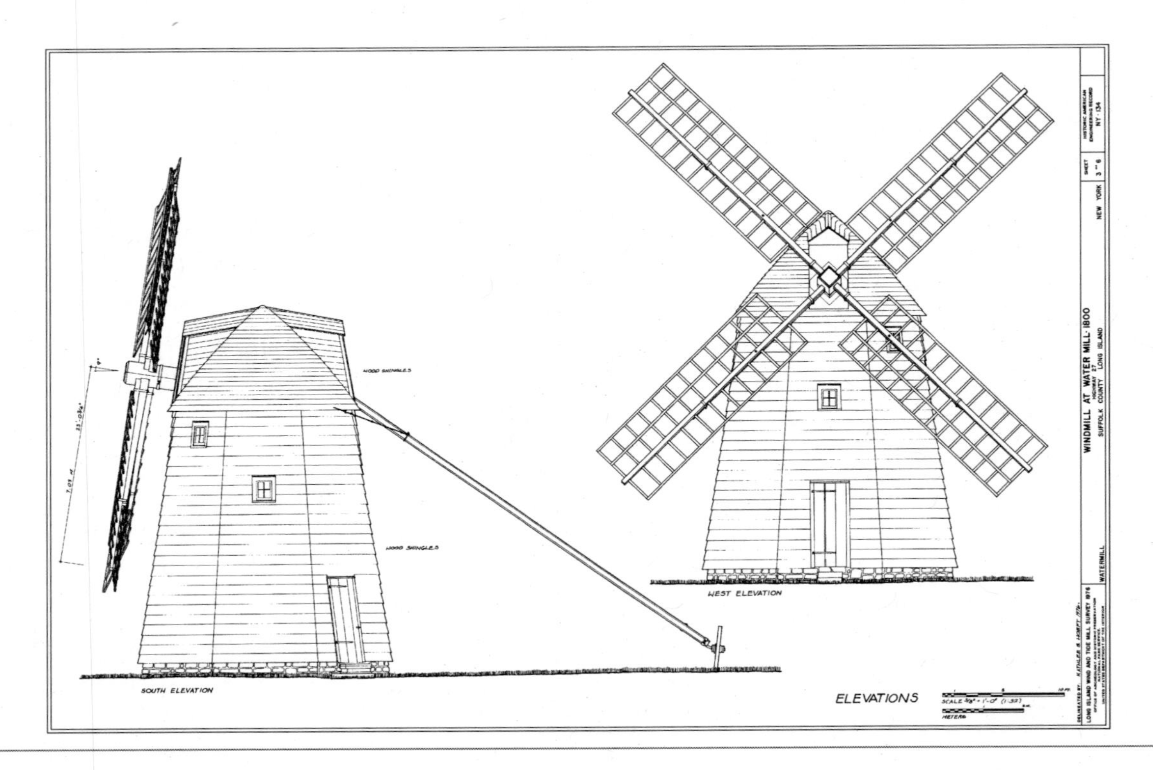

Figure 11.9. Smock mill. Windmill at Water Mill. West elevation. HAER NY-134, National Park Service. Delineated by Kathleen S. Hoeft, 1976.

wind. Most smock mills are eight-sided.[27] The mill building is composed of long timbers at each of the eight corners, each canted to give the building a smaller profile and to accommodate the pitch of the sails.[28]

Technology

The operation of the early nineteenth-century Beebe windmill in Suffolk County, Long Island, is shown in Hoeft and Long's 1976 HAER drawing (Figure 11.10). The wind hit the sails fastened on angled latticed stock and turned the polygonal windshaft. The brake wheel gear was attached to the windshaft and meshed with the wallower. In order to permit some control of the mill, the brake wheel could be slowed by using a wooden friction brake around its edge.[29] The wallower, attached to the vertical shaft, transferred the energy to it. The rotation of the vertical shaft caused the great spur gear to rotate. Paired smaller gears, known as stone nuts, meshed with the great spur gears.

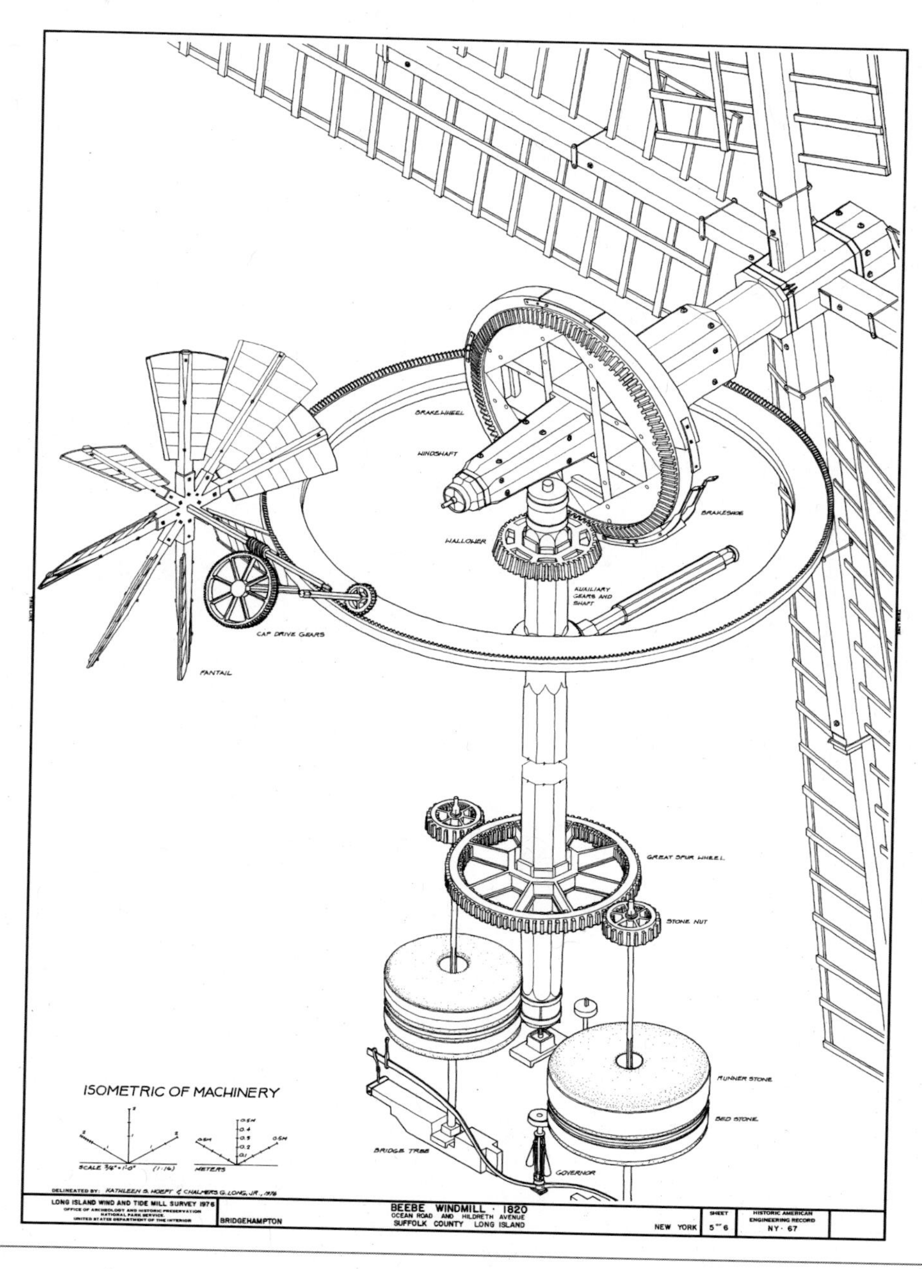

Figure 11.10. Isometric of machinery. Beebe Windmill-1820, Bridgehampton, New York, HAER NY-67, National Park Service. Delineated by Kathleen S. Hoeft and Chalmers G. Long, Jr., 1976.

The stone nuts transferred energy to the stone shafts and caused the runner stone to rotate.

Additional machinery illustrated in the isometric ensured that the windmill sails continued to face into a shifting wind. The fantail is a small windmill projecting from the back of the cap positioned at right angles to the sails. It consists of four sails or vanes set into a star-wheel of eight sockets. When the sails are directly into the wind, the vanes do not revolve. A shift in wind would cause the fantail to turn in relation to it, as the turning of its shaft causes the cap drive gears to turn, thus rotating both the cap and the sail mechanism. The fantail probably was first used in the United States in windmills in the New York area after 1800.[30]

In other mills, such as the Gardiner's Island Windmill, the sails, windshaft, frame, and cap were turned from inside the tower by means of the winding gears. The winding lever rotated a lantern pinion which in turn rotated a spur gear mounted on a vertical shaft. This shaft turned the cap gear through additional gearing.

GRAIN ELEVATORS AND FEED MILLS

With the emergence of large-scale grain production in the nineteenth century came the need to store increasingly larger amounts of grain. This led to the development of the grain elevator at mid-century. An elevator is composed of various mechanisms that perform the basic functions of receiving, shipping, and storing grain. Many of the larger elevators were constructed to receive grain brought to the site by train, boat, or barge. In the 1920s, grain elevators with the greatest total storage capacity were located in Minneapolis, Chicago, Duluth, Buffalo, and Kansas City.[31]

Wooden Elevators

Among the first grain elevators built in the United States was that erected in 1842 by Joseph Dart, a city merchant, on the Buffalo waterfront. Dart's building consisted of a series of grain bins, above which was a cupola containing weighing and spouting equipment. Incoming grain was moved to the cupola by a steam-driven belt elevator and was spouted by gravity via weighing hoppers to storage. Outgoing grain was drawn off from the bottom of the storage bins to be raised once again to the top of the cupola, where it was weighed out and spouted to barge, train, or wagon. Dart had adapted the technology of Oliver Evans's eighteenth-century gristmill for the storage of grain.

Dart's design enabled grain to be easily raised by a series of scoop-like buckets attached to a continuous belt. The design used two different varieties of elevator "legs." The "stiff leg" elevated grain in the elevator house, where it was fixed, while the "loose leg" elevated grain from ships to the elevator house. When not in use, the loose leg was stored in a raised position within the elevator house, within a distinctive tower above the cupola roof. If a ship's cargo was to be discharged, the loose leg could be lowered directly into the hold.[32]

By 1894, developments in speed and diversity of grain transfer systems prompted a corresponding evolution in building form. The application of horizontal transfer systems dramatically increased elevator capacity. In the absence of horizontal transfer systems, all bins had to be in sufficient proximity to the elevator leg to receive grain by direct gravity spouting. Horizontal conveyors permitted the transfer of grain to bins at a distance from the fixed elevator leg.

The disposition of conveying equipment affected the form of the elevator. Where conveyors were only installed above the storage bins, the classic high cupola house became typical. The high cupola accommodated the heads of a row of elevating legs together with their assembled scale and garner hoppers. Although incoming grain could be distributed to any bin by transfer along bin floor conveyors, the absence of basement conveyors required that outgoing grain, drawn from the bottom of any bin, had to be within direct spouting distance of an elevating leg so that it could be raised for weighing before shipping out.

The addition of conveying equipment to the basement floor gave the building a radically different appearance. Installation of basement horizontal transfer systems eliminated the need for elevating legs along the length of the structure. Outgoing grain could be spouted onto the basement conveying system and taken to some convenient point where elevator legs were located.

The grouping of elevator legs in the workhouse/headhouse-style of elevator dispensed with the need for a high cupola above the entire bin floor, requiring only a low cupola or gallery to house the bin floor conveying system.

Wood grain elevators were generally of timber construction. Foundations consisted of a series of concrete piers supported on piles. Timber columns were erected on piers to the height of the basement. The bin system was supported on a series of longitudinal and transverse beams spanning the columns. The bins were of laminated construction. Typically, the entire structure was clad in corrugated iron sheeting.

The wooden elevator was far from ideal as a means of grain storage. It was extremely flexible and loaded bins tended to settle. Conditions within proved to be an ideal breeding ground for vermin and grain rot. The flammable nature of the material proved to be the most serious drawback.[33]

An example of a small wooden grain elevator is the Hogan Grain Elevator in Seneca, Illinois, documented in the Historic American Engineering Record IL-25 (Figure 11.11). This elevator stored grain brought to the site by wagon.

Grain wagons were first weighed, then pulled up an inclined ramp to the dump shed. A hydraulic ram located under the floor raised one end of the wagon, dumping the grain through a grate into the funnel-shaped dump hopper below. A boot gate controlled the

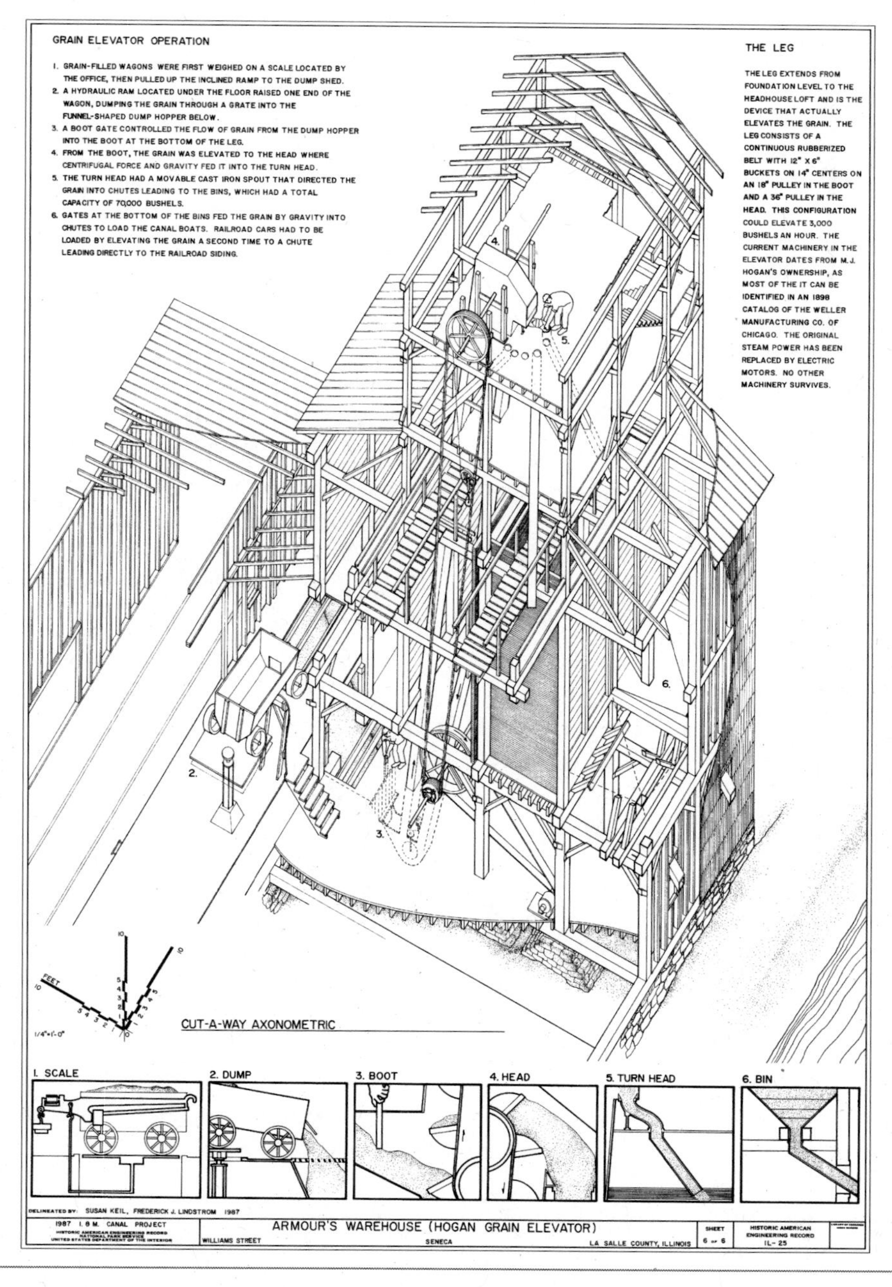

Figure 11.11. Cutaway axiometric. Armour's Warehouse (Hogan Grain Elevator), Seneca, Illinois, HAER IL-25, National Park Service. Delineated by Susan Keil and Frederick J. Lindstrom, 1987.

flow of grain from the dump hopper into the boot at the bottom of the leg. From the boot, the grain was elevated by a leg to the head, where centrifugal force and gravity fed it into the turn head. The leg extended from the foundation level to the headhouse loft and elevated the grain. It consisted of a continuous rubberized belt with buckets on 14-inch centers on an 18-inch pulley in the boot and a 36-inch pulley in the head. The leg could elevate 3,000 bushels per hour. The turn head had a movable cast-iron spout that directed the grain into chutes leading to the bins. Gates at the bottom of the bins fed the grain by gravity into chutes to load canal boats.

Steel and Tile Elevators

In 1861, a revolutionary iron grain elevator was completed in Brooklyn, New York. Constructed of wrought-iron plate cylindrical storage bins, 50 feet high and 12 feet in diameter, the elevator was supported on cast-iron basement columns and sheathed in masonry curtain walling. Later, the first elevator with steel bins was completed at Girard Point in Philadelphia.

Buffalo, New York, played an important role in the development of steel as a suitable, economical material for elevator construction. Buffalo's two pioneering steel elevators, the Great Northern and the Electric, both completed in 1897, used cylindrical bins with hemispherical bottoms. Both also pioneered the use of electric power in grain elevators.

During the first decade of the twentieth century, the use of ceramics in grain storage enjoyed a brief period of popularity. Tile bins were constructed in courses of plain and channel tiles. The channel tiles accepted horizontal tensile reinforcing bands. The bin floor was constructed of a grid of I-beams filled in with hollow ceramic book tiles. Unlike steel elevators, tile elevators were truly fireproof. These tile bins were considered obsolescent by 1913.

The first elevator to use steel grain bins on a large scale was Buffalo's Great Northern Grain Elevator noted above. This elevator was also the first to be powered by electricity.[34]

Concrete Elevators

Concrete grain elevators were first constructed in Europe. The impetus for the adoption of the concrete elevator in the United States was provided by F. H. Peavey. This Minneapolis grain dealer commissioned engineer C. F. Haglin to investigate the pioneering "ferro-concrete" (reinforced concrete) grain silos that had been constructed in Europe in the late nineteenth century.

Haglin recognized the inherent structural advantages of the cylindrical bin, particularly in view of the large volume storage required in the bulk American grain trade. Heglin also devised a new system of form work that did not need to be "struck" after every lift and dispensed with the need for full scaffolding. His forms consisted of two circular rings separated by yokes. In 1899, Haglin designed and erected a single cylindrical bin 124 feet high and 20 feet in diameter, with walls graduated in thickness from 1 foot at the base to 5 inches at the top. Following this experiment, work commenced on the first reinforced concrete grain elevator in the United States in 1900.[35]

Larger concrete elevators were constructed in major port cities such as Buffalo. One such large-scale elevator, Buffalo's Hecker or Standard Elevator, was documented in HAER drawings (NY-241; see Figure 11.12). This elevator was designed to receive grain from lake boats and rail cars and to ship grain by rail cars, canal boats, and barges. It also reconditioned grain using cleaners, clippers, washers, and driers.

Its vertical elevating systems included two marine towers with marine and receiving legs, workhouse legs for shipping, receiving, and reconditioning, and a car unloading jack leg. Horizontal grain transfer was by means of conveyors on the feed floor and bin floors equipped with loading hoppers and unloading trippers. Receiving systems consisted of garners and scale hoppers in the marine towers, rooftop "V" hoppers, and turnspouts in the bin floor. Shipping gravity-fed systems included bin spouts on the feed floor; the garner, scale hopper, and distribution floor turnspout of the west workhouse; and shipping spouts.

Grain to be shipped was spouted from the storage bins to the feed floor conveyors. Grain to be shipped by rail was transferred to the boot of either east workhouse lofting leg. Having been elevated to the top of the workhouse, the grain was weighed out in carload drafts, to be directed to the car loading spouts via the distribution floor turnspouts.

Feed Mills

Hundreds of facilities were erected throughout the country in the twentieth century to produce feed for

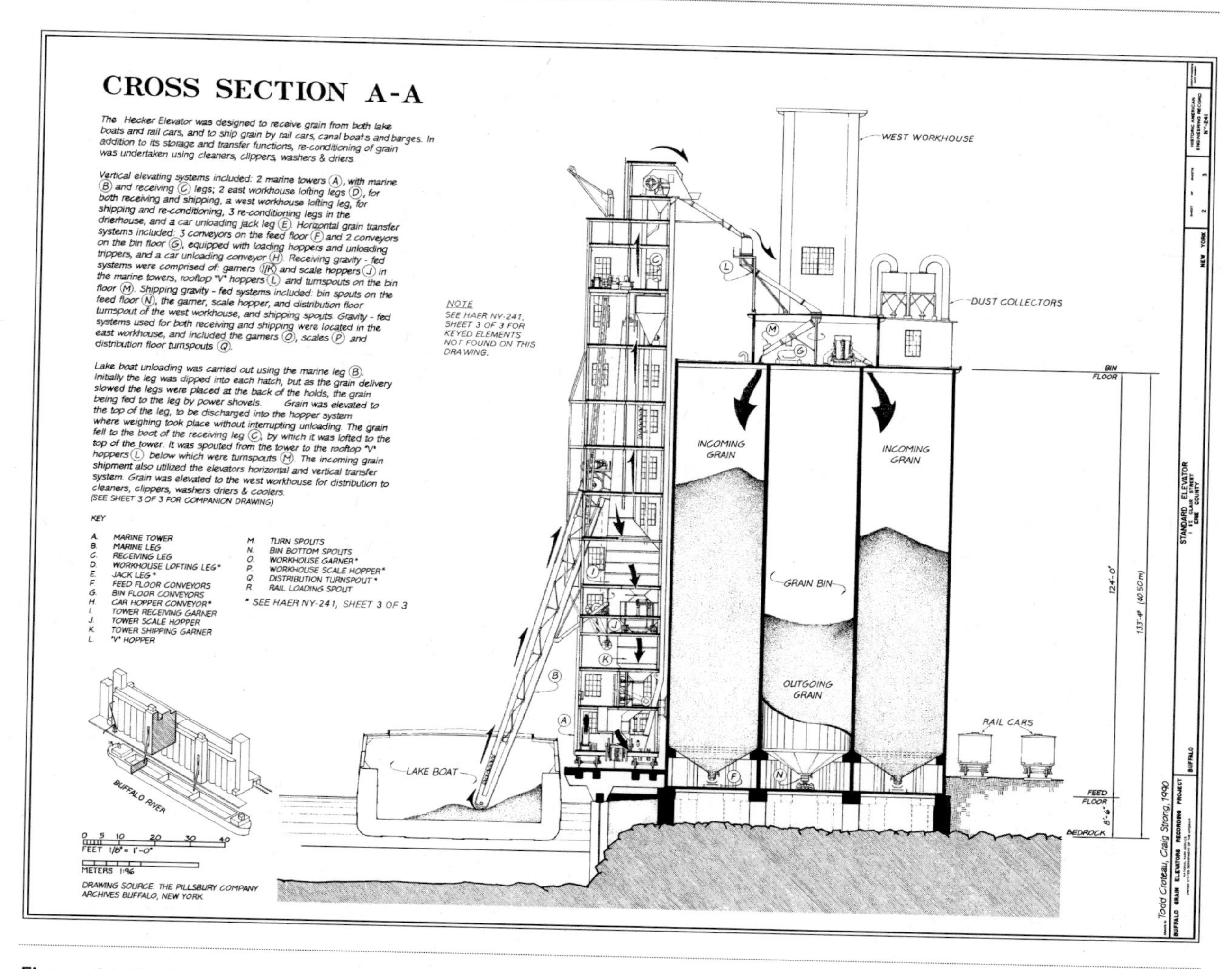

Figure 11.12. Cross Section A-A, Standard Elevator, Buffalo, New York. HAER NY-241, National Park Service. Delineated by Todd Croteau and Craig Strong, 1990.

livestock and pets. In such mills, various grains (the principal ingredients in animal feeds) were received, milled ingredients were mixed, and the mixture was then bagged. An example of one such feed mill was the Wollenburg Grain and Seed Elevator in Buffalo, a structure documented in the Historic American Engineering Record (Figure 11.13). As shown in the longitudinal section of this mill, principal ingredients of feed, corn, oats, and soybean meal were received by railroad boxcars. Cars were unloaded by a power shovel, which scooped grain from the car to a hopper, which then fed the receiving lofting leg. The lofting leg raised the grain to the top of the building, where it was poured into a scale hopper. The grain was then spouted to receiving bins or shipping bins if it only needed to be bagged. Grains needed for mixing, such as millet, buckwheat, and sunflower, were received and stored bagged.

When required for processing, grain was spouted from the base of the receiving bins to be re-elevated to the top of the house by the distribution leg. Most grains were spouted from this point to the mill. The grains that did not require milling, such as whole oats, were routed to the cleaner and oat clipper located on the scale floor. The grain was then spouted directly to the shipping bins.

Grain transferred to the mill was ground, elevated, sifted, cleaned, and deposited in the mill. Mill products to be mixed, such as rolled oats and cracked corn, were taken by hand cart to the elevator sacking floor. Special components such as sunflower seed were hand mixed into the milled ingredients. The mixture was then hand fed into the distribution leg via the trapdoor, while other components of the mixture, such as barley, were fed from the receiving legs. The elevated mixture was

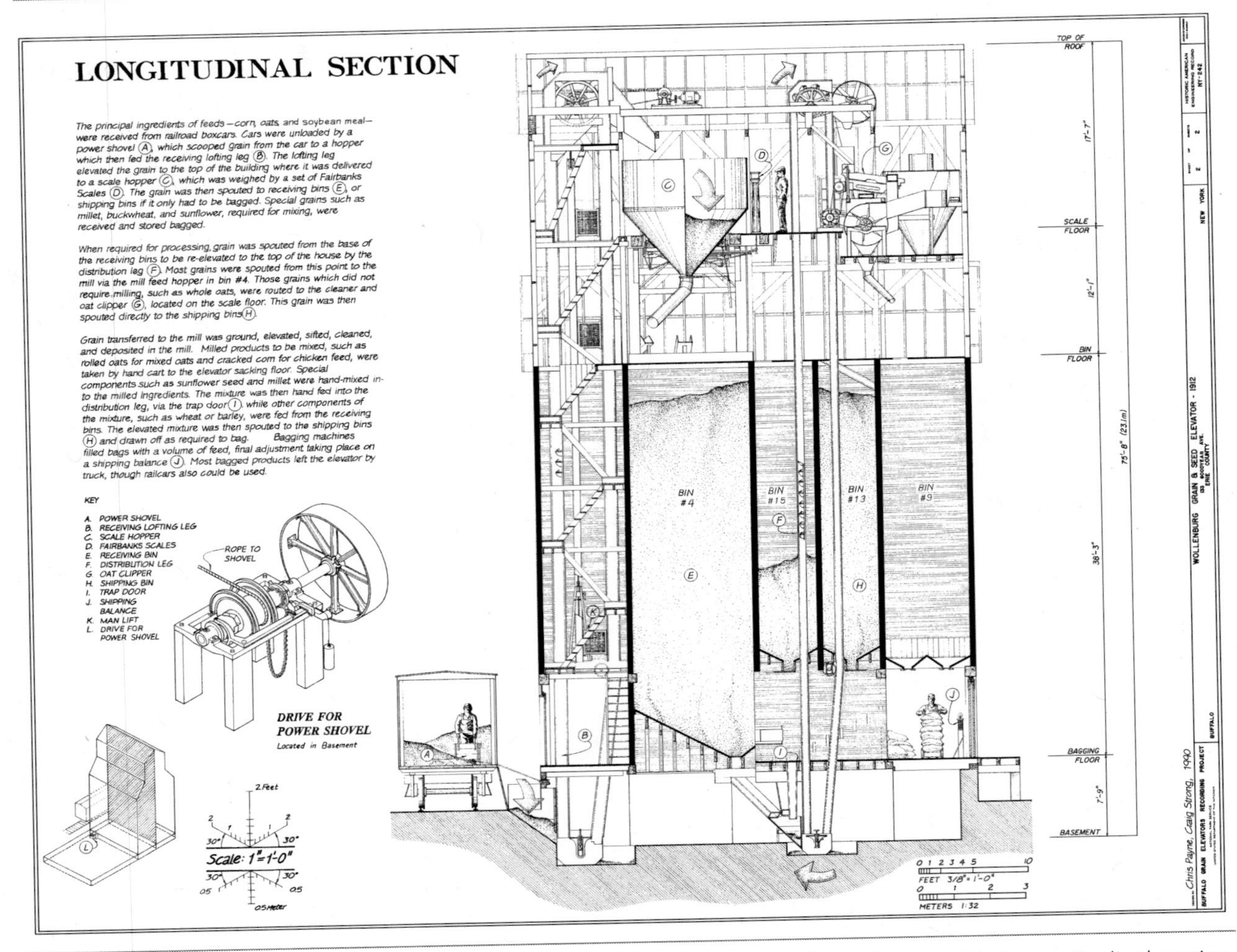

LONGITUDINAL SECTION
The principal ingredients of feeds—corn, oats, and soybean meal—were received from railroad boxcars. Cars were unloaded by a power shovel (A), which scooped grain from the car to a hopper which then fed the receiving lofting leg (B). The lofting leg elevated the grain to the top of the building where it was delivered to a scale hopper (C), which was weighed by a set of Fairbanks Scales (D). The grain was then spouted to receiving bins (E), or shipping bins if it only had to be bagged. Special grains such as millet, buckwheat, and sunflower, required for mixing, were received and stored bagged.
When required for processing, grain was spouted from the base of the receiving bins to be re-elevated to the top of the house by the distribution leg (F). Most grains were spouted from this point to the mill via the mill feed hopper in bin #4. Those grains which did not require milling, such as whole oats, were routed to the cleaner and oat clipper (G), located on the scale floor. This grain was then spouted directly to the shipping bins (H).
Grain transferred to the mill was ground, elevated, sifted, cleaned, and deposited in the mill. Milled products to be mixed, such as rolled oats for mixed oats and cracked corn for chicken feed, were taken by hand cart to the elevator sacking floor. Special components such as sunflower seed and millet were hand-mixed into the milled ingredients. The mixture was then hand fed into the distribution leg, via the trap door (I), while other components of the mixture, such as wheat or barley, were fed from the receiving bins. The elevated mixture was then spouted to the shipping bins (H) and drawn off as required to bag. Bagging machines filled bags with a volume of feed, final adjustment taking place on a shipping balance (J). Most bagged products left the elevator by truck, though railcars also could be used.
KEY
A. POWER SHOVEL
B. RECEIVING LOFTING LEG
C. SCALE HOPPER
D. FAIRBANKS SCALES
E. RECEIVING BIN
F. DISTRIBUTION LEG
G. OAT CLIPPER
H. SHIPPING BIN
I. TRAP DOOR
J. SHIPPING BALANCE
K. MAN LIFT
L. DRIVE FOR POWER SHOVEL
ROPE TO SHOVEL
DRIVE FOR POWER SHOVEL
Located in Basement
Scale: 1"=1'-0"
BIN #4
BIN #15
BIN #13
BIN #9
TOP OF ROOF
SCALE FLOOR
BIN FLOOR
BAGGING FLOOR
BASEMENT
FEET 3/8"=1'-0"
METERS 1:32
WOLLENBURG GRAIN & SEED ELEVATOR · 1912
Chris Payne, Craig Strong, 1990

Figure 11.13. Longitudinal Section, Wollenburg Grain and Seed Elevator-1912, Buffalo, New York. Longitudinal section. HAER NY-242, National Park Service. Delineated by Chris Payne and Craig Strong, 1990.

then spouted to the shipping bins and drawn off to be bagged.

GLOSSARY

Arms. (1) Spokes extending from the main shaft of a waterwheel, which in turn support the shrouding or rims of the wheel. (2) Shafts extending from the outer end of the windshaft of a windmill to which the sails are attached.

Balance rynd. A curved iron bar which crosses the eye of the runner stone, fitting into slots on either side; also called the *millstone bridge* or *crossbar*.

Bedstone. The lower, stationary stone in a pair of millstones.

Bins. Storage containers for grain, usually on the upper floor of a mill, from which grain could be fed into millstone hoppers.

Bolter. Machine used to sift flour into lots of differing textures or degrees of fineness.

Boot or *boot pit*. The lowest level of a grain elevator in which grain is initially dumped.

Brake wheel. In a windmill, the gear placed on the inner end of the windshaft that conveys momentum to the wallower.

Breastshot wheel. A waterwheel powered by a head of water striking the wheel at a point from one-third to two-thirds the height of the wheel, causing the wheel to revolve in a direction opposite that of the flow of water in the sluiceway.

Bridgetree or *bridge tree*. An adjustable beam upon which the millstone spindle is supported; may be raised or lowered to alter the distance between the grinding surfaces of the stone to produce a finer or coarser meal.

Bridging box. A housing, mounted on the bridgetree, that contains a footstep bearing supporting the millstone spindle and ensuring that the spindle will run perfectly upright.

Cap. The rotatable portion of a smock windmill located at the peak of the roof.

Cockeye. The socket at the center of the balance rynd, which serves as supporting bearing for the runner stone.

Cock head. The pivot point at the top of the millstone sprindle which fits into the cockeye.

Conveyor. In a gristmill, an endless belt structure equipped with buckets used to move grain of flour from one level to another.

Control gate. A gate at the end of the flume nearest the waterwheel, used to control the flow of water from the flume to the wheel.

Counter-gearing. A system of two-step gearing using a combination of face or spin gears and wallowers to drive millstone spindles.

Crotch. A device placed at the lower end of the millstone spindle in overdrift drive, engaging the rynd and driving the runner stone, shaped like a fork in the road.

Damsel. A square shaft, squared section of a rounded shaft, or a forked iron shaft fitted over the top of the millstone spindle, which in rotating taps against the shoe, thus feeding grain into the stones.

Driver. A cast-iron bar fitted onto the millstone spindle; the ends of the driver fit into slots in the eye of the runner stone and connect the runner stone to the spindle.

Eye. The central hole in a millstone.

Face gear or *face wheel.* A wheel with cogs mortised into its face, usually used in conjunction with a lantern pinion.

Flume. A trough or channel that carries water from the headrace to the point where the water strikes or enters the waterwheel.

Flume gate. A gate at the end of the flume nearest the headrace of a millpond, used to control the flow of water entering the flume.

Furrow. A groove cut into the grinding surface of a millstone.

Great spur wheel. (1) Spur gear used to transfer power from the main vertical spindle through lantern or spur pinions to millstone spindles in the form of two-step gearing known as spur gear drive. (2) In a smock mill, a gear used to transfer from the vertical shaft to the stone shaft(s) by means of stone nut(s).

Gristmill. A mill used for grinding grain, principally wheat or corn.

Head. The difference in level between water entering the waterwheel and that leaving the wheel (also called *fall*).

Headwater. Water entering or feeding the waterwheel.

Headrace. A channel that conveys water from the dam or millpond to the flume or directly to the waterwheel.

Hopper. An open-topped container tapered to feed grain to millstones.

Horizontal mill. A watermill whose wheel revolves in a horizontal plane.

Hub. The center of a waterwheel, into which blades are mortised in a horizontal mill.

Lands. The areas between furrows on the grinding surface of a millstone.

Lantern pinion. A pinion gear consisting of two round staves mortised between two disks, used either as a wallower or as a millstone pinion or nut.

Lay-shaft. A shaft set at right angles to the master face wheel that transfers drive to the little face wheel in counter-gearing, or a parallel-driven shaft when the master wheel is a spur gear wheel.

Leg. In a grain elevator, a steel column containing an endless chain of buckets used to move grain from a vehicle to the top of the elevator.

Little face wheel. A face wheel that transfers drive from the lay-shaft to millstone spindles in counter-gearing.

Lofting leg. In a grain elevator, a steel column equipped with buckets used to elevate grain to the top of the building.

Marine leg. A long, steel column housing an endless chain of buckets dipped into the hold of a freighter from a grain elevator to carry grain from the freighter to the elevator.

Marine tower. A movable or fixed portion of a grain elevator in which the marine leg is housed.

Master face wheel. A face wheel mounted on the waterwheel shaft in counter-gearing; used to transfer power to lay-shafts via lantern pinions.

Millpond. A body of water, usually created by damming a stream, that serves as a source of water for the waterwheel.

Neck bearing. A wooden bearing in the center of the bedstone through which the millstone spindle passes.

Nut. A pinion, located at the top of the quant, which engages the great spur wheel in overdrift drive; or a pinion mounted on the stone pinion in underdrift drive.

Overdrift drive. A method of driving or turning millstones by bringing power down from above by means of a quant or millstone spindle connected from above to the runner stone.

Overshot wheel. A waterwheel powered by a head of water striking the wheel just forward of its highest point, causing the wheel to revolve in the same direction as the flow of water in the sluiceway.

Pair. A set of two millstones, consisting of the upper or runner stone and the lower or bedstone.

Pit wheel. A large face gear wheel mounted on the waterwheel shaft.

Pitchback wheel. A waterwheel powered by a head of water striking the wheel at or just back of its highest point, causing the wheel to revolve in a direction opposite that of the flow of water in the sluiceway.

Post mill. The earliest type of windmill. It consists of a box-shaped body mounted and turning on a horizontal main shaft (post).

Quant. A shaft that serves as a millstone spindle in overdrift drive.

Rap. A block on a shoe against which the damsel strikes to ensure an even flow of grain from the hopper to the millstones.

Runner stone. The upper, moving stone in a pair of millstones.

Rynd (or *rind*). A crossbar containing the bearing on which the upper stone of a pair of millstones rests.

Shoe. A tapering trough vibrated to feed grain into the stones for grinding.

Shroud. The rim of a waterwheel which forms the sides of bucket enclosures.

Skirt. The outer edge of the grinding surface of a millstone.

Smock mill. An octagonal, wood-framed windmill in which the mill's cap is turned to face the sails into the wind.

Spindle. The shaft on which the runner stone rotates.

Spur gear (or *spur wheel*). A gear with cogs mortised or cut into its edge.

Spur gear drive. A system of two-step gearing using a system of face gears, spur gears, and lantern or spur pinions to drive millstone spindles.

Stone case. A circular wooden enclosure around a pair of millstones, also called a casing, hoop, husk, tun or vat.

Stone nut. In a windmill, a spur or lantern pinion mounted on a stone shaft that transfers motion from the vertical shaft by meshing with the great spur wheel.

Sweeper. A device attached to the runner stone that sweeps meal from between the edges of the stones and the stone case and carries the meal to the spout opening in the case.

Tail water. Water leaving the waterwheel.

Tailrace. A channel that conveys water from the waterwheel back into the millstream.

Tentering. The process of adjusting the distance between the upper and lower millstones.

Tentering staff. A beam, connected to the bridgetree by the brayer, permitting the bridgetree to be raised or lowered and thus adjusting the distance between the upper and lower millstones.

Trundle. A lantern pinion on a millstone spindle in counter-gearing; or a pinion resembling a smaller version of a face gear.

Tub wheel. A horizontal waterwheel with curved blades that revolves in a wooden or masonry hoop or tub enclosure.

Turnspouts. Movable pipes used to distribute grain into the individual hoppers of a grain elevator.

Two-step gearing. Systems of interlocking gears designed to permit several pairs of millstones to be driven from a single waterwheel.

Underdrift drive. A method of driving or turning millstones by bringing power up from below by means of a millstone spindle connected from below to the runner stone through the eye of the bedstone.

Undershot wheel. A waterwheel powered by a head of water striking the wheel at a point near the bottom of the wheel, causing the wheel to revolve in a direction opposite to that of the flow of water in the sluiceway.

Wallower. (1) The first driven gear wheel in a water mill, driven by a gear wheel from the main waterwheel. (2) In a windmill, the gear that transfers motion from the horizontal windshaft to the vertical shaft.

Windshaft. The usually polygonal wood shaft connecting the windmill sails to the brake wheel.

Workhouse. The portion of a large grain elevator in which elevator personnel work, usually located between groups of bins or atop the bins.

BIBLIOGRAPHY

Craik, David. *The Practical American Millwright and Miller.* Philadelphia: H. C. Baird, 1870.

Evans, Oliver. *The Young Mill-Wright & Miller's Guide.* Philadelphia: Oliver Evans, 1795.

Hazen, Theodore R. The Art of the Millstones, How They Work. www.angelfire.com/journal/millrestoration/millstones.html.

Hefner, Robert J. *Windmills of Long Island.* Published for the Society for the Preservation of Long Island Antiquities. New York: W. W. Norton & Company, 1983.

Howell, Charles, and Allan Keller. *The Mill at Philipsburg Upper Mills and a Brief History of Milling.* Tarrytown, NY: Sleepy Hollow Restorations, 1977.

Huls, Mary Ellen. *Water Mills: A Bibliography.* Monticello, IL: Vance Bibliographies, 1986.

International Correspondence Schools. *Water Wheels.* Scranton, PA: The Colliery Engineer Company, 1898.

Hunter, Louis C. *A History of Industrial Power in the United States, 1780–1930.* Volume 1: *Waterpower in the Century of the Steam Engine.* Charlottesville: University Press of Virginia, 1979.

Kuhlman, Charles Byron. *The Development of the Flour-Milling Industry in the United States.* Boston: Houghton Mifflin Co., 1929.

Larkin, David. *Mill.* New York: Universe Publishing Co., 2000.

Leary, Thomas E., John R. Healey, and Elizabeth C. Sholes. Buffalo Grain Elevators. HAER NY-239, 1991.

Maher-Keplinger, Lisa. *Grain Elevators.* New York: Princeton Architectural Press, 1993.

National Park Service. Tub Wheels. Harpers Ferry National Historical Park Photo Archives. www.nps.gov/hafe/waterpwr/tub.htm.

Reynolds, John. *Windmills and Watermills.* New York: Praeger, 1970.

Wailes, Rex. *English Windmills.* London: Routledge & Kegan Paul, Ltd., 1954.

Regional Guides

Illinois

Vierling, Philip E. *Early Water Powered Mills of the Des Plaines River and Its Tributaries, Illinois.* Chicago: Illinois Country Outdoor Guides, 1995.

Maryland

Cook, Eleanor M.V. *Early Water Mills in Montgomery County.* Rockville: Montgomery County Historical Society, 1990.

Missouri

Suggs, George C. *Water Mills of the Missouri Ozarks.* Norman: University of Oklahoma Press, 1990.

Montana

Haines, Tom. *Flouring Mills of Montana Territory.* Missoula: Friends of the University of Montana Library, 1984.

Nebraska

Buecker, Thomas R. *Water Powered Flour Mills in Nebraska.* Lincoln: Nebraska State Historical Society, 1983.

New Jersey

Weiss, Harry B., et al. *The Early Grist and Flouring Mills of New Jersey.* Trenton: New Jersey Agricultural Society, 1956.

Ohio

Garber, D. W. *Waterwheels and Millstones: A History of Ohio Gristmills and Milling.* Columbus: Ohio Historical Society, 1970.

Pennsylvania

Lord, Arthur C. *Water-Powered Grist Mills, Lancaster County, Pennsylvania.* Millersville, PA: A. C. Lord, 1996.

South Carolina

Batson, Mann. *Water-Powered Gristmills and Owners, Upper Part of Greenville County, South Carolina.* Travelers Rest, SC: M. Batson, 1996.

Virginia

Moffett, Lee. *Water Powered Mills of Fauquier County, Virginia.* Berryville, VA: Virginia Book Company, 1980.

NOTES

1 Bryant Franklin Tolles, "Textile Mill Architecture in East Central New England: An Analysis of Pre-Civil War Design," *Essex Institute Historical Collections* 107 (July 1971): 228–229.

2 Louis C. Hunter, *A History of Industrial Power in the United States,* volume I (Charlottesville: University Press of Virginia 1979), 54.

3 Charles Howell, "Colonial Watermills in the Wooden Age," in Brooke Hindle, editor, *America's Wooden Age: Aspects of Its Early Technology* (Tarrytown, NY: Sleepy Hollow Restorations, 1975), 134.

4 Howell, 127.

5 Joseph P. Frizzell, "The Old Time Water-Wheels of America," from *Transactions of the American Society of Civil Engineers* XXVIII (April 1893): 3, as cited on the Pond Lily Mill website (www.angelfire.com/journal/pondlilymill/frizzell.html).

6 Howell, 128.

7 Howell, 128.

8 Howell, 128, 133.

9 Howell, 133.

10 Frizzell, 4.

11 Dr. John Lovett, Jr., Tub Wheel, www.spoom.org/pdf/lovett%20tub%20wheel.pdf

12 Hunter, 307.

13 Hunter, 321.

14 Brenda Krekeler, Ohio's Old Mills Today, fpw.isoc.net/krek/Overview_04_Waterwheels.html; John Blake Campbell, *Campbell Water Wheel Company Water Wheels, Dams, Hydro-Electric Plants & Water Supply Systems*, 1932, as reproduced on Pond Lily Mill website (www.angelfire.com/journal/pondlilymill/campbell.html).

15 Theodore R. Hazen, The Art of Millstones, How They Work, www.angelfire.com/journal/millrestoration/millstones.html.

16 Hazen.

17 Howell, 137.

18 Howell, 138–140.

19 Eugene S. Ferguson, Foreword to Oliver Evans, *The Young Mill-Wright & Miller's Guide* (Wallingford, PA: The Oliver Evans Press, 1990), iii–v.

20 Howell, 143–144.

21 Charles Byron Kulhman, *The Development of the Flour-Milling Industry in the United States* (Boston: Houghton Mifflin Company, 1929), 121–122.

22 Theodore R. Hazen, How the Roller Mills Changed the Milling Industry, www.angelfire.com/journal/millrestoration/roller. html.

23 John R. Bowie, delineator, Mascot Roller Mills, Ronks vicinity, Pennsylvania, Historic American Engineering Record, PA-506, 1997–1998.

24 Mark Berry, History of Windmills, from Windmill World website, www.windmillworld.com/windmills/history.htm.

25 An exhaustive scholarly discussion of Long Island's windmills, illustrated with photographs and HAER drawings, is found in Robert J. Hefner, *Windmills of Long Island* (New York: W.W. Norton & Company, 1983).

26 Hefner, 15.

27 Berry.

28 T. Allan Comp, in Hefner, 11.

29 Berry.

30 Hefner, 81.

31 John A. Droege, *Freight Terminal and Trains*, 2nd edition (New York: McGraw-Hill Book Company, 1925), 287.

32 Buffalo History Works, Grain Elevators: A History, www.buffalohistoryworks.com/grain/history/history.htm.

33 Thomas E. Leary, John R. Healey, and Elizabeth C. Sholes, Buffalo Grain Elevators, NY-239, 5–7.

34 Leary, Healy, and Sholes, 8–13.

35 Leary, Healy, and Sholes, 15–19.

CHAPTER 12

Iron and Steel Production

Among the earliest surviving industrial remains in the United States are those associated with the iron industry. Furnaces, foundries, forges, rolling mills, charcoal and lime kilns, and blacksmith shops and their remnants are still scattered throughout much of the United States, especially in rural areas.

The small-scale metalworking industry produced many of the items of everyday life in the eighteenth and nineteenth centuries: Dutch ovens, bake plates, stew pans, teakettles, and other kitchenware; toys for children; building materials such as nails and brick binders; and farm implements and tools.[1]

Hundreds of iron furnaces were in operation, and most small towns had a small forge and blacksmith. Some blacksmiths may have extracted small amounts of iron from locally available ore, but most were consumers rather than producers of iron, obtaining iron pigs or bars from local furnaces or bloomeries.

Furnaces were typically the center of small communities known as iron plantations. These rural, self-sufficient villages included the furnace, housing for the workers, a gristmill and sawmill, croplands, and woodlands. An ironworks had to be near water and raw materials, a forest for charcoal, limestone for flux, and iron ore. During the months when the furnace was in blast, hundreds of workers were needed.[2] For instance, at Hopewell Furnace, in Pennsylvania's Chester County, between 200 and 250 workers were on the payroll in the mid-nineteenth century. With their families, they constituted a settlement of perhaps 600. Workers at a typical plantation included woodcutters, colliers, miners, molders, founders, patternmakers, and their helpers. At Hopewell, the company rolls included laborers, cabinetmakers, carpenters, masons, clerks, farmworkers, boxmakers, and wheelwrights. Some workers lived near the furnace, while others lived several miles away near the charcoal hearths or woodlots. The works itself included barns, company stores, springhouses, blacksmiths, a school, a church, housing, a carpenter's shop, forges, sawmills and gristmills, and a casting house.[3]

Today, iron plantations survive in various degrees of intactness. Some contain stone buildings used for functions such as charcoal storage. Others retain remnants of waterpower, such as the wheelpit, dam, millrace, and tailrace. Others contain walls pointing to a charging terrace or charging bridge. Others retain slag and cinder piles. A few have been restored to interpret an iron plantation to a modern audience.[4]

THE IRON FURNACE

The Early American Iron Industry

During the colonial period, ironmakers employed a variety of production techniques, some ancient, some more modern. Primitive man had discovered that chunks of iron ore heated on a stone hearth with bellows and a charcoal fire would produce a spongy mass of iron which could be refined through reheating and hammering. This metal became known as wrought iron. Slabs of wrought iron were called blooms, and the simple ironworks that produced them, a bloomery. A bloomery had to simultaneously accomplish two separate objectives: reduce the iron oxides in the ore to metal and separate the metal from the gangue in the ore. In the heat of the fire, the gangue reacts with iron oxide in the ore to make liquid slag, and carbon monoxide formed by burning charcoal reduces the rest of the iron oxide to particles of solid, metallic iron. A bloomery required only simple equipment, including a hearth or fireplace, a bellows operated by hand or animal or waterpower, and tools for manipulating and pounding the iron as it was heated and reheated. The hearth was provided with a water-cooled metal bottom-plate, and cast-iron plates lined the sides. The hearths, rectangular in shape, were about 2 feet deep and 3 feet wide and were surmounted by a tall chimney in the form of a truncated pyramid for carrying off the hot waste gases. Such works were easily erected and also easily disappeared. Many bloomeries operated in colonial America, particularly in Massachusetts, New Jersey, and Pennsylvania. Bloom smelters practiced their art in North America from the seventeenth century until the early years of the twentieth century.

A primary disadvantage of the bloomery was its limited production. It could supply small blacksmith shops with the amount of iron necessary to meet the needs of local customers but could not accommodate larger markets. In addition, a bloomer had to stop smelting when it was time to remove the loup from his hearth. By doing so, he wasted most of the thermal energy used to heat up the furnace. Because the bloomer did not melt the iron that was produced, he was unable to separate it completely from the slag.[5]

Blast Furnace

For large-scale iron production and production of cast iron, the blast furnace was developed. Unlike a bloomery, a blast furnace could run continuously. The furnace crew charged ore, fuel, and flux into the top of the furnace shaft and blew air in at the bottom.[6] Built in the form of a flattened pyramid, the early blast furnace was usually between 25 and 30 feet high (Figure 12.1). With a hollow center, it had thick stone outer walls and an inner lining of brick or other fire-resistant materials. A layer of clay or stone chips between the lining and the outer wall allowed the inner walls to expand in response to the enormous heat produced by smelting, reducing iron ore. The furnace shaft tapered outward from a small opening at the top called a throat or trunnel head to a point about two-thirds of the way down, the bosh. Below the bosh, the furnace tapered inward to support the materials with which the furnace was charged. The bottom of the furnace, the crucible, was the chamber in which the melted iron and liquid slag accumulated from the fiery mass above. A small hole, the tuyere, located near the bottom of the crucible, admitted the nozzles of the bellows that supplied the air blast. The bellows were usually powered by a waterwheel, requiring that the furnace be located near a stream. It was also typically built adjacent to a hill from which raw materials could be carried to the top of the furnace over a bridge for charging.

At ground level in front of the crucible was the hearth, a working area hollowed out from the side of the furnace. A damstone prevented the contents of the crucible from spilling out into the hearth. A part of the inner wall, the timp, was located behind the damstone, leaving a small opening through which workers could insert probing tools. This opening was normally plugged with clay, opened only when the furnace was being tapped. The melted slag, lighter than iron, collected in the upper part of the crucible and was drawn off through an opening, the cinder notch. The liquid iron sunk to the bottom and gushed out the tapping hole into a series of sand molds.[7]

The furnace was charged with three raw materials: iron ore, charcoal, and limestone. Limestone promoted the separation of iron from impurities. In many colonies, iron ore was plentiful in bogs, outcroppings, or deposits close to the ground surface.

Few above-ground remains survive from either colonial or early-republic forges. Where sites have not been reused, industrial archaeologists can often locate production wastes that provide evidence of early ironmakers' techniques.[8]

CHARCOAL MANUFACTURE

Charcoal was used to fuel furnaces because it is nearly pure carbon and burns hotter than wood. Charcoal is produced by controlled burning of wood. The burning was not allowed to progress beyond active smoldering; otherwise the wood would be consumed in flames. Properly controlled smoldering permits enough heat to burn off the spirits and pitch in the wood, leaving a material with a high carbon content.

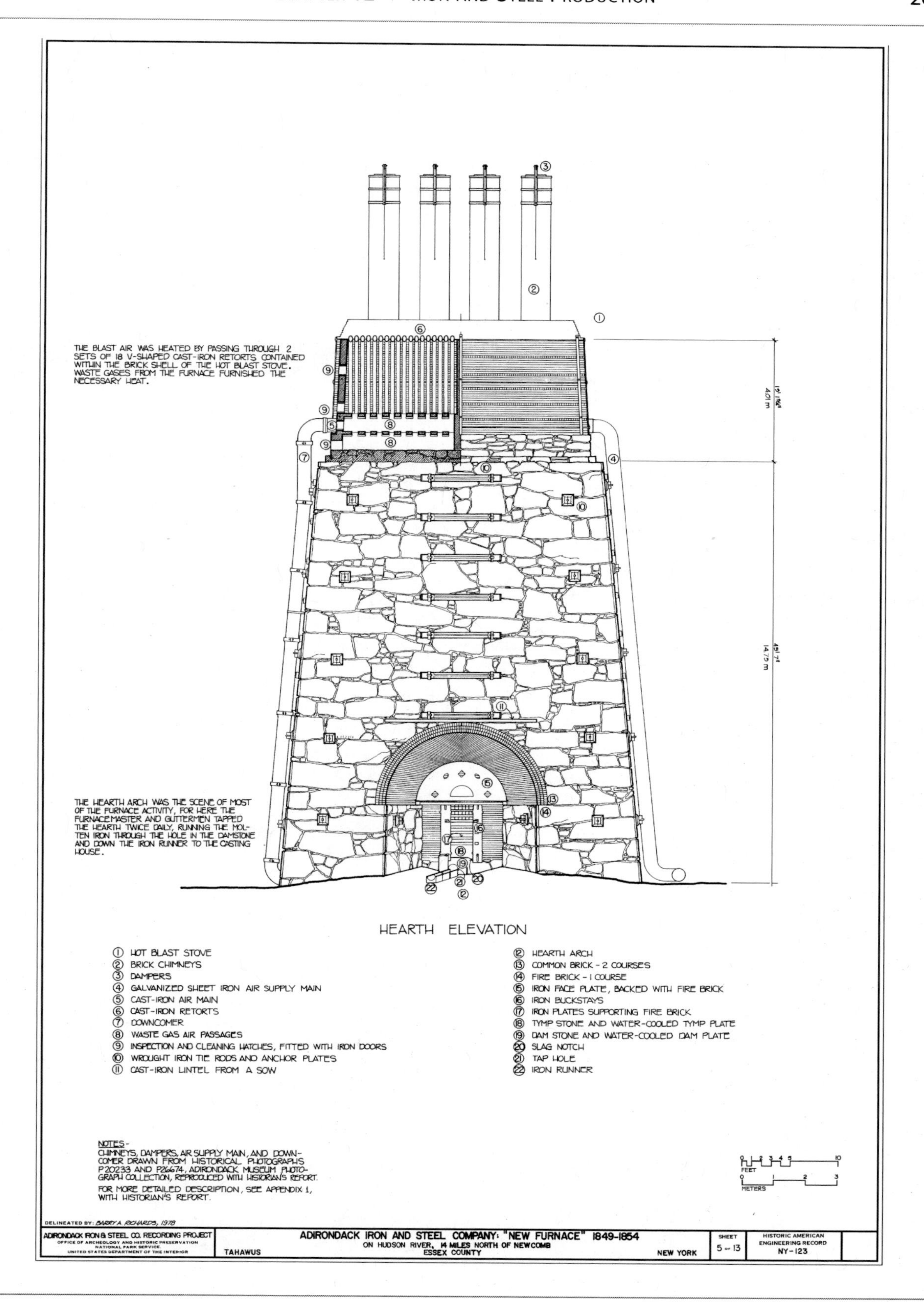

Figure 12.1. Adirondack Iron and Steel Company, New Furnace, 1849–1854, Tahawus, New York. HAER NY-123, National Park Service. Delineated by Barry A. Richards, 1978.

In the pre–Civil War period, wood was charred in earth-covered mounds. These mounds were also referred to as heaps or pits. In the mound, wood was stacked in a pile, 30 to 40 feet in diameter, with an approximately 1-foot-diameter space in the middle to function as a chimney. The wood mound was usually 10 to 14 feet high in the center and contained about 30 cords. After all the spaces were filled with smaller pieces of wood to make the mound compact, it was covered with a layer of charcoal dust and then earth and leaves. A hole was left at the top for the chimney, and small 3- to 6-inch-diameter vent holes were opened around the sides, about a foot above the ground. To start the burn, burning embers and tinder were dropped into the center chimney. The side vents were opened and closed as needed to supply a natural draft and draw the burning from the middle of the mound to the sides. It generally took about a week to fully char the wood.

After cooling, the charcoal was carefully shoveled into wagons with wooden shovels and rakes to minimize breakage and then taken to the furnace site. The mound method produced a weak charcoal, one that crumbled easily in the blast furnace. More durable charcoal was produced by permanent charcoal kilns. Typical charcoal yield of mounds was about 30 to 35 bushels of charcoal per cord of wood.[9]

CHARCOAL KILNS

Early charcoal kilns were rectangular or round, usually constructed of brick on a stone foundation. A typical rectangular kiln measured 40 to 50 feet long and 12 to 15 feet high and wide (Figure 12.2). Capacity generally ranged from 55 to 70 cords of wood. Typical yield was between 45 to 50 bushels of charcoal per cord of wood. The rectangular kiln was frequently found in the southern states and less frequently found in New England.

More common in New England were round kilns, typically 28 to 30 feet in diameter and 12 to 16 feet high. Some were constructed with vertical walls, while others had battered walls. Round kilns required two or three heavy iron bands around the walls to provide structural stability. The walls were ventilated with ankle, knee, and waist vents, and the kiln typically had a vaulted brick roof. From 33,000 to 40,000 bricks were used to construct a typical round kiln.[10]

It took four or five men one day to load the wood to fuel a single burning of the kiln. It was ignited with a long-handled torch at the bottom, in space below the wood that was left by the skids. When the charring was complete, after about 5 days, all vents were closed to suffocate the fire, followed by 5 to 6 days of cooling. The entire cycle took about 10 to 12 days.[11]

By the late 1870s conical kilns had gradually replaced the larger cylindrical kilns. Generally 25 to 30 feet in diameter at the base and 25 to 35 feet high, conical kilns were designed for 25 to 45 cords of wood.[12]

Decline of the Charcoal Industry

As increasing amounts of iron and then steel were fabricated in coke-burning furnaces, the demand for charcoal began to decline. The total number of charcoal furnaces dropped from 500 in the 1860s to fewer than 100

Figure 12.2. Rectangular-type charcoal kiln. Reprinted from Henry S. Overman, *The Manufacture of Iron* (Philadelphia: H. C. Baird, 1850), 110.

in the 1890s and to fewer than 50 after 1900. A slight revival occurred with the advent of backyard barbecue grills in the post–World War II period, but charcoal production never again came close to its nineteenth-century peak.

Remnants of Charcoal Kilns and Mounds

Mounds and kilns were constructed as close as possible to the source of their wood supply. Builders determined that hauling of heavy logs should be minimized, and lighter weight charcoal should be moved instead. Therefore, many charcoal mounds and kilns were located in higher elevations, away from present-day roads.

Because of the large number of bricks used in kiln construction, concentrations of brick are one sign of a kiln site. Another indication is charcoal itself. In general, the closer it is to the kiln, the darker the soil. Terrain is a third clue. Kilns in the northeastern states were often constructed into 15- to 20-foot-high embankments. A single kiln may have a single concave depression cut into an adjacent low hill, while an eight-kiln site may have eight such depressions. A trace of an old road may be present leading up to the kiln site, while another road trace may lead around the hill to the top of the embankment.

The kiln ruin itself may be characterized by anything from visible brick or stone walls 3 to 6 feet high, complete with vent holes, large kiln-girding hoops, and mounds of charcoal, to no walls but a low, 30-foot circular mound of thousands of bricks, or only a circle of black ground. Identifying charcoal mounds is extremely difficult because no brick or iron hardware was used in its construction. [13]

LIME PRODUCTION

Lime is produced by burning or calcining limestone without melting it. Lime is a widely used commodity, employed in the manufacture of paper, glass, whitewash, mortar, and cement, in tanning leather, in sugar refining, as a water softener and bleach, and as an agricultural neutralizer of acid soils. It is also used as a flux for iron and steel production.

Calcining limestone has a number of basic requirements. First, carbonate of lime must be brought to high heat in order to free the carbonic acid gas. The heat must be maintained for several hours to permit the gas to escape. The time needed for complete expulsion of gas is in proportion to the size of the pieces of stones heated. Large stones take a longer time, smaller stones, shorter. Large-scale commercial calcining of lime usually took place in kilns. When limestone could not be broken into small pieces, the lime burner placed the largest blocks in the center of the kiln where they were exposed to the greatest heat.[14]

Lime Kilns

One of the earliest lime kilns in the country was operated in seventeenth-century Rhode Island. Surviving documents indicate that this kiln measured 16 feet in diameter at the top, 13 feet in diameter at the center, and 10 feet in diameter at the bottom. It was 15 feet in depth.[15]

Lime kilns (Figure 12.3) were typically classified by types of operation: intermittent or continuous. Intermittent kilns were those in which each burning of limestone was a separate operation. The kiln was charged with limestone, the limestone was burned and

Figure 12.3. Limestone kiln. Reprinted from Quincy Adams Gillmore, *Practical Treatise on Limes, Hydraulic Cements and Mortars* (New York: D. Van Nostrand, 1870).

cooled, and the kiln was emptied. A disadvantage of intermittent kilns was the irregular quality of the products. These kilns were usually small and operated by local farmers as demand required.

As would be expected, farm kilns were smaller than commercial kilns, generally measuring from 6 to 8 feet in interior diameter. They were usually constructed into the sides of a low rise near an outcrop of limestone. The sides of the kiln were covered with earth, leaving an opening in front to draw out the burned lime. The top kiln opening was level with the ground at the upper level for ease in dumping limestone.[16]

Where no hillside was available, the stone structure was banked with earth, with an earthen ramp leading to the top. Heights ranged from 10 to 20 feet, and stone walls were from 18 to 24 inches thick. The wall interiors were either square or round and generally had an inside diameter or width of about 9 feet. Openings at the bottom varied in size and style. Earlier kilns had smaller openings to protect from sudden drafts that caused too rapid burning. These openings were simple holes at the bottom of the wall supported by stone lintels. Later kiln openings were arched, similar to those of a small blast furnace. An iron grate was placed a foot or so above the bottom of the shaft. Limestone was piled above the grate, while firewood was piled below.

The shaft of the kiln was egg-shaped, with the larger diameter closer to the top. The kiln was fired by placing a small pile of kindling on the grate at the bottom of the shaft. Firewood was placed on this about a foot high, then limestone in alternate layers with wood.

In continuous kilns, also known as running or perpetual kilns, limestone and fuel were charged in alternate layers. As the burning progressed, burnt lime was removed from the bottom, while fresh layers of limestone and fuel were added to the top.

The inside walls of some early kilns were protected from extreme heat by a lining of refractory stone, usually a good-quality sandstone. When kilns were worked by intermittent fires, the limestone charge rested on arches built up from pieces of stone to be burned and laid dry. The fire gradually worked its way toward the front as the draft increased. The opening was regulated to obtain the required degree of combustion and new fuel added as necessary to maintain the fire. Outside air, which entered through the fire door, carried the flames to all parts of the arch. When the upper part of the kiln was smaller in diameter than the lower, the draft sometimes drove the flames out the door.

As the character of the lime business changed from a local to a regional market, demand for both quantity and quality increased. In some places, intermittent kilns were lined with firebrick and built a little higher to meet new demands. In most places, new kilns of more advanced design were built.

The new kilns were 25 to 28 feet high, with an inside diameter of 5 to 6 feet at the top, 10 to 11 feet near the middle, and 7 or 8 feet near the bottom. The inside resembled an egg standing on its fatter end. These kilns were lined with regular brick or firebrick, set 14 to 18 inches thick. There was a 5- or 6-foot arched opening at the bottom through which the fuel was loaded and burned lime removed. About a foot above the bottom there was an iron grate on which the fuel burned.

By the 1860s perpetual kilns were introduced. These kilns are characterized by three types of operation: mixed feed (limestone and fuel fed in alternate layers), separate-feed (limestone and fuel not in direct contact), and rotary kilns. Mixed feed kilns were the most difficult to manage with certainty, although when running favorably were the most economical.

The second type, the separate-feed, became the first modern lime kiln. Remnants may be seen in its distinctive steel shell rising from 25 to 35 feet above the kiln base. The kilns rose up to a maximum of 50 feet above ground level. The insides of the shells were lined with firebrick. Around and outside the bottom of the shell were small furnaces, also known as fireplaces, in which the fuel was burned. Fuel for the furnaces was coal or wood, whose flames were directed through two large openings around the side of the shell and directly on the limestone passing downward past an opening. Fresh limestone was fed in at the top of the shell. At some plants, the limestone was carried to the top by small cars that ran on tracks directly from the quarry to trestles atop the kilns.

The third type of perpetual kiln was a rotary kiln. Rotary kilns were in essence horizontal kilns resembling a long, large-diameter pipe, which rotated slowly on driven rollers. The charging end was slightly higher than the discharge end, so the charge would move slowly through the kiln. Its chief disadvantage was high fuel consumption. Rotary kilns required that limestone be crushed to fairly even size, preferably finely ground.[17]

Kiln Sites

Surviving kiln remains range from those of farm kilns to early and later commercial kilns. Later commercial kiln remains were larger than those of farm kilns and were made with more massive stones. Some had front walls as much as 10 feet high, with ramps leading to the top from the uphill side.

Farm-type kilns were usually found at the base of a hill, sometimes just below a limestone outcrop. Road traces provide evidence of means to transport the production of the kiln.

Ruins of later commercial kilns are squat, square stone structures. Usually present are iron binding rods such as those found in blast furnaces. Early modern kiln remains have square or round concrete or stone bases with steel shells rising above the bases. They were multi-unit operations and were typically located near railroad sidings for access to markets. The locations of these later commercial kilns are often indicated on historic maps of the nineteenth and early twentieth centuries.[18]

NINETEENTH-CENTURY IRON MANUFACTURE

A typical nineteenth-century iron furnace consisted of the following components: a bridge house, a charging terrace or bench, the furnace itself, an engine shed, and a casting shed (Figure 12.4).[19]

The Process

By the nineteenth century, the reduction of iron ore took place almost exclusively in the blast furnace. William Fairbairn describes the parts of a furnace in his mid-nineteenth-century text on iron manufacture:

> The blast-furnace consists of a large mass of masonry, usually square at the base, from which the sides are carried up in a slightly slanting direction, so as to form, externally, a truncated pyramid. In the sides there are large arched recesses, in which are the openings into the furnace for the admission of the blast, and for running out metal and the cinder.[20]

To keep the furnace operating continuously for months without a break, a reliable air pump driven by water or steam power was required. At early nineteenth-century American furnaces, the blowing engine had to develop about 20 horsepower.[21]

Inside the stone furnace, iron ore was heated in burning charcoal until the metal became molten, separated from the earth and rock in which it was found, and trickled down into the hearth where it collected and remained molten until the founder let it out. A flux was added to the mixture, generally limestone, which helped the liquid iron separate and flow freely apart from impurities. These impurities, called slag, floated on top of the heavier iron and were drawn off before the iron was let out. Oxygen for the burning mixture was forced in at the furnace and exhausted through the open stack at the top. The ironmaster could adjust the blast to control the temperature inside the furnace.

Before the blast commenced, the position of the tuyures, the nozzles that carry the blast into the furnace, had to be checked, as did the blast, in order to make sure the crucible and hearth were just the right temperature.

On the casting floor, the pig bed was prepared. Workers cut runners into the sand casting floor leading from just in front of the hearth to long oblong depressions, each of which was connected to several smaller depressions. The large "sow" molds and smaller "pigs" were formed in the sand with a pattern of hardwood.

When the molten metal reached the right temperature, the founder broke out the clay plug. A white-hot ribbon of molten iron flowed across the bottom stone and into the runner. The workers urged the running metal along with iron rods. It then flowed into the sow and pig molds. As the molds filled, slag appeared at the taphole, ending the blast. The pig iron was sold to founders, who remelted and cast it into finished products, or to proprietors of finery or puddling furnaces or forges, who converted it to wrought iron.[22]

The founder then worked at the empty hearth, chipping the slag out of the corners and off the sides of the hearthstones with a long iron ringer. After scraping the nose of the burden protruding into the empty hearth, he replaced the plug in preparation for the next casting.

After the iron had cooled, the pigs were broken off the sows, weighed, and placed in a waiting cart. The metal remaining in the runners as well as spilled metal was gathered up, broken into small pieces, and then thrown into a barrel. Because this metal would require remelting at the foundry, it would sell for less than sows

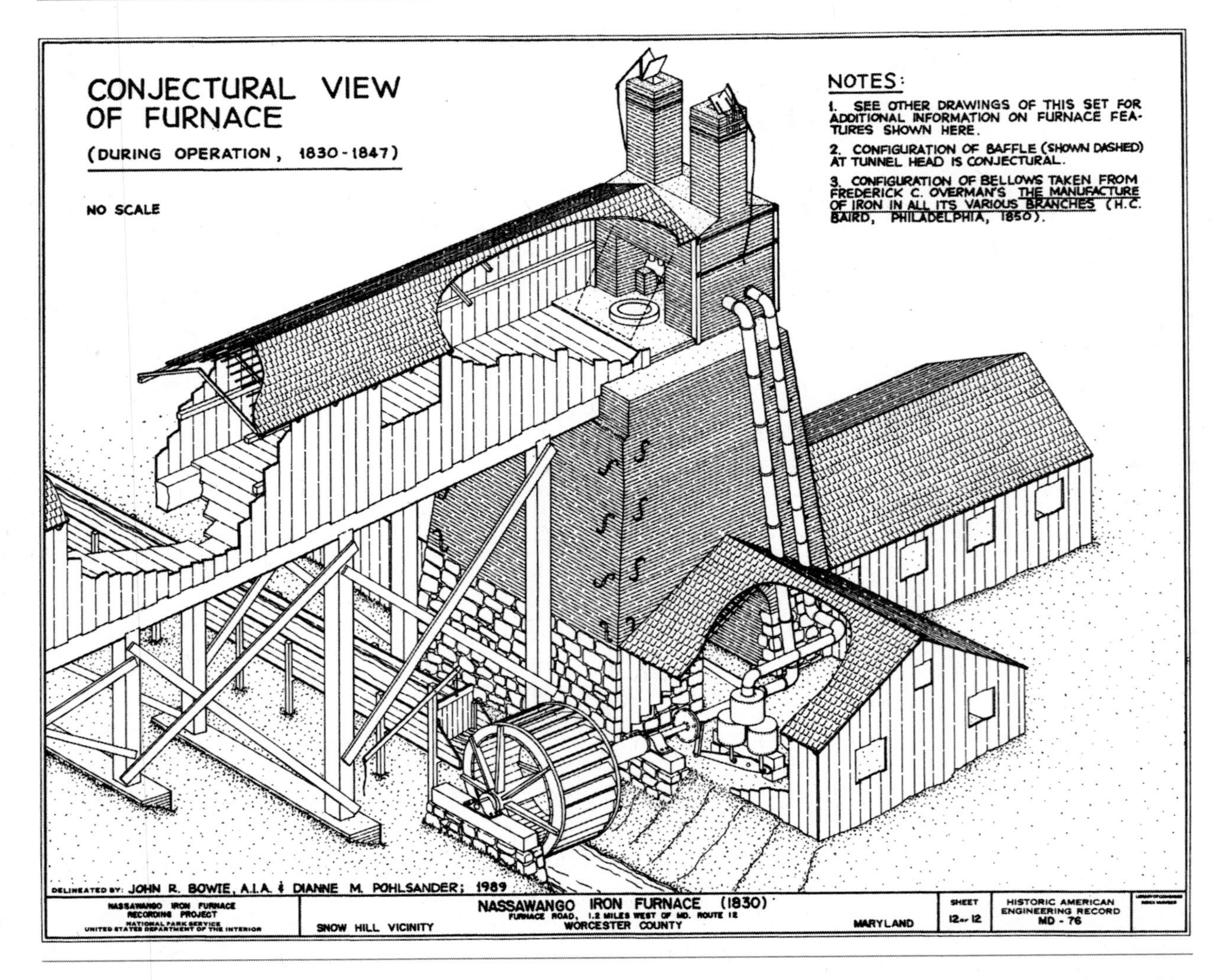

Figure 12.4. Conjectural view of Nassawango Iron Furnace (1830). HAER MD-78, National Park Service. Delineated by John R. Bowie, AIA, and Dianne M. Pohlsander, 1989.

and pigs. Then cinders would be loaded on a cart and dumped on the slag pile nearby.

The Furnace

The furnace itself consisted of an inner and outer wall. The outer wall was generally constructed of any stone close at hand, often granite, *grauwach* (a dark coarse-grained sedimentary stone), or slate. The inner wall or lining material was most critical because it had to stand up to high heat over long periods of time. Early iron-masters preferred fine-grained white limestone. Later furnaces generally used sandstone or firebrick. Between the lining and the outer wall was a space of about 8 inches filled with stone chips or broken furnace cinders. This space allowed the lining to expand and contract in response to temperature changes.

The shape of a nineteenth-century iron furnace was standardized. The furnace itself varied from about 32 to 40 feet high, with the upper portion being roughly conical in shape. Workers, standing on the charging terrace or bench, would empty wheelbarrow loads of iron ore, limestone, and charcoal into the opening at the top of the furnace, known as the throat or trunnel head. The diameter of the shaft increased below the trunnel head, reaching its greatest diameter, and then tapered inward and downward to form a small opening above the hearth. This area is called the bosh. The purpose of the bosh was to support the furnace burden while funneling the molten iron and slag down to the crucible of the hearth.

Determining the taper of the shaft, bosh, and crucible depended on the kind of ore to be smelted, whether the charcoal was made from hard or soft woods, and the kind of iron desired. When either hydrates or oxides of iron were to be smelted, as was most frequently the case in the United States, a uniform, almost straight stack was called for. The maximum stack height was generally 40 or 50 feet.

A typical hearth measured 5½ feet in height, 24 inches in width at the bottom, and 36 inches at the top. Major parts of the hearth include the bottom stone, damstone, sidestones, tuyere stones, topstone, tuyere holes, timpstone, and back stone.

In the nineteenth century, heat exchangers or hot-blast stoves were installed in many furnaces to improve efficiency, such as the one on Snow Hill, Maryland's Nassawango Furnace (Figure 12.5). Hot gas emerged from the top of the furnace and was deflected by an iron plate into the chamber containing iron pipes and passed out the stack covered with the damper. The blast engine pumped air up a sheet-iron pipe, through the iron pipe where it was heated, and down the cast-iron pipe to the tuyere.[23]

By the early twentieth century, the iron furnace began to change. Charcoal was replaced by coke and coal. Conveyors replaced the bench, charging bridge, and bridge house. The shape and proportions of the stack, bosh, and hearth changed to suit new materials. More tuyeres were added, and a bell hopper was installed at the trunnel head. Furnaces grew to 80 feet in height. However, despite changes, twentieth-century refractory furnaces are recognizable as descendants

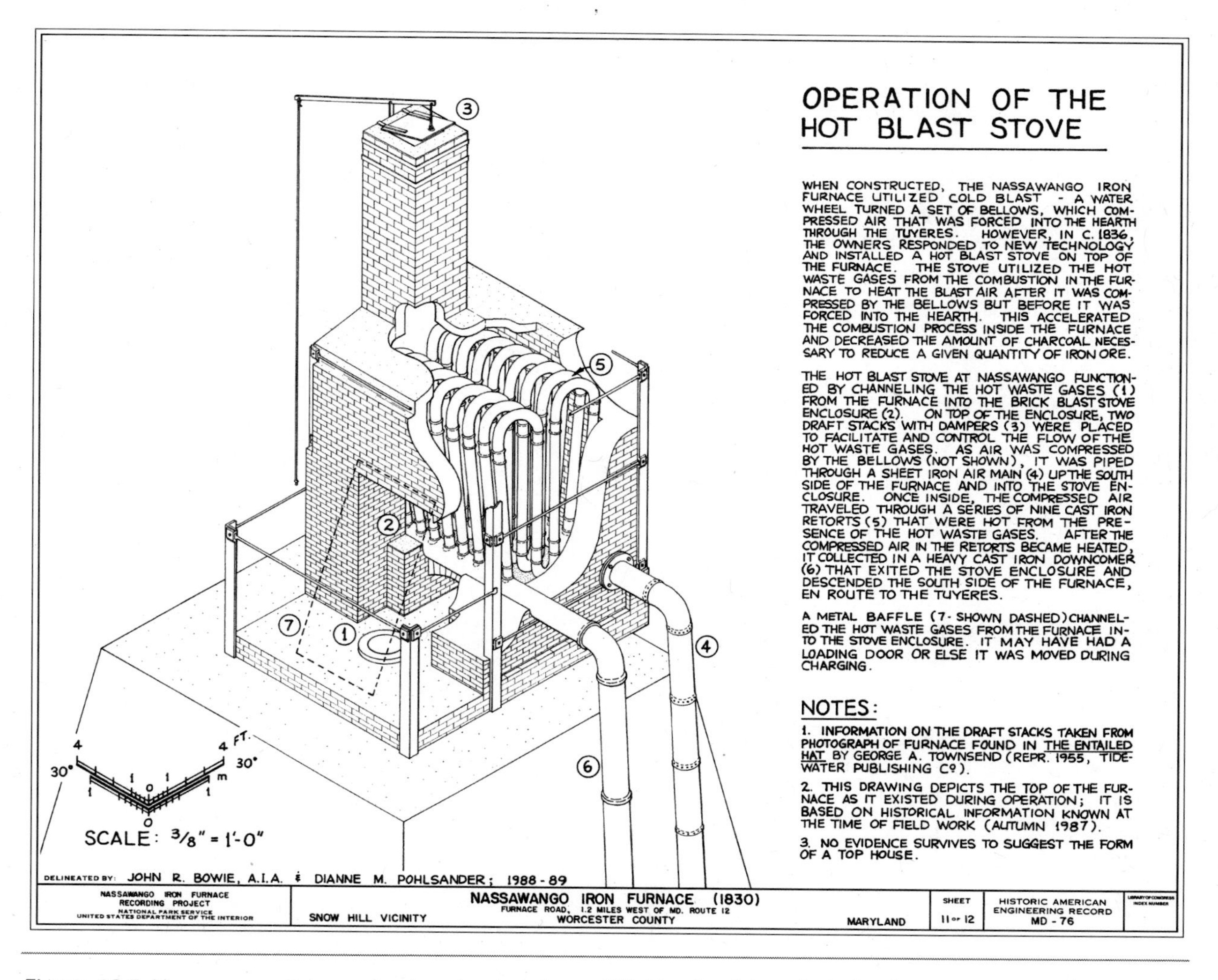
OPERATION OF THE HOT BLAST STOVE
WHEN CONSTRUCTED, THE NASSAWANGO IRON FURNACE UTILIZED COLD BLAST - A WATER WHEEL TURNED A SET OF BELLOWS, WHICH COMPRESSED AIR THAT WAS FORCED INTO THE HEARTH THROUGH THE TUYERES. HOWEVER, IN C. 1836, THE OWNERS RESPONDED TO NEW TECHNOLOGY AND INSTALLED A HOT BLAST STOVE ON TOP OF THE FURNACE. THE STOVE UTILIZED THE HOT WASTE GASES FROM THE COMBUSTION IN THE FURNACE TO HEAT THE BLAST AIR AFTER IT WAS COMPRESSED BY THE BELLOWS BUT BEFORE IT WAS FORCED INTO THE HEARTH. THIS ACCELERATED THE COMBUSTION PROCESS INSIDE THE FURNACE AND DECREASED THE AMOUNT OF CHARCOAL NECESSARY TO REDUCE A GIVEN QUANTITY OF IRON ORE.
THE HOT BLAST STOVE AT NASSAWANGO FUNCTIONED BY CHANNELING THE HOT WASTE GASES (1) FROM THE FURNACE INTO THE BRICK BLAST STOVE ENCLOSURE (2). ON TOP OF THE ENCLOSURE, TWO DRAFT STACKS WITH DAMPERS (3) WERE PLACED TO FACILITATE AND CONTROL THE FLOW OF THE HOT WASTE GASES. AS AIR WAS COMPRESSED BY THE BELLOWS (NOT SHOWN), IT WAS PIPED THROUGH A SHEET IRON AIR MAIN (4) UP THE SOUTH SIDE OF THE FURNACE AND INTO THE STOVE ENCLOSURE. ONCE INSIDE, THE COMPRESSED AIR TRAVELED THROUGH A SERIES OF NINE CAST IRON RETORTS (5) THAT WERE HOT FROM THE PRESENCE OF THE HOT WASTE GASES. AFTER THE COMPRESSED AIR IN THE RETORTS BECAME HEATED, IT COLLECTED IN A HEAVY CAST IRON DOWNCOMER (6) THAT EXITED THE STOVE ENCLOSURE AND DESCENDED THE SOUTH SIDE OF THE FURNACE, EN ROUTE TO THE TUYERES.
A METAL BAFFLE (7 - SHOWN DASHED) CHANNELED THE HOT WASTE GASES FROM THE FURNACE INTO THE STOVE ENCLOSURE. IT MAY HAVE HAD A LOADING DOOR OR ELSE IT WAS MOVED DURING CHARGING.
NOTES:
1. INFORMATION ON THE DRAFT STACKS TAKEN FROM PHOTOGRAPH OF FURNACE FOUND IN THE ENTAILED HAT BY GEORGE A. TOWNSEND (REPR. 1955, TIDEWATER PUBLISHING Cº).
2. THIS DRAWING DEPICTS THE TOP OF THE FURNACE AS IT EXISTED DURING OPERATION; IT IS BASED ON HISTORICAL INFORMATION KNOWN AT THE TIME OF FIELD WORK (AUTUMN 1987).
3. NO EVIDENCE SURVIVES TO SUGGEST THE FORM OF A TOP HOUSE.
SCALE: 3/8" = 1'-0"
DELINEATED BY: JOHN R. BOWIE, A.I.A. & DIANNE M. POHLSANDER; 1988 - 89
NASSAWANGO IRON FURNACE RECORDING PROJECT
NATIONAL PARK SERVICE
UNITED STATES DEPARTMENT OF THE INTERIOR
SNOW HILL VICINITY
NASSAWANGO IRON FURNACE (1830)
FURNACE ROAD, 1.2 MILES WEST OF MD. ROUTE 12
WORCESTER COUNTY
MARYLAND
SHEET 11 OF 12
HISTORIC AMERICAN ENGINEERING RECORD
MD - 76
LIBRARY OF CONGRESS INDEX NUMBER

Figure 12.5. Nassawango Furnace heat exchanger, Snow Hill, Maryland. Installed about 1836. HAER MD-78, National Park Service. Delineated by J. R. Bowie and Dianne M. Pohlsander, 1988–1989.

of the charcoal-fired, nineteenth-century iron furnaces.

Coke Furnace

Charcoal and early coke furnaces were similar in construction, but coke furnaces were generally larger and required a higher pressure blast. A low hearth, often made of sandstone, is a distinctive characteristic of early coke furnaces. Typical dimensions were a height of 50 feet, a bottom width of 50 feet, and a top width of 25 feet; diameter of boshes, 15 feet; hearth height, 6 feet; hearth width at bottom, 3 feet, and at top, 48 inches.

Converting Pig Iron to Wrought Iron

The fining process was brought to North America by seventeenth-century colonists. An early finer melted pig iron in a small hearth containing a charcoal fire blown with a strong air blast. The air oxidized the carbon and silicone in the pig. Similar to a bloomer, a finer made a loup, a mass of solid iron particles and liquid slag, in the bottom of his hearth. He hammered the loup to consolidate the metal and expel the slag.[24]

Ironmasters had to use charcoal fuel in fineries because the sulfur in coal contaminated the liquid iron. A finer might burn as much as 2 or 3 pounds of charcoal for every pound of iron made. Finers employed one of several processes to produce wrought iron.

Walloon Process

The original form of fining was termed the Walloon process after its place of origin (Figure 12.6). Among the forges that employed this process was the famous Valley Forge in Pennsylvania. A finer using the Walloon process passed a long pig through the opening at one side of the hearth with a downward slant so that its end was in the charcoal fire just above the tuyere. A strong air blast was needed to make his fire hot enough to melt the iron. The finer melted the end of the pig in the oxidizing flame that formed immediately in front of the tuyere. He lifted the metal that accumulated in the bottom of the hearth into the air blast until he was convinced that all silicon and carbon had been oxidized. The accumulated metal was then worked into a loup. He lifted the loup from the hearth and took it to a nearby helve hammer. With hammer blows he consolidated the iron and expelled as much slag as he could. During the hammering, he reheated the loup from time to time in a separate chafery fire to keep the slag remaining in the iron molten. The quality of the bar iron depended on the skill and care of the finer, helveman, knobbler, and chafer.[25]

Lancashire (Charcoal Hearth) Process

Later in the nineteenth century, American finers began to employ the Lancashire process, developed in Sweden by British artisans. This process made iron more nearly free of carbon and slag than could be made by the Walloon method. In the United States, this method was often termed the charcoal hearth process.

The fining hearth was constructed of water-cooled cast-iron plates. It stood about 4 feet tall, with the top of the charcoal bed about 1½ feet above the floor. The hot gases from the fire flowed through a heat exchanger to preheat the air blast and were then burned under a boiler to recover their remaining thermal energy.

As the iron sank with the burning charcoal, the metal was repeatedly lifted into the hottest part of the fire. The pigs melted in about 15 minutes. The droplets

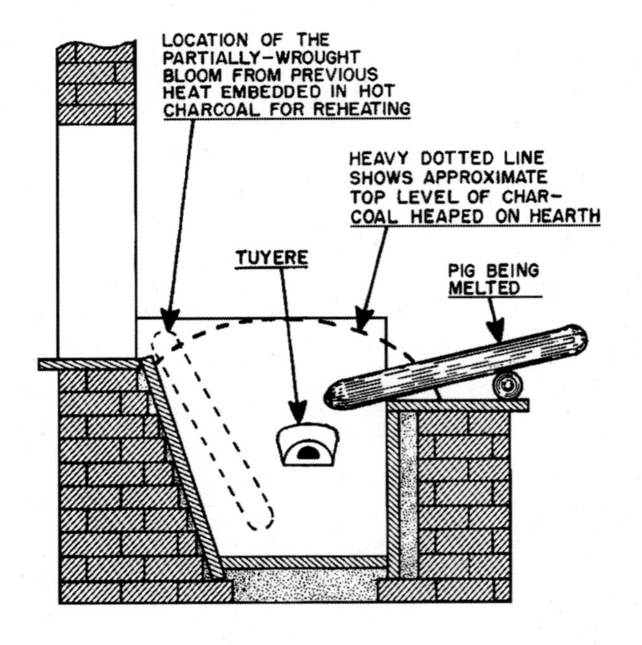

Figure 12.6. Arrangement of a Walloon hearth. Reproduced with permission from Harold E. McGannon, *The Making, Shaping and Treating of Steel,* 8th edition (Pittsburgh: United States Steel, 1964), 10.

of molten pig metal fell through the oxidizing zone of the fire, where most of the silicon and phosphorus they contained burned out, and accumulated on the bottom plate of the hearth as a semi-solid mass.

For the next stage, which lasted about 15 minutes, the helper joined in the finer's work. Forcing long bars under the accumulated mass of iron and using the edge of the foreplate as a fulcrum, they lifted the iron a few inches above the bottom plate, cutting into it and forcing in slag as they lifted. This slag oxidized the remaining silicon, phosphorus, and carbon.

In the final stage, the finer and helper broke up the lump of iron into smaller pieces which they lifted to the top of the fire. They remelted the now purified iron and allowed the droplets to accumulate in the bottom of the hearth. The finer added hammer scale, iron oxide that broke off loups during hammering, as needed, to keep the accumulating loup covered with a layer of iron oxide. After about 15 minutes, the finer had melted all the iron. He then levered the loup out of the hearth for hammering.[26]

Puddling

Ironmasters could not make wrought iron with mineral coal in a bloomery or a finery because the fuel, in contact with the metal, contaminated it with sulfur. By the mid-nineteenth century, an alternative to the blast furnace was becoming common in American ironworks: the reverberatory or puddling furnace (Figure 12.7). This furnace differed from the forge fire in that the fuel did not come in contact with the iron, but instead burned in a fireplace or grate adjacent to but separate from the hearth. The hearth was heated partly by the flame heating the walls of the furnace, but most of the heat reaching the hearth was reflected off the roof of the furnace, hence the name. Pennsylvanians attempted to replace fining with coal-fired puddling as early as 1817. By 1850, puddling had reached its heyday, with large numbers of furnaces in operation in various locations of eastern Pennsylvania.

In the 1820s, Pennsylvanians redesigned the English-style puddling furnace to burn anthracite,

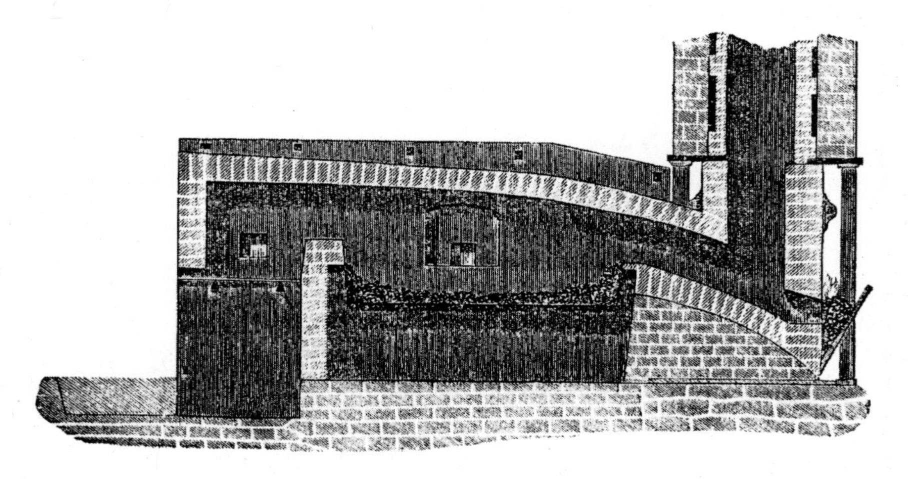

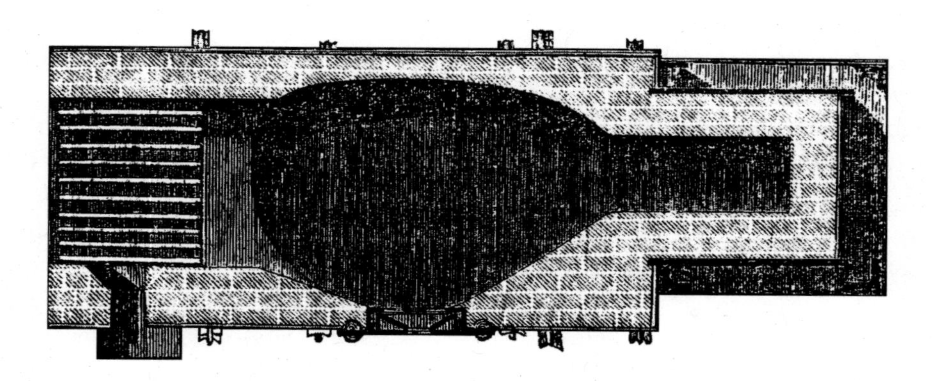

Figure 12.7 Puddling furnace (Overman 1850: 264–265).

equipping it with a larger firebox to drive the flame over the hearth. A blower forced air through a coal fire burning on a grate. Hot gas from the fire passed under the roof and out the flue to the furnace stack heating the hearth.[27]

Although puddlers carried out the same basic chemical and physical processes as finers, they worked their metal on a hearth separated from the fuel, where they could always see it. They handled about 600 pounds of iron in a heat, three times as much as a finer. Standing in front of the working door of his furnace, the puddler had the 10-inch-square firing hole to one side and the chain for adjusting the stack damper on the other. His tools rested in a tub of water (the bosh). Pig and coal were piled on the cast-iron floor plates in front of the furnace.

To start a heat, the puddler shoveled in about 50 pounds of mill scale. Then an assistant placed about 600 pounds of pig iron on top of the mill scale, using broken pigs. The puddler closed and sealed the furnace door with coal dust and put the upper and lower heat shields in place while the helper built up the fire.

The puddler removed as much of the silicon, manganese, and phosphorus from the molten metal as possible. To remove phosphorus, he oxidized the silicon and manganese but not the carbon, in a process called "clearing the iron." As soon as the iron was cleared, the puddler brought on the boil. He initiated oxidation of the carbon without allowing the phosphorus to return to the metal. As the temperature increased, making the bath of slag and iron more fluid, rapid decarburization took place, giving off large volumes of carbon monoxide. Whenever lumps of solidifying iron appeared through the slag, the puddler stirred them under so that they would not be oxidized by the flame. When the carbon was nearly eliminated, the boiling subsided and the apparent volume of the slag diminished. The grains of iron, enveloped in slag, formed clusters from which the excess slag drained off.

The puddler began the turning and balling. Using a paddle, he lifted the lumps of iron so that they would not adhere to the cold bottom of the hearth and turned them into a single mass. Punching and prying, he separated a lump of about 200 pounds, which he shaped into a ball. Puddling balls were then sent to the squeezer, who pressed out the liquid slag and shaped it into a rough cylinder. Before squeezers, ironmasters used helves to consolidate the iron and expel the slag from puddle balls.

FORGING

Forging is the process of producing wrought iron from blooms and rods by hammering. The simplest machine by which iron was forged in the nineteenth century is the German forge-hammer, also known as the tilt-hammer. The cast-iron hammer varied in weight (from 50 to 400 pounds) according to the purposes for which it was designed. Lighter hammers were sufficiently heavy for drawing small iron and nail rods, while for forging blooms from 60 to 100 pounds in weight, a hammer weighing 300 to 400 pounds was employed. The hammer struck on an anvil, a cast-iron block about the weight of the hammer. The helve or handle was constructed of sound, dry hickory or more commonly of white oak. The fulcrum, a cast-iron ring, was tightly wedged upon the helve. A wrought-iron ring fastened to the upper side of the helve received the taps of the cam. On the lower side it struck against a vibrating piece of lumber to increase (by recoil) the force of the hammer. The shaft was commonly made of wood. If the motive power was water, a waterwheel was directly fastened upon it. If the motive power was steam, a flywheel was attached and the power conveyed by leather straps or belt. The cast-iron wheel, either round or octagonal, had cams fastened in it.[28]

In 1833, Scotsman James Nasmyth patented an improved forge hammer powered by steam. Steam hammers began to be widely used in English and French forges in the 1840s.[29] By the mid-nineteenth century, a version of this hammer had proved useful in the United States for shingling blooms or slabs. One such hammer, a double-acting type, constructed by Messrs. B. & S. Massey of Manchester, England, was displayed at the 1876 Centennial Exposition in Philadelphia. Its operation is described in an exhibit guide:

> The arrangement for working the valves in these hammers . . . is a combination of self-acting and hand-worked gearing, and it is different from that ordinarily employed, being without the usual cams or sliding-wedge. As the hammer rises and falls when in action, a hardened roller on the back of the head slides on the face of a curved lever, which rotates about a pin near its upper end, and is held by a spiral spring always in position against the

> roller. At every movement of the hammer this lever operates a valve-spindle and regulating-valve, the length traveled by the hammer being controlled by another lever attached to the fulcrum-point of the curved lever, and by which this pin may be raised or lowered by hand, and the points at which the steam is admitted or allowed to escape varied at pleasure.

The illustrated hammer could strike as many as 250 blows per minute with a pressure between 40 and 60 pounds.[30]

STEEL PRODUCTION

To make steel, an artisan had to get the right amount of carbon in his iron. A steelmaker could start with pig iron and remove all the silicon and most of the carbon, or he could begin with wrought iron and add the desired amount of carbon. Although Americans experimented with steelmaking in bloomeries and fineries, they had little success in making a useful product, due to the difficulties in controlling the carbon distribution.

Initially, steel produced in America was "blister steel" made by a cementation process in which carbon was diffused into wrought-iron bars in cementing furnaces. Because the carbon diffused in from the surface and did not always reach the center of the bar, blister steel was not homogeneous.

To make blister steel homogeneous and free of slag inclusions, artisans had to melt it in crucibles and cast the liquid in ingot molds, a process pioneered by Benjamin Huntsman of Sheffield, England, in 1742. Americans began to have success with crucible steelmaking only after 1860.

In addition to problems with technique, American crucible steelmakers had difficulty getting the high-purity iron they needed for starting material. American importers of iron competed with Sheffield firms who had contracts to buy all of the best grades.

Crucible Steel

Makers of crucible steel depended on three highly skilled specialists: the melter, the puller-outer, and the teemer. The melter oversaw the charging of the pots and their placement in the furnace. The furnace typically held 24 pots. With the anthracite-fired furnaces initially used in the United States, melting took three to five hours. The melter's most difficult task was judging when the steel was ready for casting. While this was occurring, the teemer adjusted the molds to the angle he liked to pour into. He and the puller-outer wrapped multiple layers of water-soaked sacking around their legs and thighs. When the steel appeared ready, the melter signaled the puller-outer to start his work.

Soon after they started making crucible steel, Americans diverged from the English practice by adopting gas-fired furnaces. This made the melter's job easier because he could regulate the temperature by manipulating valves.

DEVELOPMENT OF MODERN STEELMAKING

Pig iron consists of iron combined with numerous other elements, the most common of which are carbon, manganese, phosphorus, sulfur, and silicon. Depending on the composition of the materials used in the blast furnace and the manner in which the furnace is operated, pig iron may contain 3.0 to 4.5 percent of carbon, 0.15 to 2.5 percent or more of manganese, as much as 0.2 percent of sulfur, 0.025 to 2.5 percent of phosphorus, and 0.5 to 4.0 percent of silicon. In refining pig iron and converting it to steel, all five of these elements must either be removed or reduced drastically in amount.

Modern steelmaking uses either an acid process or a base process. Carbon, manganese, and silicon can be removed with relative ease by either acid or base processes. The removal of phosphorus and sulfur requires basic processes. Oxidation is employed to convert pig iron into steel. Each steelmaking process has been devised primarily to provide a means where controlled amounts of oxygen can be supplied to the molten metal undergoing refining.

Coke, the residue from the destructive distillation of bituminous coal, is necessary for the production of steel. Two processes, the beehive process and the by-product or retort process, are used to manufacture metallurgical coke. These two processes are described in the chapter on extractive industries (Chapter 13).

The beehive process has an advantage for certain peak requirements where the high investment cost of a by-product cannot be justified because of long inoperative periods. Beehive coke is usually made near a mine that supplies coal that can be successfully coked in this type of oven. In the steel industry of the United States, the transition from beehive to by-product ovens was accelerated with the start of World War I, when the construction of by-product coke ovens was begun in many locations in the eastern and central states.[31]

The Blast Furnace

The purpose of a steelworks blast furnace is to chemically reduce and physically convert iron oxides into liquid iron, also termed hot metal. The furnace itself is a large steel stack lined with refractory brick where iron ore, coke, and limestone are dumped into the top, and preheated air is blown into the bottom. After six to eight hours, the raw materials descend to the bottom of the furnace, where the products are liquid slag and iron ore. Once a blast furnace is started, it typically runs continuously for four to ten years, with only short interruptions for maintenance (Figure 12.8).

The Blast Furnace Plant

An ore storage yard is commonly located close to the blast furnace. Each piece of material—ore, pellet, sinter, coke, and limestone—is dumped into separate storage bins. The various raw materials are weighed prior to use by rail-mounted scale cars or by computer-controlled weigh hoppers. The weighed materials are then dumped into a winch-powered skip car that rides on rails up the inclined skip bridge to the receiving hopper at the top of the furnace.

Materials are held at the top of the furnace until a charge has accumulated. The materials are charged into the blast furnace through two stages of conical bells, which seal in gases and distribute the materials evenly around the circumference of the furnace throat. The top of the furnace also includes four "uptakes," where hot, dirty gas exits the furnace dome. A series of mechanisms, which include a scrubber and cooler, cleans the gas so that it is ready for burning. The clean gas pipeline is directed to the hot blast "stoves." Usually three or four cylindrical stoves are placed in a line adjacent to the blast furnace.

Large volumes of air generated by a turbo blower flow through the cold blast main to the stoves. The cold blast then enters a stove that has been previously heat-

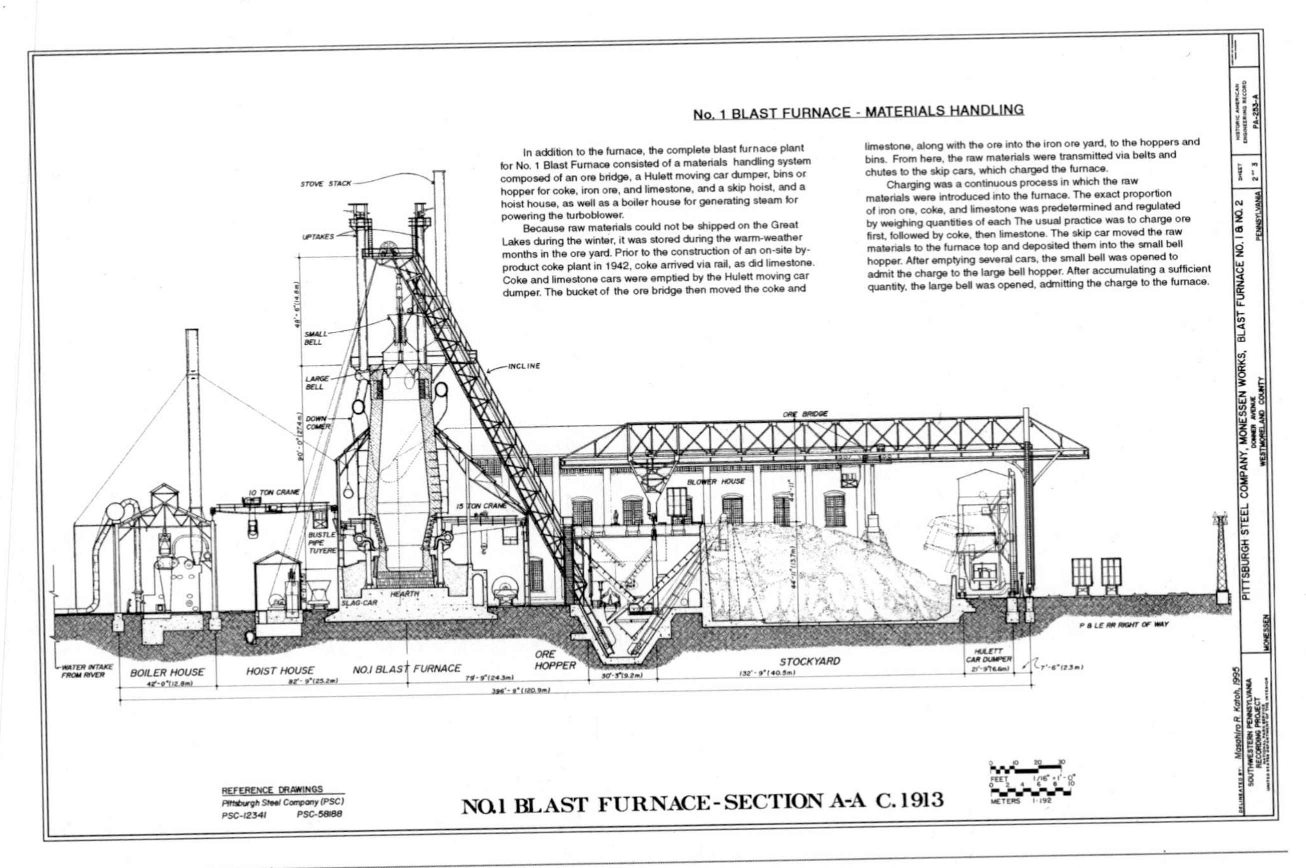

Figure 12.8. No. 1 Blast Furnace. Pittsburgh Steel Company, Monessen Works, Monessen, Pennsylvania. HAER PA-253A, National Park Service. Delineated by Masahiro R. Katoh, 1995.

ed, and the heat stored in the refractory brick is transferred to the cold blast to form hot blast. This heated air (whose temperature could range up to 2300 degrees) exits the stove in the hot blast main which runs up to the furnace. The hot blast main enters into a ring-shaped pipe that encircles the furnace, called the bustle pipe. Tuyeres placed in the bustle pipe direct the hot blast into the furnace.

The molten iron and slag drip past the tuyeres on the way to the furnace hearth. A drill mounted on a pivot base called the taphole drill swings up to the iron notch and drills a hole through the refractory clay plug into the liquid iron. Once the taphole is drilled open, the liquid iron and slag flow down a deep trench called a trough. Set across and into the trough is a block of refractory called a skimmer. The hot metal flows through this skimmer opening over the iron dam and down what is called the iron runner. The liquid iron flows into refractory-lined "ladles" known as sub or submarine cars because of their shape.[32]

Bessemer Steelmaking

The dawn of modern steelmaking was proclaimed by Englishman Henry Bessemer in his 1856 paper read before the British Association for the Advancement of Science, "Manufacture of Malleable Iron and Steel without Fuel."

The concept of pneumatic steelmaking was independently developed by Bessemer and American William Kelly of Eddyville, Kentucky. Kelly developed the concept at Suwanee Iron Works, which he co-owned with a brother. Kelly eventually perfected his approach to steelmaking in employment at the Cambria Steel Company in Johnstown, Pennsylvania. It was here that he fabricated his first tilting converter.

The Kelly Process Company, organized to operate under his patents, built a converter at an ironworks in Wyandotte, Michigan. Here the first automatic process steel ever made in the United States was blown in 1864.

Meanwhile, another American engineer, Alexander L. Holley, obtained the right to manufacture steel under Bessemer's patents. He erected a plant in Troy, New York, and began making steel in 1865. The following year, the two companies merged, producing steel by what became known as the Bessemer process.

The essential part of the Bessemer process is the blowing of air through molten cast iron to remove the metalloids by which cast iron differs from steel and wrought iron. These metalloids are removed by oxidation. In the Bessemer process, the oxygen blown through the molten metal directly oxidizes or burns out the carbon, silicon, and manganese. The extremely rapid oxidation of these furnishes the heat.

After about three to five minutes, half of the silicon and manganese is burned out. The remainder of these elements and all the carbon are removed in the following five to six minutes. After some minutes, the flame begins to drop. The converter is then turned down, and a ladle is run in the track above. This brings into the mixture just enough carbon, manganese, and silicon to produce the whole of the molten metal in the converter needed to make the steel of the composition desired.

By the early twentieth century, large steel plants used "bottom-blown" converters (Figure 12.9a). The bottom-blown converter has been the principal type used in both the acid and basic air-blown pneumatic processes for the production of steel ingots. Two or three vessels working together with proper metal from the "mixer" could produce an immense amount of steel each day. The mixer is a large vessel or furnace holding and keeping hot 75 to 300 or more tons of metal from the furnace. It mixes and equalizes irons of various compositions, giving these converters the advantage of uniform and hot metal with which to work.

One alternative to the bottom-blown converter is the side-blown converter (Figure 12.9b). This converter can be arranged for either submerged or surface blowing. The submerged side-blown converter presents several difficult maintenance problems, and the design has not been popular. The most extensively used side-blown converters are acid-lined vessels arranged for surface blowing, used chiefly in casting foundries.

The top-blown converter (Figure 12.9c) was used in the basic oxygen process method of making steel. In this process, an oxygen lance was lowered through the top of the vessel with the tip of the water-cooled lance about 5 to 6 feet above the metal bath.[33]

After recarburization, as the process is called, the steel and the slag are poured out into a ladle waiting below from which the steel is "teemed" (or poured) through a hole in the bottom into ingot molds arranged on trucks on the rail tracks which run through the building.

When the molds have been filled and a crust develops on the steel, the cars are pulled to the "stripper" where the molds are removed, leaving the white hot ingots standing on the cars. From the stripper, the ingots

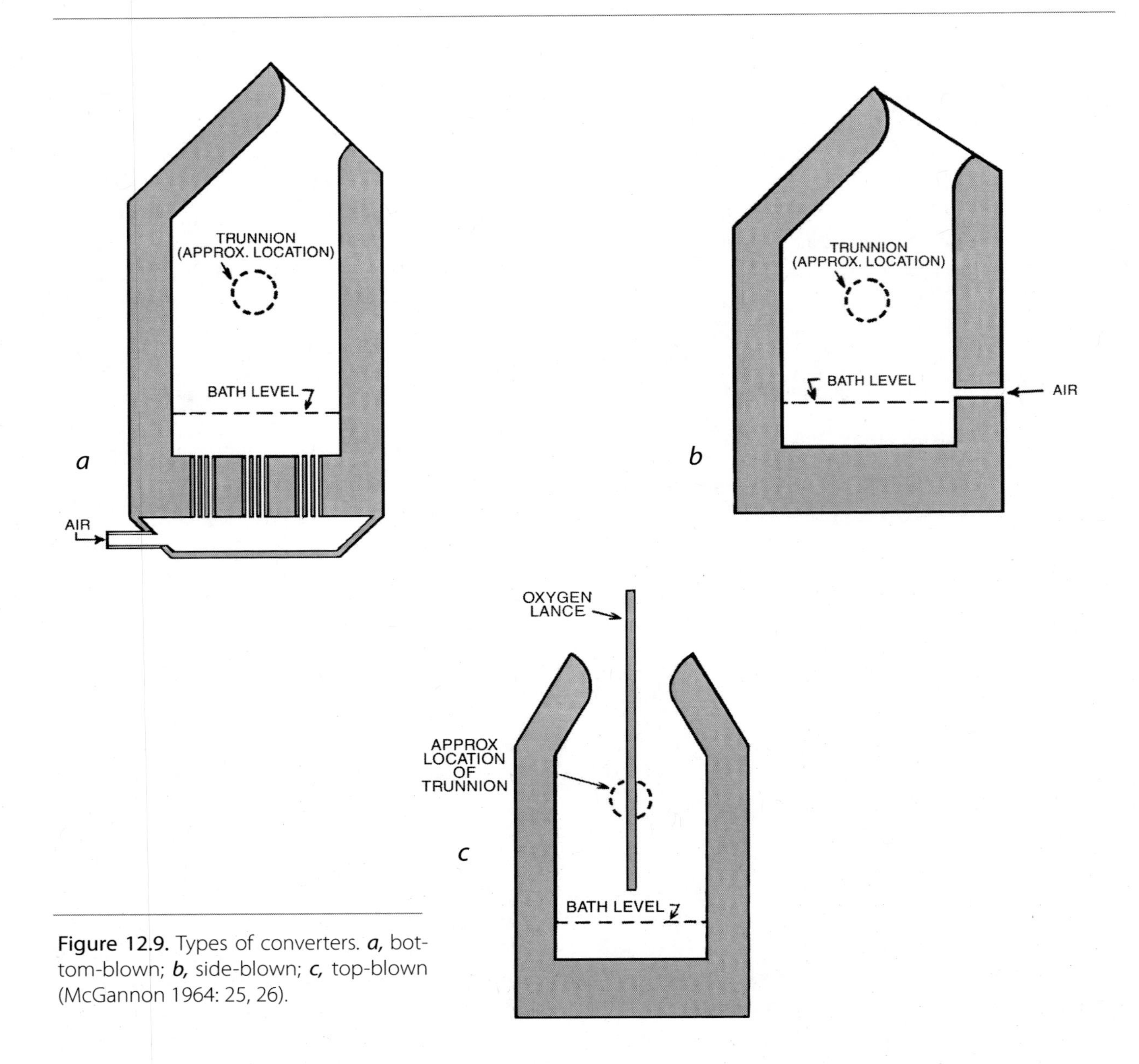

Figure 12.9. Types of converters. *a,* bottom-blown; *b,* side-blown; *c,* top-blown (McGannon 1964: 25, 26).

go to the gas-fired soaking pits, where the molten interiors of the ingots gradually solidify by cooling while the outer crusts are reheated. After the temperatures of exteriors and interiors are equalized, the ingots are white hot and ready for rolling.

Historically, Bessemer steel was largely used for the manufacture of rails, rod, wire, pipe, and merchant bar.

Open-Hearth Steelmaking

Open-hearth steelmaking can be traced back the mid-nineteenth century. In 1845, John Marshall Heath received a patent for a process of making steel patterned after puddling iron. Practical application of his idea was limited until a furnace was developed that would supply the required heat.

In 1860, C. W. Siemens solved this problem with the invention of the regenerative process. The first successful open-hearth furnace was used in Birmingham, England. The advantages of this new process were soon apparent. It could use pig iron of varied character and composition. Because no air is blown through the metal and little comes in contact with it, the conversion occurs quietly and smoothly, with little loss by oxidation.

The slowness of the conversion proved an advantage, as control was very easy.

The melting in an open-hearth furnace is done largely by indirect or radiated heat. The bath must remain hot enough to remain molten after purification of the metal.

In the early twentieth century, two chambers built up with a checkerboard of firebrick were placed under each end of the rectangular stationary open-hearth furnace. Each had one chamber for air and one chamber for gas.

An open-hearth furnace occupies a sort of hollow square. The furnace proper is on one side, the regenerative chambers on two sides, and the chimneys and flues on the remaining side. Reversing valves force the incoming gas and air to each travel through its respective hot regenerative chamber up through the ports and into the furnace where they unite and burn with a very hot flame. The hot gases leave through similar ports in the other end of the furnace and in the process heat the checker-work in the regenerative chamber. Every 15 to 20 minutes, the valves are reversed and the direction of the flow changes. In this way, the incoming gas and air are preheated, and the furnace can burn with a much hotter flame than it would with cold air.

The original scheme was to melt pig iron and burn out the silicon, manganese, and carbon by action of the flame and the addition of iron ore, the process developed by Siemens in England. In France, P. and E. Martin altered the method, diluting molten pig iron in the Siemens furnace by melting and dissolving in it steel scrap. Modern steelmaking uses the combination of the two processes, hence the name, Siemens-Martin.

Basic versus Acidic Process

In the early twentieth century, steel was made either by the basic open-hearth process or by the acid open-hearth process. In the basic process, the phosphorus, as well as the silicon, manganese, and carbon, are eliminated. In order to remove the phosphorus, lime in the form of calcium oxide or calcium carbonate is added.

At the beginning of charging, limestone or burnt lime is shoveled into the hearth of the white-hot furnace. When cold metal is charged, the pigs of iron are conveyed into the furnace by the melter and his helpers by means of long-handled iron tools called peels. This is followed by charging some or all of the scrap or iron which is to be made part of the charge. More modern facilities use machine charging.

During the melting down of the pig iron with the charged scrap, the air and flame burn out about half of the silicon and manganese in the metal. To remove the remainder of these and the carbon, more ore is added to keep the bath boiling. The carbon is partially removed in carbon monoxide gas formed from the oxygen of the iron ore and the carbon of the metal. The covering of slag that forms, protecting the bath from flame, transfers oxygen from the furnace gases to the bath and helps to burn out the carbon. The lime charge unites with the phosphorus and takes it into the slag. Samples are frequently taken to determine the composition of the bath. If samples show the bath to have the desired composition, it is poured.

When ready to tap, a big ladle is suspended from a crane under the spout of the furnace. With a tapping bar, the plug of clay is removed from the tap hole, and the molten steel gushes out into the ladle. The slag is the last to drain out. Later furnaces have been built as "tilting." These furnaces can be tipped to pour the metal into the ladle.

While the furnace is again charged, the steel teems through the nozzle of the big ladle into the waiting ingot molds. These ingots are then moved to the stripper, to the soaking pits, and then to the rolling mills.

In the acid-lined furnace, no attempt is made to reduce the phosphorus. Therefore, the materials charged have to be very low in phosphorus and sulfur. No lime is added. The flame simply melts down the pig iron and scrap. Iron oxide is added from time to time to keep up the boil until test bars show that the carbon, silicon, and manganese have been eliminated. The metal is then tapped.

Originally, open-hearth steel was used for plate, boiler tubes, structural shapes, and axles. Rails were later formed from open-hearth steel.

The Layout of the Modern Open-Hearth Steel Plant

Open-hearth steel plants generally employ stationary furnaces of over 200 tons of individual capacity which produce steel from hot-metal and scrap charges. The furnaces are usually housed in large steel buildings, usually over 300 feet long, arranged to facilitate the charging of the furnaces, the making of the heats, and the handling of the finished molten steel (Figure 12.10).

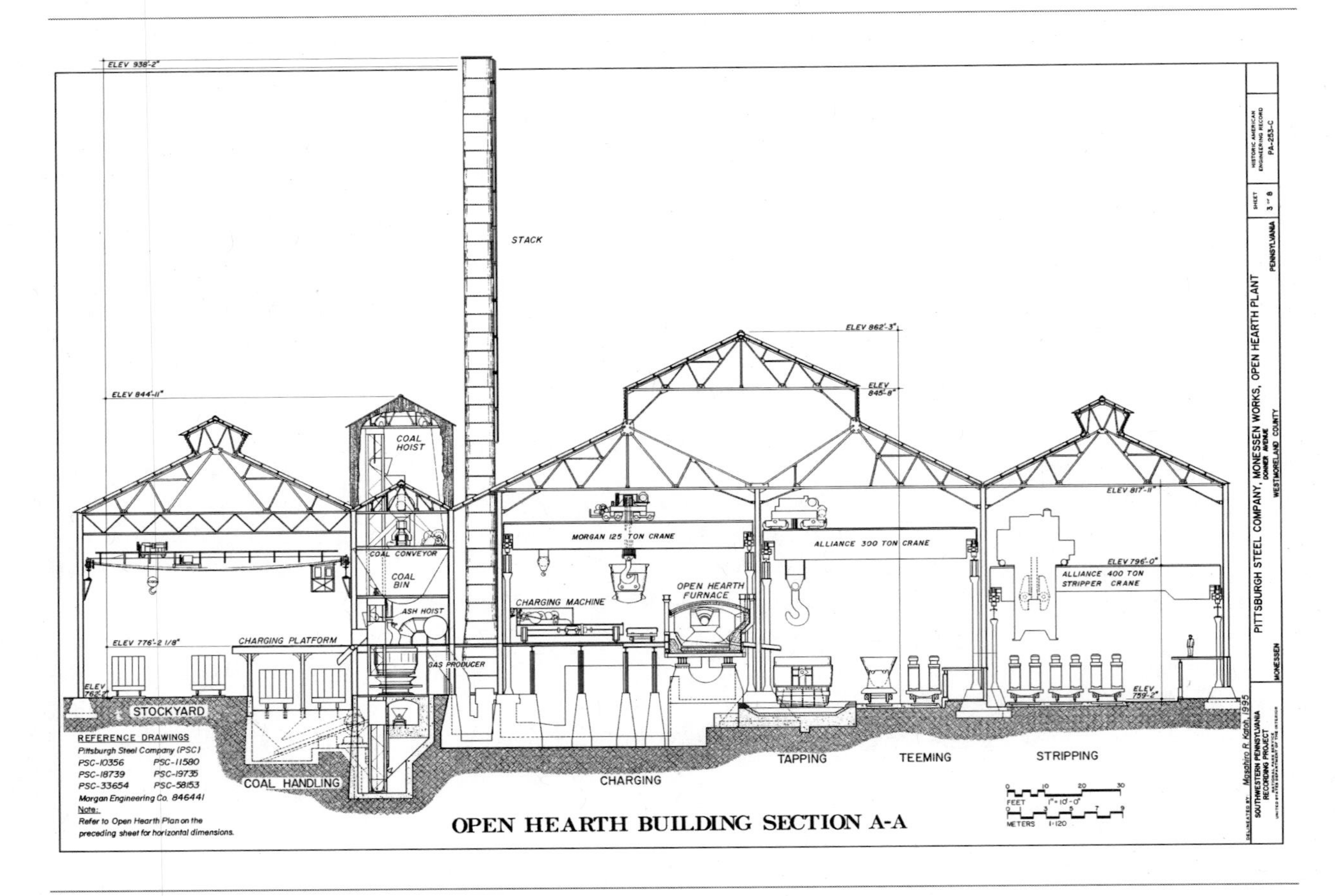

Figure 12.10. Pittsburgh Steel Company. Monessen Works, Open Hearth Plant, Monessen, Pennsylvania. HAER PA-253C, National Park Service. Delineated by Mashahiro R. Kotoh, 1994.

Furnaces are typically arranged lengthwise along the centerline of the building. In general, a maximum of twelve furnaces are placed in one row. Between each pair of furnaces is a vertical, composite steel column. The tall central section of the column rises to a height sufficient to support the main roof truss, while the two shorter heavy steel sections support one of two tracks of electric overhead traveling crane runways that extend the length of the building.

The part of the floor in front of the row of furnaces is called the charging floor and is elevated above the general yard level of the plant. The charging floor is usually 75 to 85 feet wide. The remainder of the floor space of the building, located on the tapping side of the furnace, is called the pouring floor or pit side, and is at general yard level.

On the charging side of the furnaces and close to them, a broad-gauge track is laid, on which locomotives move buggies that carry the charging boxes loaded with scrap and other solid materials to be charged into the furnaces. Parallel to this track is a special track of very wide gauge on which the charging machines operate. These machines are equipped to pick up the charging boxes one at a time from the buggies, thrust the boxes through an open door into the furnace, and turn them to dump their contents on the hearths.[34]

Electric Arc Steelmaking

The first electric arc furnace for steelmaking was used in the United States in 1906. The technology of electric arc steelmaking had been developed by Heroult in France in 1899, and the first shipment of electric steel was made in 1900. The first Heroult furnace in the United States was installed in the Syracuse, New York, plant of the Halcomb Steel Company. This furnace, designed to make steel ingots, was of the single-phase, two-electrode rectangular type of 4-ton capacity. In 1909, a 15-ton, three-phase round furnace, then the largest electric steelmaking furnace in the world, was installed by United States Steel in Chicago. The first electric furnace for the production of steel for commercial castings was that of the Treadwell Engineering Company in Easton, Pennsylvania. It was first operated in August 1911.[35]

An advantage of the electric arc furnace over other types is that most grades of steel can be produced. The furnace can be more economical than other methods when carbon and low-alloy steel production requirements are insufficient to justify use of the blast furnace/open-hearth combination to produce steel; facilities are installed in industrial areas of high scrap-steel availability but distant from natural sources of coke, limestone, and high-grade iron ore; and the nature of subsequent processing is such that steel-production requirements are intermittent.[36]

Furnaces

All furnaces of the Heroult type (Figure 12.11) are designed to tilt in two directions, one for pouring and the other for slagging. Heroult furnaces are mounted on toothed rockers which rest on, and intermesh with, toothed rails. They are tilted by a motor-driven rack-and-pinion mechanism, generally placed underneath the furnace.

Modern electric arc furnaces are constructed with inside shell diameters ranging from 7 to 25 feet and a steel capacity ranging from 8,000 to 420,000 pounds. The shell is cylindrical in shape with a dished bottom and is of welded and bolted construction. The bottom of the furnace may be laid first with a layer of clay, magnesite, or silica brick. Modern furnaces are of the swing type, top-charged design. In this type, the roof and supporting structure for the electrode masts are lifted and swung to one side by motor-driven or hydraulic equipment. Scrap charges of as much as 100 tons can be charged into large furnaces with one bucket.[37]

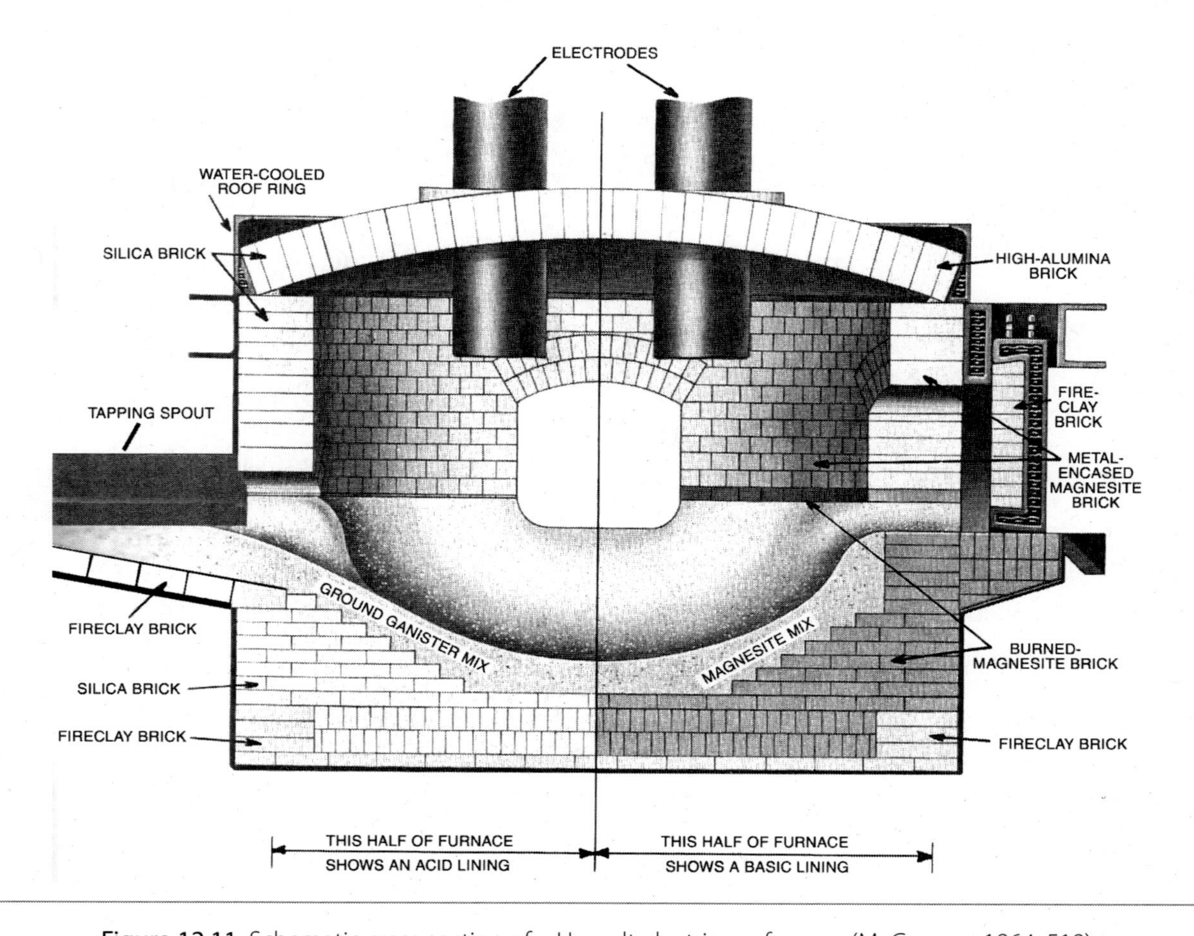

Figure 12.11. Schematic cross section of a Heroult electric arc furnace (McGannon 1964: 519).

Facilities

The modern arc furnace facility incorporates a lower-level stockhouse located as an extension to the end of the furnace building or adjacent and parallel to the furnace building. At these locations, drop-bottom charging buckets are loaded at ground level and moved by overhead crane to the furnace floor. The different lots of alloy scrap are kept in separate bins, often in outdoor stocking areas.[38]

Production Method

In charging the furnace, the power is turned off and the roof and electrodes are moved out of the way. Light or medium scrap is charged in a thin layer on the bottom, because this type of scrap melts faster than large pieces. Heavy scrap is charged in the area within or adjacent to the triangle or delta formed by the electrodes. After this, light or medium scrap is piled high around the sides of the furnace to protect the roof and side walls from the arc.

When the charging has been completed, the main circuit breaker is closed, an intermediate voltage is selected, and the arcs are struck under automatic control. After about 15 minutes, maximum voltage and current are applied for the fastest melting of the scrap. The electrodes melt the portion of the charge directly underneath them and continue to bore through the metallic charge, forming a pool of molten metal on the hearth. Oxygen for this and other oxidizing reactions is obtained by oxygen gas injected into the bath or oxides of alloying elements added in the furnace.

In the double-slag method, the original slag with its oxidation products is slagged off or largely removed from the surface of the bath by cutting off electric power to the electrodes, raising the electrodes, back-tilting the furnace slightly, and then raking the slag out. Materials used in making up the second slag are burnt lime, fluorspar, and silica sand, with powdered coke to supply carbon.

In tapping a heat, the electrodes are raised to the maximum height after the power is shut off, the taphole is opened, and the furnace is tilted by a control mechanism so that the steel is drained from the furnace into a ladle set on the pit side of the furnace. The slag comes after the steel and serves as an insulating blanket during tapping.[39]

SHAPING IRON AND STEEL

As the final task, iron- and steelworkers shape the metal into the forms desired by their customers. They hammer or roll wrought iron or steel into bars, rods, or sheets sized to the buyer's specifications.

Makers of wrought iron could use the slow, heavy blows of a helve hammer to shape a rough billet. Iron- and steelmakers turned to tilt hammers to make smoothly finished bars. Millwrights developed tilt hammers with a light wooden beam with cams on a rotating drum to lift and release the beam.

For heavier tasks, such as making iron axles for railway cars, artisans have preferred steam hammers because they could precisely regulate the force of each blow. Steam hammers could deliver heavier blows than any tilt hammer, and ironmasters increasingly adopted them as they undertook forgings for railways and marine engines.[40]

Tube Mill

Steel tubing is extensively used for industrial piping to boiler tubes, bedsteads, flagpoles, utility poles, columns, automobile axle housings, airplane fuselages, motor mounts, bicycle frames, and many other uses.

Welded Tube

The first shop in the United States for manufacturing butt-welded pipe was established in Philadelphia by Morris, Tasker, and Morris. The idea that pipe could be butt-welded continuously was conceived by John Moon in 1911. In 1921 and 1922, the Fritz-Moon Tube Company was formed, and continuously butt-welded pipe was made on a production basis.

The butt-welding process is schematically shown in Figure 12.12. The raw material used to produce tubes, termed skelp, is formed from steel slabs which are then heated in a furnace. A pushout bar and a pair of pull-out rolls remove the heated slabs from the furnace. The skelp mill itself consists of roll stands, scale-breaking edging stands, and edging stands. The final product of the skelp mill is coiled skelp.

The skelp is transformed into welded tube by a sequence of operations, including uncoiling, rolling and leveling, flash welding, heating, threading, forming and welding, cooling, and descaling.[41]

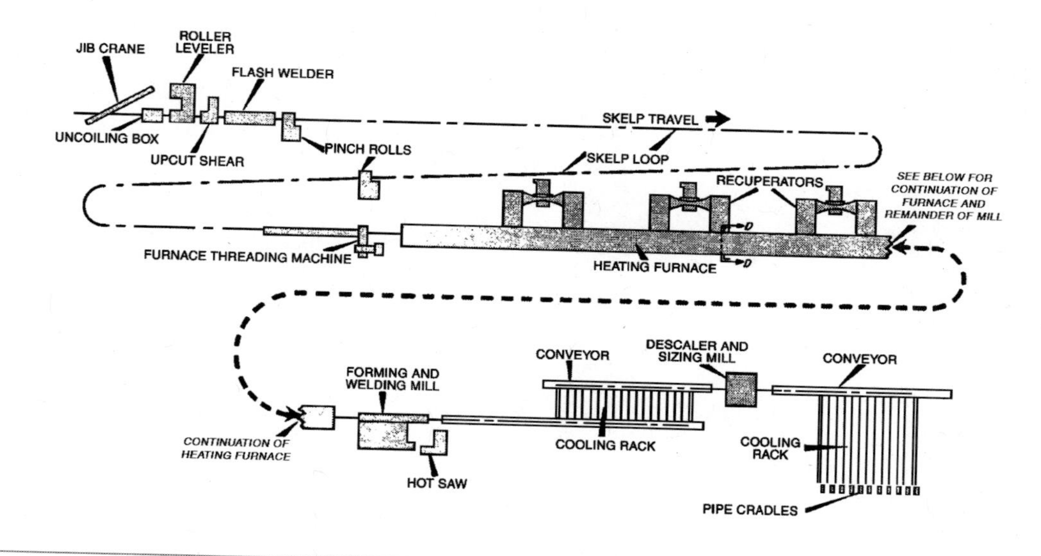

Figure 12.12. Schematic plan of the layout of a typical continuous butt-weld pipe mill (McGannon 1964: 830).

Seamless Tube

Prior to 1920, seamless-tube mills in the United States were only able to make pipe about 6 inches in diameter and no longer than 25 feet. In 1925, United States Steel developed the double piercing process, which consisted of piercing a heavy-walled shell and expanding this in a second piercing operation to a larger diameter tube with a lighter wall. This permitted the production of tubes up to 16 inches in diameter. Later developments enabled the maximum diameter to be increased to 26 inches.[42]

In modern seamless pipemaking, a series of processes is followed. For example, in the 1960s at the Lorain Works of United States Steel, the sequence for producing 3½- to 16-inch-diameter seamless pipe included the first piercing mill, the second piercing mill, the reheating furnace, the plug rolling mill, and the reeling machine. The piercing mill pierces the solid billet to create a shell, and the shell passes to a second piercer where it is rolled to a larger-diameter, thinner structure. The plug rolling mill further reduces the tube diameter and thickness.[43]

Rolling Mill

A rolling mill shapes metal by an entirely different process. The motion of two rolls turning toward each other squeeze the metal and cause it to flow over the surfaces of the rolls in their direction of rotation, extruding it out the gap between them. Throughout the eighteenth century, American ironmakers used small rolling mills to convert hammered bars into plates that could pass through slitting rolls to make nail rods.

Rolling Rails and Beams

With the development of the railroad and T-shaped rail in the 1830s came the need for mills to make them. The eventual solution proved to be the development of a universal mill, one with both vertical and horizontal rolls, developed by New Jersey engineer William Borrow.

Whether destined to be sold in intermediate shapes or further rolled down into finished products, ingots have to be "cogged" or broken down into intermediate-sized slabs, blooms, or billets. The cogging or first rolling is generally accomplished in the reversing mill. The white-hot ingot from the soaking pit is run back and forth through rolls forced closer and closer together so that the piece becomes thinner and longer with each pass.

Plate mills are usually three-high, with tables of small rollers on each side which tilt to feed the plate into the rolls and receive it on the other side, from which it is fed in again. For wide plate, the rolls may be

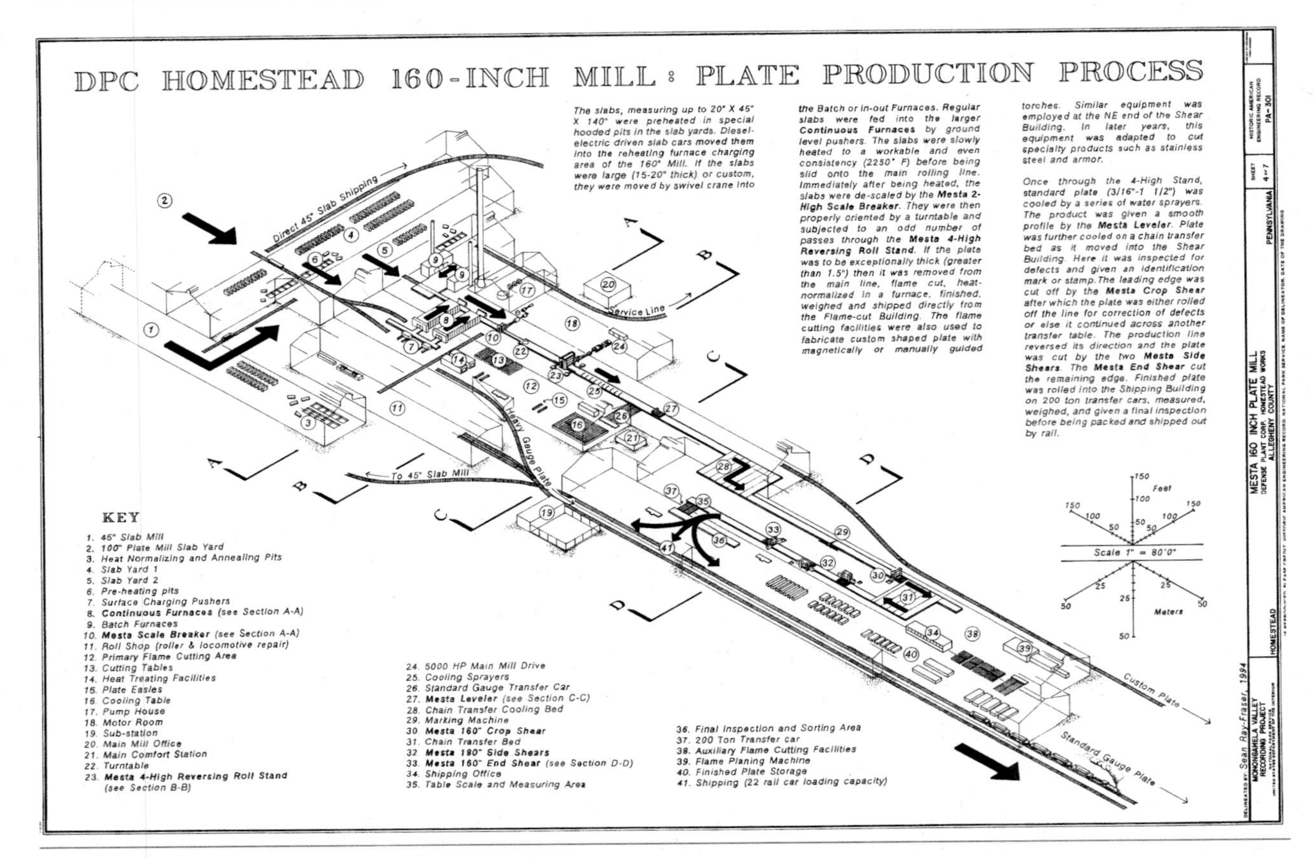

Figure 12.13. Schematic layout of plate mill at Defense Plant Company's Homestead Works, Pennsylvania. HAER PA-301, National Park Service. Delineated by Sean Ray-Fraser, 1994.

140 inches or more long and as much as 3 feet in diameter (Figure 12.13). The rolls are flooded with water to keep them cool. The steel is rolled down from 3-inch-thick slabs to the width required.

After the final pass, the plates go upon the "hot bed" where they are marked with the boundaries of the smaller plates or pieces into which they will be cut. Cranes with magnets or hooks convey the long plates to the shears where a square-edged knife cuts them.

Rolling of Iron and Steel Rods

Artisans found iron and steel softer and easier to form when hot. When they tried to hot-roll a long bar, it would cool off between passes through the rolls. In the mid-nineteenth century, English engineers developed the looping mill to solve the problem. The catcher grasped the bar emerging from one pass, looped it around himself, and inserted it in the next pass. Time saved between passes, coupled with the mechanical work done by the rolls, helped keep the iron at the hot working temperature.

Steel rods are formed from long, approximately square billets of steel. The principal types of rolling mills used for the rolling of steel rods are referred to as two-high, three-high, and four-high mills (Figure 12.14). As the names suggest, the classification is based on the manner of arranging the rolls in the housings.

The first bars and rods were rolled in two-high mills. After each pass they were pulled back over the top roll and inserted into the next groove by the roller. The three-high mill simplified the rolling process.

Long, thin bars or rods are quite pliable when at white heat. To facilitate rolling, the catcher catches the forward end of the bar as it comes through and, giving it a quarter twist, inserts it in the proper return groove without waiting for the whole bar to run through.

Four types of mills have been developed specifically for roll rods: the Morgan or continuous rod mill; the Garrett or looping rod mill; the combination or combined continuous and single Belgian rod mill; and the double Belgian rod mill.

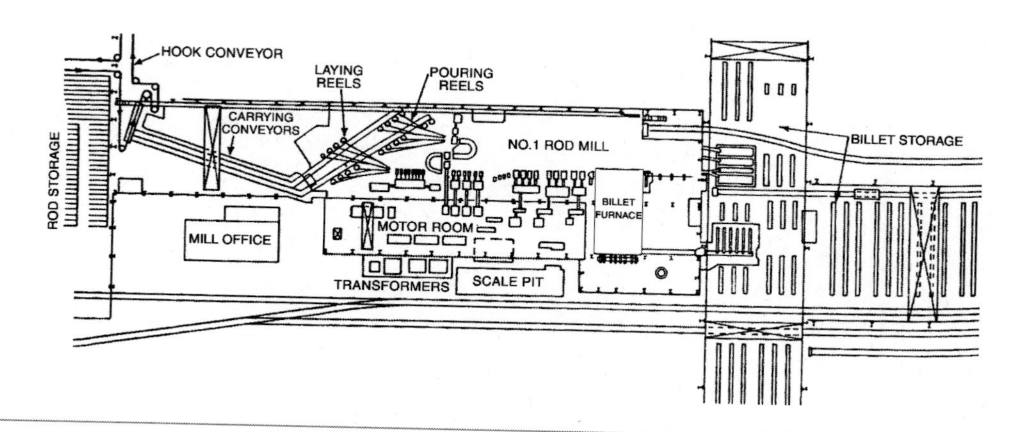

Figure 12.14. Rod rolling mill (McGannon 1964: 780).

The first continuous rod mill in the United States was erected in 1869 by Washburn and Moen in Worcester, Massachusetts. This mill consisted of a number of pairs of horizontal rolls and an equal number of vertical rolls, arranged in a series so that the first, third, and succeeding stands of an odd number were horizontal and the second, fourth, and succeeding stands of even number were vertical. This arrangement of rolls overcame the necessity of giving each piece a quarter turn between passes. These mills had the disadvantage of the vertical rolls frequently falling out of adjustment.

C. H. Morgan, general manager for Washburn and Moen, eliminated the need for vertical rolls by the development of the twisting guide. This was a closed delivering guide in which grooves were cut in a spiral so that after any pass, the piece was forced through it and twisted a quarter turn before it entered the next pass.

Since the time of Morgan's first mill, improvements have been made to auxiliary equipment. These improvements have enabled the production of more accurate rods in greater tonnages at a delivery speed of up to 6,000 feet per minute.

Until about 1882, the continuous rod mill had no competitor for speed, length of rod, and tonnage produced. William Garrett, superintendent of plant of the Cleveland Rolling Mill Company, conceived a plan where the looping mill could be modified to roll No. 5 rod in long lengths, directly from 4 by 4 inch billets without reheating. Garrett inserted between the sets of rolls looping troughs which guided the forward ends of rods around and into the next groove in the rolls. This mill could roll rods from billets of larger diameter and greater weight than had been previously possible and could give rods a more uniform shape and diameter.[44]

Preparation for Drawing

The primary object of rod rolling is to put the steel into a shape that can be most efficiently cold-drawn into wire. To prepare the rod for drawing, preliminary treatment is necessary. This consists of acid cleaning and coating, often followed by a heat treatment. The heat-treating is usually one of the following: patenting, annealing, or normalizing.

All mill scale, oxide, or dirt must be removed before the rods are drawn into wire. This is accomplished by placing the rods in a solution of hot, dilute sulfuric acid for 15 to 30 minutes. Following the acid bath, the rods are thoroughly rinsed in a spray of high-pressure water. The next step is to give the rods a suitable coating, usually lime. Borax is another coating that has been used. After the coating is applied, the rods are baked in an oven to dry the coating.

The rods are then delivered to the wire-drawing equipment. There are two processes for drawing wire: dry drawing and wet drawing. Mechanically, the two are the same; the wire is drawn through a die and wound up on a block. The difference in the processes is in the coating applied to the wire and the lubricant used.[45]

Wire Mills

The first wire-drawing mill in the United States was built in 1775 by Nathaniel Miles at Norwich, Connecticut. Other early mills were established in succeeding years, but none was very prosperous. By 1820, practically no wire was manufactured in the United States. In 1831, the industry was reinvigorated in the United States with the establishment of the firm of Washburn, Moen and Company in Worcester, Massachusetts, by Ichabod Washburn and Benjamin Goddard.

In 1840, the first wire rope in the country was manufactured by John A. Roebling at Saxonburg, Pennsylvania. In 1851, the first American-made wire nail machine was built by Thomas Merton. In 1875, the first steel wire nails were made, and in 1888, the production of wire nails exceeded those of cut nails.

Rolling the Wire Rod

The slender rods or bars of metal from which wire is drawn are known as wire rod. It is made in various shapes and sizes, but for common and fine wires, 7/32-inch round rod is the standard. As the rod comes from the rolling mill, it is wound into coils. Each coil represents the rod made from a single billet, and its weight is determined by the weight of the billet used.

After it has been made according to specifications, the steel for wire is cast into ingots and rolled on the blooming mill. In most cases, the blooming mill is succeeded by a billet mill, and the blooms are rolled into billets 4 inches square or less (Figure 12.15).[46]

Several primary tools are used in wire drawing: the die and the block. The die has no moving parts and it does not remove any of the metal, but it uniformly reduces the cross-sectional area of the steel and improves its finish. In early wire drawing, chilled iron, steel plates, and alloy steel were all used, but by the 1920s tungsten carbide emerged as the principal die material.

Wire was first drawn in very short lengths and was pulled through the die in straight pieces. As the lengths grew longer, a means of storage became necessary. The block serves this purpose and consists of a steel casting in the form of a cylinder, the sides of which have a slight taper.

Several types of drawing machines are used. These include the drawbench, bull blocks, motor blocks, and continuous machines. A drawbench is a mechanism used to give a single draft to heavy material. It handles the largest sizes drawn. The machine consists of a horizontal framework 50 to 100 feet long, along the centerline of which runs a heavy roller chain driven by heavy sprocket wheels. The die through which the material is drawn is located at the opposite end of the frame from the drive. A carriage mounted on wheels is

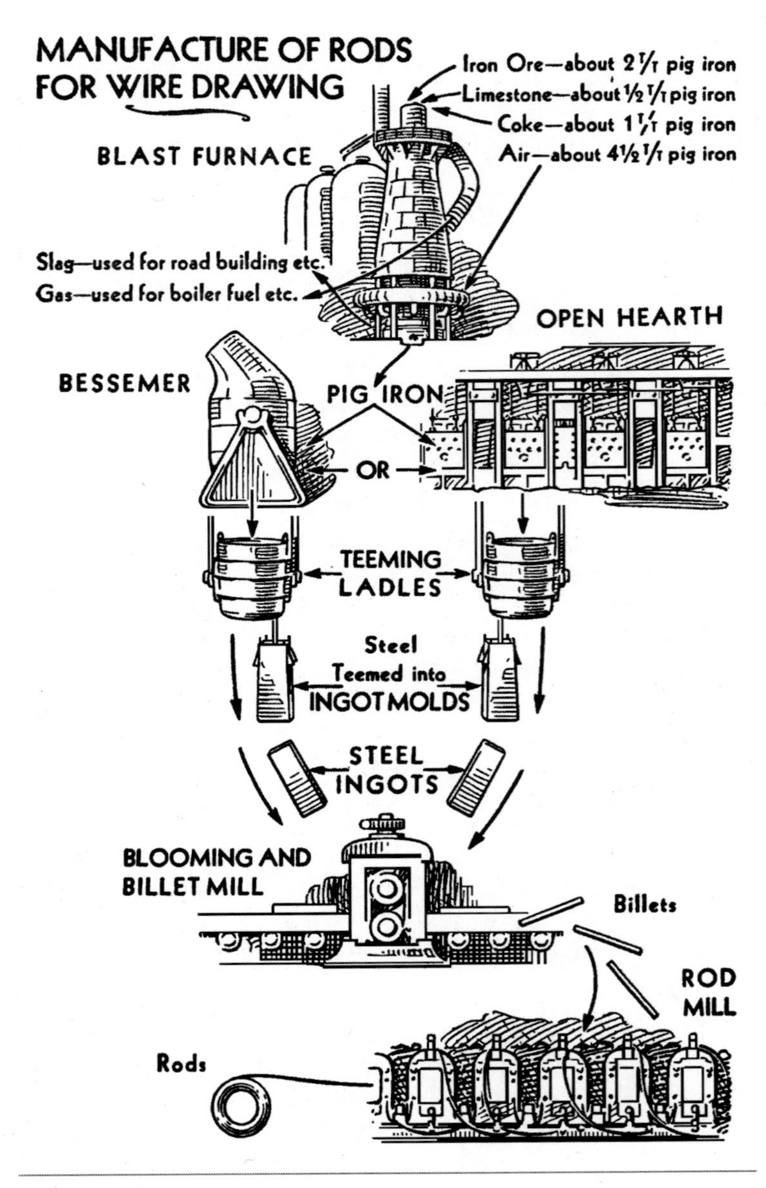

Figure 12.15. Manufacture of rods for wire drawing. Reprinted from American Steel and Wire Company, *Making Steel and Wire* (Cleveland: American Steel and Wire Company, 1934), 29.

arranged to travel along the upper surface of the frame or bench.

Sizes ½ inch to 1 inch are usually drawn on horizontal blocks called bull blocks, very heavy machines built to pull these coarse sizes. They are driven by an individual variable-speed motor. Motor blocks are also driven by an individual motor but also have a vertical spindle. Wire that requires three or more drafts from the rod is usually drawn on continuous machines.[47]

GLOSSARY

Air furnace. A horizontal reverberatory furnace in which the metal is melted by the flame from fuel burning at one end of the hearth, which passes over the iron toward the stack at the other end.

Anneal. Heating iron to above a critical range, holding it at that temperature for a required time, and slowly cooling it to make the iron less brittle.

Arch bricks. Bricks that surround fire holes or arches on kilns.

Bellows. A box with flexible sides, often constructed of leather, in which expansion and contraction draws air through a side valve and expels it through a nozzle.

Bessemer process. A process of making steel from cast iron through burning out carbon and impurities through the bottom of the molten metal. By the early twentieth century, it was largely supplanted by the open-hearth process.

Billet. Iron stock with a round-cornered square or rectangular cross section, able to be further processed by forging or rolling.

Blast. A continuous blowing, to which the charge of ore or metal is subject in a furnace.

Blast furnace. A tall-shaft variety of furnace operated by forced draft.

Bloom. A mass of wrought iron from a puddling furnace or bloomery.

Bloomer. A worker in a bloomery or puddling furnace, also called a *puddler*.

Bloomery. A forge that makes wrought-iron blooms directly from ore, or less frequently, from cast iron. Its product was generally of poor quality, as it never was molten.

Blowing tubs. Blast-producing machinery consisting of waterwheel-driven pistons and cylinders. This machinery replaced bellows.

Bosh. The bottom inward-sloping surfaces of the furnace cavity.

Breast. The part of the furnace lining connecting the spout with the bottom and made up with a taphole for every heat.

Calcining. Heating to high temperatures without melting to remove volatile matter.

Carburize. To add carbon to low-carbon steel by heating it to above critical range while in contact with carbonaceous matter such as charcoal or coke.

Cast iron. Iron containing so much carbon, usually above 1.7 or 2 percent, that it is not usefully malleable at any temperature.

Cementation. A steelmaking process in which wrought-iron bars were held at high temperature in a furnace in the presence of charcoal. If left long enough, the carbon would penetrate the iron, converting it to steel.

Charcoal furnace. A blast furnace that uses charcoal as its fuel.

Charcoal kiln. A round, rectangular, or conical structure made of brick and stone used for charring wood.

Charge. A given weight of metal, stone, and/or fuel used in a furnace or a kiln.

Clamp. A temporary-type brick kiln.

Coke. The residue of coal remaining after high-temperature heating, used as fuel in furnaces and forges.

Cold blast. A blast furnace that operates with air at room (outside) temperature.

Crucible. A vessel with a refractory lining used for melting or calcining iron or steel.

Crucible steel. A type of steel made by melting steel made in cementation furnaces in clay crucibles.

Cupola. A type of blast furnace for melting metal that consisted of a vertical cylinder lined with refractory material and equipped with openings for the entrance of a blast. In this furnace, metal was melted in direct contact with the fuel.[48]

Draft. The stream of air blast delivered by the bellows or air pump that maintains combustion in the furnace.

Drop hammer. A forging hammer that drops vertically onto the work piece, usually relying on a powered cylinder to lift the hammer head and to add to the force of the downward stroke.

Dross. The solid scum that forms on the surface of a metal when molten or melting.

Electric arc furnace. The standard furnace for melting steel in large quantities since World War II. Arcs pass among electrodes hung from the lid of the furnace and the metal of the furnace hearth to produce intense heating in the center of the furnace.

Firebridge. A short wall inside a reverberatory furnace, made of highly refractory material, which separates the fire from the molten metal.

Flume. An inclined channel for conveying water to supply power to a waterwheel or turbine.

Flux. The basic material added to the furnace charge that unites with sand, ash, and dirt during melting to form slag.

Forge. A general term that includes furnaces or a shop with a hearth where wrought iron is produced directly from the ore.

Foundry. A factory in which metals are molded or forged for the production of castings.

Gangue. Non-iron mineral matter mixed with the ore.

Gate. The end of the runner where molten metal enters a mold.

Hearth. The floor or sole of the furnace.

Helve hammer. A name given to the heavier types of water-powered hammers, especially in the United Kingdom. The helve is the main beam of the hammer pivoted at the tail end.

Hot blast. Furnace blast preheated to 400 degrees or more.

Ingot. The casting from which rolled or forged iron is to be produced.

Knee vents. The middle row of vents in a charcoal kiln, set at knee height.

Lime kiln. Round or square stone or brick structure for calcining limestone.

Long ton. The usual unit of measurement for iron: 2,240 pounds.

Malleable. Capable of being hammered or rolled without breaking.

Pig bed. Small, open sand molds, made in the floor of the foundry near the furnace, to hold the over iron and other waste metal.

Pig iron. Cast iron that has been run into pigs directly from the blast furnace. The iron in the sand molds resembles a sow with suckling pigs.

Puddle. To work metal, while molten, into a desired shape with a long iron tool.

Puddling. The process developed in the late eighteenth century in which melted cast iron is converted into wrought iron. It ultimately superseded the refining method associated with a charcoal furnace, as a cheaper and faster method of making wrought iron.

Puddling furnace. A small reverberatory furnace in which cast iron is converted into wrought iron.

Race. A natural or man-made waterway that conveys water to power a waterwheel or turbine.

Refinery. A furnace with a shallow hearth for converting or refining pig iron to wrought iron.

Refractory. Material capable of enduring high temperatures without fusing, corroding, or deforming.

Reverberatory (or *reverbatory*) *furnace*. A hearth furnace where the flame is drawn over a firebridge and sweeps through the chamber to the chimney. Burning gases heat the stock, roof, and side walls of the furnace, and the radiated heat melts and superheats the metal.

Rolling and slitting mill. A foundry in which wrought-iron bars were rolled into plate iron, then passed through cutters that sheared the plate into long thin rods used primarily by nail makers.

Runner. An enlarged pouring basin or deep channel connecting with gates to bring metal to them.

Shingling. The process of forging wrought iron by using a hammer to squeeze out most of the slag.

Sinter. Ore that has been burned, not melted, to separate its elements.

Skip car. A small car that carries ore, fuel, and flux from storage bins to the top of a modern blast furnace.

Skip hoist. The inclined track on which a skip car is hoisted to the top of a blast furnace.

Slag. A by-product of heating or melting of iron and steel consisting of oxides and other impurities that are generally unwanted. The process of forming slag is essential in removing impurities during iron and steel production.

Slag hole. An aperture in the furnace slightly above the level of the molten iron through which the slag is drawn off.

Spout. A channel casting bolted to the furnace which, when lined with refractory material, forms a continuation of the bottom of the furnace and carries the molten metal from the taphole to the ladles or molds.

Steel. A compound of iron containing beween 0.15 and 1.35 percent of carbon. Modern alloys include chromium, lead, manganese, molybdenum, nickel, vanadium, and tungsten.

Taphole. Opening in the furnace breast through which molten metal is allowed to run to the spout.

Tapping. Opening the aperture at the spout to permit the molten metal to run from the furnace.

Tunnel head. The top of the furnace shaft at its smallest inside diameter. Sometimes called the *trunnel head*.

Tuyere. A nozzle, usually constructed of iron, through which blast is provided for the furnace.

Waist vents. The upper row of vents in a charcoal kiln, at waist height.

Wrought iron. A malleable iron, aggregated from particles without subsequent fusion. It contains so little carbon (generally less than 0.15 percent) that it does not harden usefully when cooled rapidly.

BIBLIOGRAPHY

American Iron and Steel Institute. Steelworks. The online resource for steel. www.steel.org.

American Steel and Iron Company. *Making Steel and Wire*. Cleveland: American Steel and Wire Company, 1934.

Camp, J. M., and C. B. Francis. *The Making, Shaping and Treating of Steel*. Pittsburgh: Carnegie Steel Company, 1920.

Chard, Jack. *Making Iron and Steel: The Historic Processes, 1700–1900*. Ringwood, NJ: North Highlands Historical Society, 1995.

Fairbairn, William. *Iron: Its History, Properties, & Processes of Manufacture*. Edinburgh: Adam and Charles Black, 1865.

Gillmore, Quincy Adams, *Practical Treatise on Limes, Hydraulic Cements and Mortars*. New York: D. Van Nostrand, 1870.

Gordon, Robert B. "Material Evidence of Ironmaking Techniques." *Journal of the Society for Industrial Archeology* 21:2 (1995): 69–80.

—. *American Iron, 1607–1900*. Baltimore: Johns Hopkins University Press, 1996.

Lewis, W. David. *Iron and Steel Production*. Wilmington: The Eleutherian Mills-Hagley Foundation, 1976.

Lovis, John B. *The Blast Furnaces of Sparrow Point: One Hundred Years of Ironmaking on the Chesapeake Bay*. Easton, PA: Canal History and Technology Press, 2005.

McGannon, Harold E. *The Making, Shaping and Treating of Steel*. 8th edition. Pittsburgh: United States Steel, 1964.

National Park Service. *Hopewell Furnace: A Guide to Hopewell Furnace National Historical Site, Pennsylvania*. Washington, DC: National Park Service, 1983.

Osborn, H. S. *The Metallurgy of Iron and Steel, Theoretical and Practical: In All Its Branches, with Special Reference to American Materials and Processes*. Philadelphia: Henry Carey Baird, 1869.

Overman, Henry S. *The Manufacture of Iron in All Its Various Branches*. Philadelphia: H. C. Baird, 1854.

Rolando, Victor R. *200 Years of Soot and Sweat: The History and Archeology of Vermont's Iron, Charcoal, and Lime Industries*. Manchester Center, VT: Vermont Archeological Society, n.d.

Spring, LaVerne W. *Non-Technical Chats on Iron and Steel and Their Application to Modern Industry*.

Vukmir, Rade B., M.D., J.D. *The Mill*. Lanham, MD: University of Press, 1999.

NOTES

1 David Weitzman, *Traces of the Past: A Field Guide to Industrial Archaeology* (New York: Charles Scribner's Sons, 1980), 134.

2 Weitzman, 134

3 Weitzman, 140.

4 Reconstructed or restored iron plantations are located throughout the United States. Links to many of these museum properties are available on the Tannehill Ironworks website (www.tannehill.org/links.html).

5 Robert B. Gordon, *American Iron, 1607–1900* (Baltimore: The Johns Hopkins University Press, 1996), 100.

6 Gordon, 100.

7 W. David Lewis, *Iron and Steel Production* (Wilmington, DE: Eleutherian Mills-Hagley Foundation, 1976), 11.

8 For details about information that can be gathered from early iron wastes, see Robert B. Gordon, "Material Evidence of Ironmaking Techniques," *The Journal of the Society for Industrial Archeology* 21:2 (1995): 69–80.

9 Victor R. Rolando, *200 Years of Soot and Sweat: The History and Archeology of Vermont's Iron, Charcoal and Lime Industries* (Manchester Center, VT: Vermont Archeological Society, 1992), 149–152.

10 Rolando, 154.

11 Rolando, 156–158.

12 Rolando, 158.

13 Rolando, 167–168.

14 Rolando, 206.

15 Rolando, 210.

16 Rolando, 207.

17 Rolando, 208–209.

18 Rolando, 225.

19 See also illustration of the Buckeye Furnace in Weitzman, 142–143.

20 William Fairbairn, *Iron: Its History, Properties, & Processes of Manufacture* (Edinburgh: Adam and Charles Black, 1865), 55.

21 Gordon, *American Iron*, 100.

22 Gordon, *American Iron*, 100.

23 Gordon, *American Iron*, 113.

24 Gordon, *American Iron*, 126.

25 Gordon, *American Iron*, 128.

26 Gordon, *American Iron*, 129–131.

27 Gordon, *American Iron*, 137.

28 Henry S. Overman, *The Manufacture of Iron* (Philadelphia: H. C. Baird, 1854), 334–337.

29 The American Blacksmith: Steam Hammers. www.blackiron.us/steam-hammers.html.
30 American Blacksmith.
31 Harold E. McGannon, editor, *The Making, Shaping and Treating of Steel* (Pittsburgh: United States Steel, 1964), 98–99.
32 John A. Ricketts, How a Blast Furnace Works, Steelworks website (www.steel.org).
33 American Society of Mechanical Engineers, *Oxygen Process Steel-Making Vessel, Trenton, Michigan, A National Historic Mechanical Engineering Landmark*, 1985.
34 McGannon, 459–462.
35 McGannon, 29.
36 McGannon, 518–519.
37 McGannon, 522.
38 McGannon, 535.
39 McGannon, 537–538.
40 McGannon, 562–563.
41 McGannon, 826–831.
42 McGannon, 843–844.
43 McGannon, 858–859.
44 McGannon, 777. LaVerne W. Spring, *Non-Technical Chats on Iron and Steel and Their Application to Modern Industry* (New York: Franklin A. Stokes Company, 1917), 282–283.
45 McGannon, 782.
46 McGannon, 774–775.
47 McGannon, 788–789.
48 Fairbairn, 57.

Chapter 13

Extractive Industries

The remnants of extractive industries are found in many regions of the United States. In New England, remnants include those associated with stone quarrying. In the Appalachian region from Pennsylvania to Alabama, many of these remnants and active industrial complexes are associated with the coal industry. In northern New Jersey, zinc was mined. The upper Midwest, including the Upper Peninsula of Michigan and northern Minnesota, has remnants and active copper and iron mines. In the Southwest, other types of mining, including lead, molybdenum, precious metals, and diamonds, have left their marks on the landscape. This chapter is an introduction to some of the types of mining that have left their traces on the landscape of the United States.

Coal Mining

History of Coal Mining

Pennsylvania was among the most important early centers of coal mining in the United States. By the early nineteenth century, substantial commercial exploitation of coal deposits had begun. Bituminous and anthracite coal production increased in the nineteenth century as trade and population grew. Soft (bituminous) coal was primarily used in industry for generating steam or was processed to form coke for steel manufacturing. Hard (anthracite) coal was primarily used for domestic heating. Because the anthracite fields of northeastern Pennsylvania were closer to coastal population centers, they were the first to be exploited on a large scale. Waterways such as the Schuylkill Canal and the Lehigh Canal aided in moving coal to population centers. By 1832, the annual output of Pennsylvania anthracite coal had reached 501,951 tons.

Large-scale exploitation of the bituminous fields of western Pennsylvania began after the extension of railroads to the area in the last quarter of the nineteenth century. Bituminous production equaled anthracite in 1870 and surpassed it in every succeeding year. Pennsylvania's bituminous coal industry reached its peak in 1918 (when 178 million tons were mined) and declined in following years. In part, this decline was due to increased competition with other coal-producing states, particularly Kentucky and West Virginia.

Techniques of Coal Mining

The type of coal mine created depended on the location of the coal seam. When coal outcropped to the surface, drift mines were used. To open a mine, a main heading or "drift" was dug horizontally into the side of a mountain to reach coal. Panel headings were driven to act as cross streets in the mine. Side entries were then opened off of the major headings to serve as alleys. A drift mine was less expensive to open than other types, because coal could be retrieved from the time excavation began. There was no need for elaborate transportation systems or ventilation systems. In the late nineteenth and early twentieth centuries, miners walked into the entrance and horses or mules dragged the coal from the mine.

Slope mines are defined as those in which the opening proceeds downward at an angle of less than 90 degrees. This type of mine was used where the coal seam was near the surface but not at a great depth. Early slope mines required more machine power or manpower than drift mines. The twentieth-century use of locomotives and conveyors made mining easier.[1]

Shaft mines have a vertical opening. Initial costs are higher at these mines because substantial preliminary digging through rock and shale is required before the buried coal seam is reached. Miners and equipment are lowered to the work site with hoists or elevators.

The most common historic method of working the coal was the room-and-pillar system. Rooms were opened approximately every 50 feet off of side entries. These rooms could reach a length of 300 feet and were often up to 30 feet wide. For the first 30 feet or so, the rooms were driven narrowly, only about 9 feet wide, to create a thick pillar in the area of the entry. When all the rooms along a section of a side heading had been opened, the pillars of coal between them would be removed. This method increased the amount of coal recovered from a field and allowed the roof to settle, relieving pressure. The drawback of this method was that each room was isolated, and as mines grew, supervision, distribution of cars and equipment, and gathering of coal became increasingly difficult.

One approach to address this problem was the short-wall method. In this method, a continuous sawtooth face was worked off of each side entry. Man and equipment could be concentrated in the area of the working face.

A second major mining technique was long-wall mining. This technique produced more coal than the room-and-pillar method but required proper equipment and structural conditions. In this technique, two parallel tunnels were driven to one side, hundreds of feet apart. The area between the tunnels was mined by a large gang of workers who moved forward and used movable hydraulic pillars to support the roof as they proceeded. No pillars were left to support the roof; all was taken out of the mine.[2]

In early mines, a skilled miner would laboriously pick a narrow wedge several feet long and as deep as possible at the bottom of a coal seam. He would then drill holes into the hole at several locations. To drive headings and rooms, twentieth-century miners often made a cut with compressed-air cutters, similar to jackhammers (and known as "punch" or "pick" machines), at the base of the coal face, boring two or three holes at the top to hold explosives. Ideally, this undercut allowed the coal to break cleanly from the seam.

By the early twentieth century, electrical undercutting machines began to replace the compressed-air units. Technology advanced further with low-vein, short-wall chain breast cutting machines, which cut coal with an endless chain fitted with sharp bits. The chain rotated around a flat, tongue-like guide that could slice 10 feet into the bottom of a coal seam.

Once the coal was cut and shot free, miners would push cars to the face on tracks laid along either side of the room. The loose coal was hand-loaded, and the car was then pushed back to the side entry. Later, mechanical loading was introduced, with the scraper loader consisting of a bucket or scoop attached to an electric hoist. Eventually, face conveyors were used in many mines. When mine cars were loaded, they were collected and carried to the surface by the main haulage locomotive that ran along the main opening.

During the early twentieth century, steadily increasing demand for coal led to the opening of coal fields in areas previously considered inaccessible. Because coal in these isolated regions was often located in steep slopes, coal miners adopted special techniques for moving men and materials. One such technique was the use of inclines to move miners and supplies to the mine opening. Such a system was employed at West Virginia's Kay Moor Mine.[3]

Mine Ventilation

Ventilation has been a continuing vexing problem for underground coal mine operators. Lack of ventilation causes the buildup of noxious or explosive gases, often with disastrous results.

The standard method of mine ventilation has been to force a current of air through the mine and use heavy doors to direct the draft through each heading. This method has often been inefficient because the doors in the mine may not be tightly sealed, resulting in sections of the mine deprived of fresh air.[4]

Another method is the installation of mechanical fans, located near or at the portal, pushing fresh air into the mine. Other mine fans pull air through the mine to exhaust it via the ventilation opening. Typically, the fan used employs a disk or axial-flow design similar to an airplane propeller. Its primary purpose is to force air directly into mine shafts below. An example is the Baltimore fan shown in Figure 13.1 which was used at the Dorrance Colliery in Wilkes-Barre, Pennsylvania.

In the mid-twentieth century, two types of mine fans were in general use: the centrifugal and the propeller or axial fan. The propeller fan had begun to dominate on account of its adaptability to most mine problems. A commonly used type of fan was the Aerodyne, made by the Jaffrey Manufacturing Company. These fans were available in sizes ranging from 5,000 to 600,000 cubic feet per minute and had seven possible blade positions. The ability to adjust the pitch of the blades permitted a maximum efficiency over a wide change in the mine characteristics.[5]

A system of airways and airshafts was also designed to improve ventilation. For example, in Berwind-White's Eureka No. 40, the airway split into two parts at the main heading. On opposite sides from one another, the two airways ran parallel to the main heading for its entire length. At each side or panel heading, the main airways would split again, with one passage bridging the heading and the other turning off to run parallel to the lateral entry. At the limit of the workings, the side airway entered the side heading. The air then followed the heading back to the main haulage road and from there returned to the drift mouth. When initially planned, the main airways had a cross-sectional area of 50 square feet. The company usually left a 35-foot pillar between the main heading and the main airways. The overcasts were usually built of wood. As the mine grew, the company enlarged the airways and rebuilt most of the original overcasts in brick and concrete. The company also sank airshafts near the working faces of its mines to serve as exhausts, thus reducing the distance the air circulated.[6]

The remnants of the fan house and fan setting for Eureka Mine No. 40 exist near Berwind, Pennsylvania. It used a 16-foot, 200-horsepower rope-driven fan. The fan ventilated the workings through a 10-foot-square airshaft which reached a depth of 45 feet.[7]

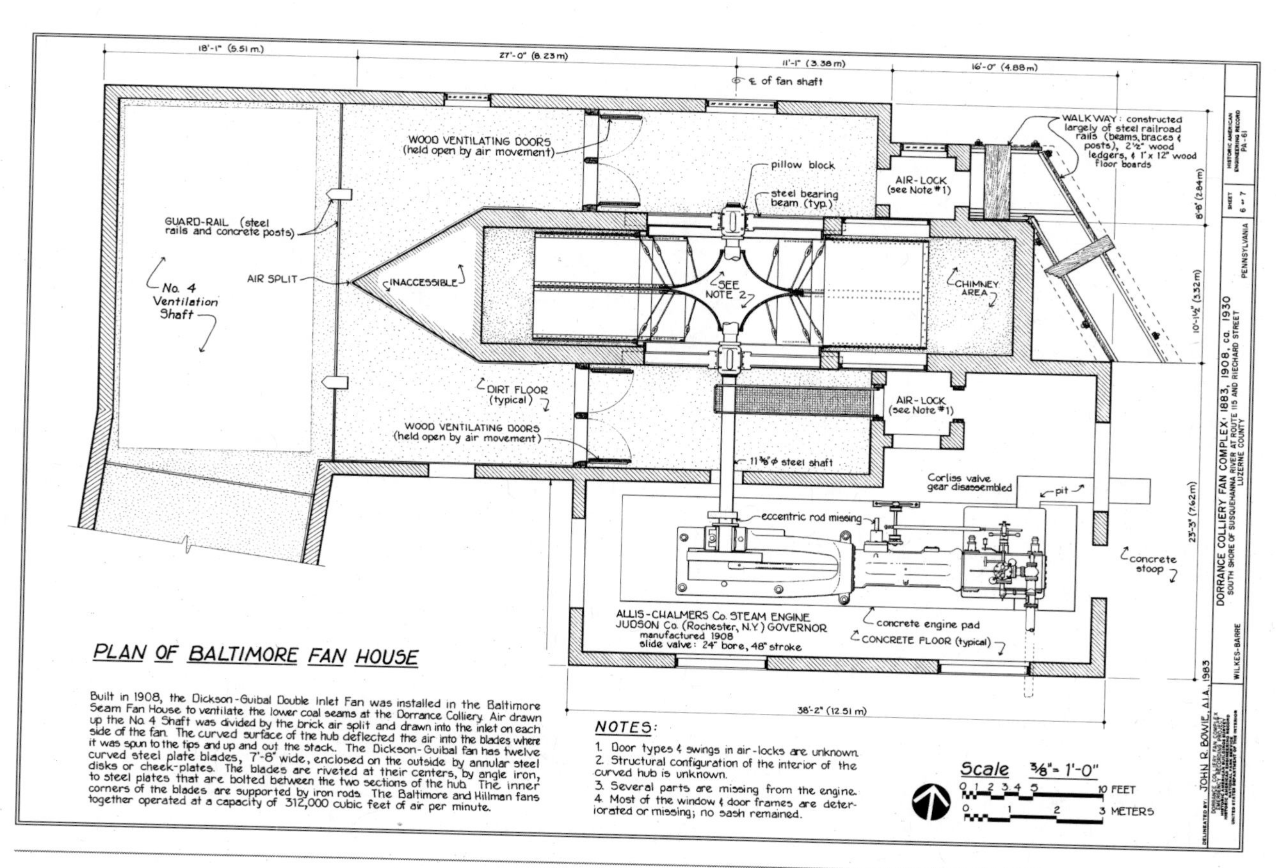

Figure 13.1. Baltimore Fan. Dorrance Colliery Fan Complex, 1893, 1908, ca. 1930. HAER PA-61, National Park Service. Delineated by John R. Bowie, AIA, 1983.

Coal Camps

Coal camps were unplanned, temporary settlements initially filled with tents and shanties. Later, small dwellings could be built, but a coal camp lacked the amenities and services found in some coal towns. When the coal was depleted, the camp was abandoned. These sites are ephemeral, and the shanties and structures do not last long; their only trace is often ruined foundations and debris.

Coal Company Towns

In his book on the Big Sandy River Valley, George Torok describes a typical southern Appalachian coal company town:

> These coal towns were primitive, congested places that offered little more comfort than earlier camps. Monotonous rows of poorly constructed houses were built with timber cleared for the railroad. Hogs dug through the garbage, and waste and runoff ran into the creeks.[8]

With the arrival of large corporations came the construction of standardized housing in company towns. Most dwellings were simple detached, wood-framed houses, with three or four rooms, a porch, and an outdoor privy. Beginning in the late nineteenth century, a few large coal companies built model company towns offering an improved standard of living. Among the earliest were Stonega, Virginia; and Holden and Gary, West Virginia.[9] These model company towns featured such amenities as landscaped streets, parks, sewer systems, garbage collection, a large company store, recreational facilities, and modern conveniences.

The most common surviving building type in both the anthracite and bituminous coal regions is the house. Most such houses are box frame constructions with a simple, square lumber frame. Interior walls were generally bare wood.[10]

Among common house types were a standard rectangular, three-room, single-story dwelling with a front porch; the shotgun house; the basic I, with its end facing a gabled roof, two rooms long but only a single room deep; L-shaped dwellings; and two-story shotguns. Later, prefabricated houses became common in some communities.[11] As would be expected, houses for coal operators and managers were larger than those for workers, often Victorian or bungalow designs.[12]

A comparison of southwestern Pennsylvania coal town residences to residential architecture of other coal communities is included as a chapter in Margaret M. Mulrooney's *A Legacy of Coal: The Coal Company Towns of Southwestern, Pennsylvania*.[13]

Company Stores

The company store was usually the most prominent building in the center of coal towns. It was typically a three-story rectangular building of wood or brick construction, with the first level used for shipping and receiving, the second for retail, and the third for offices and storage.[14]

Collieries

A colliery is the entire mining plant of a mine, including surface improvements and the underground workings. The underground workings typically include several mines and several types of mines. The surface structures were designed to support the underground operations and prepare coal for shipment. A typical colliery might include a breaker or tipple, railroad tracks, boiler house, hoisting house, ventilating fans, powerhouses, lamp house, wash house, and refuse dumps.

Generally the largest structure in either an anthracite or bituminous colliery was the facility where coal was processed and prepared for shipment. In an anthracite colliery, this structure was termed a breaker, while in an anthracite colliery, it was termed a tipple.

Breakers

The breaker was the central structure on the surface of an anthracite colliery (Figure 13.2). Its purpose was to facilitate the preparation of coal for shipment to market. The term breaker initially referred to the crushers or rollers used to break coal into smaller sizes. Soon, however, the name was given to structures that housed not only rollers but other mechanical devices, including conveyors, shakers, picking tables, washers, elevators, and separators to break, wash, and size coal, remove impurities, and load the coal for shipment.

Breakers were necessary for anthracite mining because of the character of coal as it emerged from the ground. Coal came from the mine in a variety of sizes, from huge chunks to powder mixed with impurities.

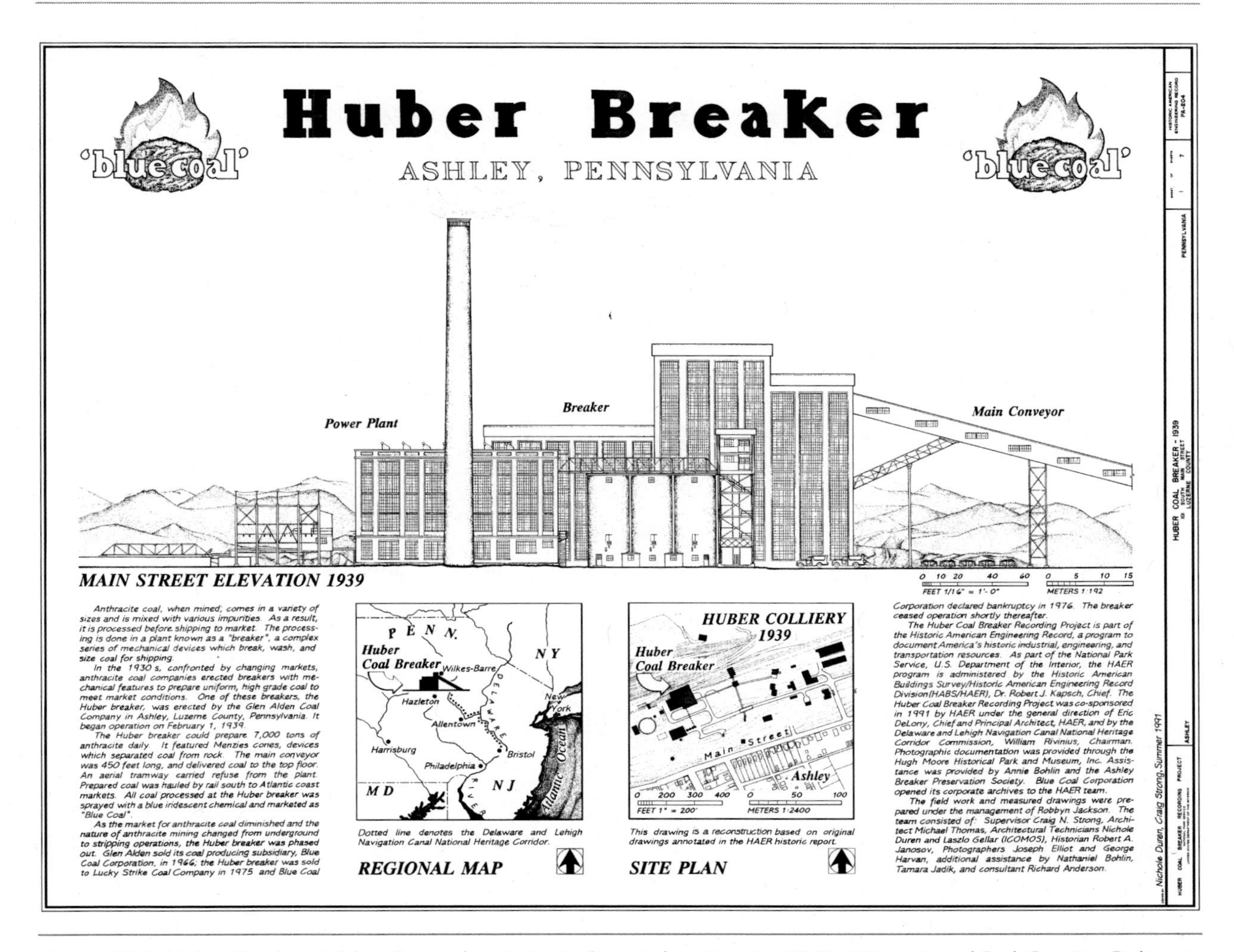

Figure 13.2. Huber Breaker, Ashley, Pennsylvania. Main Street elevation. HAER PA-204, National Park Service. Delineated by Nichole Duren and Craig Strong, 1991.

Because of size variety and impurities, mine-run coal was not marketable and had to be processed to meet consumer expectations. As the industry evolved, changing market conditions meant the development of an increasingly complex preparation process which required more complex breakers

In the earliest days of mining, most coal preparation was done underground by hand by miners who broke up large lumps of coal. In the 1840s, simple sheds housing coal hoppers, rollers, and screens began to appear in the anthracite region. The first of these was erected at Wolf Creek colliery near Minersville, Pennsylvania, and used a system of rolls and screens for breaking and sizing coal developed by Joseph Batten of Philadelphia.

During the 1850s and 1860s, changes in roller technology, the addition of more screens and chutes, and the introduction of washing systems resulted in an increase in the size of breakers. All breakers depended on gravity, and as the preparation process became more complex, the height of breakers increased to permit gravity flow of coal through various appliances. By the late nineteenth century, breakers ranged in height from 60 to 115 feet and capacities ranged from 1,000 to 4,000 tons per day. Breakers were generally wood-framed structures, usually constructed of pine, hemlock, or oak timber and had a large number of windows to provide illumination for the picking process.

By the second decade of the twentieth century, new building materials were introduced to breakers. These new structures were constructed of steel and concrete, and many older wooden ones were remodeled, replacing the original wood framework with steel beams. Among the first of these new breakers was the Loomis in Hanover Township, Luzerne County, Pennsylvania, erected in 1914 of a combination of reinforced concrete and steel.

In a modern coal breaker, coal was conveyed to the foothouse. In the foothouse, the coal received a preliminary screening and hand picking before it was transported to the top of the breaker. This transport was often by means of a chain scraper conveyor, which was preferred over a belt because of a lower rate of coal breakage.

After the coal reached the top floor of the breaker, it began to be washed with calcium hydrate–treated mine water circulated in the breaker. Dirty water was pumped to a thickener, which removed the silt, and the clean water was recirculated. In the Huber Breaker in Ashley, Pennsylvania, the coal then went through five separate, yet interconnected, processing stages: mine-run coal process, coarse-coal process (Figure 13.3), the fine-coal process, the Menzies cone process, and the prepared-coal loading process. Throughout the processes, sizing and separating devices were employed in the breaker.

Picking shakers or tables were used at the preliminary, mine-run stage to separate refuse from the coal by hand. In the mine-run coal process, the stream of coal passed over a double-screened "bull" or lump-and-steamer shaker. The smaller pieces fell through the bull-shaker screens to the third deck and ran through a set of grate, egg, stove, and nut shakers to picking shakers, where debris was picked by hand. The lump and steamer sizes proceeded to picking shakers on the second deck, where impurities were picked by hand.

After the picking on the second deck, the lump and steamer coal flowed to a pair of rollers which crushed it to grate, egg, and smaller sizes. The coal then

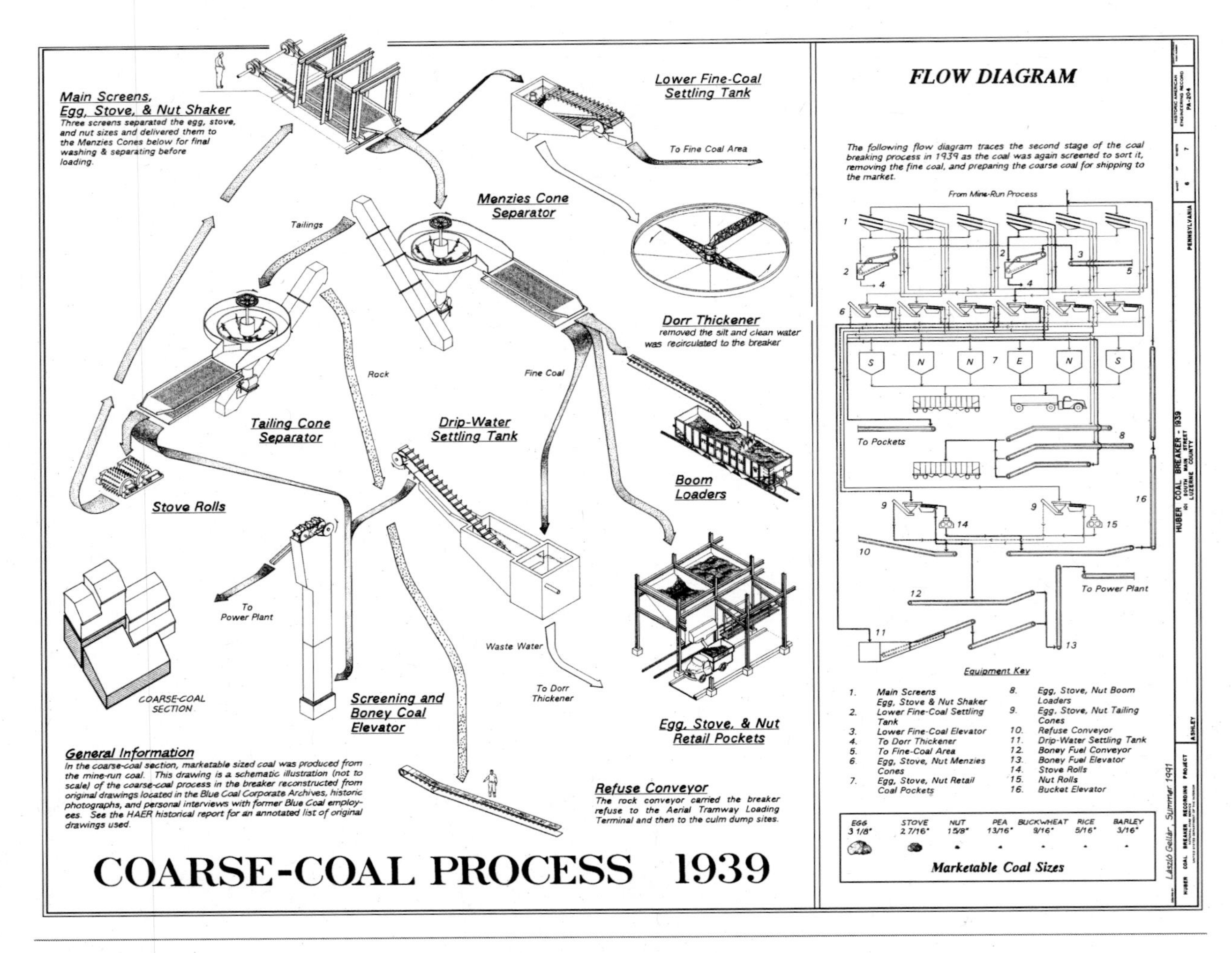

Figure 13.3. Huber Company Breaker. Ashley, Pennsylvania. Coarse-Coal Process 1939. HAER PA-204, National Park Service. Delineated by Laszlo Gellar, 1991.

dropped to the third deck, where along with coal from the third-deck picking process, it was broken in either egg or stove rollers.

The coal then entered the second, coarse-coal process stage. It was screened and rolled again to prepare the coarse sizes for shipment. The fine coal that fell through the shaker screens entered a fine-coal settling tank prior to entering the fine-coal section of the breaker. The coarse-coal sizes passed from the shakers to Menzies cone separators for final washing and separating before loading.

In the fine-coal process, the final cleaning and sizing stage occurred for coal smaller than chestnut size. Two conveyors carried the particles to an elevator. The fine-coal shakers contained a set of four screens representing each of the fine-coal sizes. After sizing, the coal was conveyed to Menzies cones for final washing. The cleaned fine coal proceeded to storage pockets to await loading.

By the 1930s, the Menzies cone method of coal separation began to be employed in breakers. In this method, coal was separated from refuse using circulating upward current of high-pressure water. The machine worked on the principle that coal has a lower specific gravity than its contaminants. Once the lighter coal was fed into the cone, the upward current of water carried it over the top, leading the impurities behind.

Loading and shipping occurred at ground level. There, four railroad tracks, each for a different size of coal, allowed empty cars and trucks to enter the breaker for loading. The loading took place using long belt booms lowered into the cars to prevent coal breakage.[15]

Tipple

A tipple is the term used by the coal industry to describe the structure or group of structures where the mine run of finished coal is processed and loaded into railroad hopper cars for customer delivery.[16]

A tipple is the equivalent of the breaker in an anthracite colliery. Early tipples were simple wooden structures located alongside or over railroad tracks. After coal was dumped into the tipple, it was channeled through a chute to railroad cars below. In early mines, workers hand-sorted coal in the tipple.[17]

At a shaft mine, the tipple and headframe were usually combined in one structure, if possible. Where the headframe formed part of the tipple, self-dumping cages were frequently used for hoisting. The coal passed over the various chutes and screens which separated it into sizes, and the different sizes were delivered to railroad cars standing on tracks beneath the tipple.[18]

The arrangement of a tipple and the amount of machinery contained within it depended largely on whether the coal was weighed in the tipple as run-of-mine coal or as screened coal and on the amount of screening and sizing the coal needed before it was shipped. The principal preparation of bituminous coal for market was separation into lump and slack by bars or screens. Bituminous coal was usually cleaned in the mine.

An example of a twentieth-century tipple is the reinforced concrete structure and separator facility erected for Pennsylvania's Berwind-White Company at Eureka No. 40 in 1928 (Figure 13.4). The main conveyor ran from the dump, located at the northernmost end of

Figure 13.4. Eureka Mine No. 40 Tipple and Separator. HAER PA-184, National Park Service. Photograph by Jet Lowe, 1988.

the plant, to the main screen and wet cleaning plant, located at the angle of the L.

Mine cars would arrive in a covered area in front of the tipple, where they were uncoupled and then entered a rotary dump. Beneath the rotary dump, a mechanical flapgate would direct the load of coal or rock to its respective conveyor. The coal was carried to the main screenhouse and discharged onto the top deck of a shaking screen. The top deck removed coal larger than 4 inches round, while the second deck sized coal down to 1.5 inches. The third deck separated the finest sizes of coal.

Coal sized larger than 4 inches was hand picked and discharged directly down a chute to railroad cars and run-of-mine product. Coal passing over the second deck was taken to the wet-cleaning plant. The smallest sizes were carried to the separator building for air cleaning. Coal too small to be hand-cleaned and too large for dry cleaning was processed in the Menzies hydrotator and classifier located in the main screenhouse.[19]

Few historic tipples remain in operation, but the remains of wood tipples may sometimes be found rotting alongside railroad tracks.[20]

Headframe

The selection of a headframe (Figure 13.5) depended on a large number of factors, including tonnage, hoisting speed, method of hoisting ore and waste, and height at which the core would be discharged. There were three general types of headframes in use: the A-type, the four-post type, and the six-post type. Four- and six-post headframes, whose names refer to the front posts, were installed mostly in coal mines.

The choice between the A-frame and the other types depended on the shaft layout. A rectangular shaft could be equipped with an A-frame type, while square shafts were usually equipped with the four-post type. Circular and elliptical shafts were generally equipped with four- or six-post headframes. Headframes could be made of wood, steel, or reinforced concrete. Wooden frames were usually not as high as steel headframes, nor were they subject to as high a hoisting speed or required to supply as large a tonnage.

Two hoisting methods were employed with headframes: skip, and cage and car. For large tonnages at metal mines, the skip was used almost exclusively. With skip hoisting, the skips dumped directly into large storage bins or into smaller transfer bins.[21]

Boiler House

The boiler house was centrally placed in proximity to the principal places for which it provided steam. At a shaft or slope mine, the best location was usually near the hoisting engine. The size of the boiler house depended on the type of boilers used, the method of firing, and the requirements for cleaning and repair.[22] Anthracite boiler houses were usually of wood-framed construction, although stone, brick, or steel structures were also used.

A mine power plant produced the compressed air needed to power coal cutters and also the power necessary to operate the ventilation fan or fans. At Kay Moor Mine in West Virginia, steam from three boilers (with a capacity of 150 horsepower each) was used to power the compressor. Shortly after construction, the powerhouse was enlarged to include a new steam engine and dynamo. An additional boiler was also installed. This improved plant was able to produce electricity for a haulage locomotive and haulage motor, and electric lights. Later improvements were aimed at increasing its capacity and reliability and included installation of new boilers and generators and a rotary converter to transform AC power into DC.

An article in a 1904 issue of *Engineering and Mining Journal* describes the planned new power plant in Berwind-White's Eureka mine:

> The new power-plant will be in brick buildings. There will be in the boiler-house twelve 200 h.p. Stirling boilers in four sets of 600 h.p. each to which the coal will be fed by Roney stokers. In the power-house . . . there . . . are being installed, two duplex Ingersoll-Sergeant compressors, each with a capacity of 3,000 cu. ft. of air per minute, and two General Electric alternating-current 300-k.w. generators. The current will be stepped up to 6,600 volts, carried by lead-covered cables a longest distance of about three miles, and stepped down to the working mine pressure of 500–550 volts by a rotary transformer.

By the 1920s, the old cylindrical boilers of many mine plants had been replaced by modern vertical or horizontal tubular and watertube boilers.

Engine House

The size of an engine house depended on the size of the engine and what other machinery was placed in the

Figure 13.5. Headframe. Maryland Shaft No. 2, Wilmere vicinity, Pennsylvania. HAER PA-182, National Park Service. Photograph by Jet Lowe, 1988.

building. Typically, the engine house contained the air compressor, dynamos, and the hoisting engine. In smaller plants, the engine house was often sheathed in sheet iron, while in larger plants it was made of brick, tile, or concrete.

Electric and Compressed-Air Power Stations

Electricity was used in anthracite collieries for haulage, pumping, lighting, and operating the breaker. If electricity was to be provided for a single colliery, the generator could be installed in a separate building erected for the purpose or in one of the engine rooms about the plant. At collieries in which compressed air was used for power, the power stations were often arranged with a central station serving several neighboring collieries.

A typical electrical power plant at an early twentieth-century mine was contained in a building 30 feet by 40 feet and served machinery generating from 200 to 300 horsepower. When power was also used for haulage, coal cutting, pumping, and lighting, the power generation need was from 600 to 800 horsepower, and the building could be up to 60 by 100 feet in size. The power plant was not necessarily very near the mine opening. By the 1950s, electric powerhouses were abandoned, and coal operators purchased power from utility companies.[23]

Air compressors had to be placed where the air they used was as free of dust and as cool as possible. Where the hoisting engine was separate from the boiler house, the air compressor could be placed under the same roof as the engine. The air compressor building, if separate from the engine house, typically had sufficient room to house at least two compressors.

At the air compressing end, connections were made with the outside of the building for a fresh air inlet. Connection with a water supply was necessary to furnish water to the water jackets of the compressors for cooling the air.

Hoisting Engines

In the early twentieth century, hoisting engines were usually of the horizontal, high-pressure, slide-valve type. At a few mines, hoisting was done by Corliss valve engines. The location of the hoisting engine depended on the kind of opening, whether it was a shaft or a slope.

With a shaft opening, the distance between the hoisting engine and the headframe was such that the rope would coil regularly on the drum. It was also located so that it was unnecessary to put carrying pulleys between the drum and the headframe. At collieries where coal was raised through a slope and the slope was connected to the main structure through an inclined plane, the preferred location of the engine was on line with the slope at some point back of the breaker.[24]

Wash House

The wash house provided a safe storage place for the miners' street clothes during a shift and permitted the miners to shower before going home.

Lamp House

The lamp house was one of the most important surface facilities from the miners' point of view. In this building, mining safety lamps were used to detect potentially explosive gas accumulations underground and were stored and serviced daily.

Inspection House

Prior to leaving the colliery, the loaded railroad cars passed through a coal inspector's house. Coal samples were taken from different points in randomly selected cars. The samples were weighed and tested against size and impurity specifications in the coal testing laboratory.

Blacksmith Shop

The blacksmith shop, typically constructed of sheet iron or masonry, ranged from about 15 feet square to about 20 by 30 feet. It was here that miners' tools requiring sharpening and mine cars requiring repair were taken. The blacksmith shop contained at least one forge and anvil. Blast was furnished by a blower driven by electricity or steam.

Tool House

Often placed close to or adjoining the blacksmith shop, the tool house usually measured about 15 feet square. The house was used to store the company's shovels, picks, bars, sledges, drills, and other equipment.

Carpenter Shop

The carpenter shop was typically at least 20 by 30 feet in size, located close to the blacksmith and machine shops and near the lumber yard. Its primary purpose was to repair mine cars.[25]

Machine Shop

Often the machine shop was collocated with the blacksmith shop and carpenter shop, placed between the two. It was frequently equipped with a 15-horsepower upright engine connected by line shafting with the carpenter and blacksmith shops in order to run the saw in the first and the forge blower in the second. The machine shop generally did not make machine parts. Instead, it functioned as a repair shop, replacing mass-produced machine parts.[26]

Storehouse

The size of the storehouse depended on the nature of the operations, the kinds of machinery, and the variety and quantity of supplies needed. The stock usually included miners', carpenters', and blacksmiths' tools, iron and steel, nails, screws, bolts, nuts, spikes, mine car fittings, fittings for machinery, belting, and steam and water pipes.

Oil House

At large operations where oil was purchased in carload lots, a separate house was erected for its storage. Here the clerk dealt out lamp oil to miners and lubricating and other oil to company hands.

Refuse Disposal

Mining and coal preparation produced hundreds of tons of breaker refuse and mine rock daily. This waste was sent to waste dumps known as either culm or gob piles. These piles are familiar landscape features in many coal-mining areas and are often the most visible above-ground remnant of mining. In the case of the Huber Breaker, the refuse was conveyed to the dump by an aerial tramway system.

COKE PROCESSING

Coke is the residue of destructive distillation of bituminous coal. Not all bituminous coals form coke. There are three principal kinds of coke, classified in accordance with the methods by which they are manufactured. Coke was made in one of three ways from bituminous coal: in a beehive oven, in by-product ovens, or in gas retorts.

Until the 1930s, most coke was produced near the mine in beehive ovens. The ovens were constructed of hemispherical chambers of firebrick and were usually 12 feet 6 inches in diameter (Figure 13.6) and held a charge of about 6 tons of coal that was coked in 48 to 72 hours. Modern beehive ovens were constructed of masonry, brick, and tile. The space between the lining and the outside walls was filled with waste brick to prevent the loss of heat to the exterior. The heat required to start the coking process was supplied by that retained in the walls of the oven from the previous charge. Beehive ovens were built at the mine to be near the coal supply, their product shipped to distant markets. Few beehive coke ovens remain. According to George Torok, some stone remains of these ovens could be found at Stonega, Virginia, in 2004.[27]

By-product coke was manufactured in rectangular firebrick chambers or retorts. These ovens were from

Figure 13.6. Leetonia Iron and Coal Company beehive coke oven, Leetonia, Ohio. From Titchenal family website (www.titchenal.com).[9]

35 to 40 feet long, 16 to 20 inches wide, and 8 to 12 feet high, and were capable of holding a charge of about 15 tons of coal that was coked in 12 to 21 hours. The heat for coking was provided by burning gas in vertical or horizontal flues in the walls of the oven. No air was admitted to the oven. By-product ovens were built where the coke was consumed, the coal being shipped to them from the mines. After the 1920s, by-product coke became the most common type in the United States. Most was used in making pig iron.

Gas-house coke was the residue obtained by heating gas in small, closed retorts to make illuminating gas. The amount of coke produced was small, and because gas-house coke was soft and porous, it was of little use except for domestic heating.

Metal Mining Districts

Among the principal metals mined in the United States are copper, iron, and lead. The landscape of underground metal mining and processing features several distinctive buildings and structures. Headframes, shafts, shaft-rockhouses, hoist houses, ore bins, concentrators, and smelters form the major elements of the mine landscape.

Headframes

In the past, the simplest way to get miners, waste rock, ores, and supplies out of mines was by manpower or animal power. For example, at the mid-nineteenth-century Ophir Mine, a narrow flight of steps led to the surface. Slightly more complex was the windlass, a hand-operated winch. Yet another solution was use of a small steam engine to pull ore cars up an inclined ramp.

By the late nineteenth century, the headframe hoisting system predominated (Figure 13.7).[28] Headframes mark the locations of mines. The gallows-like structures, topped by sheave wheels, served to hoist men, equipment, and ore from the mines. Like other features associated with mining, the earlier sorts are more likely to have been constructed of wood rather than metal. The headframe's cable ran over the top of the sheave wheel and was powered by an engine in the hoist house.

Shafts

Vertical shafts may be rectangular or circular in cross section. Rectangular shafts are most common in the United States, and circular shafts are most common in

Figure 13.7. Soudan Iron Mine, Minnesota. Headframe. Tower vicinity, Minnesota. HAER MN-30, National Park Service. Photograph by Jet Lowe, 1987.

Europe. Rectangular shafts of shorter and wider cross shafts became increasingly popular in the twentieth century (Figure 13.8). Generally, two skip roads, paths for mine lifts, occupy one end of the shaft, while one or two cage compartments and the pipe and ladder compartments are placed at the opposite ends. In the skip compartments, about 3-inch clearance was left between the shaft timber and the front and back of the skip, and 3 inches or more from the face of each guide to the side of the skip box. A typical cage compartment was about 6 feet wide and 10 feet 10 inches long. This size was large enough to accommodate a 28-man single-decked cage.[29]

Shaft-Rockhouses

At Quincy Mine, Michigan, and other below-ground copper mines, shaft-rockhouses had three purposes: to provide a building for a crusher to break up the ore hoisted from the mine; to house storage bins for the crushed ore so that the train that carried it to the mill could be loaded quickly; and to hold the sheave carrying the rope between the mine car or skip and the hoisting engine. The sheave had to be high enough so

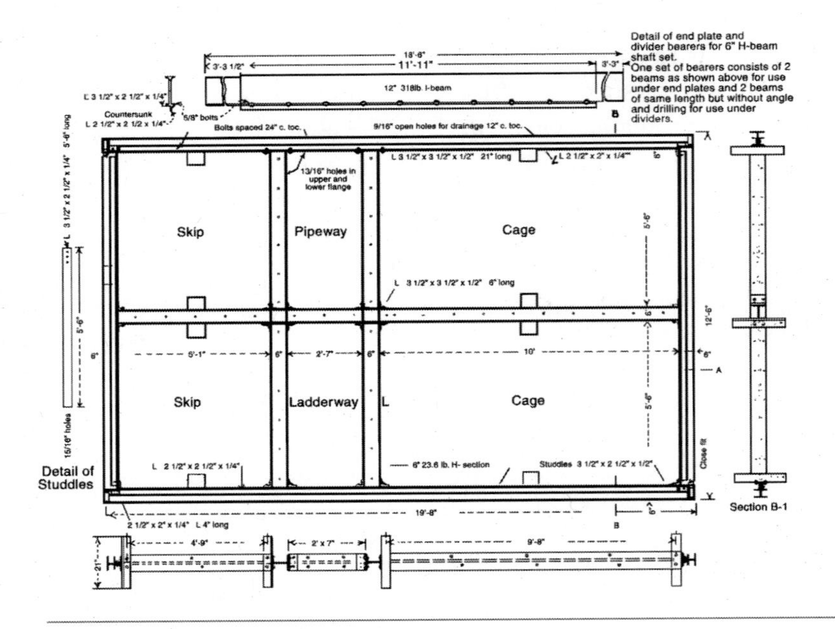

Figure 13.8. Cross section of Eureka Mine shaft, Ramsey, Michigan. Reprinted from Lucien Eaton, *Practical Mine Development and Equipment* (New York: McGraw-Hill Book Company, Inc., 1934), 12.

that when the skip load of ore was dumped, it would flow by gravity through the crusher and into the storage bins. The Quincy No. 2 house is 120 feet high (Figure 13.9).[30]

Hoists

Hoisting systems can be characterized as unbalanced, counterweighted, partly balanced, or balanced. A hoist

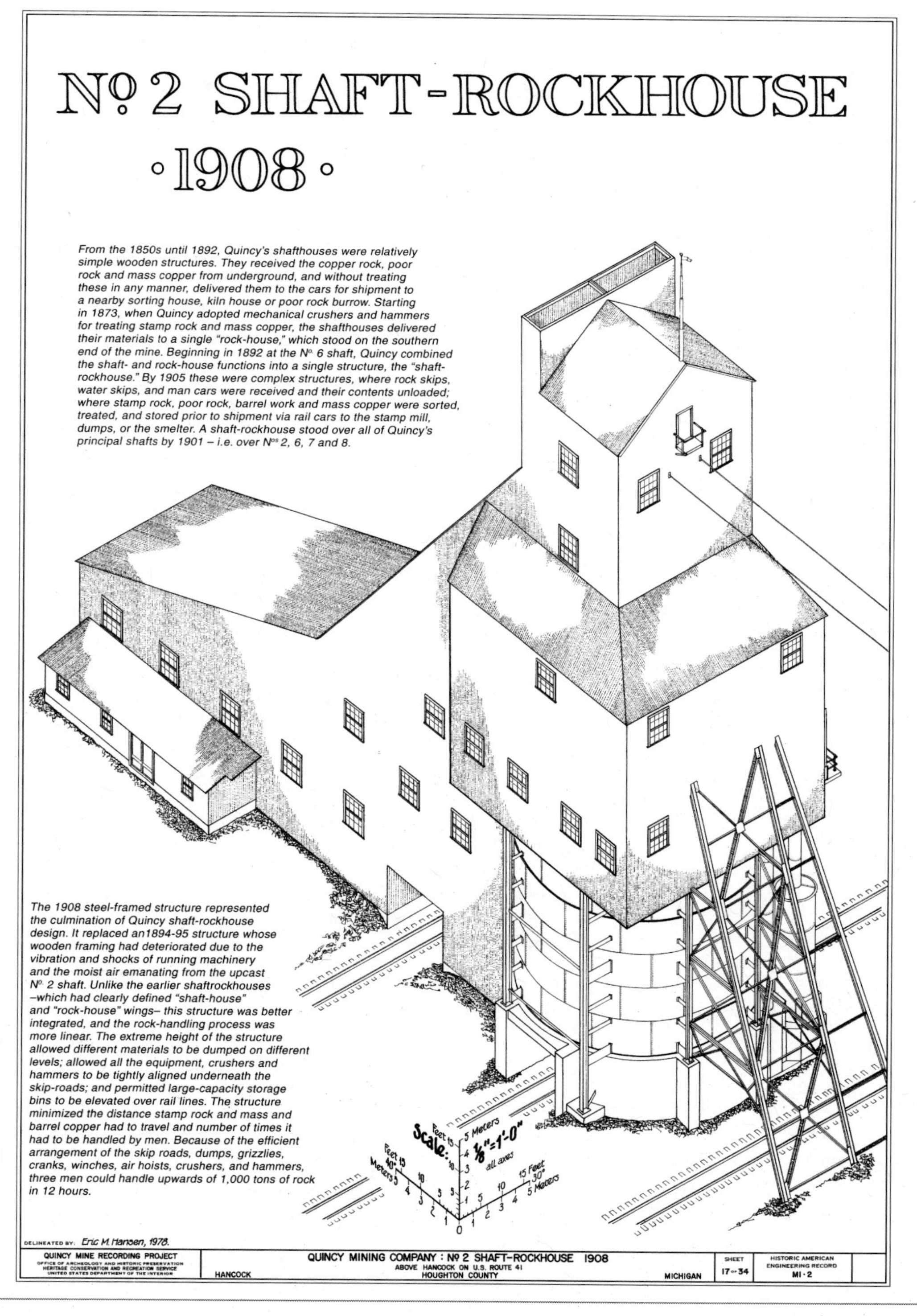

Nº 2 SHAFT-ROCKHOUSE
·1908·
From the 1850s until 1892, Quincy's shafthouses were relatively simple wooden structures. They received the copper rock, poor rock and mass copper from underground, and without treating these in any manner, delivered them to the cars for shipment to a nearby sorting house, kiln house or poor rock burrow. Starting in 1873, when Quincy adopted mechanical crushers and hammers for treating stamp rock and mass copper, the shafthouses delivered their materials to a single "rock-house," which stood on the southern end of the mine. Beginning in 1892 at the Nº 6 shaft, Quincy combined the shaft- and rock-house functions into a single structure, the "shaft-rockhouse." By 1905 these were complex structures, where rock skips, water skips, and man cars were received and their contents unloaded; where stamp rock, poor rock, barrel work and mass copper were sorted, treated, and stored prior to shipment via rail cars to the stamp mill, dumps, or the smelter. A shaft-rockhouse stood over all of Quincy's principal shafts by 1901 – i.e. over Nºs 2, 6, 7 and 8.
The 1908 steel-framed structure represented the culmination of Quincy shaft-rockhouse design. It replaced an1894-95 structure whose wooden framing had deteriorated due to the vibration and shocks of running machinery and the moist air emanating from the upcast Nº 2 shaft. Unlike the earlier shaftrockhouses –which had clearly defined "shaft-house" and "rock-house" wings– this structure was better integrated, and the rock-handling process was more linear. The extreme height of the structure allowed different materials to be dumped on different levels; allowed all the equipment, crushers and hammers to be tightly aligned underneath the skip-roads; and permitted large-capacity storage bins to be elevated over rail lines. The structure minimized the distance stamp rock and mass and barrel copper had to travel and number of times it had to be handled by men. Because of the efficient arrangement of the skip roads, dumps, grizzlies, cranks, winches, air hoists, crushers, and hammers, three men could handle upwards of 1,000 tons of rock in 12 hours.
Scale: 1/8"=1'-0" all axes
Feet
Meters
DELINEATED BY: Eric M. Hansen, 1978.
QUINCY MINE RECORDING PROJECT
QUINCY MINING COMPANY : Nº 2 SHAFT-ROCKHOUSE 1908
ABOVE HANCOCK ON U.S. ROUTE 41
HOUGHTON COUNTY
HANCOCK
MICHIGAN
SHEET 17-34
HISTORIC AMERICAN ENGINEERING RECORD
MI-2

Figure 13.9. No. 2 Shaft-Rockhouse, 1908. Quincy Mine Company. Hancock, Michigan. HAER MI-2, National Park Service. Delineated by Eric M. Hansen, 1978.

can be balanced either by a tail rope or a particular drum shape. The totally unbalanced system was chiefly used in shaft-sinking operations. The counterweighted system found its greatest application in service hoists. Most common was the partly balanced. Hoists were driven by electric power, steam, compressed air, or internal-combustion engines.[31]

Hoist Houses

Hoist houses are typically rectangular, often sheathed in board and batten siding. Double board siding became more common after about 1895, and corrugated galvanized iron siding had become common by about 1900. The 1918–1920 No. 2 Hoist and Engine House at Quincy Mines in Michigan is of reinforced and brick veneer construction (Figure 13.10).

Ore Bins

Ore bins range in capacity from a few tons to a few thousand tons. The shape varies from square to round. The material from which the bins were constructed depended, to a large extent, on the capacity. Smaller-capacity bins were generally made of wood, while larger-capacity bins were made of steel or concrete or a combination of the two.[32]

Mine Drainage

Because many mines were substantially below the water table, draining them became a significant issue. The most simple and most common method of draining western hard rock underground mines was to dig adits in the hillside below the flooded mine. A low-tech solution was to bail out the mine by using buckets on a windlass to carry water. Yet another alternative was to use a water skip, which could be pulled up an inclined ramp.[33]

The most efficient solution to drainage in deeper mines was a pump. Mine pumps can be divided into three general classes: those with an underground pump and motive power applied from the surface, those in which motive power is applied at the pump, or air lifts

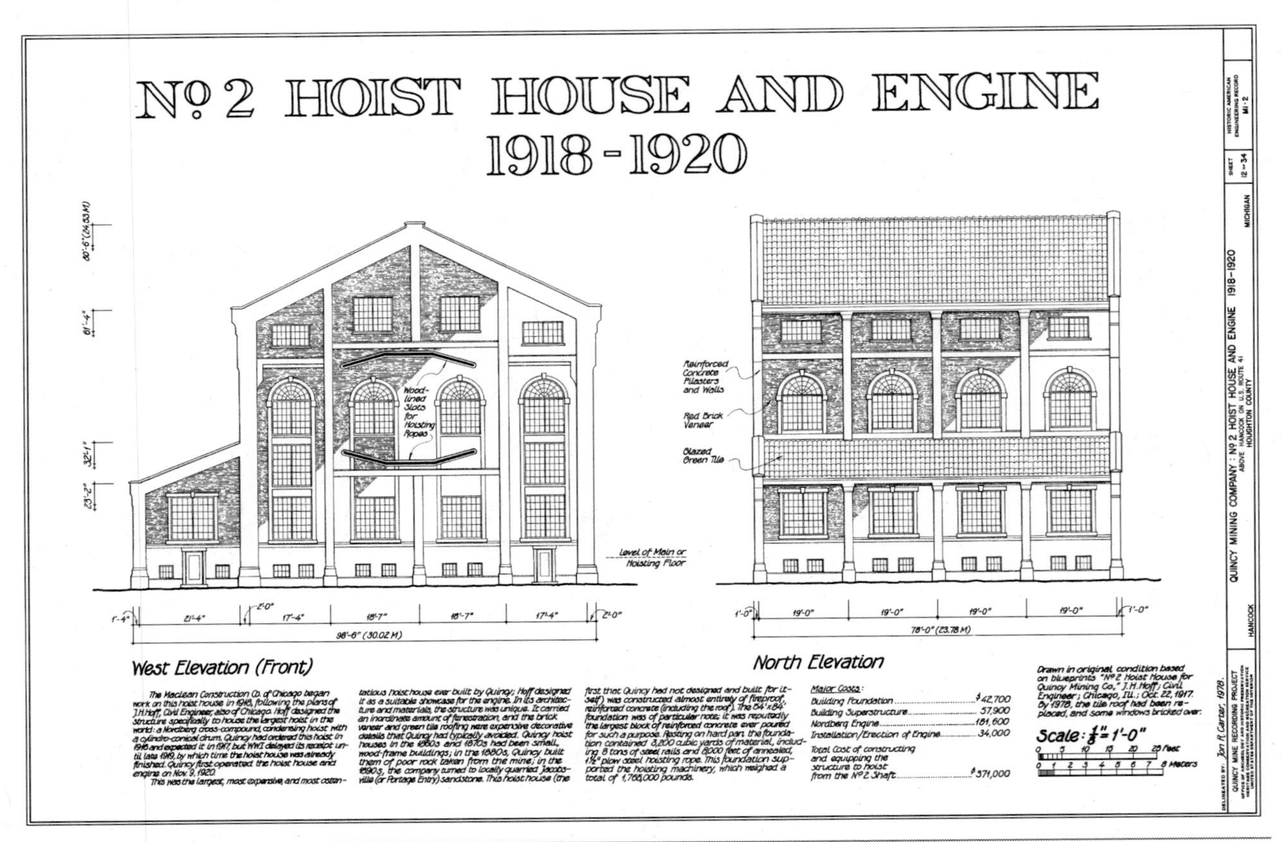

Figure 13.10. Quincy Mining Company: No. 2 Hoist House and Engine, 1918–1920. Hancock, Michigan. HAER MI-2, National Park Service. Delineated by Jon R. Carter, 1978.

and inspirators. The first class includes Cornish pumps, such as those installed in the 1860s and 1870s at Nevada's Comstock mines, and centrifugal pumps. The second class includes reciprocating and centrifugal pumps. In these cases, when a permanent pumping plant was installed, it was usually placed at the bottom level of the mine.[34]

Concentrators

The milling-concentrating process increases the concentration of the metals by removing waste rock. By doing so, the ore is transformed into a higher-grade commodity that can bear the cost of transportation to the smelter. The milling process involves stages of crushing, in which ore is reduced to a size where valued products are more easily recovered. One technique is by means of a stamp mill, which is among the most distinctive features associated with hard rock mining. A stamp mill crushes ore to the consistency of sand by a series of weights driven by camshafts.

Engineers categorize crushing equipment into two types: rock breakers, which crush the ore received from the mine to approximately ¼ inch in diameter, and fine crushers, which reduce the broken ore to "milling size," generally a coarse sand or finer.[35]

The typical stamp mill was constructed on a sloping site and had several levels, each with a stone foundation. It was identifiable by its long, sloping roof and its stepped foundation. The roofline, sometimes pierced by gable windows or skylights, followed the same gradient as the entire building. Mills constructed between about 1870 and 1890 usually had board and batten siding. Corrugated metal siding became common about 1900.[36]

A schematic rendering of the stamp mill process used at the Quincy Mining Company is shown in Figure 13.11. This process yielded an average of 30 pounds of copper per ton of rock stamped.

Smelting

The smelter is typically one of the largest buildings in the mining district and features a tall stack to disperse noxious smoke away from populated areas. Smelting ore is a multistep process, with the exact steps dependent upon the ore being processed. It takes place in a series of buildings and structures connected by conveyors. Many of these buildings have clerestoried or monitor roofs to let in light and dissipate heat. Slag piles are usually located close by.[37]

For example, in smelting of lead at Utah's early twentieth-century Tooele Smelter (Figure 13.12), ore was received from the mine in railroad cars and was dumped into the receiving bins. The ore was conveyed from the receiving bins to the sample mill where it was crushed to 3/8 inch or less in diameter, and a small sample was chemically analyzed. The crushed ore was conveyed to the storage bins.

Conveyors then carried the crushed ore to the roasters. The roasted ore, termed calcine, dropped into hoppers that emptied into calcine cars. The cars were unloaded at the charge bins. The next step was sintering. Calcine, ore, and flux were combined and fed into sintering machines, where the mixture was ignited. The burning fused the material into a low-sulfur, porous product known as sinter. The sinter was then fed into railroad cars.

These cars were pulled back to the charge bins where the sinter was mixed with coke and fluxing materials as it was dumped into charge cars. These were then pulled to the top of the blast furnace. Smelting occurred in the blast furnace. The charge melted, forming three layers: slag, matte, and lead bullion. Matte was sent to the copper smelter, the slag to the slag dump, and the molten lead bullion was tapped into ladles and then taken to the drossing plant. In the drossing plant, the molten lead was poured into drossing kettles and allowed to cool so that impurities could be skimmed from the surface. The bullion was then cast into ingots, loaded, and taken to a refinery.[38]

Transportation

Large mine complexes and districts relied on railroads to connect mine to mill to waste dumps and smelters. A maze of trackage connected these parts.

Some mining districts also included inclines designed to traverse slopes too steep for railroads. With a slope of 20 to 40 degrees, inclines depended on cables to haul specially designed ore cars. Inclines, which typically conveyed ore from mine to tipple or ore bins, may be seen as straight lines on a hillside.

In many western mining districts, materials were also moved by water, and flumes are an important part of the landscape. Flumes are usually open structures by

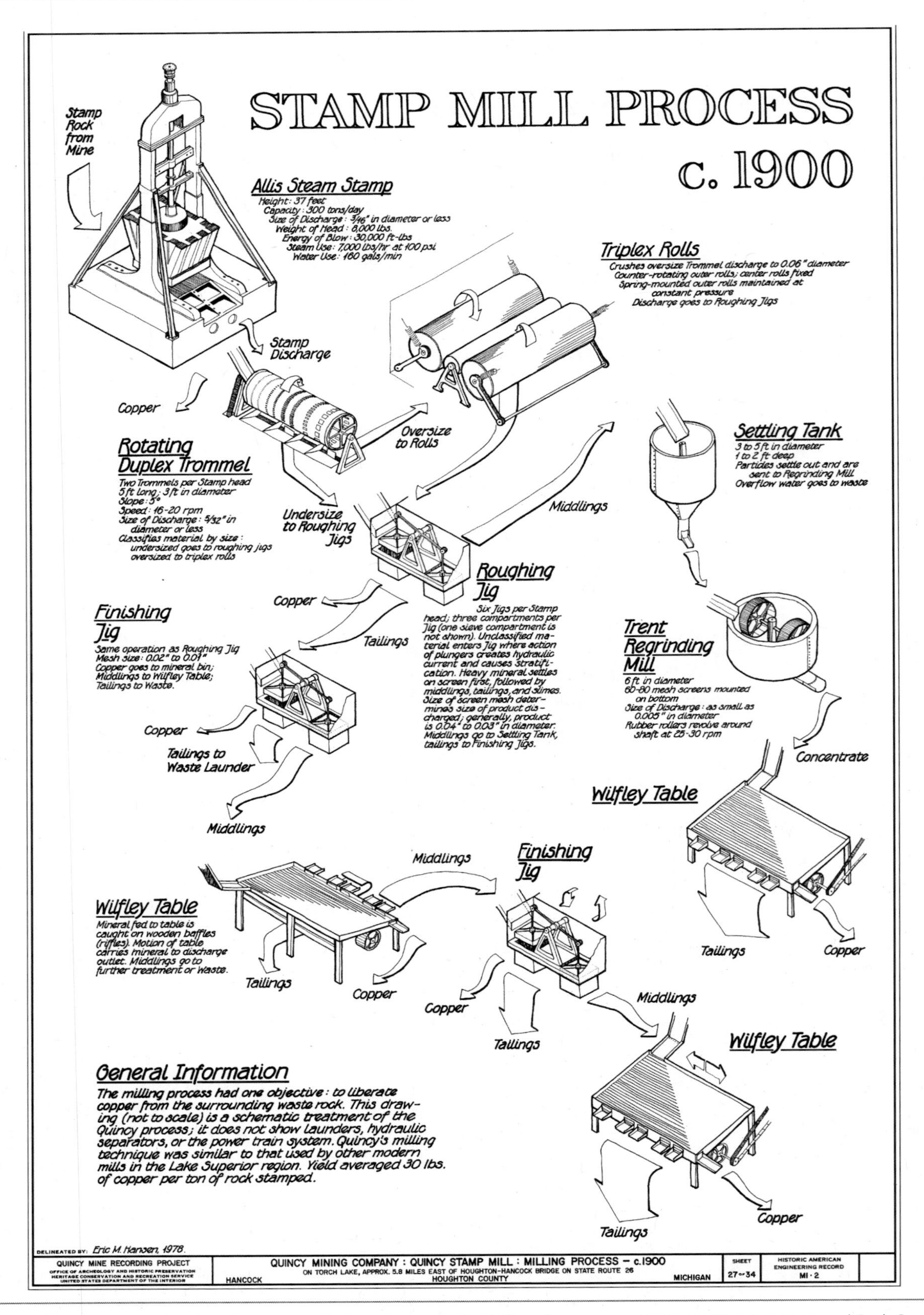

Figure 13.11. Quincy Mining Company: Quincy Stamp Mill: Milling Process ca. 1900. HAER MI-2, National Park Service. Delineated by Eric M. Hansen, 1978.

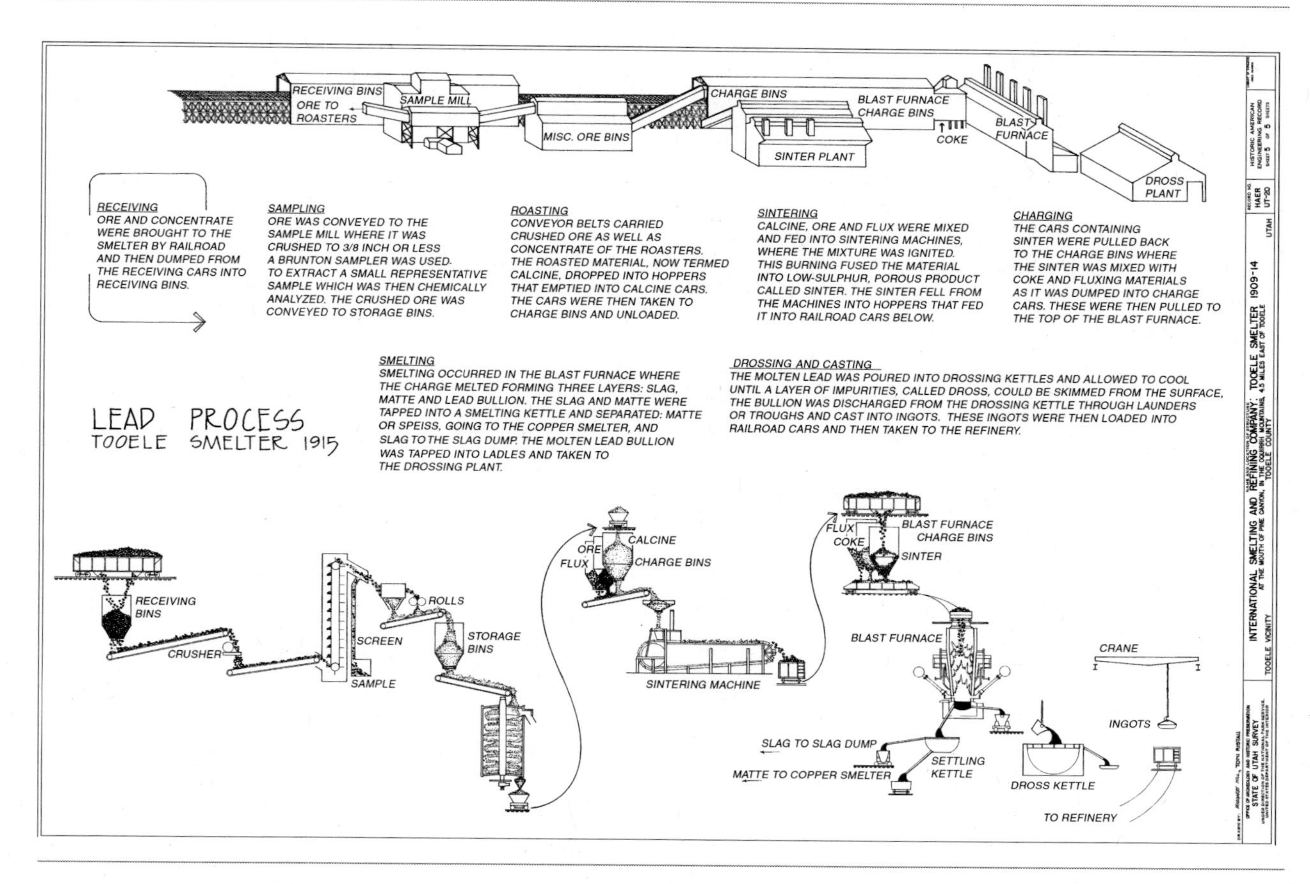

Figure 13.12. Diagram of International Smelting and Refining: Tooele Smelter, Lead Process, 1915. HAER UT-20. Delineated by Margaret Hill and Toni Ristau, n.d.

which tail water carries the waste from the concentrating process to tailing dumps or ponds.

In areas of particularly rugged topography, aerial tramways were sometimes used. These consisted of a continuous cable from which ore buckets hung suspended by towers and pylons.[39]

Housing

Often the most distinctive element of a mining district is housing. In her study of residential architecture of the mining town of Park City, Utah, Deborah Lynn Randall determined that the majority of the older housing could be classified as one of three main types: the rectangular miner's cabin, the T-plan cottage, and the pyramidal roofed cottage. Although these houses are found in many parts of the Midwest and West, a concentration is an indication of a mining community.

Company towns present a different picture. Typically these communities contained rows of standardized housing lacking high-style ornamentation. Company housing varied from small cottages to large and long attached row houses.[40]

Iron Mining and Processing

While in the mid-1990s, the Lake Superior area of Minnesota and the Upper Peninsula of Michigan had a near monopoly on United States iron ore production, the initial mining of iron in the eighteenth and nineteenth centuries was much more widely dispersed, with iron mines located in proximity to iron processing facilities.

Early iron mining in locations such as New Jersey and Pennsylvania took place in either open pit or underground mines. When magnetic testing revealed deposits of iron ore close to the surface, open pit or trench mining was used. The surface rock and soils were removed, and the underlying deposit was worked in levels or terraces. The side walls of the pits or trenches were sloped inward or braced with timbers to prevent the rock from crumbling and the pit from collapsing.

Early underground mining relied on black powder for blasting, and picks and shovels for gathering the loosened ore. Drilling into the body of the ore was done either by a miner using a short hammer and a short, hardened steel drill, or by two miners, one with a long drill and the other driving with a long hammer.

Several methods were used to remove the ore from an underground mine. The simplest methods were by human or animal power, carrying the ore and waste rock up a ladder, or using a hand-operated winch or windlass. Some mines used wood or metal cars powered by men or animals on narrow-gauge iron rails. Later steam engines replaced the hand power. Eventually the headframe came to predominate. A drum or sheave was located at the top of the headframe, over which passed a rope, cable, or chain. An engine pulled the cable to raise or lower a cage for miners, ore, and/or equipment.

After the ore was mined, mechanical crushing and grinding methods such as a stamping mill were used to reduce the size of the ore and remove waste rock. The ore was also roasted and heated to drive off impurities. These operations were usually performed near the site of the blast furnace rather than at the mine.

Large-Scale Iron Mining

Iron mining in northern Michigan and Minnesota began in the mid-nineteenth century, eventually becoming concentrated in the Mesabi Range of Minnesota and the Marquette and Menominee ranges of Michigan and the Gogebic Range of Michigan and Wisconsin. As in eastern mines, the early mines were small, shallow ore pits which were often soon exhausted. By 1875, underground mining had begun.

The first Michigan iron ores to be mined were soft ones, hematites, and they were mined near the surface. These were soon exhausted, and later mines were developed up to 4,000 feet deep. An underground iron mine, like other types of underground mines, was topped by a headframe or shafthouse at ground level.[41]

At the Cliff Shaft Mine in Ishpeming, Michigan (Figure 13.13), reinforced concrete headframes were built in 1919 to replace two badly deteriorated wooden structures. The structures, designed by the Condron Company with architect George W. Maher as consultant, have a rectangular base and taper upward, rising 88 feet 9 inches to the eaves and 96 feet 9 inches to the tip of the roof.[42]

Copper Mining

Michigan dominated United States copper production in the late nineteenth century. Output reached 50 million pounds per year by the early 1880s, when the state contributed 75 percent of the nation's production. By the 1960s, Michigan copper mining had basically ceased. Copper mining in the state took place in a belt at the western end of the Upper Peninsula.[43]

Copper mining in Michigan used deep shaft mines. Surface equipment of these mines included the headframe; a hoist house, first equipped with a steam-powered hoist but often later electrified; a boiler house to supply steam to the hoist; a machine shop and/or blacksmith shop; a dryhouse, where miners changed clothes;

Figure 13.13. Cliff Shaft Mine Headframe, Ishpeming, Michigan. HAER MI-3. Photograph by Jet Lowe, 1978.

and warehouse facilities.[44] Larger mines also included specialized facilities such as a drill shop, where drills were manufactured and sharpened; general offices; an engine house; and a gear house, where gears were fabricated and stored.

Production Processes

The production of copper includes three major stages: mining, milling, and smelting.[45]

At the turn of the twentieth century, copper was typically mined with shafts placed at intervals, with levels every 100 feet down. The mine shafts at Michigan's Baltic Mine were 9 feet high by 23 feet wide and were divided into three sections. Two of the sections were used to transport rock skips and man cars, while the third section contained utilities.[46]

Two copper mining complexes documented in the Historic American Engineering Record illustrate two methods of processing the rock gathered from the mine. In the Quincy Mine in Hancock, Michigan, the skip transported rock to the shaft-rockhouse (Figure 13.14).

In the rockhouse, a skip dumped the copper rock and mass copper where it first passed across grizzlies which separated out the smaller rock, while the larger rock was reduced in size with a drop hammer. The rock was then crushed and deposited into a stamp rock bin. From this bin, the rock was released into hopper cars, and from there transported to the stamp mill.[47]

Open-Cut Mining

By the mid-twentieth century, open-pit copper mining largely supplanted shaft mining (Figure 13.15). Mineable material consists of vast masses of low-grade ore containing only about 0.8 percent copper. Often the zones of ore are overlain by waste rock with little or no metal content. Mining involves both the disposal of waste rock and the recovery of copper ore. Pits are dug in the form of amphitheaters, with various levels or steps cut into the sides. On the top of these steps or benches, huge electric shovels eat into the walls of the pit which have been loosened by blasting. The bench tops serve as haulways to transport both ore and waste

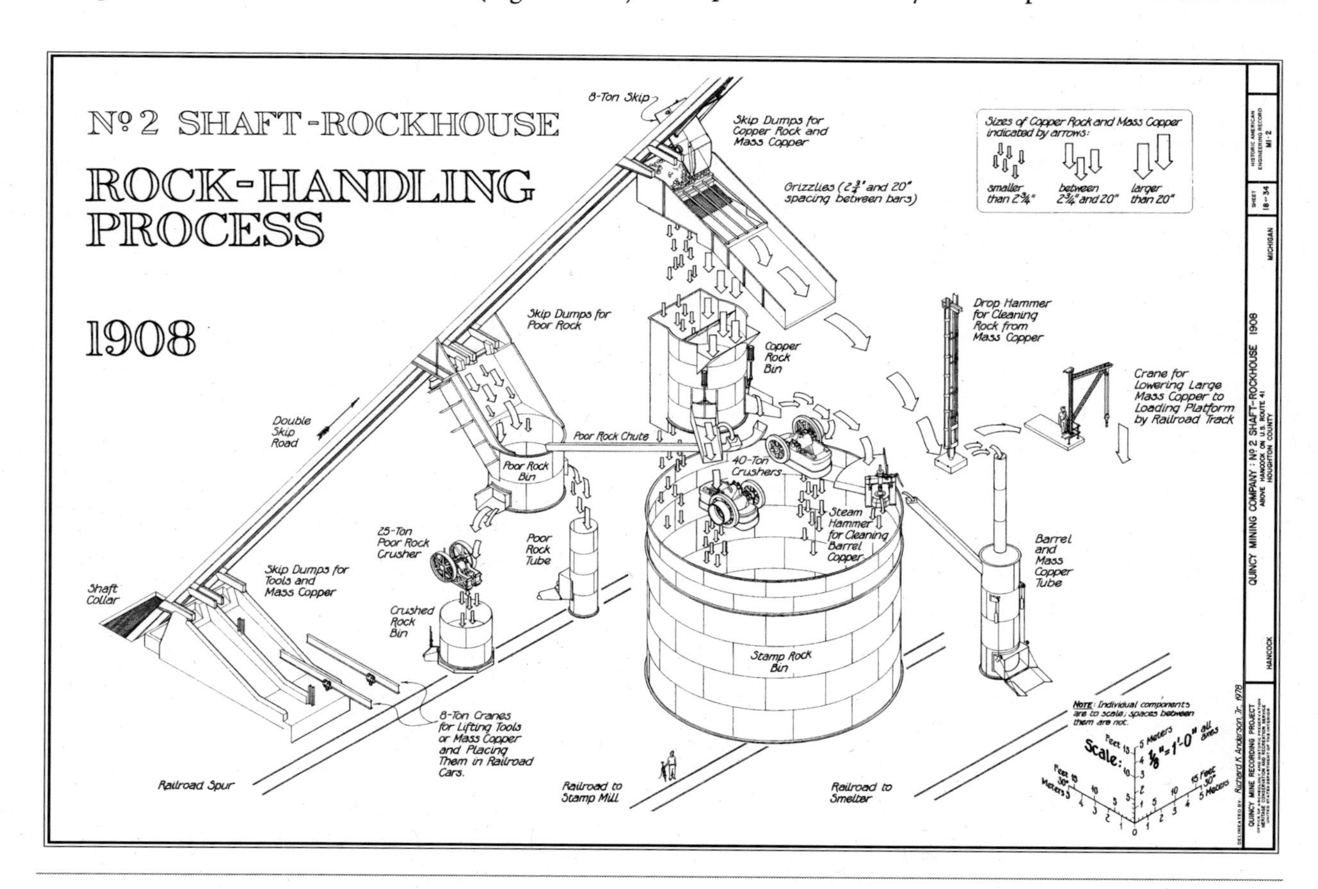

Figure 13.14. Quincy Mining Company No. 2: Shaft-Rockhouse, 1908. Hancock, Michigan. Rock-handling process. HAER MI-2, National Park Service. Delineated by Richard K. Anderson, Jr., 1978.

Figure 13.15. Electric shovel in a surface mine. Utah Copper Company, Bingham Canyon Mine, Bingham Canyon, Utah. HAER UT-21, National Park Service. Photograph by Jack E. Boucher, 1972.

and are connected by switchbacks or arranged in a spiral form around the periphery of the pit. Each bench is wide enough to carry the car tracks and catch any loose material that falls from the bench above, commonly about 50 feet. For metal mines, banks of up to 225 feet in height have been mined, but these banks can be a source of danger and are preferably replaced by several smaller banks, usually from 12 to 75 feet in height. The ore is delivered to the crusher and mill by electric train or diesel truck.[48]

An alternative to the power shovel is the dragline excavator (Figure 13.16). At the end of a long boom is a sheave over which a hoisting rope passes to a bucket, which is hauled toward the machine by a second rope, thus picking up its load like a drag scraper. As soon as the bucket is filled, it is lifted into the air; the excavator is turned until its boom is above the dump, and the bucket is discharged. The excavator digs below the level on which it stands. In large sizes the booms are from 40 to 250 feet long, and the buckets hold from 5 to 20 cubic yards.[49]

At the Kennecott Copper Mill in Kennicott, Alaska, mine tramways delivered rock to the concentration mill/leaching plant, a complex of wood-framed buildings that sprawled up a hillside. The first step was crushing. The tram delivered up to 1,200 tons of ore per day, which first passed through a grate, termed a grizzly, with 3.5-inch openings. Oversized ore was crushed with a hand-held sledge. Buchanan jaw crushers then reduced to ore to golf-ball size in preparation for the secondary crushers. High-grade ore was hand-picked out and sent by conveyor for direct shipment. The remaining ore was sent to a Symons disk crusher which reduced it to pebble size. The crushed ore was transported by elevator to the vibrating screens in the upper floor of the mill. Finer ore was screened for chuting to the gravity concentration department, while coarser ore was directed to the Traylor roller mill for re-crushing. This roller crusher continued the milling until sand was produced. Sands and gravels passed via the vibrating screens to the gravity concentration department.[50]

Copper Processing

In modern copper mills, large ore is first crushed to pieces less than an inch in diameter. These pieces are then ground in the concentrating mill to pieces less than 1/64 inch in size. Further grinding reduces the ore to the tiny size of copper-bearing particles. Water is added, and the particles in the resulting slurry are

Figure 13.16. Dragline excavator. Drummond Coal Company Cedrum Mine, Townley, Alabama. HAER AL-44, National Park Service. Photograph by Jet Lowe, 1993.

ground so fine that they are less than .0001 inch in diameter. The ore is then sent to the flotation section.[51]

Milling at Michigan's Quincy Mining Company liberated copper from the surrounding waste rock, which took place in the stamp mill. Rock entered the mill and initially passed through the Allis steam stamp, which used a 4-ton head to crush the ore. The discharge passed through the rotating duplex trammel that classified material by size. Oversized material passed to triplex rolls which crushed it further and discharged material with a diameter of 0.06 inch. This material and undersized material from the trammel then passed to the roughing jig, which stratified the input. Heavy minerals settled first, then middlings, tailings, and slimes. The material then passed to the finishing jig: copper then went to the mineral bin; middlings went to the Wilfley table; and tailings to waste. Mineral fed to the table was caught on wooden baffles (riffles). The motion of the table carried mineral to a discharge outlet, while middlings went for further treatment or to waste.[52]

The flotation department at Kennicott, Alaska (Figure 13.17) received 3 or 4 percent copper from the concentration tables. During the flotation process, reagents were added that removed more copper in a frothing action. Following settling, filtering, and drying, a concentration was produced that contained 32 to 35 percent copper, ready to be shipped for smelting.[53]

In mid-twentieth-century mills, pulverized ore went through the flotation section, where chemical reagents were added to the slurry. One chemical created bubbles in the mixture, while another coated the mineral particles and caused them to adhere to the bubbles. The bubbles, with some waste still attached, rose to the top of the flotation cells as copper concentrate containing from 15 to 35 percent copper. Material not floating to the top of the cell was called tailings and was disposed of as a waste product. To prepare the concentrate for smelting, most of the water was removed in thickening tanks and by filtration.[54]

Fluxing materials, lime and silica, were added to the copper concentrate, and the resulting mixture was charged into reverberatory furnaces. At a temperature of 2700 degrees F, chemical reactions induced by the flux occurred in the melted mass, and a fluid waste material, slag, formed and rose to the top. Heavier iron and copper sulfides settled to the bottom in an impure mixture called matte. The molten matte was charged into the converter, where silica was again added and air was blown through the hot liquid mass. Eventually nothing remained but about 98 percent pure molten copper, which was then conveyed to the refining furnace.

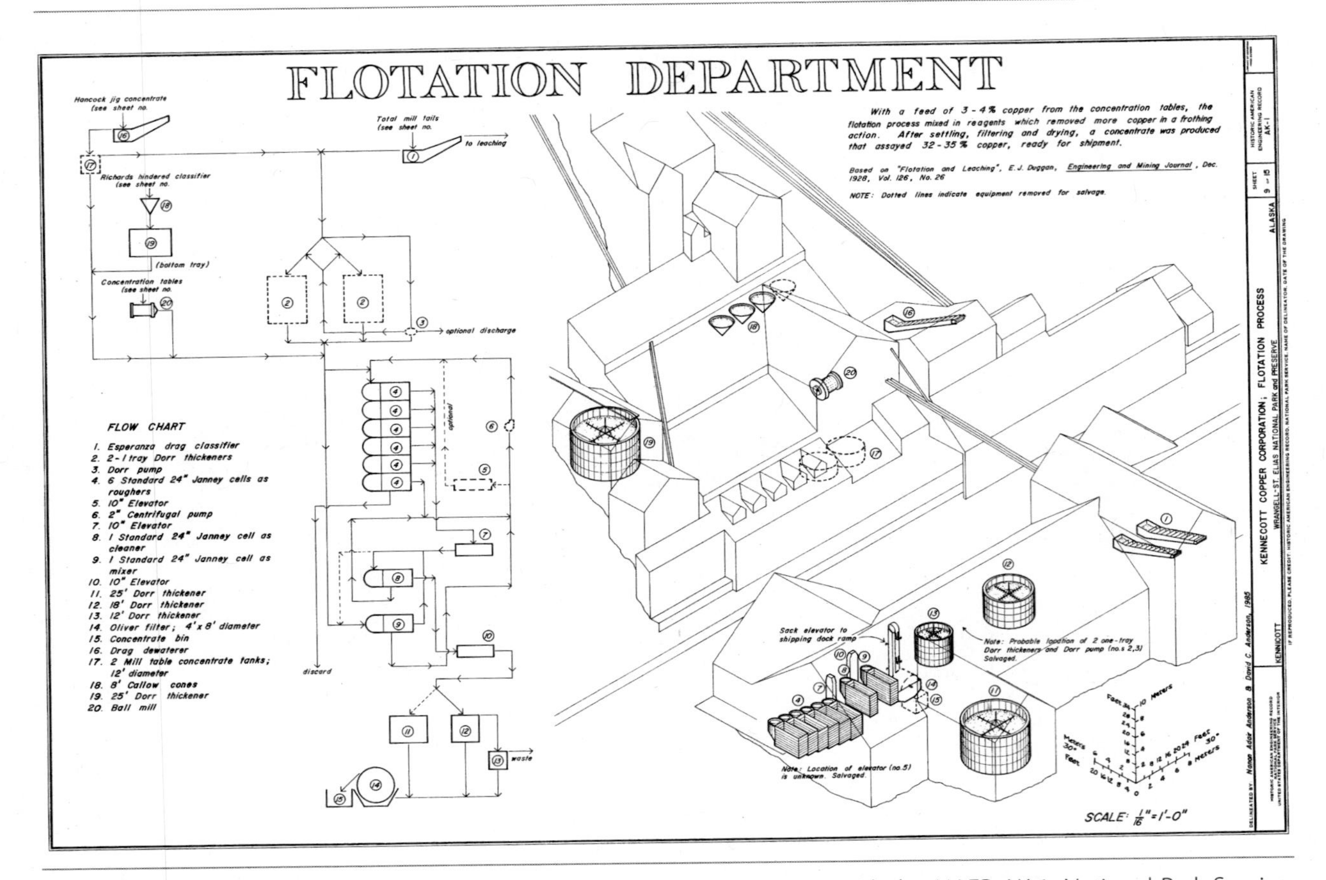

Figure 13.17. Kennecott Copper Corporation: Flotation Process. Kennicott, Alaska. HAER AK-1, National Park Service. Delineated by Nanon Adair Anderson & David C. Anderson, 1985.

Mining Waste

At the Lavender Pit in Bisbee, Arizona, a total of 380 million tons of material was removed, including 94 million pounds of copper ore, 111 million tons of leach material, and 175 million tons of waste rock. In Bisbee, waste material was disposed in three major areas. Waste rock or overburden was hauled to one side of the pit and dumped in a huge ridge. Material to be leached or piled so that water could be passed through it to remove further concentrations of copper was conglomerated into a leach dump. The third area was the tailings pond, where tail water (from which nearly all metals have been removed) was piped from the copper concentrator and dumped into huge ponds to evaporate. The tailings dams had terraces about 20 feet high. The light-colored, symmetrical terraces are distinctive features of mining regions of the western United States.

In lead and zinc mining areas, waste is concentrated in huge piles of chat, sandy wastes from the concentrating process that tower above the landscape. In coal mining areas, a characteristic feature is the culm bank or gob pile, consisting of slate and other waste rock that is separated from the coal. In metal mining districts, a characteristic feature is the slag pile or slag heap that results when ores are smelted. Often outlasting the smelters themselves, they remain as dark or black, steep-sided hills.[55]

Zinc and Lead Mining

Among the earliest important centers for zinc mining and production was the Ogdensburg area of New Jersey. The first zinc mines in the United States were opened in 1838 in Sussex County, and commercial mining began ten years later with the creation of the Sussex Zinc and Iron Mining Company, later known as the New Jersey Zinc Company. Mining was concentrated on Mine Hill in Franklin and Sterling Hill in Ogdensburg. Ores mined in the area consisted primarily of zincite, willemite, and franklinite.

By 1900, the United States was producing nearly one-third of the world's zinc, most of it from New Jersey mines. The world's largest zinc smelter was built in the Lehigh Valley of Pennsylvania to process zinc from New Jersey mines.[56]

New Jersey's Sterling Hill Mine, now a museum, was accessed by a 10-foot-wide horizontal adit that extended to a shaft. The adit was equipped with massive wood air doors. The shaft descended into the earth at a 53-degree angle and was divided into five compartments, two of which carried skips filled with ore and two which carried 40 men and materials. The remaining compartment held cables, pipes, and ladders. The shaft connected 18 levels approximately 100 vertical feet apart.

To free the zinc ore, a series of holes were drilled into a tunnel wall and were filled with explosives for blasting. After the ore was freed, slusher machines scraped the fallen ore into a 4-foot-diameter shaft sunk into the tunnel floor.

The shaft dumped ore into a car on the level below, and then the car dumped the ore into the orepass, an angled pit that ended with a rock crusher. A huge mesh-like screen, colloquially known as a grizzly, caught the larger rocks that might clog the crusher. The crusher, which was located more than 1,000 feet underground, pulverized the ore, and then dropped it to a lower level into skips hauled to the surface.[57]

By the mid-twentieth century, New Jersey's Zinc mine had been depleted. Mines remained active in New York, Virginia, Tennessee, Kansas, Missouri, and Oklahoma, as well as in Idaho, Utah, and Washington State, among others.[58]

Zinc and Lead Processing

Since ores removed from the ground contained only relatively small percentages of zinc, the metal-bearing portions had to be separated from waste rock prior to further treatment. As with other metals, this was done by milling, followed by either gravity concentration or flotation.[59]

When delivered to the smelter, concentrate contained zinc in the range of 48 to 60 percent. Depending on the process used and the character of the concentrates, the percentage recovered could be over 90 percent. Most zinc concentrates contain zinc in the form of sulfide, and the concentrate first had to be put through a roasting step in which most of the sulfur was burned. Zinc could then be recovered by either reduction with carbon or by an electrolytic process.

Carbon reduction requires the use of heat in excess of 1,000 degrees C. This process was undertaken by the horizontal retort process, the vertical retort process, or the blast furnace. In the electrolytic process, the zinc content of the roasted ore was leached out with dilute sulfuric acid. The zinc-bearing solution was filtered and purified, and zinc was recovered from solution using lead anodes and sheet aluminum cathodes.[60]

Lead Production

In the early twentieth century, extraction of lead from ore was based on three principles: reduction of lead oxide by carbon or carbon monoxide; the reaction between lead sulfide and lead sulfate or oxide; and the decomposition of lead sulfide by metallic iron. The first is known as the roast-reduction method, the second is the basis of the roast-reaction method, and the third is the basis of the precipitation method.

Lead smelting in the United States began with log and ash furnaces employed in Missouri prior to 1850. Subsequent generations of smelters were erected in the Missouri lead belt. Modern blast furnace smelting of silver-lead ore was begun chiefly in the reduction of the ore of a single mine or group of mines, as at Eureka, Nevada, Cerro Gordo, Colorado, and several places in Utah. The charges were generally rich in lead. The function of the blast furnace is to reduce the metal slag so that a separation can be made.

In smelting lead ore, the sulfur must be burned off and impurities must be combined in a slag. Several methods have been used: roasting, employing a hand-raked reverberatory furnace; roast-reaction; precipitation; and lime-roasting.[61]

WESTERN MINING SITES

Donald Hardesty, writing about Nevada hard rock mining sites, notes:

> Unfortunately . . . surviving technology and buildings are not common at mining sites. Rather they are rich in trash dumps, residential house foundations, privies, and other remains of the miners themselves.[62]

The illustration of the Homestake Mine site (Figure 13.18) depicts what is found at one Nevada hard rock mine site. Mine shafts associated with the lower rock dump are present at the site. More extensive

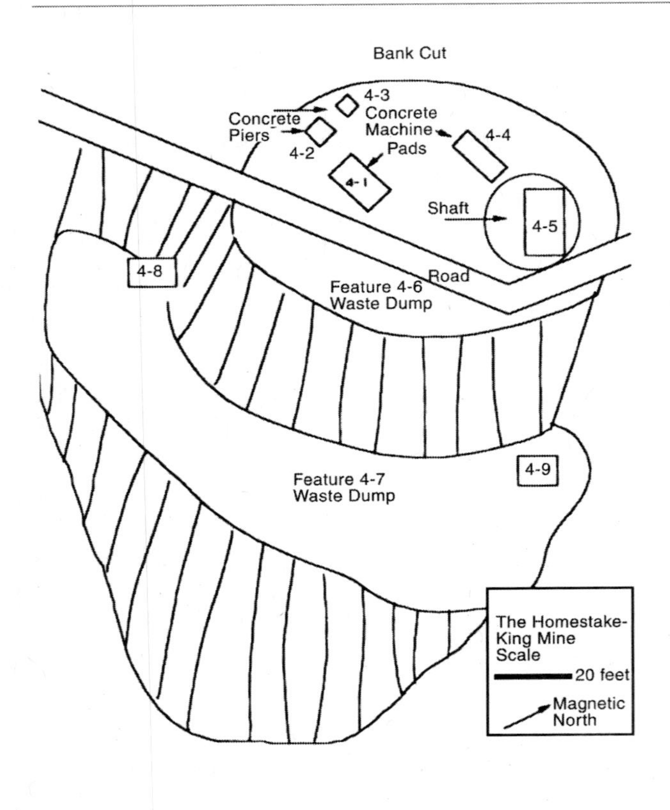

Figure 13.18. Homestake Mine Site. Reprinted with permission from Donald L. Hardesty, *The Archaeology of Miners and Mining: A View from the Silver State,* Special Publication Series No. 6 (Society for Historical Archaeology 1988), 36.

remnants may be found in other precious metal mining areas—for example, in California and South Dakota.

Gold and Other Precious Metal Mining and Processing

The initial technique used in the United States for recovery of gold was the placer method, in which gold eroded from ores close to the surface was found in riverbed gravel. The heavier metallic content of this gravel was recovered by panning. Placer gold ranges from nuggets to fine flakes.[63]

Dredging

River gravels and flat-lying placers of large extent were most suitable for dredging. The depth of gravel was generally not greater than 75 feet. The lower end of the digging ladder of a dredge was supported by steel cables passing over the sheaves on the front, from which they ran down to winding drums or winches by which the ladder was raised or lowered (Figure 13.19). Digging was done by a line of buckets which moved over rollers on top of the digging ladder. The material from the buckets fell into a hopper, and from the hopper, the material passed into a revolving screen, through the openings of which the gold and fine sand dropped to the gold saving tables below.[64]

Figure 13.19. Mining dredge. Solomon River Gold Dredge. Nome vicinity, Alaska. HAER AK-2, National Park Service. Photograph by Walter Smalley, Jr., 1987.

Hydraulic Mining

An elaboration of placer mining was hydraulic mining, which employed water under pressure to dislodge material or move sediment. It was developed in the region around Nevada City, California, in the mid-nineteenth century. Hydraulic mining often applied water under very high pressures, developed by bringing water from High Sierra locations to holding ponds several hundred feet above the surface to be mined. By the early 1860s, while hydraulic mining was at its height, it replaced small-scale placer mining. While generating great revenues, hydraulic mining was devastating to the riparian environment and agricultural areas of much of California.[65]

Initial underground exploration for gold occurred by driving tunnels and drift and sinking shafts. Shafts for exploratory work were usually sunk in the vein, cutting the vein transversely. Drifts extended horizontally through the vein. In the early twentieth century, excavation of exploratory shafts had begun to be supplanted by churn and core drills with regularly spaced holes.[66] In underground mines, workers selectively mined the highest-grade ore possible to maximize values per ton. Mine cars deposited crude ore into large bins from which it was fed into a mill.

Assaying

In silver and gold assaying, the initial step was to crush the ore into a powder. It was then mixed with a flux in a crucible, melted in a furnace, and poured into a cast-iron button mold. After cooling, the glassy slag outer surface was chipped off, leaving a silver-lead button. The button was cold-hammered into a cube, put into a bone-ash cupel, and heated in a muffle furnace. Upon melting, the lead was oxidized and changed to a lead oxide, which was absorbed by the bone-ash, leaving free silver or gold.

Assay houses were built close to the mill or mine but sufficiently distant to be unaffected by vibrations. Distinctive archaeological remains include a scatter of crucible and cupel fragments, slag and charcoal, and nails, window glass, and lumber. The site may also include a concrete pad on which the scales were placed.[67]

Milling and Concentrating

On-site milling was particularly advantageous because of the high monetary value of the metals, their location in remote regions, and the fact that they could be reduced to bullion by simple processes. In the typical California gold mill, a jaw crusher performed the initial task of breaking rock from the mine. The device pushed a movable plate toward a fixed plate, with ore introduced between the two plates. The repetition of this action broke the ore until it could pass through a gap between the plates.

For the fine crushing stage, a gravity stamp mill was used. This device, known to have been in use by the fifteenth century, was modified by the use of a sturdier frame construction that enabled the operators to increase the weight of individual stamps and the drop rate from 90 to 100 drops per minute.[68]

For contraction, three methods were commonly used in nineteenth- and twentieth-century gold mills: amalgamation, gravity concentration, and cyanidation. The amalgamation process extracted gold chemically by introducing mercury to crushed ore. The contact of free gold formed a "pasty" amalgam, an alloy containing approximately one-third gold by weight. Millwrights recovered the alloy from the rest of the material by dressing the mercury to a fixed metal surface on which the amalgam accumulated. Improved techniques included amalgam machines or the direct addition of mercury and water into crushing machinery.[69]

Gravity concentration separated mineral grains according to differences in specific gravity. In practice, most methods sorted particles by relative weight. Ore was introduced into a water trough equipped with a shaking bed. The oscillation of the bottom grate graded the feed, with heavier particles (including metals or metal-rich minerals) resting on the bottom. Shaking tables shook ore over a slightly inclined tabletop fitted with riffles (wooden slats). Heavier particles caught against the riffles and were shaken by the table's motion to one end of the table to be bagged as concentrate.[70]

Cyanidation recovered gold by its solubility in potassium- or sodium cyanide solution (Figure 13.20). In general operation, a cyanide mixture was introduced to the milled ore in a settling tank, which dissolved the gold and other metallic minerals into solution over a period of several days. The metal-rich solution was then siphoned from the tanks and precipitated using zinc or aluminum dust.[71] The concentration method selected for a mill rested largely on the character of the ore and available finances. Frequently, a combination of methods was used, typically amalgamation and gravity concentration or amalgamation and cyanidation.[72]

Figure 13.20. Closeup of cyanidation tubs, Donovan's Mill, Nevada. Silver City, Nevada. HAER NV-3. Photograph by Marty Stupich, 1980.

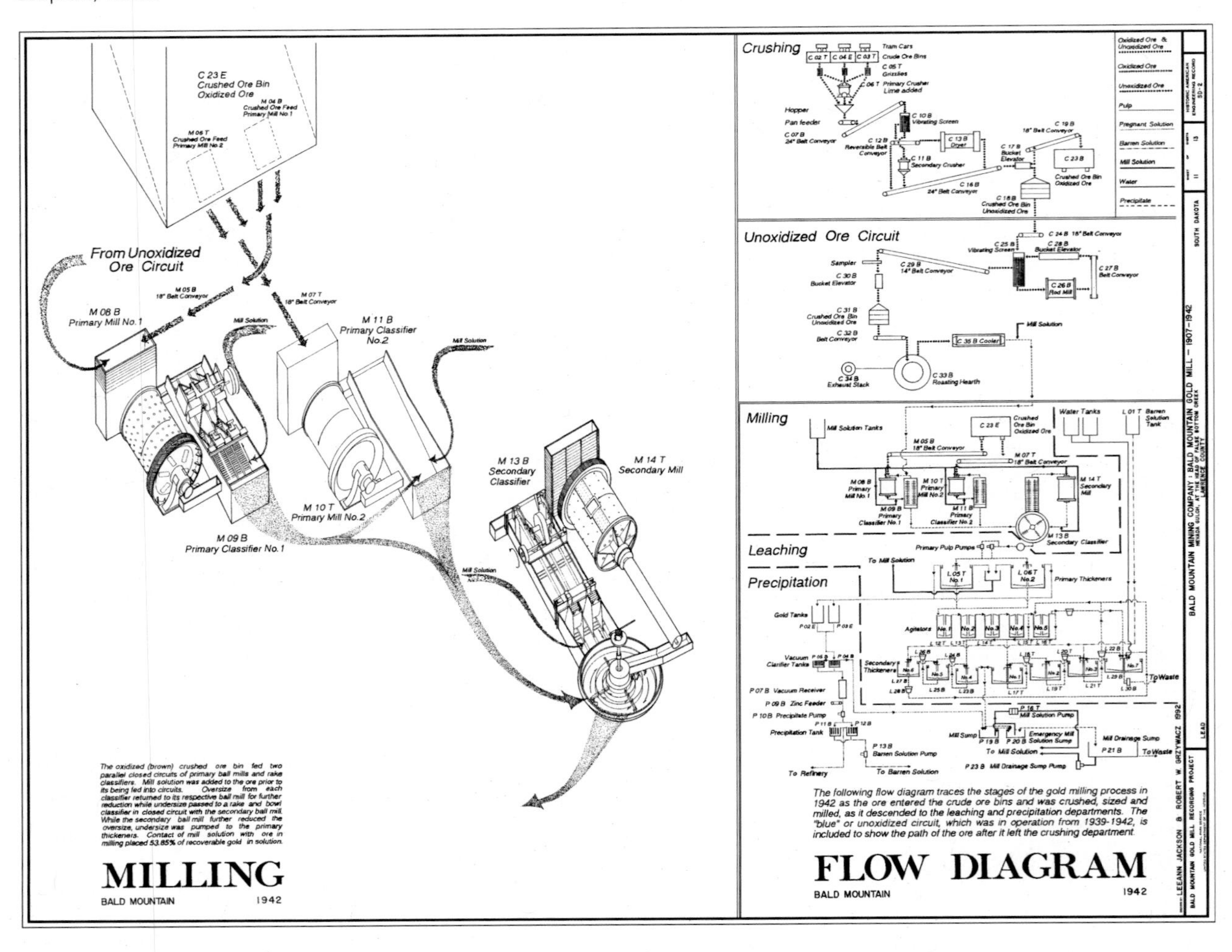

Another process, chlorination, was used in concentration in the Black Hills of South Dakota. In this process, developed in Germany in the mid-nineteenth century, finely crushed ore was mixed with chlorine and sulfuric acid diluted in water, and placed in chlorination barrels. Chlorine gas was produced and gold was dissolved. The resulting gold precipitate was separated from the chlorine solution and placed in filters, where it was pressed until the fluids passed through and solids were deposited in the bags.[73] This process, from milling to precipitation, is graphically shown in Figure 13.21.

Flotation

By the twentieth century, technology had been developed to separate several metals from raw ore through a process known as flotation. Flotation was developed as a method to process low-grade ore after many of the richest deposits had been tapped out. It is a process in which ore is immersed in a water-and-chemical solution and valuable minerals are floated away from the other material, assisted by their difference in weight.

In the 1860s, it was established that ore ground into fine particulate and mixed with water and oil would attach to the oil. When agitated, the minerals separated out more readily. In a modern flotation mill, a metallurgist crushed a sample of the ore, analyzed the content, and worked out the optimum chemical process to extract the metals.

Flotation circuits were tailored to specific metals: zinc, copper, lead, gold, and silver. A circuit could be adjusted to isolate one mineral one day and another the next. The raw ore arriving from the mine was crushed into small particles and then mixed with water. This mixture was sent through a series of jigs, with the heavier lead and gold settling out. The rest continued on to the flotation cells: long rectangular troughs equipped with agitators. A mineral-rich froth formed at the top.[74]

RESEARCHING THE MINING DISTRICT

In researching the history of a former mine or district, several types of sources are frequently used.

Mine History Collections

Although mining collections are available at archives throughout the country, a few collections are particularly notable. These include the Russell L. and Lyn Wood Mining History Archives at the Colorado School of Mines, the Michigan Technological University Archives and Copper Country Historical Collections, and the Anaconda Collection at the American Heritage Center, University of Wyoming. Information about these and other collections is available online.

In addition, mine museums, such as Pennsylvania's Anthracite Heritage Museum, and local and county historical societies in mining regions have at least limited collections of materials about their specific types of mining.

Graphic Documentation

As indicated by Alan Alanen, some of the earliest depictions of northern Michigan communities are artist's sketches and reporters' verbal descriptions of mining villages that appeared in the pages of *Harper's New Monthly Magazine* in the 1850s.[75] During the remainder of the century, *Harper's* and other publications included articles that depicted mines and quarries in many parts of the United States. As indicated elsewhere in this volume, the nineteenth-century issues of *Harper's* have been digitized and are available on Cornell University's *Making of America* website. Other nineteenth-century periodicals are indexed in *Poole's Periodical Index*.

The standard periodical sources on mining technology include the *Mining and Scientific Press* and *The Engineering and Mining Journal*. A useful textbook on mining technology is Peele's *Mining Engineers Handbook* (1918).[76]

During the late nineteenth and early twentieth centuries, street scenes and mining activities in mining communities proved a popular subject for professional photographers. Many collections of these images are available in university special collections, local history collections of public libraries, and state and local historical societies. One such collection, containing many images of mining in Colorado, in the collection of the

(ON FACING PAGE): **Figure 13.21.** Bald Mountain Mining Company: Bald Mountain Gold Mill, showing milling, leaching, and precipitation process, 1907–1942. HAER SD-2. Delineated by Lee Ann Jackson & Robert W. Grzywacz, 1992.

Denver Public Library, may be accessed through the American Memory website of the Library of Congress.

An online source of copyright-free images of mines (among other subjects) is the United States Geological Society Earth Science Photographic Library Archive, available at the USGS website. A second source of copyright-free historic images of some facets of mining is the Archival Research Catalog of the National Archives and Records Administration. This database includes a substantial number of digitized historic images, and searches may be limited to digitized images. Another source of historic mine photographs, emphasizing Colorado mines, is the Wood Mining History Archives at the Colorado School of Mines.

Maps

A primary source of information about mine and former mine locations is current and historic USGS topographic maps. Current and historic topographic maps for specific states are often found at state libraries, research university libraries, and major public libraries. United States Government Document Depository Libraries have a wider geographic range of topographic maps but may not retain historical editions.

On the current USGS map series, the familiar cross pickax symbol indicates a quarry or open-pit mine. Earlier maps used this symbol to indicate any kind of mine. A gravel, sand, clay, or borrow pit is indicated by crossed shovels. A mine shaft is indicated either by a small cross or by a small, half-darkened square, depending on the date of the map. Tailings are indicated on standard edition maps by beige, tightly hachured, scabby symbols, and on new or replacement maps by finely dotted beige symbols. A mine dump is denoted by brown hachure marks. On provisional maps yet to show symbols,"mine dumps" or "tailings" may be written in.[77]

Other governmental maps include townsite surveys for some western states issued by the General Land Office and available at the National Archives in College Park, Maryland.[78]

Information about the proposed layouts of speculative mine community developments, companies, towns, and model villages is usually available on plat maps filed with deeds with the county recorder in most states and with the municipal clerks of Connecticut and Vermont.

Mine District Reports

Important information to be used in the compilation of a mine history is contained in mine reports issued by state and federal government agencies. For example, in Pennsylvania, the Pennsylvania Bureau of Mines issued annual reports beginning in the late nineteenth century which listed all of the active coal mines in the Commonwealth. Many of these reports emphasize statistics and mine injuries, but some indicate improvements made to a particular mine during the period of the report. Another source is the *Annual Report of the Geological Survey of Pennsylvania*.

On the national level, the USGS issued historic reports for entire mining districts. Nearly every mining district was the subject of at least one definitive report that described the geological structure and the development of mines. The reports were done by economic geologists writing when the districts were in production. Well-illustrated and well-written, these reports frequently describe mining and treatment of ores and may make reference to mining and milling techniques. Many contain panoramic photographs that indicate the various mines and geological formations. Citations for some of these documents, including USGS numbered series beginning as early as 1880, are indexed online at USGS's Publications Warehouse. Some publications, including some important historic reports, are available with full content, while bibliographic citations are given for other publications.

BUILDING STONES

The chief building stones quarried in the United States have included sandstone, limestone, granite, marble, and slate.

Limestone

Limestone has historically been used for three primary purposes in the United States: for the production of lime, as a flux in the production of iron and steel, and for building and roadmaking. Production of limestone for use as a flux has historically been concentrated in major steel- and iron-producing states such as Pennsylvania, Ohio, and Alabama, while limestone production for the remaining two primary uses has been geographically dispersed. In 1894, the leading producers of limestone for building were Illinois, Indiana, Ohio, New York, and Pennsylvania.[79]

Indiana, whose limestone graces many prominent buildings, established its first organized quarry in 1827 near Stinesville. In 1929, at the zenith of production, Indiana furnished 12 million cubic feet of dimension stone. In recent years, production has averaged about 2.7 million cubic feet.[80]

Sandstone

In the late nineteenth century, sandstone had a variety of uses in the United States: for building elements, such as window sills and lintels, wall cladding, and belt courses; for street work, including paving blocks and curbing; as an abrasive; in engineering structures such as canals and bridge abutments; and in miscellaneous materials such as firebrick, fluxing, and sand for glass. At that time, the leading states in sandstone production, based on value of production, were Ohio, New York, Connecticut, Massachusetts, and Missouri.[81]

Marble

During the late nineteenth century, the leading states in marble production were Vermont and New York, with lesser amounts produced in Tennessee, Maryland, Georgia, and other states.[82] The marble of Vermont was present in irregular beds, extending north and south, with the leading counties in order of importance being Rutland, Bennington, Franklin, and Addison. Most Vermont marble was bluish-gray in color.

Granite

In 1894, the leading producers of granite in terms of monetary value were Massachusetts, Maine, Rhode Island, Vermont, and Maryland. Much of late nineteenth-century production of granite was used for paving blocks.[83]

Slate

During the late nineteenth century, slate was used for roofing, billiard tables, mantels, floor tiles, steps, flagging, and in the manufacture of writing slates. Slate for the last-mentioned purpose was almost entirely produced in Pennsylvania and Vermont.[84]

Commercial quarrying in the Vermont-New York region became feasible when the railway was established in the area, about 1850. The first quarry workers were largely Welsh immigrants. Initial quarrying of slate from the region coincided with the introduction of eclectic architectural styles into the United States. The color varieties of the stone made it well suited to the polychromy that characterized much architecture of the period.[85]

Quarrying Building Stones

Initially limestone was cut in the quarry using two-man crosscut saws. These were subsequently replaced by gang saws. The introduction of the channeling machine enabled the industry in many states to double and triple production. Modern quarries typically use diamond belt saws to dimension quarry blocks.[86]

In quarrying sandstone, a channeling machine was sometimes used, but the primary method was deep drill holes made by a simple machine driven by cranks charged with heavy blasts of powder. Large blocks were freed from the quarry and then broken by picks and iron wedges.

In the late nineteenth century, building stones were quarried from open-pit quarries. The technique used to obtain the stone depended on its type and hardness. The principal object was to obtain large and well-shaped blocks with the least outlay of time and money.

Because of the greater hardness of granite, there was less to fear from the use of explosives than when quarrying either sandstone or marble. In Maine and Massachusetts quarries, no machinery was used other than the steam drill and hoisting apparatus. Drills were used to open a lewis hole or a series of holes. They were then charged and fired simultaneously. In the Hallowell, Maine, quarries, the sheets of granite were separate from one another, and all that was necessary was to loosen the blocks from the quarry.

Beginning in the late nineteenth century, the Knox system of blasting began to replace the lewis hole in larger quarries. A round hole was first drilled to the required depth. Into this a reamer was drilled, producing V-shaped grooves at opposite sides to the entire depth of the hole. The charge was then inserted and, instead of driving the tamping down upon the top of the charge, an air space or cushion was reserved between the charge of powder and the tamping, giving the explosive the greatest possible chance for expansion. The force of the explosion was directed in the line of the grooves, and no shattering of rock occurred.[87]

In opening a marble quarry after initial testing, the surface material was stripped by blasting. Derricks were then placed into position, and channelers, drills, and gadders began to cut the stone. A channeler cut two grooves or channels across the grain of the stone about 5 feet apart. The stone separated from the rest was called the key course.

Slabs of slates were raised from the quarry and cut into rough blocks with a circular saw. The stock was then split into individual slates while the stone was kept moist. Splitting occurred by dividing the block along the cleave, which is parallel to its length. Division of the block continued until the pieces reached the desired thickness. A second worker trimmed the slates or dressed the sheets with a machine equipped with a rotary blade. The worker took a slate and rapidly trimmed two adjacent edges with what would be the exposed side of the roofing slate facing down, so that the beveled edge produced by trimming would show on the roof. A final step was to punch the slates for nail holes using a foot-powered puncher.[88]

Figure 13.22. Clark & McCormack Quarry. Derrick. Rockville, Minnesota. HAER MN-48, National Park Service. Photograph by Jet Lowe, 1990.

Equipment and Facilities

Equipment in a typical quarry included machinery to lift the stone and a means to transport it to sites where the stone was worked, usually termed quarry sheds.

Stones were typically lifted by a crane or derrick, often the quarry crane (Figure 13.22). This machinery consisted of a central mast, supported by strong guys or backstays, and bearing a movable beam, the jib, that carried the pulley block, the power being applied to the winch at the base of the mast. In the mid-nineteenth century, the crane was improved by a method that allowed the weight and the jib to be operated in unison. By the late nineteenth century, the crane was further improved by compound gearing, in which the effective radius of the crane could be altered by raising or lowering the jib. These cranes were capable of carrying loads of half a ton to 10 or 12 tons. These cranes had the advantage of being portable.

For quick action and heavy loads, a steam derrick crane was preferred to the quarry crane. By the early twentieth century, electric cranes had been introduced in some quarries. Several other types of cranes were frequently used in the stoneyard or workshop. The gantry crane, used in the yard, consists of a bridge supported at either end and on movable piers. The overhead traveling crane consists of a beam atop which rode a movable carriage on tracks.[89]

The stone was transported to the processing facility by one of various methods. Some quarries used narrow-gauge railways, while others used hoist conveyors, cableways, gravity inclines, or aerial tramways.

The character and size of the quarry shed varied depending on the quarry. For example, at a late nineteenth-century quarry on Massachusetts's Cape Ann, the sheds were located near the water's edge. Of wood construction, some of them were open on one side, some of them with doors and windows of cotton cloth.[90] A typical Vermont granite industry shed was that erected for the Grearson & Lane Company in Barre in about 1900. The timber-framed shed rested on a concrete block foundation and was of timber-frame construction. It was sided in clapboard. The main block of the building was a rectangle that measured 60 feet by 194 feet. Equipment included cutting lathes, polishing lathes, a grindstone, saws, boring machines, and a diamond core drill.[91] In the same city sat the unusual 16-sided wood-framed shed with a cupola built for E. L. Smith and Company in 1889.[92]

Cutting

In the early twentieth century, a small pneumatic tool was typically used to make a small channel across the stone in the direction in which a split was to be made. The stone was then placed upside down in a frame, and a breaking hammer with a rounded head, operated by a foot lever, gave the stone a few light blows and then a heavy blow by which the stone was split along its hem. The stone was then taken to the cleaving hammer, placed in right angles to the previous position, and then struck with another blow.

By the early twentieth century, motor-operated saws had been introduced to facilitate stone cutting. One type, a gang saw, was developed by the F. R. Patch Manufacturing Company of Rutland, Vermont. Another saw, designed to cut granite, the Chase saw, had been introduced by the American Granite Saw Company. In this design, the cut, instead of being a straight line, was composed of arcs. Yet another design was the circular diamond saw, also manufactured by the F. R. Patch Company. Later in the twentieth century, these blade saws had been largely replaced by wire saws (Figure 13.23).

Another method used by stonecutters to shape blocks was to split the block to about the right size by the plug and feather method. The block was then brought to a plane surfaced on one side, accomplished by knocking off overhanging edges and projections with a spalling hammer or spalling tool. Drafts or ledges were then chiseled along two opposite edges. The whole face was then worked down to this plane with tools necessary for the required fineness of finish. The point was used for removing rougher projections, and this was followed by the peen hammer. If a smoother surface was required, it was made by bush hammering.

Another step employed with some stones was planing. In the early twentieth century, a circular planer was introduced that could plane a block measuring 4 feet high and 6 feet 10 inches wide.[93]

Figure 13.23. Clark & McCormack Quarry. Wire saw. Rockville, Minnesota. HAER MN-48, National Park Service. Photograph by Jet Lowe, 1990.

Polishing

After bush hammering, the block was transported to the shop or mill to receive further smoothing. The surface to be worked was arranged horizontally and ground smooth with an abrasive material mixed with water and moved about by a revolving iron or steel disk perforated with holes or made of concentric rings. The disk revolved by an upright shaft, and power was communicated by a main shaft running overhead. The abrasive material used was either chilled iron globules, steel emery, or crushed steel.

By the early twentieth century, mechanical stone polishing had become the norm, with machinery such as Barre granite surfacing and polishing machines and lighter Concord surfacing machines.

Finishing

Granite for columns, balusters, round posts, and urns was worked primarily in lathes, some of which were made large enough to handle blocks 25 feet long and 5 feet in diameter. Instead of using sharp cutting instruments, as in a wood lathe, granite was turned or ground away by the wedge-like action of thick steel disks.

The final step in preparing stone for shipment was preparation of the surface. Some of the more common surface treatments are shown in Figure 13.24. Rock face keeps the natural face of the rock as broken from the quarry but is slightly trimmed down by use of the pitching tool. Pointed face trims the natural face of the rock by means of a sharp-pointed tool called a point. Tooth chiseled is produced by means of a steel chisel with an edge toothed like a saw and has been used primarily on limestones, marbles, and sandstones. Square drove has a surface made with a wide chisel with a smooth edge, called a drove. This style has frequently been used as a border of a rock face or pointed surface. Patent hammered stone is produced by striking blows on the smooth surface with a rough-faced implement called a patent hammer.

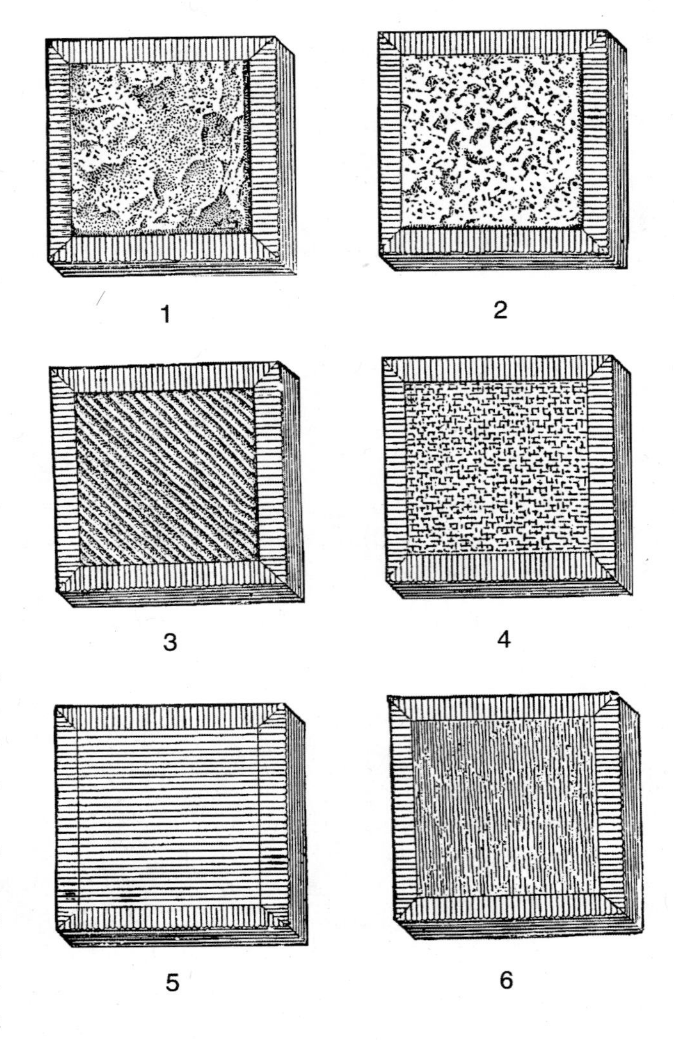

Figure 13.24. Surface treatment of stones. ***1,*** rock face; ***2*** and ***3***, pointed face; ***4,*** tooth chiseled; ***5***, square drove; ***6,*** patent hammered. Reprinted from George P. Merrill, *Stones for Building and Decoration* (New York: John Wiley and Son, 1891), Plate X.

GLOSSARY[94]

Active workings. Any place in a mine where miners are normally required to work or travel.

Adit. A tunnel driven horizontally into a hillside to provide access to a mineral deposit.

Air split. The division of a current of air into two or more parts.

Airway. Any passage through which air is carried.

Back. The roof or upper part of any underground mining cavity.

Bed. A stratum of coal or other sedimentary deposit.

Belt conveyor. A looped belt on which coal or other materials can be carried and which is generally constructed of flame-resistant material.

Boney. The broken waste that accumulates during the mining of coal. Called *gob* in some areas such as western Maryland.

Bottom. Floor or underlying surface of an underground excavation.

Brattice cloth. A fire-resistant fabric or plastic partition used in a mine passage to confine air and force it into a working place.

Breakthrough. A passage for ventilation that is cut through pillars between rooms of a mine.

Breast. The face or wall of a quarry.

Breccia. Fragmental stones, the individual particles of which are large and angular in form.

Cage. In a mine shaft, a device, similar to an elevator car, used for hoisting personnel and materials.

Car. A railway wagon, particularly a wagon adapted to carry coal, ore, and waste underground.

Car-dump. A device used for unloading a loaded railroad car.

Chain conveyor. A machine in which the material is moved along troughs by the action of scraper crossbars attached to powered chains.

Channeler. A quarrying machine capable of cutting slots in stone at any angle. It is used for cutting dimension stone off the quarry face without explosives.

Chat. The finely crushed gangue remaining after the extraction of lead and zinc materials.

Check curtain. Sheet of brattice cloth hung across an airway to control the passage of air current.

Coal washing. The process of separating undesirable materials from coal based on differences in densities.

Conglomerates. Fragmental stones composed of large, rounded fragments.

Continuous miner. A machine designed to remove coal from the face and load it into cars continuously without the use of cutting machines, drills, or explosives.

Crib. A mine roof support of prop timbers or ties, laid in alternate cross-layers, log-cabin style.

Cross entry. A mine entry running at an angle to the main entry.

Cupel. A small bone-ash cup used in gold or silver assaying with lead.

Cyanidation. A process for the extraction of gold from finely crushed ores, concentrates, and tailings by means of cyanide of potassium or sodium used in dilute solutions. The gold is dissolved by the solution and subsequently deposited upon metallic zinc or other materials. Also known as *leaching*.

Dimension stone. Natural stone that has been selected and fabricated to specific sizes or shapes.

Dragline. A large excavation machine used in surface mining to remove overburden covering a coal seam. The dragline casts a wire-rope-hung bucket a long distance, collects the dug material by pulling the bucket to itself on the ground with a second wire rope or chain, elevates the bucket, and dumps the material on a spoil bank, in a hopper, or on a pile.

Drift. A horizontal underground mine passage.

Drift mine. An underground coal mine in which entry or access is above water level and usually on the slope of a hill, driven horizontally.

Entry. An underground horizontal or near-horizontal passage used for haulage, ventilation, or as a mainway.

Face. The exposed area of a coal bed from which coal is being extracted.

Face conveyor. Any conveyor used parallel to a working face which delivers coal into another conveyor or into a car.

Feeder. A machine that feeds coal evenly onto a conveyor belt.

Flotation. A process for concentrating materials based on the selective adhesion of certain minerals to air bubbles in a mixture of water and ground-up ore.

Gadder. In quarrying, a small car or platform carrying a drilling machine in order to make a straight line of holes along its course to get out dimension stone.

Gravity separation. Recovery of gold or other heavy minerals from crushed rock or gravel by using the mineral's high specific gravity to separate it from lighter material.

Grip car. A flatcar with standard couplings, flanged wheels, and a pair of vise-like jaws used to haul mine cars over an inclined section of track.

Haulage. The horizontal transport of ore, coal, supplies, and waste.

Haulageway. Any underground entry or passageway that is designed for transport of mined material, personnel, or equipment, usually by installation of a track or belt conveyer.

Headframe. The structure surmounting a mine shaft that supports the hoist rope pulley.

Hoist. A drum on which the hoisting rope is wound in the engine house or hoist house.

Hoisting. The vertical transport of mine material.

Hoppit. A large bucket used in sinking a mine shaft for hoisting men, materials, and tools.

Incline. An entry to a mine that is neither vertical nor horizontal.

Intake. The passage through which fresh air is drawn or forced into a mine or a section of a mine.

Larry car. A mine car with high sides and a trapdoor on one side; used to move and dump mine refuse materials.

Lewis hole. An opening consisting of a series of two or more holes drilled as closely together as possible and then connected by knocking out the partition between them, thus forming one wide hole.

Main entry. A main haulage road.

Manway. An entry used exclusively for personnel to travel from the shaft bottom or drift mouth to the working section.

Menzies cone separator. A device consisting of a 60-degree cone with a short cylindrical top section provided with a stirring shaft located in its vertical axis, carrying several sets of horizontal arms with rings of nozzles projecting through the sides for the required admission of water currents. It is used to separate sizes of coal during processing.

Muffle furnace. A furnace in which heat is applied to the outside of a refractory chamber containing the charge.

Natural ventilation. Ventilation of a mine without the aid of fans or furnaces.

Open pit. A surface mine, open to daylight, such as a quarry.

Ore. A mixture of mineral and gangue from which at least one of the metals can be extracted at a profit.

Outcrop. An exposure of rock, coal, or mineral deposit that can be seen on the surface.

Overcast. An enclosed airway that permits one air current to pass over another without interruption.

Pillar. An area of coal left to support the overlying strata in a mine.

Placer. Deposits of ore or minerals that have accumulated in quantities of economic import through the natural processes of weathering and concentration.

Preparation plant. A place where coal is cleaned, sized, and prepared for market.

Prop. In a coal mine, a single post used as a roof support.

Return. The air or ventilation that has passed through all the working faces of a split.

Riffle. The raised portions of the deck of a concentrating table that serve to trap the heaviest particles.

Run-of-mine ore. Uncrushed ore as it is retrieved from a mine without processing.

Seam. A stratum or bed of coal.

Shaft. A vertical passageway to an underground mine for moving personnel, equipment, supplies, and material.

Short-wall. An underground mining method in which small areas are worked by a continuous miner with the use of hydraulic roof supports.

Skip. A guided steel hoppit, usually rectangular, used in vertical or inclined shafts for hoisting coal or minerals.

Slag. The waste product of smelting.

Slope mine. An underground mine with an opening that slopes upward or downward to the coal seam.

Split. Any division or branch of the ventilating current, or the workings ventilated by one branch.

Surface mine. A mine in which the material lies near the surface and can be extracted by removing the covering layers of rock and soil.

Tipple. Originally the place where mine cars were tipped and emptied of coal; now more generally used to designate the surface structures of a mine, including preparation plant and loading tracks.

Undercut. To cut below or undermine a coal face by chipping away the coal by pick or mining machine.

Underground mine. A mine generally several hundred feet below the earth's surface in which material is removed mechanically and transferred by shuttle car or conveyor to the surface.

Upcast shaft. A shaft through which air leaves a mine.

Working face. Any place in a mine where material is extracted during a mining cycle.

Workings. The entire system of openings in a mine for the purpose of exploitation.

REFERENCES

Alanen, Arnold R. "Documenting the Physical and Social Characteristics of Mining and Resource-Based Communities." *Bulletin of the Association for Preservation Technology* 11, no. 4 (1979): 49–68.

American Zinc Institute, Inc. *Zinc: A Mine to Market Outline.* New York: American Zinc Institute, 1958.

Benedict, C. Harry. *Lake Superior Milling Practice: A Technical History of a Century of Copper Milling.* Houghton, MI: Michigan College of Mining and Technology Press, 1955.

Bergstresser, Jack. Kay Moor Coal Mine. Historic American Engineering Record, HAER No. WV-38.

Brown, Sharon A. *A Historic Resource Study; Kaymoor, New River Gorge National Park, West Virginia.* Washington, DC: U.S. Department of the Interior, 1990.

Cooley, A. B. "Machine Shop, Bath House, Houses and Other Accessory Buildings at New Gallup American Mine." *Coal Age* 24, no. 12 (20 September 1923): 427.

Crane, Walter R. *Gold and Silver.* New York: John Wiley & Sons, 1908.

Crawford, Margaret. *Building the Workingman's Paradise: The Design of American Company Towns.* New York: Verso, 1995.

DiCiccio, Carmen. *Coal and Coke in Pennsylvania.* Harrisburg: Pennsylvania Historical and Museum Commission, 1996.

Eaton, Lucien. *Practical Mine Development and Equipment.* New York: McGraw-Hill Book Company, Inc., 1934.

Enman, Aubrey. "The Shape, Structure and Form of A Pennsylvania Company Town." *Proceedings of the Pennsylvania Academy of Science* 42 (1968).

Fields, Richard A. *Range of Opportunity: A Historic Study of the Copper Range Company.* Hancock, MI: Quincy Mine Hoist Association, 1997.

Flanagan, Joe. "Mining Majesty: Seeking the Silver Lining in a Prospecting Past." *Common Ground* (Spring 2006): 26–37.

Francaviglia, Richard V. *Hard Places: Reading the Landscape of America's Historic Mining Districts.* Iowa City: University of Iowa Press, 1991.

Greenwell, Allan, and J. Vincent Elsden. *Practical Stone Quarrying, A Manual for Managers, Inspectors, and Owners of Quarries, and for Students.* New York: D. Appleton & Company, 1913.

Hall, R. Dawson, and J. H. Edwards. *Coal Age Mining Manual.* New York: Coal Age, 1934.

Hamill, R. H. "Design of Buildings in Mining Towns." *Coal Age* 11 (16 June 1917): 1045–1048.

Hardesty, Donald. *The Archaeology of Mining and Miners: A View from the Silver State.* Special Publication Series No. 6. Society for Historical Archaeology, 1988.

Hess, Demien. Eureka No. 40. Historic American Engineering Record, HAER No. PA-184. 1988.

Huebner, A. F. "Houses for Mine Villages." *Coal Age* 12 (27 October 1917): 717–720.

Hyde, Charles K., director. *The Upper Peninsula of Michigan: An Inventory of Historic Engineering and Industrial Sites.* Washington, DC: Historic American Engineering Record, 1978.

Ingalls, Walter Renton. *Lead and Zinc in the United States.* New York: Hill Publishing Company, 1908.

International Correspondence Schools. *A Textbook on Metals Mining.* Scranton, PA: International Textbook Company, 1899.

—. *Preparation of Coal...Surface Arrangements at Bituminous Mines, Coal Washing, Manufacture of Coke. . . .* Scranton, PA: International Textbook Company, 1924.

Janosov, Robert A. Huber Coal Breaker (Ashley Breaker). Historic American Engineering Record, HAER PA-204.

Kennecott Copper Corporation. *All About Kennecott: The Story of Kennecott Copper Corporation.* New York: Kennecott Copper Corporation, 1962.

Lewis, Robert S. *Elements of Mining.* New York: John Wiley and Sons, Inc., 1933.

Merrill, George P. *Stones for Building and Decoration.* New York: John Wiley & Sons, 1891.

Moore, Elwood S. *Coal: Its Properties, Analysis, Classification, Geology, Extraction, Uses and Distribution.* New York: John Wiley & Sons, Inc., 1922.

Mulrooney, Margaret M. *A Legacy of Coal: The Coal Company Towns of Southwestern Pennsylvania.* Historic American Buildings Survey/Historic American Engineering Record. Washington, DC: National Park Service, 1989.

Park, John R. *Missouri Mining Heritage Guide.* South Miami, FL: Stonerose Publishing Company, 2005.

Parrish, Wanda C., Connie Torbeck, and Mary Daughtrey. Tower Hill No. 2 Mine. Historic American Engineering Record, HAER No. PA-424.

Reps, John W. "Bonanza Towns: Urban Planning on the Western Mining Frontier." In *Patterns and Process: Research in Historical Geography,* edited by R. H. Ehrenberg. Washington, DC: Howard University Press, 1975.

Reynolds, Terry. "Iron in the Wilderness: The Michigan Iron Industry Museum." *Technology and Culture* 30/1 (January 1989): 112–117.

Rohe, Randall. "The Geography and Material Culture of the Western Mining Town." *Material Culture* 16/3 (Fall 1984): 99–120.

Silva, Dorothy Allen. Marvine Collier. Historic American Engineering Record, HAER PA-183.

Staley, W. M. *Mine Plant Design.* New York: McGraw-Hill Book Company, Inc., 1949.

Stevens, Horace J. *Copper Handbook.* Houghton, MI: no publisher, 1900.

Torok, George D. *A Guide to Historic Coal Towns of the Big Sandy River Valley.* Knoxville: University of Tennessee Press, 2004.

U.S. Department of the Interior. National Park Service. *A Coal Heritage Study: A Study of Coal Mining and Related Resources in Southern West Virginia*. Philadelphia: National Park Service, 1993.

U.S. Geological Survey. *Sixteenth Annual Report of the United States Geological Survey*. Part IV. Washington, DC: U.S. Geological Survey, 1894.

White, Joseph H. *Houses for Mining Towns*. Washington, DC: U.S. Department of the Interior, 1914.

Zeier, Charles D."Historical Charcoal Production near Eureka, Nevada: An Archaeological Perspective." *Historical Archaeology* 21/1 (1987): 81–101.

NOTES

1 George D. Torok, *A Guide to Historic Coal Towns of the Big Sandy River Valley* (Knoxville: University of Tennessee Press, 2004), 61.
2 Torok, 63.
3 Jack Bergstresser, Kay Moor Coal Mine, HAER WV-38, 1990, 15.
4 Demian Hess, Eureka No. 40, Scalp Level, Pennsylvania, HAER PA-184, 1988, 32.
5 W. W. Staley, *Mine Plant Design* (New York: McGraw-Hill Book Company, Inc., 1949), 509.
6 Eureka No. 40, 32–33.
7 Eureka No. 40, 48.
8 Torok, 71–72.
9 Torok, 72–73.
10 Torok, 78.
11 Torok, 79–80.
12 Torok, 83.
13 Margaret M. Mulrooney, *A Legacy of Coal: The Coal Company Towns of Southwestern Pennsylvania* (Washington, DC: HABS/HAER, 1989,) 125–144.
14 Torok, 89.
15 Robert A. Janosov, Huber Coal Breaker, PA-204, 1991, 11–20.
16 Raymond P. Washlaski and Peter E. Starry, Jr.,"Coal Tipple & Coal Washer at Salem No. 1 Mine," http://patheoldminer.rootsweb.com/salem2.html.
17 Torok, 91.
18 National Park Service, Blue Heron Mining Community, Tipple and Bridge (http://www.nps.gov/biso/historyculture/blueheron.htm.
19 Eureka No. 40, 55–56.
20 Torok, 93.
21 Staley, 105–107.
22 International Textbook Company, *Preparation of Coal...Surface Arrangements at Bituminous Mines, Coal Washing, Manufacture of Coke* (Scranton, PA: International Textbook Company, 1924), 66–67.
23 Torok, 95.
24 International Textbook Company, 17–18.
25 International Textbook Company, 66:10.
26 International Textbook Company, 66:10–11.
27 Torok, 34–94; Elwood S. Moore, *Coal: Its Properties, Analysis, Classification, Geology, Extraction, Uses and Distribution* (New York: John Wiley & Sons, Inc., 1922), 318.
28 Donald Hardesty, *The Archaeology of Mining and Miners: A View from the Silver State*, Special Publication Series No. 6 (Society for Historical Archaeology, 1988), 27.
29 Lucien Eaton, *Practical Mine Development and Equipment* (New York: McGraw-Hill Book Company, Inc, 1934), 7–9, 12.
30 American Society of Mechanical Engineers, Quincy No. 2 Mine Hoist (1920) (New York: American Society of Mechanical Engineers, 1984), 2.
31 Staley, 279, 283.
32 Staley, 187.
33 Hardesty, 29–30.
34 Eaton, 326–327, 329.
35 Paul J. White, Skidoo Mine, Death Valley National Park, California, HAER No. CA-290, 9.
36 Richard V. Francaviglia, *Hard Places: Reading the Landscape of America's Historic Districts* (Iowa City: University of Iowa Press, 1991), 51, 53.
37 Francaviglia, 54.
38 Margaret Hill and Toni Ristau, International Smelting and Refining Company: Toele Smelterr, 1909–14. HAER UT-20, n.d.
39 Francaviglia, 55–58.
40 Francaviglia, 46–48.
41 Charles K. Hyde, director, *The Upper Peninsula of Michigan: An Inventory of Historic Engineering and Industrial Sites* (Washington, DC: Historic American Engineering Record, 1978), 31.
42 Hyde, 35–36.
43 Hyde, 1–2. Extensive remnants of the copper mining industry still exist in Michigan's Keewenaw Peninsula and vicinity. For information, consult the National Park Service personnel at Quincy Mines in Hancock.
44 Hyde, 4.
45 Richard A. Fields, *Range of Opportunity: A Historic Study of the Copper Range Company* (Hancock, MI: Quincy Mine Hoist Association, 1997), 15.
46 Fields, 22.
47 Richard K. Anderson, Jr., Quincy Mining Company: No. 2 Shaft-Rockhouse, 1908. HAER MI-2. 1978.
48 Kennecott Copper Corporation, *All about Kennecott: The Story of Kennecott Copper Corporation* (New York: Kennecott Copper Corporation, 1962), 9–10; Robert S. Lewis, *Elements of Mining* (New York: John Wiley & Sons, Inc., 1933), 242–244.
49 Lewis, 242–243.
50 Nanon Adair Anderson and David C. Anderson, delineators, Kennecott Copper Corporation: Crushing Department, Kennicott, Alaska. HAER AK-1, 1985.
51 Kennecott, 11.
52 Eric M. Hansen, Quincy Mineral Company: Quincy Stamp Mill, Hancock, Michigan, HAER MI-2, 1978.
53 Nanon Adair Anderson and David C. Anderson, Kennecott Copper Corporation: Flotation Process, Kennicott, Alaska, AK-1, 1985.
54 Kennecott, 11–12.
55 Francaviglia, 22–27.

[56] Andrea C. Dragon, "Zinc," in *Encyclopedia of New Jersey*, edited by Maxine N. Lurie and Marc Mappen (New Brunswick, NJ: Rutgers University Press, 2004), 893.
[57] Mary Jasch, Sterling Hill Mine: The Essence of Fluorescence, http://www.njskylands.com/atsterhill.htm.
[58] American Zinc Institute, *Zinc: A Mine to Market Outline* (New York: American Zinc Institute, Inc., 1958), 13–15.
[59] American Zinc Institute, 20.
[60] American Zinc Institute, 19–21.
[61] Walter Renton Ingalls, *Lead and Zinc in the United States* (New York: Hill Publishing Company, 1908), 37–42.
[62] Hardesty, 17.
[63] David Eve, "Bald Mountain Gold Mill," HAER No. SD-2, 18; Lewis, 225.
[64] Lewis, 235–236.
[65] "Hydraulic mining," Wikipedia website: en.wikipedia.org/wiki/hydraulic_mining
[66] Walter R. Crane, *Gold and Silver* (New York: John Wiley & Sons, 1908), 417.
[67] Hardesty, 38–39.
[68] White, 12.
[69] White, 9.
[70] White, 10.
[71] White, 10.
[72] White, 10–11.
[73] Arthur Stewart Eve, *Applied Geophysics in the Search for Minerals* (Cambridge, England: The University Press, 1929), 19.
[74] Joe Flanagan, "Mining Majesty: Seeking the Silver Lining in a Prospecting Past," *Common Ground* (Spring 2006): 29–32.
[75] Alan R. Alanen, "Documenting the Physical and Social Characteristics of Mining and Resource-Based Communities," *Bulletin of the Association for Preservation Technology*, vol. 11, no. 4 (1979): 53.
[76] Hardesty, 6.
[77] Francaviglia, 28–29.
[78] Hardesty, 6.
[79] U.S. Geological Survey, *Mineral Resources of the United States* (Washington, DC: Government Printing Office, 1894), 494–498.
[80] John R. Hill, Indiana Limestone, Indiana Geological Survey website, http://igs.indiana.edu/geology.minRes/indianaLimestone/index.cfm; Indiana Limestone Institute of America, A History of the Indiana Limestone Industry, website: http:// www.iliai.com/index.php?pageId=11.
[81] U.S. Geological Survey IV, 482–486.
[82] George P. Merrill, *Stones for Building and Decoration* (New York: John Wiley & Sons, 1891), 116.
[83] Walter B. Smith, Methods of Quarrying, Cutting, and Polishing Granite, on Stone Quarries and Beyond website: http://quarriesandbeyond.org/quarries/articles_and_books/us_geo_survey1894_granite_2.html.
[84] Merrill, 298.
[85] Philip C. Marshall, "Polychromatic Roofing Slate of Vermont and New York," *Bulletin of the Association for Preservation Technology* 11 (no. 3): 77, 79.
[86] Indiana Geological Survey.
[87] Smith.
[88] Marshall, 82–84.
[89] Allan Greenwell and J. Vincent Elsden, *Practical Stone Quarrying* (New York: D. Appleton & Company, 1913), 329–349.
[90] http://quarriesandbeyond.org/articles_and_books/the_cape_ann_quarries.html.
[91] Holly K. Chamberlain, Grearson & Lane Company, Granite Turning Shed, Barre, Vermont. HAER VT-9, 1987.
[92] Holly K. Chamberlain, E. L. Smith & Company, Sixteen-sided Granite Shed, HAER VT-8, 1987.
[93] Greenwell and Elsden, 475–476.
[94] Selected coal mining terms have been taken or paraphrased from the Kentucky Coal Council's online coal education glossary (www.coaleducation.org/glossary.htm) and from the Pennsylvania Department of Environmental Protection's glossary of mining terms (www.dep.state.pa.us/dep/deputate/minres/district/glossary.html). Other mining terms have been taken or paraphrased from Mining Life's mining glossary (www.mininglife.com/Glossary/a.asp) and the U.S. Bureau of Mine's dictionary of mining, mineral and related terms: http://xmlwords.infomine.com/wordsearch.asp?searchstring=mining&start=1.

CHAPTER 14

Aviation

Typical larger civilian or military airports are a mixture of building types and periods. Many represent the evolution of a World War II–era military airfield and still retain buildings and structures from that period. Others had their genesis in pre–World War II flying fields and contain one or more buildings, especially hangars, from the early era of aviation. These older buildings may have been joined or supplanted by more recent buildings and structures that testify to the growth and development of the airport. Yet others began in the economic expansion that followed the end of the Second World War.

TERMINALS

In the early years of aviation, the terminal building consisted of a corner of a hangar, partitioned from the rest, heated with a pot-bellied stove, and containing a desk, a few chairs, and a telephone.[1] At that time, airport planning began with the layout of the runways. The "leftover" areas were where terminals and other buildings were located. The size and function of terminals depends, of course, on the type and size of the airport.[2]

General Aviation

In the small, general aviation airport, the terminal building was typically a simple, one-story building, boxlike in form, rectangular in plan. This building generally included a central lobby, a field manager's office, an operations office, toilets, locker rooms, showers, and a heating plant. Larger terminals could include a larger lobby, a snack bar, a room for the local flying club, additional toilet facilities, and flying school rooms and office.

Non-Scheduled Commercial Airports

In the initial stage of terminal construction, the terminal of a non-scheduled commercial airport would typically include an airport operations office, shipping and receiving space and offices, communications and weather facilities, a pilots' room, a control tower, and toilet and locker facilities. As it grew, often executives' lounges and toilets were added. Finally, a large industrial airport building might include a central lobby, an operations office, a manager's office, service rooms and offices, and a heating plant.

Scheduled Commercial Airports

Pre-1970 terminals have become increasingly rare at scheduled airports because they are unable to accommodate today's planes and security requirements. At a few airports, such as Houston's Hobby, the original terminal has been retained and converted to another use (in Houston's case, a museum).

In the United States, two types of passenger terminals developed during the 1920s. The "depot hangar" or "lean-to hangar" combined a waiting room, offices, and a hangar in a single building. Newark (New Jersey), Chicago, Wichita (Kansas), and Los Angeles built terminals like this. A somewhat later example of this type

Figure 14.1. Municipal Hangar. Reading Regional Airport, Berks County, Pennsylvania. Photography by the author, 2007.

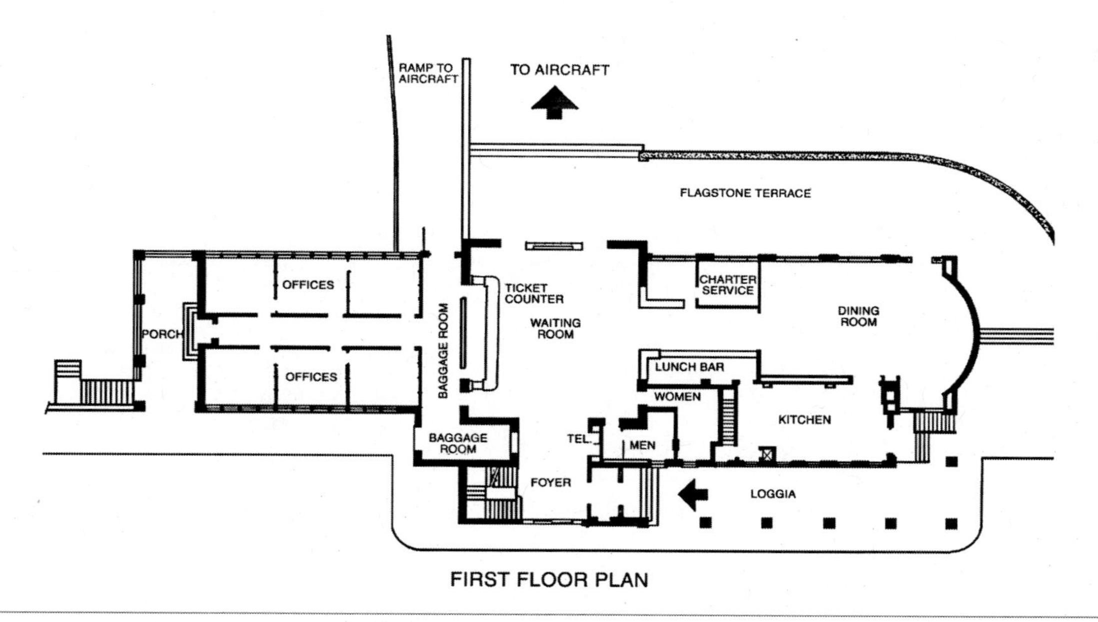

Figure 14.2. Winston-Salem Airport. First floor plan (Froesch and Prokosch 1946: 172).

is the 1930s hangar/terminal at the Reading (Pennsylvania) Airport shown in Figure 14.1.

Historically, in planning the development of commercial airport terminals, two approaches have been taken: centralization or decentralization. In a centralized plan, all passengers, baggage, and cargo are funneled through a central building and then dispersed to the plane gates. In the decentralized plan, the passenger and baggage arrive at a point near the departing plane, and all airline functions are carried on adjacent to the plane. In today's airport, centralization has become the norm, although separate international terminals or commuter terminals are often used.

In an early initial-stage centralized airport terminal (Figure 14.2), the following functions were typically present: central lobby and waiting room, manager's office, airline counter space, airline operations office, airmail and express room, snack bar, heating plant, building service, and control tower. In a second stage, often a restaurant, post office, Civil Aeronautics

Administration (CAA) office, Weather Bureau station, concessions, and additional toilets were added. In the third stage, a separate lobby and waiting area were often constructed, an observation deck was sometimes added, as was an airline cargo and baggage room, pilots' lounge, and incoming baggage room. The largest centralized terminal was usually two stories, with the following facilities: waiting room, ticket counters, baggage rooms, station manager's office, reservations facilities, operations office, radio and communications rooms, pilots' room, flight attendants' room, apron service personnel facilities, fleet service, minor repair shops, toilets, and concessions.

After World War II, new airports were built in what is called a connection or transport design, with planes parked on the tarmac and passengers walking out to them. As larger planes parked farther from the terminal, shuttle buses or mobile lounges began to transport passengers to the planes. And as jets were introduced, this became even more essential.

"Pier finger" and star-shaped terminals appeared in the 1950s in the United States. Passengers would congregate in a central area and then move out into the fingers or points of the star to depart. Chicago's O'Hare and St. Louis's Lindbergh airports were built using this design. Planes could load passengers directly from these fingers, and "moving sidewalks" often helped passengers reach their departure gate. Two levels separated arriving and departing passengers.

This design evolved into decentralized satellites or jetways, which are covered corridors that telescope out from the main terminal to meet the plane. Another design is the "linear" or "gate arrival" terminal, a long but shallow corridor with appendages coming off it where the planes tie up and passengers board and deplane.

The rise of terrorism in the 1970s increased airports' safety and security requirements. Airports intentionally added "bottlenecks" to divide "secure" regions from "open" areas. Arrival and departure areas, which had been as close to the airplanes as possible, became centrally located, resulting in longer walks for passengers.

CONTROL TOWERS

Modern airports usually feature freestanding control towers located to provide a clear view of the entirety of the tarmac, taxiways, and runways. By contrast, smaller, older airports often incorporated the control tower as part of another building, such as the terminal or a hangar. The control tower was sited to be centrally located and higher than any nearby structure. A CAA design for a rooftop control tower is shown in Figure 14.3. A tower incorporated into the terminal building had the advantage of proximity, greater stability in construction, ease of heating and air conditioning, and lower construction costs. The typical tower was 14 feet square.

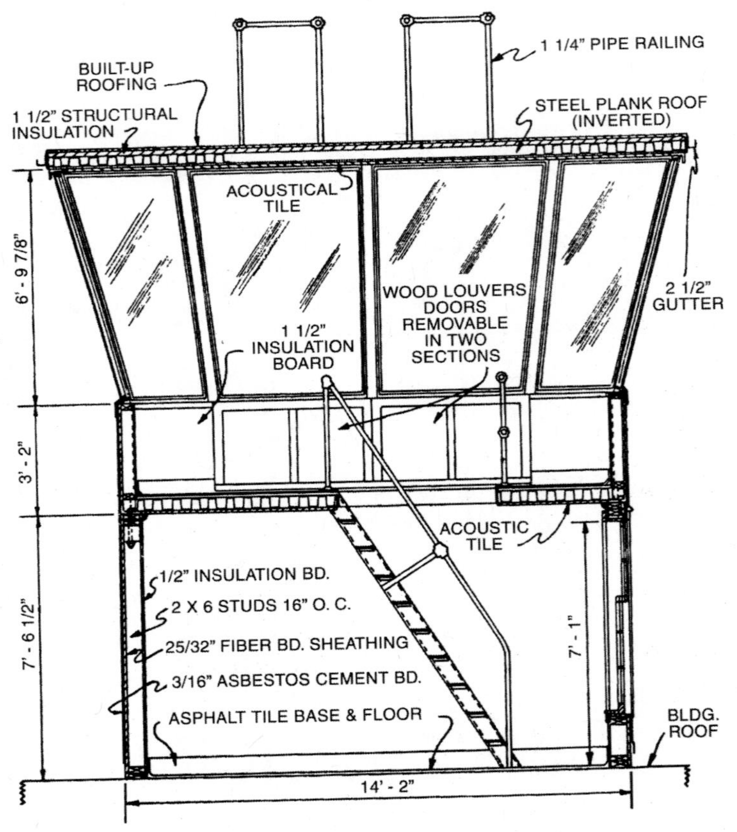

Figure 14.3. CAA Traffic Control Tower Design. Reprinted from Horace K. Glidden, Hervey F. Law, and John E. Cowles, *Airports: Design and Construction* (New York: McGraw-Hill Book Co., 1946), Figure XII.

During World War II, control towers were erected at air bases using temporary or mobilization construction. Resembling fire towers of the same period, these structures were built with open, heavy steel-framed towers with a fire-flight metal stairway. The control room was a metal-framed, square room surrounded by an observation deck. Due to the temporary character of construction, few, if any, examples of this building type remain.

In the 1940s, the CAA constructed some free-standing control towers, including one at the Roanoke, Virginia, Airport. The standard design was five stories, 52 feet in height. The first floor was designed to provide room for the heating unit, fuel storage, standby generator, and storage space. The second floor was the maintenance office, the third was the chief controller's office, the fourth incorporated toilet facilities, and the tower itself occupied the fifth floor.[3]

In the 1950s, design of military control towers reflected earlier civilian practices. An example of a control tower of this period is Building 559 at the Selfridge Air National Air Guard Base in Michigan. A six-story, steel-framed and concrete structure, it is capped by a hexagonal, glass-enclosed observation deck surrounded by a deck with a metal pipe railing.

HANGARS

Hangars may be divided into two classes: small general aviation hangars and larger scheduled, military or cargo aircraft hangars. The former are used for storage of aircraft between flights, while the latter are used for aircraft maintenance, overhaul, and repair rather than storage.[4]

General Aviation

The simplest general aviation hangar is the open shed, similar to that erected on farms to store equipment (Figure 14.4). These sheds usually are supported by timber center posts, sometimes recycled telephone poles, feature regular spaced posts along the eaves, lack walls, and often have gabled roofs sheathed in metal.

Figure 14.4. Aircraft storage shed. Delaware Airpark, Cheswold, Delaware. Photograph by the author, 2004.

These inexpensive shelters offer only scant protection against the weather.

A wood-framed hangar, measuring 50 by 60 feet, was depicted in a publication by the National Lumber Manufacturers Association. Essentially a larger version of an automotive garage, its facade featured 12-foot-high, nine-light, two-panel wood doors, extending across the wall. Other older general aviation hangars include the T-hanger and the larger private plane hangar or club hangar.

The T-hangar, whose design was initially developed in the 1940s, provides a series of "garages" for private planes (Figure 14.5). These structures benefited both the airfield operator and airplane owner by providing economy, privacy, accessibility, and freedom from damage. Planes are alternately tailed in and nosed in.[5]

As their names suggest, club hangars were built for flying clubs and contain sufficient interior space for several private planes. These hangars differ in construction materials and structural system but are unified by being designed to house small, single-engine, private planes rather than multi-engine commercial planes. An example of this hangar type is the Club Hangar, a 60-foot-wide, 40-foot-deep building erected at Capital City Airport in East Lansing, Michigan (Figure 14.6).

A third type of general aviation hangar is the fixed-base operations (FBO) hangar. In some cases, the FBO was housed in the original terminal/hangar of the airport, while in other cases, a former World War II hangar was adapted for this new use. In still other cases, a steel-framed, concrete, or wood-framed hangar was specifically built for the FBO.

Figure 14.5. T-hangars. Reading Regional Airport, Berks County, Pennsylvania. Photograph by the author, 2007.

Figure 14.6. Club hangar. Capital City Airport, Lansing, Michigan. HAER MI-320-F, National Park Service. Photograph by Dietrich Floeter, 1992.

Commercial and Military Hangars

Although many, though not all, older commercial airport hangars began as military hangars, the military branches developed standard hangar types, some of which have been adapted for commercial aviation use. A typology of military hangar construction is contained in the United States Army's Construction Engineering Research Laboratory study of the Department of Defense's hangar inventory. This study (Webster et al. 2001) is now available online.[6] The classifications included in this study have been adapted for use in this chapter.

Webster et al. divide hangars by materials, subdivide by structural system, and further divide by cross section and approximate date of construction to enable the surveyor to identify the standard plan type, when applicable.

Steel Materials

The largest number of hangars represented in the military hangar database are of steel construction. Webster et al. attribute the popularity of steel in part to its high strength-to-weight ratio, its ability to be easily assembled into members that span long distances, and its ability to be easily transported.[7]

Steel hangars may be divided into truss, girder, and long-span joist construction. Truss construction has historically predominated. Among the earliest steel truss hangars in the United States was the 1920 Fort Worden Balloon Hangar, with its concrete pier foundation, closed gambrel profile, and steel truss at the apex of its interior (Figure 14.7).

Prefabricated Hangars

In the 1930s, demountable prefabricated steel hangars were developed. One such structure, fabricated by the American Rolling Mill Company, consisted of arched rib framing spaced 17 feet 4½ inches on center. Roof purlins consisting of interlocking formed sections of sheet steel ran from rib to rib. Vertically placed panels formed the side walls for lean-to area and hangar doors. These hangars were available for dimensions up to 160 by 202 feet.[8]

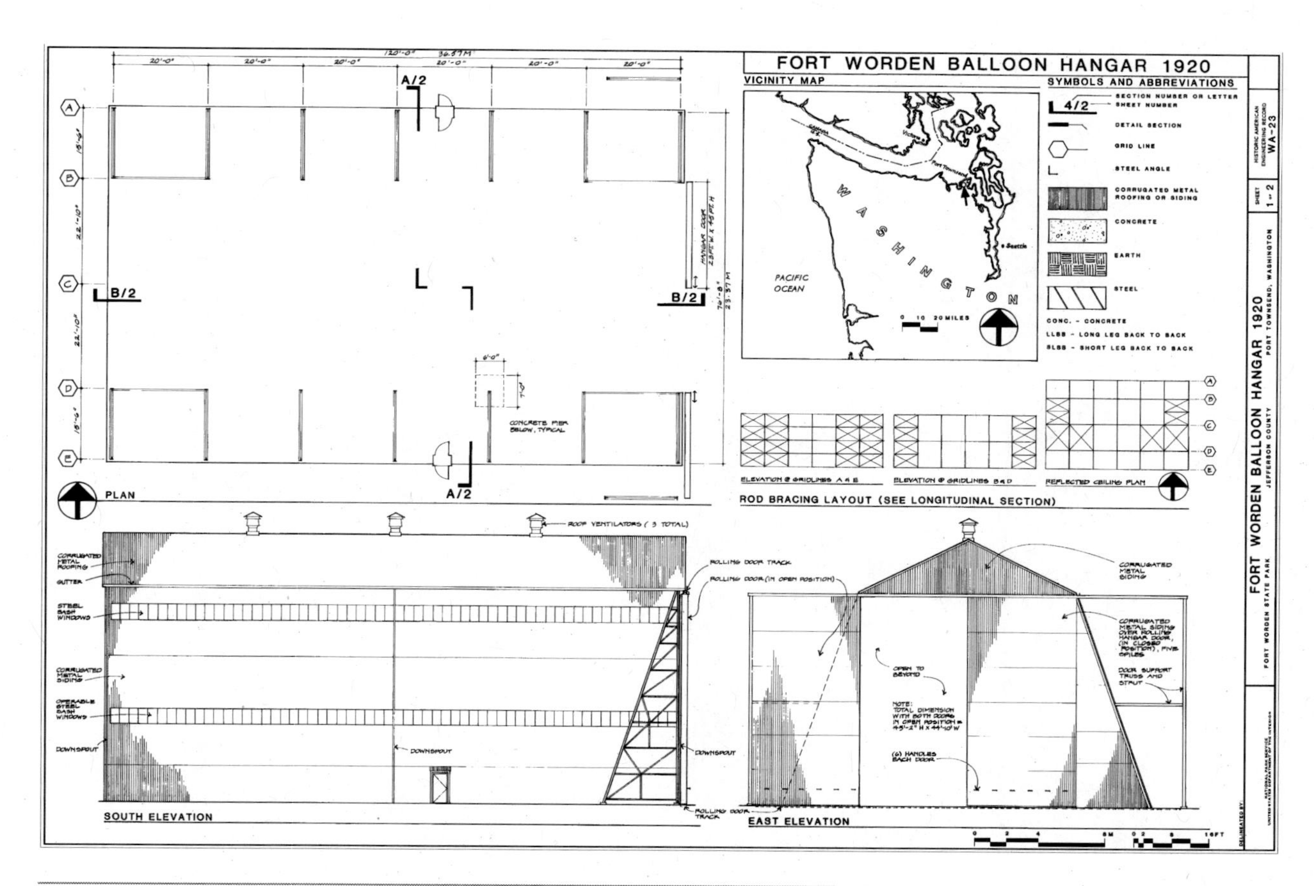

Figure 14.7. Fort Worden Balloon Hangar, 1920. Port Townsend, Washington. HAER WA-23, National Park Service.

Air-Supported Structures

Air-supported structures used a series of thin sheets of steel, welded together and sealed at all joints to the foundation. The sheets were then inflated to become a vaulted structure. By means of continuous air pressure in excess of atmospheric pressure, the structure is maintained. These hangars were designed to provide large, unobstructed, enclosed areas.[9]

Truss Construction

The truss technology used in hangar construction is based on bridge design, making it suitable for long-span design. Cross sections of steel truss hangars include closed arch, open arch (Figure 14.8), closed flat gable, open flat gable, closed gable, open gable, gable offset, gambrel A-frame, open shed, external structure, closed gambrel, sawtooth, monitor, flat, and flat exterior/arched interior.

Girder Construction

Girders are the main horizontal members of a post-and-beam structural system. Girder construction is preferred in prefabricated structural elements. Cross sections employed in military steel girder hangars include open flat gable, open gable, open gambrel, and external structure.

Long-Span Joists

Long-span joists are similar to trusses in that they are comprised of structural triangles, but they differ in scale. Joists are lighter in weight, formed of smaller structural triangles, and have lower weight limits. To compensate for lower strength, joist spacing is tighter. The use of long-span joists in the design of hangars began in the mid-1950s. Long-span joists have been used in open-shed cross-section hangars erected beginning in the 1950s.

Wood Materials

Wood hangars encompass hangars whose structural trusses are of heavy timber of lumber construction. First used in the military during World War I due to steel shortages, wooden hangar construction continued through World War II. Many used trussed structural members. Most surviving wood-framed hangars were sheathed in contemporary materials such as cement-asbestos or aluminum or sheet metal siding.

Figure 14.8. Open steel arch hangar. Aircraft Operations Hangar. Hill Field. Layton, Utah. HAER UT-85-A-6, National Park Service. Photograph by Richard Dockendorf, 1995.

Figure 14.9. Building 501. Reading Regional Airport. 1943 standard Army Air Forces wood-framed hangar. Photograph by the author, 2007.

Figure 14.10. Marine Corps Air Station, Tustin Blimp Hangar, Tustin, California. HABS CA-2707-A-1, National Park Service.

Four types of wood hangars were common in military construction: the open arch, the closed arch, the closed flat gable, and the closed gambrel. The last design was used only during World War I. The earliest surviving hangar in the United States, old Hangar Nine at Brooks Air Force Base, completed in 1918, uses the gambrel-roofed design.[10]

The increased popularity of wood hangars during World War II (Figure 14.9) has been attributed to three factors: the development and application of metal connectors for joining wood members, which permitted more accurate and efficient design criteria to be developed for wood joints, with a consequent reduction in the cross section of timber required; the development of effective synthetic wood adhesives; and the rapid development of chemicals for preserving wood and rendering it fire resistant.

Wood was used in clear spans up to 200 feet. The Lamella truss was most economically constructed of wood.[11] The largest timber hangars were the twin U.S. Marine Corps blimp hangars at Tustin, California, arched buildings with a ceiling height of 178 feet and a length of over 1,000 feet, constructed in 1942–1943 (Figure 14.10).[12]

Concrete or Terra-Cotta Materials

An early example of a masonry block hangar was Hangar No. 9 at Hamilton Field (Figure 14.11), erected in 1933, in which the walls were constructed of square terra-cotta block coated in cementitious stucco. The hangar roof is a riveted steel truss system.

Concrete hangars were constructed using poured-in-place (monolithic) concrete, precast concrete, and concrete masonry units (CMUs). Some hangars employ non-load-bearing concrete exterior walls attached to structural components. These are more appropriately classified by the material of the primary structure. Concrete military hangar construction was common during the 1940s and 1950s.

Two types of concrete hangars were common in military construction, open arch and transverse arch. In addition to truss-type hangars, reinforced concrete hangars may also be built using shell construction. The two types of shell construction are the Z-D and Whitney designs.

In the Z-D design for small spans, no ribs are employed. The roof area consists of a reinforced concrete shell of a thicker section at the spring line of the arch, tapering to a few inches at the top. In large spans, the design uses ribs 25 to 30 feet on centers, with a tin slab flush at the bottom of the ribs spanning between.

The Charles S. Whitney design employs many of the same principles as the Z-D, except that the roof slab is placed at the approximate neutral axis of the rib. A lighter and more flexible rib is used in the Whitney design. This reduces the thrust at the ends of the arch. The thrust from the arch can be taken into the lean-to rigid frame and dissipated into the ground by spread footings or battered piles. The entire area of the roof slab is reinforced with steel mesh.[13]

Two of the most prominent examples of the Z-D hangar type are engineer Anton Tedesko's designs for hangars at Loring Air Force Base (Maine) (Figure 14.12) and Ellsworth Air Force Base (South Dakota). These two have the largest monolithic concrete roofs built in the United States.

For illustrated cross sections, dates of construction, plan descriptions, and plan numbers of common military truss types, see Webster et al. (2001: 6.13–6.17).

Hangar Roof Trusses

Four main types of roof are employed for hangars: a flat or arched truss supported on columns, a segmental arch rib truss, the Lamella system, and trussless hangars.[14]

Flat or Arched Truss

A flat or arched truss, supported on columns, generally employs a Pratt, Warren, or bowstring truss. Wooden bowstring trusses were built to span 200 feet, and steel parallel chord types, 250 feet. This roof has the advantage of providing full headroom along the sides, with a clear area for lean-tos and office or shop space. They also occupy less ground area for given clearance requirements.[15]

Segmental Arch Rib Truss

The clear span of this truss is practically unlimited. Side thrust is taken care of by either tie rods placed below

Figure 14.11. Hamilton Field, Hangar No. 9 and Air Corps Shops. Northwest elevation. HAER CA-2398-F, National Park Service. Photograph by David G. De Vries, 1993–1994.

Figure 14.12. Loring Air Force Base, Arch Hangar, Building 8250, Limestone vicinity, Maine. HAER ME-64-B, National Park Service. Photograph by Jeff Bates, 1998.

Figure 14.13. Trussless hangar. Goodyear Airdock. Akron, Ohio. HAER OH-57:10, National Park Service. Photograph by Jet Lowe, 1985.

the floor slab or by heavy concrete buttresses or battered piles. If the arch springs from the ground, the headroom at that point is zero. For spans of 100 feet or more, this type is economical. The principal objection is the loss of floor space at each truss from grade up to clear headroom height. Designs include a segmental arch rib with tie rod at floor level, and a two-hinged segmental arch with tie rod at floor level.[16]

Lamella System

This truss, marketed by the Lamella Roof Company of Houston, is similar to the segmental arch. Instead of consisting of separate trusses spaced apart, the entire roof structure comprises an interwoven system of similar-sized sections, creating a series of diamonds that transmit the roof load evenly to the adjoining lower member on down to sill beams and buttresses. The truss, through its interwoven system, absorbs wind pressure. The interior of a hangar framed by this method has a cleaner appearance than one framed by either the flat or arched truss.[17] Lamella roof hangars have been designed for spans of up to 150 feet.

Trussless

These hangars employ dome-like roofs of either steel or concrete. They are built of curved slabs with or without stiffening ribs.[18] The best-known example of this technology is the original Goodyear Airdock built in Akron, Ohio, in 1929 and designed by Dr. Karl Arnstein of the Wilbur Watson Engineering Company of Cleveland (Figure 14.13). This hangar, designed to house the Goodyear blimp, was constructed of sheet metal panels

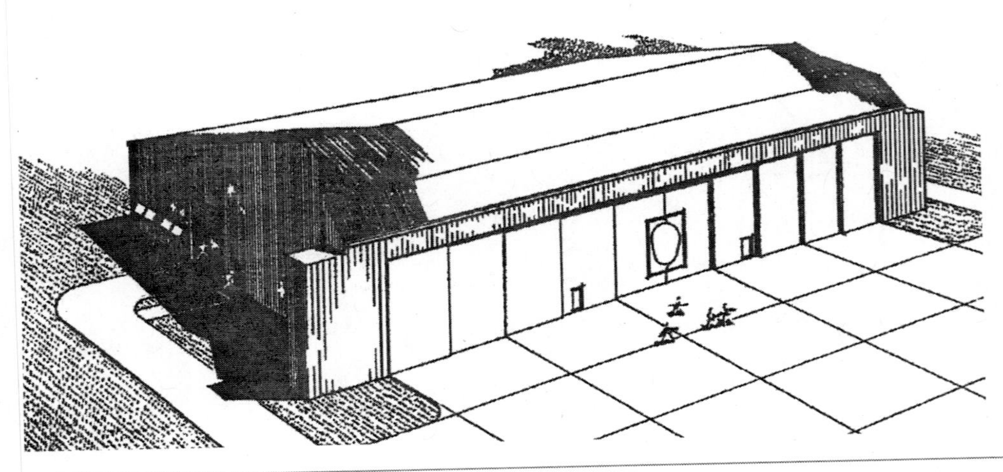

Figure 14.14. Offset gable hangar. SAC Dispersal Fuel Cell Maintenance Hangar at Wright-Patterson Air Force Base, Ohio. Reprinted from Julie L. Webster, Principal Investigator, *Historical and Architectural Overview of Military Aircraft Hangars: A General History, Thematic Typology, and Inventory of Aircraft Hangars Constructed on Department of Defense Installations* (Champaign, IL: United States Army Construction Engineering Research Laboratory, 2001), Figure 5-5.

Figure 14.15. Double cantilever hangar, Loring Air Force Base, Limestone vicinity, Maine. HAER ME-64C, National Park Service. Photograph by Jeff Bates, September 1998.

attached to 11 parabolic arches spaced at 80-foot intervals connected by the vertical and horizontal trusses.[19]

Cold War–Era Hangar Designs

Several hangar designs are characteristic of this period and were erected at Air Force bases and Naval air stations. One type, constructed as general maintenance hangars at Strategic Air Command bases, OCE Standard Plan 39-05-12, is distinguishable by its offset gable configuration (Figure 14.14).

A second design, the Miramar, features two 150 by 240 foot arched hangar bays separated by a 120 by 240 foot open shop area. This design was constructed singly and in pairs on bases such as Naval Air Station Pensacola and NAS Miramar.

The third standard type, used at Air Force installations to house multiple bombers, is the double cantilever (DC) (Figure 14.15). This construction features longitudinal arched trusses with trusses cantilevering at both ends from their central shop supports.

The fourth standard type features an exposed cantilever system on the exterior, making it simple to discern. This design, which employs concrete masts rising from its roof connecting to structural cables, was used in Naval and Air Force hangars, as well as in a small number of commercial aviation hangars, including the now-demolished TWA Maintenance Hangar at Philadelphia International Airport (Figure 14.16).

Hangar Doors

Four types of doors are commonly used for hangars: canopy, vertical-lift, jackknife, and side-sliding.

Canopy

The canopy door, which tilts upward, is counterbalanced by means of a series of cables attached to weights (Figure 14.17). An aircraft of low height is able to enter the hangar without the door being completely raised. When the door is completely open, a portion projects beyond the building wall, affording protection against rain or snow or intense sunlight. The main disadvantage is that it is the most expensive of the door types.[20]

Two types of canopy doors were used in older hangars. The unbraced canopy was used for openings not exceeding 26 feet high and 100 feet long. The braced canopy could be used for openings up to 55 feet high.

Vertical-Lift

Vertical-lift doors were designed and developed for hangars with extremely high and wide door openings. They are constructed of several horizontal panels, each of which is offset from the adjacent one (Figure 14.18).

Figure 14.16. TWA Maintenance Hangar. Philadelphia International Airport. Tinicum Township, Pennsylvania. HAER PA-561:3. Photograph by John Herr, 1999.

Figure 14.17. Canopy hangar door. Reprinted from Truscon Steel Company, *Truscon Steel Windows and Hangar Doors* (Youngstown, Ohio: Truscon Steel Company, 1945), 43.

Figure 14.18. Three-section, vertical-lift hangar door (Truscon 1945:42).

Each panel travels at a different rate of speed, so all panels arrive at their destinations simultaneously. The bottom leaf rises vertically until it clears one-half of the opening height, and then both leaves swing out and up to form a canopy, leaving the entire opening unobstructed. The door can be raised to any desired height, depending on the plane seeking entry. The main disadvantage is that two trusses must be provided instead of the one used in other types, the first outside the door, the second immediately inside.[21]

Side-Sliding

Side-sliding doors are the least costly door type. They may be divided into two types. One type operates on an accordion principle, using leaves about 4 feet wide that fold into door pockets. The other has each leaf offset from the adjacent one; each one travels in own track and at a different speed from its neighbor, so that all arrive at the open or closed position simultaneously. Typically, they are operated by cables on winches. The side-sliding door was almost universally used in arch-rib and Lamella hangars.[22]

AIRPORT LANDING AIDS AND LIGHTING

The earliest airport landing aids were people-powered. Flagmen provided aircraft separation and direction control by waving red, green, or white cloths that told pilots if they were approaching a clear field at the correct angle. Green meant all clear, while red waved the pilot off. The first radio tower was installed at an airport operated by Archie League in St. Louis in the early 1930s. Some airports followed League's lead, while others replaced the flagman by a system of red and green airport lights to show the runway threshold and side, and to allow the pilot to judge remaining distance and angle of approach. Airport lights had been introduced in the 1920s when fields were marked with rotating lights so that they could be found after dark. In the early 1930s, airports installed the earliest forms of approach lighting indicating the correct angle or descent (glide path) and whether the pilot was on target. Eventually the lights and their rates of flash were standardized worldwide, based upon International Civil Aviation Organization standards.[23]

General Aviation Airports

Beacon

The rotating beacon at an airport, the beam of which is visible for miles, indicates that the airport is equipped for night landings. This beacon is mounted higher than any surrounding obstruction and rotates at six revolutions per minute. In 1948, it was specified that this beacon should have a 36-inch light equipped with clear and green lenses.[24]

Boundary Lights

Boundary lights are placed to indicate the true outline of the entire landing area of the field. A unit is placed at each angle of the boundary, and units are placed approximately 300 feet apart.

Runway Lighting

Runway lights are typically mounted on 200-foot centers along both sides of the runway. The lights are symmetrically arranged in pairs to be directly opposite each other, except at runway intersections. The color sequence is the same for all runway lights.

Runway lights are either flush or elevated. Flush lights produce a beam of medium candlepower, do not project more than 4 inches above the ground, and are designed to withstand being run over by aircraft. Early flush lights were difficult to keep watertight and were easily obscured by vegetation. Elevated lights, also known as "snow lights," are of medium candlepower. Extending well above the ground, their small size and breakable fiber or aluminum column mounting prevent them from being a serious obstruction to aircraft.

High-intensity runway lights were introduced by the Armed Forces before 1943. These lights produce a high candlepower beam. Located in a row on each side of the runway, they are placed approximately 10 feet out from the paved width of the runway to reduce the chance of being struck by aircraft.

Range Lights

Range lights with green globes are used to indicate the ends of runways and to code them. Set in a line at right angles to the centerline of the runway, they are located 100 feet beyond the end of the runway.

Approach Lighting

Approach lights are an additional device to permit the incoming pilot to line up properly with the designated runway. The standard approach lighting configuration used in the United States employs lights beginning 3,000 feet from the end of the runway.

Taxiway Lights

Taxiways are typically lighted with elevated or flush lights with blue filters. They are mounted either on both sides of a taxiway or on the side that will be at the left of an incoming airplane. They are placed between 100 and 200 feet apart based on traffic volume.

Wind Cone or Wind Sock

A wind cone or wind sock is installed at airports and other aircraft landing locations to indicate wind direction. Typically 12 feet long and 36 inches wide at the mouth to meet Federal Aviation Administration (FAA) standards, modern wind socks are usually orange or white/red nylon construction, are mounted on metal or fiberglass masts, and are illuminated with spotlights. An alternative to the wind cone is the wind tee, which is similar to a weathervane.

Commercial Airports

Commercial airports employ the following types of lights: approach lighting, runway threshold lighting, runway edge lighting, runway centerline and touchdown zone lights, and taxiway edge centerline and clearance bar lights.

Approach Lighting

The United States Approach Light System (ALS) starts at the landing threshold and extends into the approach a distance of 2,400 to 3,000 feet for precision-instrument runways and 1,450 feet for nonprecision-instrument runways (Figure 14.19). It includes one crossbar 1,000 feet from the runway threshold. Roll guidance is provided by bars 14 feet in length placed at 100-foot centers on the extended centerline of the runway. The bars consist of five closely spaced lights to give the effect of a continuous bar of light. Approach lights are normally on pedestals of varying heights (Figure 14.20a).[25]

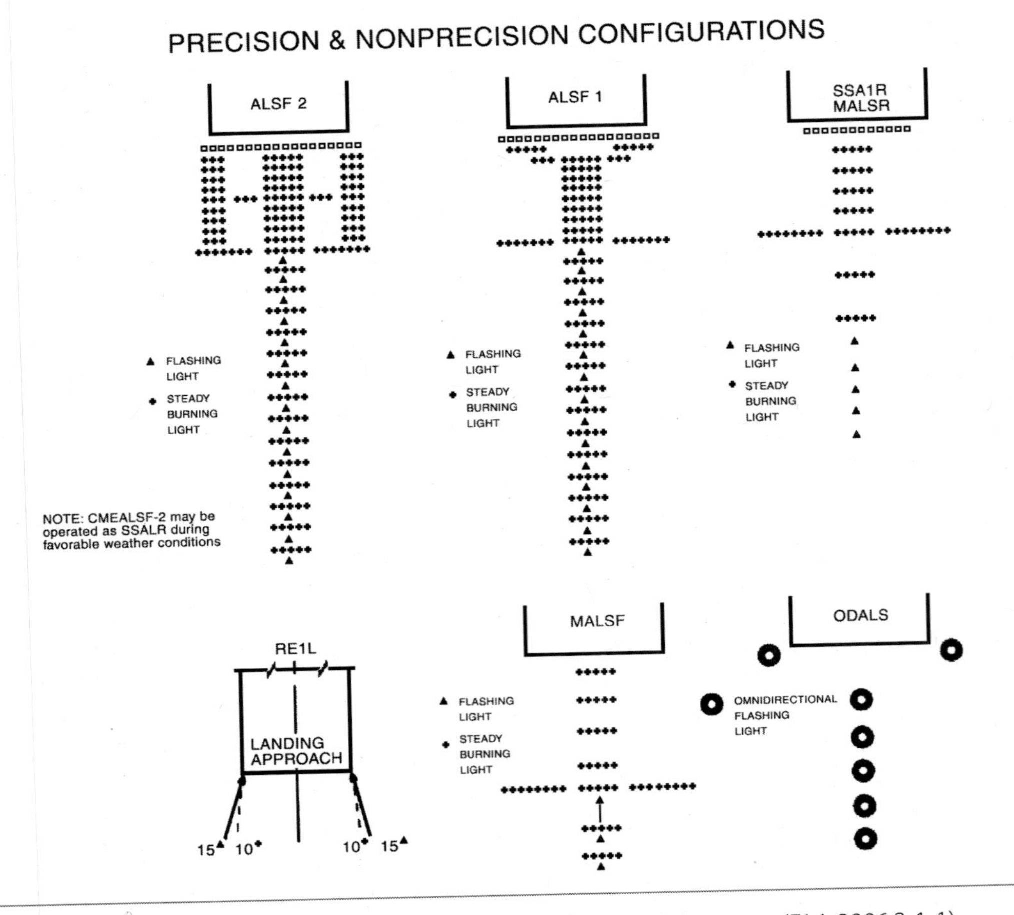

Figure 14.19. Approach lighting system, commercial airports (FAA 2006:2-1-1).

Runway Threshold Lighting

The identification of the threshold is a major factor in a pilot's decision to land or not to land. For this reason, the region near the threshold is given special lighting consideration. At large airports, the threshold is identified by a complete line of green lights extending across the entire width of the runway and at small airports by four lights on each side of the threshold. When these lights are extended across the runway, they are of the semi-flush type, while those at the sides of the runway may be elevated. Threshold lights in the direction of landing are green, while in the opposite direction, they are red to mark the end of the runway.[26]

Runway Edge Lighting

Runway edge lights (Figure 4.20b) are usually elevated units mounted on frangible fittings, projecting no more than 30 inches above the surface and located not more than 10 feet from the edge of pavement. Longitudinal spacing is not more than 200 feet. The lights are white, although the last 2,000 feet of an instrument runway are yellow to indicate a caution zone.[27] In pavement, runway lights are flush with the pavement (Figure 14.20c). In addition, to facilitate ground maneuvering, runway locations are indicated by backlit box signs (Figure 4.20d).

Runway Centerline Lights and Touchdown Zone Lights

Runway centerline and touchdown zone lights have been installed only at those airports equipped for instrument operations. Touchdown zone lights are a three-bulb unit on either side of the runway centerline and extend 3,000 feet from the runway threshold. They are spaced at intervals of 100 feet.

Figure 14.20. Types of airfield lights: ***a,*** approach lights; ***b,*** runway edge light; ***c,*** flush runway light; ***d,*** runway box sign; ***e,*** taxiway light. Photographed at Reading Regional Airport by the author, 2007.

Taxiway Edge, Centerline Lights, and Clearance Bar Lights

Taxiway edge lights (Figure 14.20e) outline the edges of taxiways during periods of darkness. These fixtures emit blue light. Taxiway centerline lights are steady burning and emit green light. Clearance bars are installed at holding positions on taxiway. Clearance bars consist of three in-pavement, steady-burning yellow lights. For ground maneuvering, taxiways, like runways, are marked by backlit box signs.

AIRPORT PAVING

While most of the infrastructure of many airports has been replaced recently due to changing technology, aviation demands, or security requirements, one of the oldest remaining elements at many airports is a portion of the airport paving: runways, tarmac, aprons, and/or taxiways. For example, several years ago, a portion of the paving of Pennsylvania's Reading Regional Airport remained from the original airport construction in the 1930s.

Often paved areas have been enlarged, runways and taxiways widened or lengthened. One way to gather evidence of original paving is to compare current pavement configuration to that shown on an older airport plan and to look for telltale seams in the present pavement.

Runways

The first paved runway at a major U.S. airport was constructed at Newark Airport, New Jersey. The typical mid-twentieth-century airport runway was subject to an impact at landing speed far in excess of the design for highways. The supporting foundation for runways of this period was typically constructed of gravel, broken stone, or soil and gravel composition. In *The Construction of Roads and Pavements*, Thomas Radford Agg illustrates typical cross sections for airport runway surfaces used in 1940.

Modern airport runways are subject to considerably higher weight loads and volumes than those envisioned by Agg. To accommodate the load of heavy transports and jumbo jets, two types of pavement are used: flexible pavements and rigid pavements (Figure 14.21).

The standard subbases for rigid pavement specified by the CAA in 1948 included 4 inches of a P-154 subbase course. The subbase material consists of hard durable particles or fragments of granular aggregates mixed or blended with fine sand, clay, stone dust, or other similar binding or filler materials.[28]

Two rigid base courses were acceptable: P-301 and P-302. P-301, a soil cement base course, consisted of Portland cement, water, and soil, in which a 6-inch course was produced by uniformly mixing together the soil, cement, and water and spreading, shaping, and compacting the mixture. P-302, a lean mix cement aggregate base course, consisted of Portland cement, selected granular materials, and water, proportioned and mixed in a central plant, and spread on the underlying course. The rigid pavement wearing surface, designated P-501, consisted of either plain or reinforced Portland cement concrete pavement. In some cases, the wearing surface was then covered with either a bituminous prime coat or bituminous surface treatment.[29]

For flexible base courses, a variety of materials were acceptable to the Civil Aeronautics Administration: P-201, bituminous base course; P-204, mixed-in-place base course; P-205, dry-bound macadam base course; P-206, water-bound macadam base course; P-208, aggregate base course; P-209, crushed aggregate base course; P-210, caliche base course; P-211, lime rock base course; P-212, shell base course; P-213, sand-clay base course; P-214, penetration macadam base course; P-215, emulsified asphalt aggregate base course; P-216, mixed-in-place base course; and P-217, aggregate-turf pavement. The components and method of construction of each of these base courses was indicated in the CAA's publication *Standard Specifications for Construction of Airports* (1948 and other editions).

Flexible-surface courses included P-401, bituminous surface course; P-405, key stone mat course; P-408, blended natural limestone and sand asphalt surface; P-410, asphalt concrete surface course; and P-411, surface treatment with limestone and sand asphalt mixture.[30]

Runway Layouts

One key to a desirable airport layout is to provide the shortest taxiing distance from the terminal area to the takeoff ends of the runway and to reduce the taxiing distance for landing aircraft as much as practicable. Another key is to orient runways so that wind conditions allow

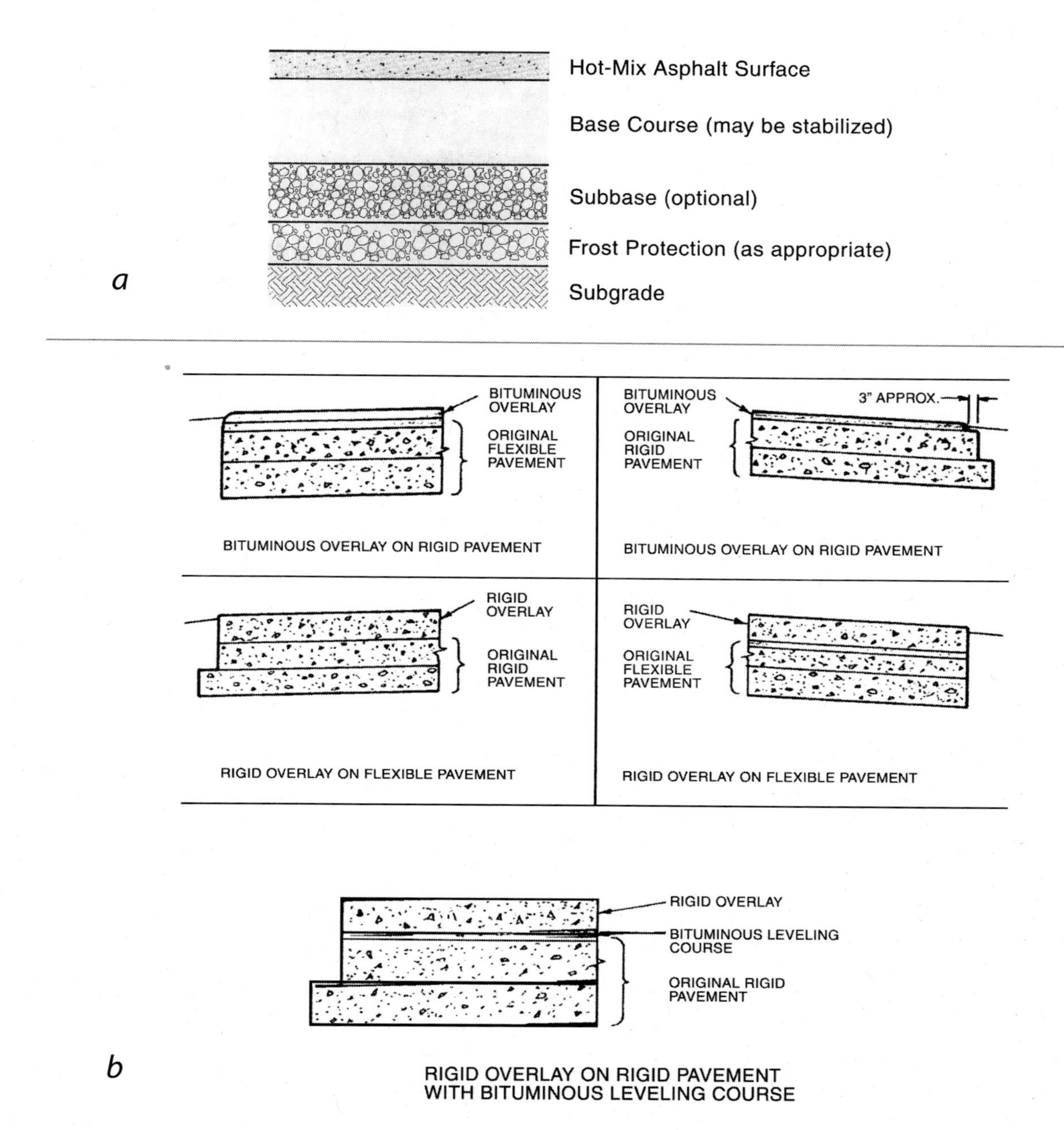

Figure 14.21. Typical airport runway cross sections: ***a,*** rigid pavement structure; ***b,*** flexible pavement structure. From *Guidelines and Procedures for Maintenance of Airport Pavements*. Advisory Circular No. 150/5380-6A.

landings and takeoffs in either direction, a desirable characteristic in high-volume airports.

The most common modern runway configurations are single runways, parallel runways, intersecting runways, and open-V runways. Close parallel runways are spaced from a minimum of 700 feet for air-carrier airports to less than 2,500 feet. Intersecting runways are necessary when relatively strong winds blow from more than one direction, resulting in excessive crosswinds when only one runway is provided.[31]

Aprons

Loading aprons are used for parking aircraft during loading and unloading of passengers, mail, and cargo.

The apron layout depends directly on the way aircraft gate positions are grouped around the building.[32]

Holding Aprons
Holding aprons, also known as run-up pads or holding bays, are placed adjacent to the end of runways as storage areas for aircraft prior to takeoff. They are designed so that one aircraft can bypass another whenever this is necessary. Space for four average aircraft is usually provided.[33]

Taxiways

The primary function of taxiways is to provide access from the runways to the terminal area and service hangars. Taxiways are arranged so that aircraft that have just landed do not interfere with aircraft taxiing for takeoff. Taxiway pavement standard widths range from 25 feet to 75 feet, depending on the character of air traffic.[34]

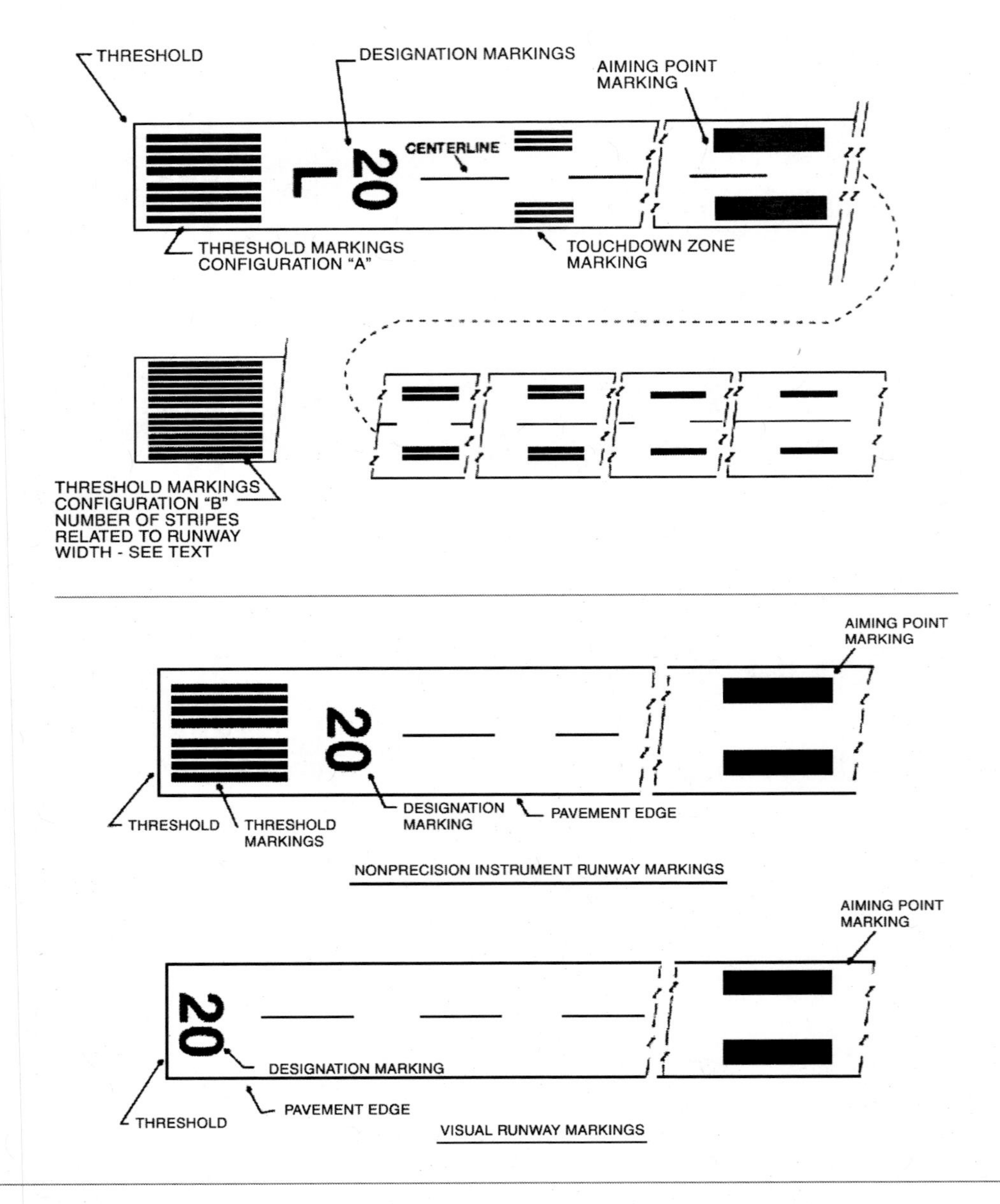

Figure 14.22. Runway markings. Reprinted from Federal Aviation Administration, *Aeronautical Information Manual: Official Guide to Basic Flight Information and ATC Procedures*, 2007.

Runway and Taxiway Markings

To assist pilots in taxiing, takeoff, and landing, runways are marked with lines and numbers (Figure 14.22). These markings, though not mandatory, are recommended by FAA standards. White is used for marking runways, while yellow is used for taxiways and aprons. The end of each runway is marked with a number, which indicates the magnetic azimuth (clockwise from magnetic north) of the runway in the direction of operations. The marking is given to the nearest 10 degrees, with the last digit omitted. For instance, the east end of a due east-west runway would be marked 27 (for 270 degrees), and the west end would be marked 9 (for 90 degrees). Should two parallel runways exist, these runways would be designated 9L-27R or 9R-27L, to indicate the direction of each runway and their position. A third parallel runway would be marked C for center runway. These markings are 60 feet tall and vary in width from 5 feet to 23 feet wide.[35]

Required markings for a visual runway are its designation, centerline, threshold for runways used by international commercial transports, and aiming point for runways 4,000 feet or longer used by jet aircraft. The markings required for a nonprecision-instrument runway are the same as for a visual runway. The required markings for a precision-instrument runway are the runway designation, centerline, threshold, fixed distance marking, touchdown zone, side strips, and holding positions.

To prevent soil erosion, many airports provide a paved blast pad 150 to 200 feet long adjacent to the end of the runway. Some airports have a stopway designed to support aircraft during rare aborted takeoffs.

Taxiway markings include the centerline and a holding line when the taxiway intersects a runway. When the edge of the taxiway pavement is not immediately apparent, it is marked by two 6-inch-wide yellow stripes.

Night Navigation Lighting

Following the First World War, the U.S. Post Office began to operate a series of air mail routes in the eastern United States. On August 20, 1920, the Transcontinental Air Mail Route was opened, extending from New York to San Francisco. Since pilots typically relied on landmarks for navigation, night flight along the route was precluded, and, as a result, air mail was often no faster than that carried by rail. Experiments were conducted in an attempt to rectify this problem. A 1923 experiment showed that pilots could navigate at night using rotating light beacons.

Beacons were positioned every 10 miles along the airway. At the top of a 51-foot steel tower was a 1 million candlepower rotating beacon, visible for 40 miles (Figure 14.23). The top of the tower also included two color-coded course lights that pointed up and down the airway. Green signified an adjacent airfield, while red signified no airfield. The course lights also flashed a Morse code letter corresponding to the number of the beacon.

Beacons also assisted daytime navigation. Each tower was erected on an arrow-shaped concrete slab painted yellow. The arrow pointed to the next higher-numbered beacon. An equipment/generator shed adjacent to the tower had the beacon number and other information painted on the roof.

By the fall of 1924, the lighted segment extended from Rock Springs, Wyoming, to Cleveland, Ohio, and by the following summer, it extended all the way to New York. Later the lighted airways program was turned over to the Commerce Department's Bureau of Lighthouses, and an improved version of the beacon was fielded in 1931. In 1933, the Federal Airway System operated by the Airways Division included 18,000 miles of lighted airway, including 1,550 rotating bacons and 236 intermediate landing fields.[36]

In the 1960s, navigation technology had advanced so quickly that many pilots felt the beacon system had become antiquated. In a cost-cutting effort, the FAA began decommissioning beacons In Montana, the responsibility for system maintenance was transferred to the state, and it remains the only state that operates its own lighted airway beacon system in its mountainous western third. In 1966, the Montana Aeronautics Commission selected 12 beacons along with 8 federally maintained beacons. In 2002, the state operated 14 airways beacons, 3 obstruction beacons, and 4 airport beacons. The airway beacons are located at Lookout Pass, St. Regis, Alberton, University Mountain, Bonita, Avon, MacDonald Pass, Spokane, Strawberry, Hardy, Wolf Creek, Whitetail, Homestake, and Canyon Resort. The beacons are 24-inch dome-type lights of at least 2 million candlepower.[37]

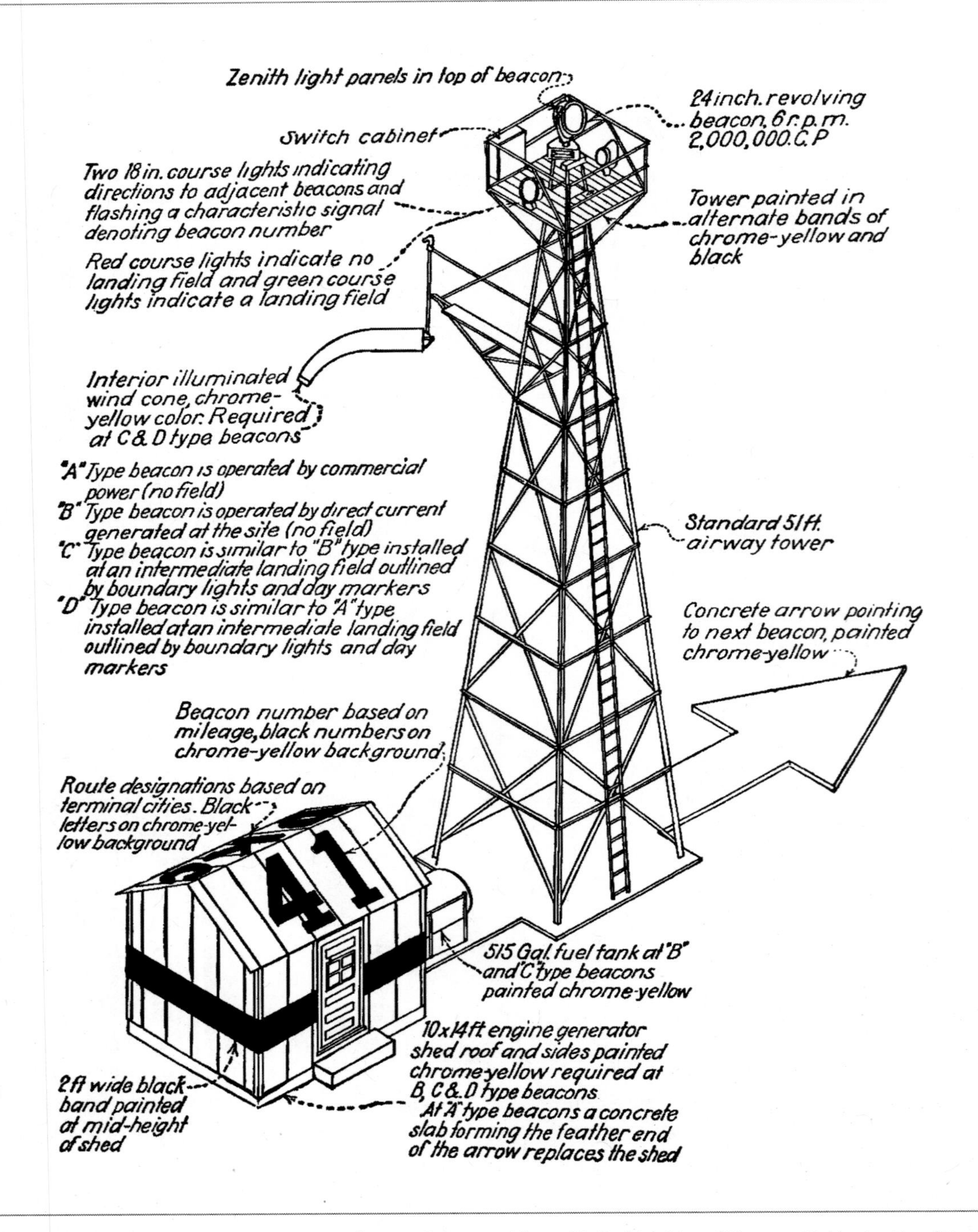

Zenith light panels in top of beacon
24 inch. revolving beacon, 6 r.p.m. 2,000,000 C.P
Switch cabinet
Two 18 in. course lights indicating directions to adjacent beacons and flashing a characteristic signal denoting beacon number
Tower painted in alternate bands of chrome-yellow and black
Red course lights indicate no landing field and green course lights indicate a landing field
Interior illuminated wind cone, chrome-yellow color. Required at C & D type beacons
"A" Type beacon is operated by commercial power (no field)
"B" Type beacon is operated by direct current generated at the site (no field)
"C" Type beacon is similar to "B" type installed at an intermediate landing field outlined by boundary lights and day markers
"D" Type beacon is similar to "A" type installed at an intermediate landing field outlined by boundary lights and day markers
Standard 51 ft. airway tower
Concrete arrow pointing to next beacon, painted chrome-yellow
Beacon number based on mileage, black numbers on chrome-yellow background
Route designations based on terminal cities. Black letters on chrome-yellow background
41
515 Gal. fuel tank at "B" and "C" type beacons painted chrome-yellow
10 x 14 ft. engine generator shed roof and sides painted chrome-yellow required at B, C & D type beacons
At "A" type beacons a concrete slab forming the feather end of the arrow replaces the shed
2 ft wide black band painted at mid-height of shed

Figure 14.23. Airway navigation beacon installation. Reprinted from Philip Van Horn Weems, *Air Navigation* (New York: McGraw-Hill Book Company 1943), 80.

The airway beacons of at least one state, Wyoming, have been surveyed. The most common remnant is the concrete arrow. In at least one site, some remnants of the old generator system and a portion of a building foundation still exist.[38]

Code beacons and course lights are remnants of "lighted" airways that predate the current electronically equipped federal airways system. Few of these beacons exist today and mark airway segments in remote mountain areas. The code beacon flashes the three- or four-character airport identified in Morse code six to eight times per minute with green flashes for land airports and yellow flashes for water airports. Course lights, which can be seen from only one direction, are used in conjunction with the rotating beacons. Two course lights, back to back, direct coded flashing beams of light in either direction along the airway.[39]

Airway Maps and Information

An efficient way to determine the whereabouts of airway beacon sites is by reference to old aeronautical charts. One source of a limited number of historical aeronautical charts, both for viewing and for downloading, is the National Oceanic and Atmospheric Administration Historical Map and Chart project website.[40]

Other sources include map collections of large research libraries. For example, the Map Collection of the central research library of the New York Public Library has a large collection of United States strip airway maps, most dating from 1929 though the 1930s. The maps include not only visual references for daylight flying, but many have a nighttime map depicted on the reverse side. The night maps are printed in black and white, with red references showing beacons, airfields, and other references.[41] Two series were issued in about 1930: those of the Department of Commerce and those of the Army Air Corps. Typical segments of the strip maps include Milwaukee–St. Paul-Minneapolis (1930), Uniontown (Pennsylvania)–Dayton (Ohio), and Portland (Oregon)–Spokane (1931).

Aeronautical charts, like other maps, use standardized symbols to depict airfields and airway beacons. An Army, Navy, or Marine Corps field is depicted with a heavy circle, while a commercial or municipal airport is depicted with a gear symbol. An emergency landing is depicted with a cross. Beacons are also illustrated. A rotating beacon is shown by an open five-point star, while a rotating beacon with course lights has added arrows, a number, and dots indicating the pattern of the light. A rotating beacon with flashing code beacon is indicated by an open multiple-point star.

GLOSSARY[42]

Air-navigation facility (historic). Any airport, emergency landing field, light, or other signal structure, radio-directional-finding facility, radio or other electrical communications facility, and other structure or facility used as an aid to air navigation.

Air traffic control tower. A facility that uses air and ground communications, visual signaling, and other devices to provide air traffic control services to aircraft operating in the vicinity of an airport or in the movement area.

Airport (historic). Any locality, of either water or land, that is adapted for aircraft landing and takeoff and provides facilities for shelter, supply, and repair of aircraft; or a place used regularly for receiving or discharging passengers or cargo by air.

Airway (historic). An air route between air traffic centers over terrain best suited for emergency landings, with landing fields at intervals equipped with aids to air navigation, and a communications system for the transmission of information pertinent to the operation of aircraft.

Aperture door. To accommodate large aircraft, the doors of some hangars were modified with an adjustable opening that closed around the tail of the plane. This opening is called an aperture door because it is similar in design to the adjustable aperture in a camera lens.

Approach light. One of several lights used to indicate a favorable direction for aircraft landing on a runway.

Apron. The paved area surrounding a hangar.

Boundary light. Any one of the lights designed to indicate the limits of the landing area of an airport or landing field.

Dirigible. A lighter-than-air craft capable of being propelled and steered for controlled flight.

Door pocket. The enclosure that receives open hangar doors.

Draft curtains. Metal partitions that run transversely and longitudinally along the ceiling of a hangar bay to prevent the spread of fire.

Fixed-base operator (FBO). A commercial operator supplying maintenance, flight training, and other services at an airport.

Gate position. The space allotted to one airplane for parking at a terminal building.

Guide slope. The angle between the horizontal and the glide path of an aircraft.

Hangar. An enclosed structure for housing aircraft.

Instrument landing system (ILS). A radar-based system allowing appropriately equipped aircraft to find a runway and land when clouds are as low as 200 feet.

Intermediate or *emergency landing field* (historic). A locality, either on water or land, adapted for aircraft landing and takeoff, which was located along an airway and was intermediate to airports connected by the airway, but which was not equipped with facilities for shelter, supply, and repair of aircraft and was not used regularly for the receipt or discharge of passengers or cargo by air.

Landing strip (historic). A narrow and comparatively long area forming part of an airport or an intermediate or auxiliary field, which is suitable for aircraft landing and takeoff under ordinary weather conditions.

LTA. Lighter-than-air craft, usually referring to powered blimps and dirigibles.

Maintenance hangars. Buildings providing a protected area for aircraft maintenance rather than garaging.

Marker, boundary. A painted cone, solid circle, or dish denoting the boundary of the available area for landing on an airfield.

Nose hangar. A hangar designed to provide shelter for the forward section of an aircraft. Typically, the hangar had a cantilevered roof that overhung the nose and wings of the aircraft with either a canvas flap or a canopy-type door fitted with cutouts to accommodate the fuselage.

Nose pocket. A clear space centered at the back of a hangar bay and often flanked by office modules. It accommodates the nose of the aircraft during maintenance.

Obstruction light. A red light used to indicate the position and height of an object hazardous to flying aircraft.

Overhead tail door. In addition to the large doors on a hangar facade, a hangar often has an overhead tail door that allows clearance for the tail of large aircraft. Tail doors may be of the rolling and swing-up varieties.

Range light. One or more lights placed at the ends of a runway to indicate the end and relative length of the runway.

Seaplane. A water-based aircraft with a boat-hull fuselage, often amphibious.

Sound-deadening boards. A special interior finish often used in hangars to reduce noise reverberation caused by loud aircraft engines.

Taxiway. A prepared strip to enable an aircraft to taxi to and from the end of a runway.

Tetrahedron. A polyhedron of four faces with weathercock characteristics located on a landing field to indicate the direction of the wind.

Wash rack. A facility used to wash aircraft to prevent fuselage deterioration.

Wind rose. A diagram of the points of the compass with lines indicating the relative velocity and direction of winds.

BIBLIOGRAPHY

Air Transport Association of America. *Airline Airport Design Recommendations*. Washington, DC: Air Transport Association of America, 1946.

Civil Aeronautics Administration. *Standard Specifications for Construction of Airports*. Washington, DC: U.S. Government Printing Office, 1948.

Froesch, Charles, and Walther Prokosch. *Airport Planning*. New York: John Wiley & Sons, Inc., 1946.

Glidden, Horace K., Hervey F. Law, and John E. Cowles. *Airports: Design and Construction*. New York: McGraw-Hill Book Co., 1946.

Horonjeff, Robert, and Francis X. McKelvey. *Planning and Design of Airports*. New York: McGraw-Hill, Inc., 1983.

National Lumber Manufacturers Association. *Airplane Hangar Construction*. Washington, DC: National Lumber Manufacturers Association, 1930.

Truscon Steel Company. *Truscon Steel Windows and Hangar Doors*. Youngstown, OH: Truscon Steel Co., 1945.

Webster, Julie L. Webster, Principal Investigator; written by Michael A. Pedrotty, Julie L. Webster, Gordon L. Cohen, and Aaron R. Chmiel. *Historical and Architectural Overview of Military Aircraft Hangars: A General History, Thematic Typology, and Inventory of Aircraft Hangars Constructed on Department of Defense Installations*. Champaign, IL: United States Army Construction Engineering Research Laboratory, 2001.

Wood, John Walter. *Airports: Some Elements of Design and Future Development*. New York: Coward-McCann, 1940.

NOTES

1 Charles Froesch and Walter Prokosch, *Airport Planning* (New York: John Wiley & Sons, Inc., 1946), 154.

2 Froesch and Prokosch, 154.

3 Horace K. Glidden, Hervey F. Law, and John E. Cowles, *Airports: Design, Construction and Management* (New York: McGraw-Hill Book Company, 1946), 234–244.

4 Froesch and Prokosch, 189.

5 Froesch and Prokosch, 200. For a discussion of types of early T-hangar types, see J. F. Woerner, "T-Type Personal Hangars," *Aero Digest* 53:3 (September 1946): 44–45, 96.

6 http://www.cecer.army.mil/techreports/webster98/webster98_idx.htm.
7 Webster et al., 6-2.
8 Froesch and Prokosch, 200.
9 Froesch and Prokosch, 200.
10 Old Hangar Nine, Brooks Air Force Base, ASCE History and Heritage of Civil Engineering. http://live.asce.org/hh/index.mxml?lid=130.
11 Froesch and Prokosch, 193.
12 Blimp Hangars, ASCE History and Heritage of Civil Engineering, http://live.asce.org/hh/index.mxml?lid=14.
13 Froesch and Prokosch, 198.
14 Froesch and Prokosch, 192–193.
15 Froesch and Prokosch, 192.
16 Froesch and Prokosch, 192.
17 Froesch and Prokosch, 192–193.
18 Froesch and Prokosch, 193.
19 Bill Lebovich, Goodyear Airdock, Historic American Engineering Record Site Data Form, HAER OH-57. 1987.
20 Froesch and Prokosch, 200, 202.
21 Froesch and Prokosch, 202; Truscon Steel Company, *Truscon Steel Windows and Hangar Doors* (Youngstown, OH: Truscon Steel Company, 1945), 42.
22 Froesch and Prokosch, 202.
23 Roger Mola, "Aircraft Landing Technology," on Centennial of Flight website, http://www.centennialofflight.gov/index.cfm.
24 Civil Aeronautics Administration, *Standard Specifications for Construction of Airports* (Washington, DC: U.S. Government Printing Office, 1948), 472.
25 Robert Horonjeff and Francis X. McKelvey, *Planning and Design of Airports* (New York: McGraw-Hill, Inc., 1983), 509–510.
26 Horonjeff and McKelvey, 510, 513
27 Horonjeff and McKelvey, 513.
28 Civil Aeronautics Administration, 65.
29 Civil Aeronautics Administration, 216, 224, 288, 315, 332.
30 Civil Aeronautics Administration, 233, 248, 255, 268, 280.
31 Horonjeff and McKelvey, 203–204
32 Horonjeff and McKelvey, 387.
33 Horonjeff and McKelvey, 323.
34 Horonieff and Mckelvey, 200, 300.
35 Federal Aviation Administration, *Aeronautical Information Manual: Official Guide to Basic Flight Information and ATC Procedures*, 2007, 658–659.
36 John Schamel, "The Development of Night Navigation in the U.S.," www.atchistory.org/History/nightnav.htm.
37 Debbie K. Alke, Administrator's Column, *Montana and the Sky* 53:7 (July 2002): 2.
38 Mel Duncan, Wyoming Airway Beacons, www.atchistory.org/History/FacilityPhotos/WY/RadioBeacons/Wyoming-BeaconHistory.htm
39 FAA 2007, Chapter 2-1.
40 http://historicalcharts.noaa.gov/historicals/historical_zoom.asp4.
41 Strip airway maps (ca. 1932). Website: www.atchistory.org/History/Maps/Maps.htm.
42 This glossary is adapted in part from Webster et al. (2001) and National Advisory Committee for Aeronautics, "Nomenclature for Aeronautics," *Report 474* (Washington, DC: U.S. Government Printing Office), 193.

Chapter 15

Industrial Building Construction

Early American industrial buildings generally resembled other building types or represented conversion of buildings to new uses. Blacksmith shops, barns, and sheds were converted to manufacturing use, as were dwellings, commercial buildings, and even schools and churches.

Buildings adapted to or erected for industrial use were often only slightly bigger than gristmills, residences, and institutional buildings such as town halls, schools, or churches and were built of the same materials: wood, brick, and stone.

Industrial Buildings: 1793–1930

The building recognized as the country's first factory, Slater Mill in Pawtucket, Rhode Island, was erected in 1793 for Samuel Slater's textile manufacturing business (Figure 15.1). This was the first of many textile factories erected on the rivers and streams of New England. In his study of industrial architecture of the region, Richard E. Greenwood characterized these early fabric mills:

> Like the common gristmills, sawmills, and fulling mills of the eighteenth century, the textile mill was conceived of as a timber-framed container of machinery and industrial processes. Although the earliest textile mills were domestic in scale, the large size and number of the various machines and the goal of mass production predisposed manufacturers toward structures with a large, open interior.[1]

Architectural historian William Pierson notes the English had established a basic architectural vocabulary for the early textile mill: a rectangular edifice, long and narrow in proportions, with several stories, many windows, and an unbroken, uncomplicated interior space.[2]

To supplement the natural light provided by large windows, early mill owners turned to supplementary attic windows. The earliest type of such windows was the trapdoor monitor, used on the Harrisville Upper Mill shown in Figure 15.2. This is a narrow strip of windows that takes the appearance of a raised section of roof, similar to a trapdoor in a floor.

A major constraint in the design of early industrial buildings was the power transmission system: the arrangement of gears, shafts, pulleys, and belts that transferred the energy from the revolutions of the waterwheels to the machinery. Each machine was connected by a drive belt to a revolving overhead shaft. This resulted in a linear arrangement of machinery in parallel aisles and an elongated, rectangular floor plan. Because the factory owner relied largely on natural light, the building was narrow to allow inner rows of machinery to be lit by windows. Since unencumbered open space was viewed as advantageous, most early factories had few partitions and only one or two rows of posts running the length of the factory floor. In some early mills, the top floor of the mill was hung from the roof frame, allowing the floor below to be free of posts.[3]

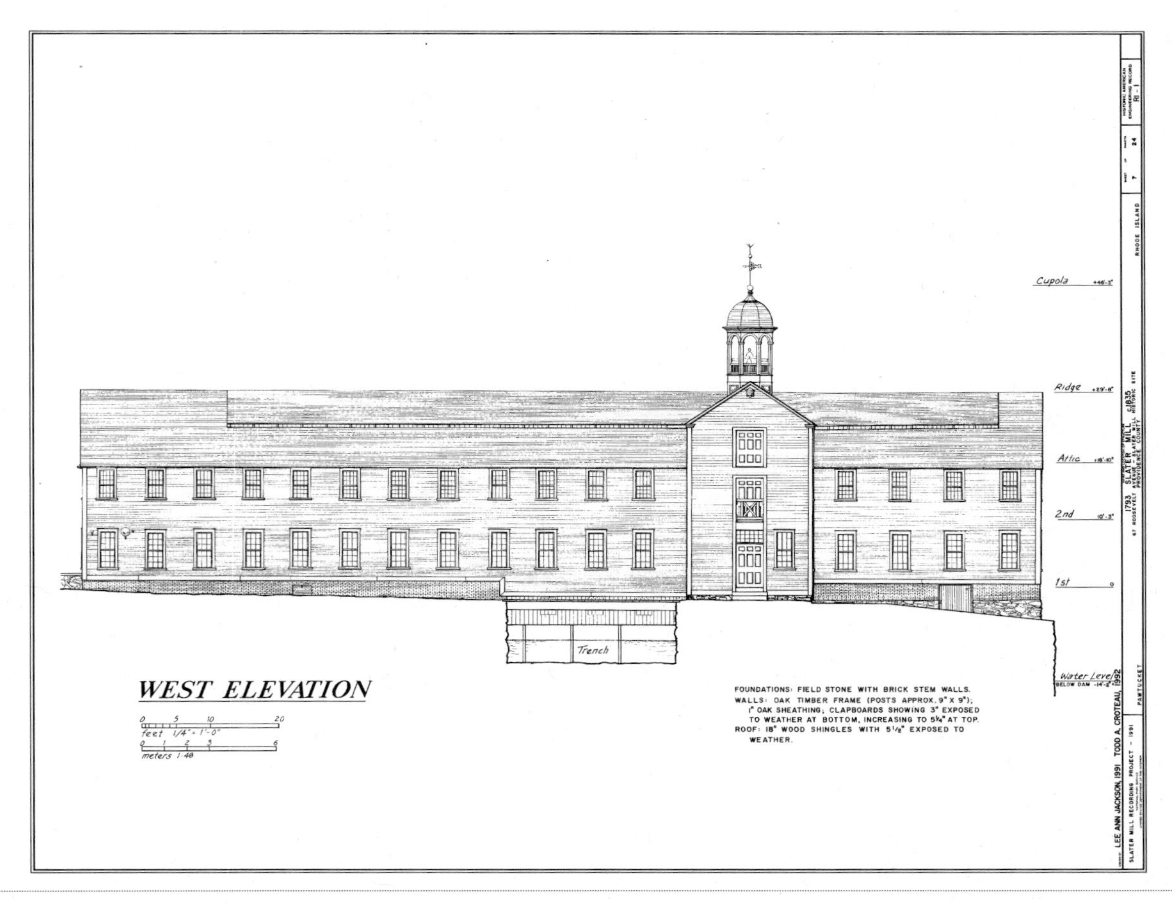

Figure 15.1. Slater Mill. Pawtucket, Rhode Island. West elevation. HAER RI-1, National Park Service. Delineated by Lee Ann Jackson, 1991, and Todd A. Croteau, 1992.

Figure 15.2. Harrisville Upper Mill. Harrisville, New Hampshire. Photograph by the author, July 2006.

The pre-1850 New England mill drew upon regional building traditions. Its skeletal plank and timber method of flooring and side-support was used in colonial houses, barns, and ships. Wood was employed for framing because it was inexpensive and readily available, it was light and easy to handle, it lent itself to the creation of unobstructed interior spaces, and its tensile strength allowed it to absorb machinery vibrations. Early mills, such as Slater's, used a repetitive post and beam and joist pattern that could be multiplied without basic change in construction method. This system prevailed until about 1810, when bearing-wall construction became common in factory design.[4]

By the early nineteenth century, New England mills had begun to grow in size. An example of this new type of mill is that of the Lippitt Company in West Warwick, Rhode Island. Three stories in height plus attic, this building is taller, wider, and longer than the Slater mill. A new design feature also became common. Along the ridgepole, the roofline was broken and the level raised to create an extended clerestory monitor. This was a way of admitting more light than was possible with the earlier trapdoor monitor.[5]

In about 1810, the wooden-box form began to be supplanted with the box structure of stone and rubble or brick masonry. In most instances, these new buildings were comparable in size and scale to their wooden predecessors. Among the first American mills to be built with stone walls was the original Georgia Mill in Smithfield, Rhode Island. An example of this new style is the Harris Company Upper Mill in Harrisville, New Hampshire.[6] Stone mill construction was concentrated in limited areas of New England, particularly Fall River, Massachusetts; Willimantic, Connecticut; western Massachusetts and northwestern Connecticut; and Rhode Island.[7]

Locally available stone such as granite, schist, sandstone, and limestone were noncombustible and inexpensive building materials. The style of masonry employed in mill walls ranged from roughly cut and coursed stone to coursed ashlar. In some cases, the facade of the building was constructed of finely finished stone, while the remaining sides had a rougher finish.[8]

Another early development in mill construction came with the removal of the interior stairway to an outside tower. This configuration provided more space for machinery and also offered some fire protection. By 1815, the first bell cupola was built to crown a stair tower.[9]

Stone mills of the 1820s through 1840s were of more finely detailed construction. An example is the Granite Mill or Cheshire Mill in Harrisville, New Hampshire. Comparatively low and narrow and of moderate length, they retained earlier rectangular floor plans, clerestory monitors, and an outside beam and hoist serving each level through loading doors. Greek Revival stylistic details were frequently employed, and bell towers were often placed over the center of roof ridges.

Between 1830 and 1850, New England mills evolved into the massive blocks that began to be constructed in textile centers, including Lowell, Lawrence, Haverhill, New Bedford, and Fall River, Massachusetts, and Manchester, New Hampshire. One such mill is the Amoskeag Mill in Manchester.

A common characteristic of these mills was greater horizontal and vertical size than earlier buildings. In many instances, a flatter roof replaced the steeper roof. In pitched roofs, dormer windows or skylights often replaced eyebrow or clerestory monitors. The outside stair tower gained prominence.[10]

As mid-century approached, buildings rose in height to as tall as six stories. Simplified brick construction often replaced stone, and "slow-burning" construction of interconnected timber members, heavy floor planking, and wooden columns and beams was employed. Regularly placed window apertures featured granite lintels (Figure 15.3).[11]

By mid-century, a new standard of mill construction developed that employed brick or stone bearing walls, interior columns of cast iron, and floor beams and roof rafters or trusses of wood.[12] In brick construction, strength of exterior walls was concentrated in brick piers that projected from adjacent sections of panel walls as pilasters and supported the heavier roof-framing members.[13]

Iron was introduced incrementally into the fabric of American industrial buildings. Its first use was in tie rods that braced brick structures and as elements in predominantly timber roof trusses. Subsequent steps included use of iron roof trusses and cast-iron bearing plates and columns to support machinery loads in multistory lofts. Combination framing systems used cast-iron columns in compression and wood columns in tension. Mill engineers experimented with using cast-iron

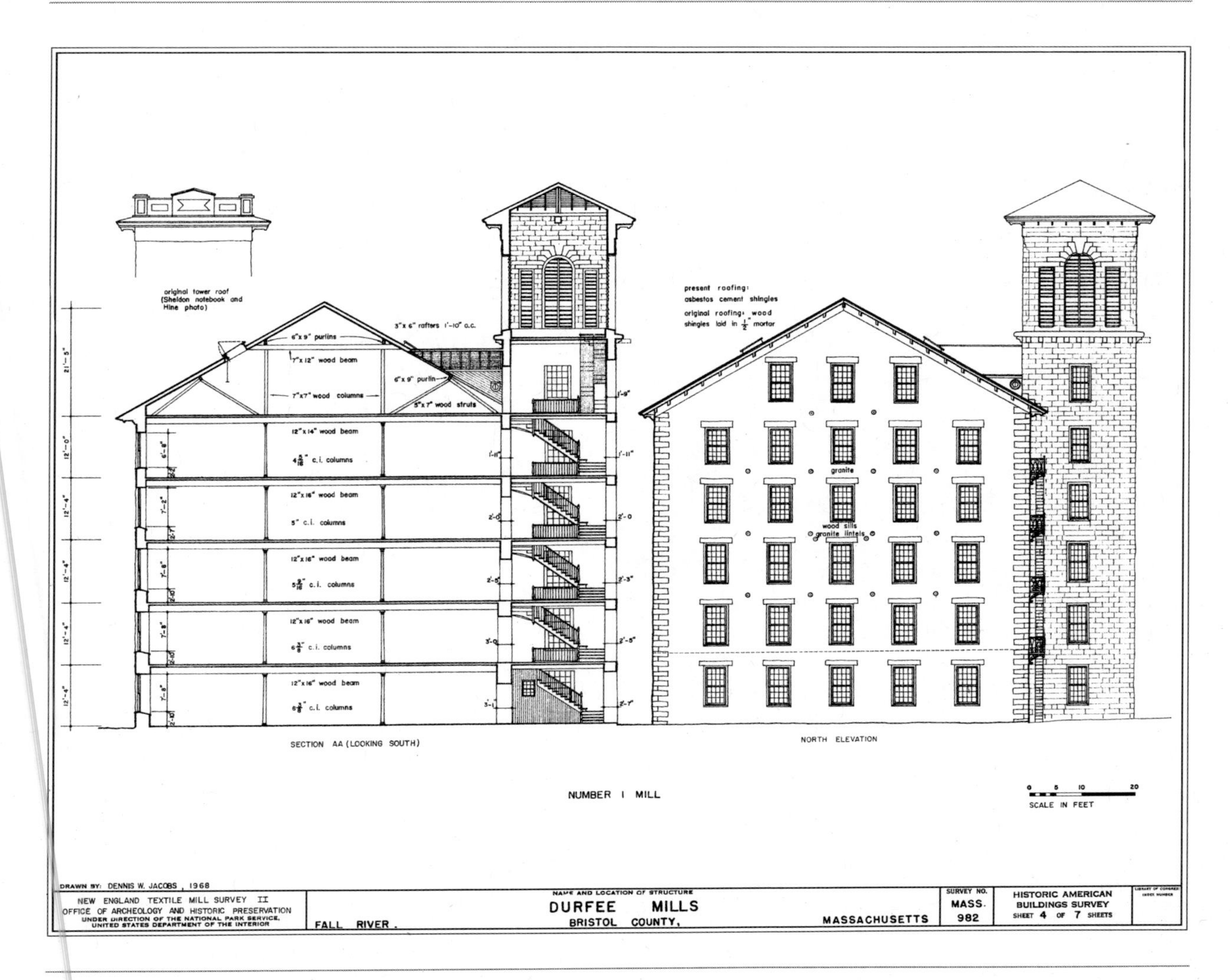

Figure 15.3. Mid-nineteenth-century mill construction. Durfee Mills, Fall River, Massachusetts. HABS MA-982, National Park Service. Delineated by Eric DeLony, 1968.

columns in textile mills as early as the mid-1820s. By the 1870s, buildings framed entirely of wrought iron and sheathed in corrugated sheet metal, with clear spans up to 150 feet, were recommended for railroad or industrial use.[14]

During the late nineteenth century, cast iron, wrought iron, and steel were all employed in industrial building construction. Steel rapidly increased in popularity in industrial building construction because its tensile and compressive strength provided the stability and spans needed in industrial buildings. Steel-framed loft buildings were expensive because their steel members had to be protected with fireproofing materials. Therefore, steel-framed industrial lofts had to be justified by programmatic and economic considerations. Early steel-framed industrial buildings included a machine shop at the Jones & Lamson works in Springfield, Vermont, and another at Pratt and Whitney in Hartford, Connecticut.[15]

By the late nineteenth century, the advantages and disadvantages of multistory factory buildings were being debated in the engineering and architectural press. One source indicated that vibrations in multistory New England mills could have an adverse effect on delicate machinery such as carding machines and could result in the uneven working of fiber.

Such sources recommended two-story mill buildings with machinery on the second floor, with belts running from pulleys below. Recommended construction featured strong dividing brick walls extending through the roof, with iron sliding doors for communication. This design not only reduced vibration but also prevented the spread of fire.[16]

Despite the promotion of lower mill buildings with greater floor area, taller textile mills continued to be built, particularly in New England. The Factory Mutual Insurance Association developed a standardized mill design in 1890. Of brick construction, this oblong building was four stories in height, clad in brick with a stair tower appended to the middle of one of the longer sides. A two-story end block was used as a picker shed. Versions of these standard mills were erected in Fall River (in stone) and in other cities.[17]

Building materials used in factories also changed in the late nineteenth century. Iron, stone, and brick had been almost universally substituted for wood. Some iron manufacturers, such as Connecticut's Berlin Iron Bridge Company, could furnish complete drawings and specifications for erection of iron manufacturing buildings.[18]

As Betsy Hunter Bradley indicated,[19] buildings designed and erected as factories reflected both the demands of machines and workers. As the scale of manufacturing increased during the late nineteenth century, the layout of manufacturing space was determined in part by the largest area that could be effectively supervised by a single foreman. Sturdy building framing was needed to resist the vibration and oscillation that accompanied equipment operation. Positioning of machinery was affected by the location of power distribution. Need for daylight in work areas also played an important role in building design. In long, narrow manufacturing spaces of the nineteenth century, workbenches lined exterior walls or were placed perpendicular to them next to windows. The center bay of the factory, typically divided by two rows of columns, where lighting conditions were poorest, was used for storage and as a transportation corridor.

By the turn of the twentieth century, the wall construction most commonly employed in mill buildings included solid brick walls, iron columns with brick columns, or iron columns and purlins covered with corrugated iron. Corrugated iron was the cheapest construction but could not be used for buildings that were heated. Brick walls had the advantage of rigid construction suited to withstand the action of cranes and heavy machinery. In the case of very thick walls under trusses, the most economical construction was thought to be iron columns with curtain walls.[20]

Despite the common use of other materials, in some areas of the country such as New England and the Pacific Northwest, timber framing continued to predominate into the twentieth century. As Henry Tyrell noted in 1911, in the New England states more mills were framed of wood than all other materials combined. Fire resistance was improved in twentieth-century timber-framed mills by use of metal beam hangers to anchor floor timbers bearing on walls and by the separation of belt towers from the working floors by brick partitions with fire doors.[21]

Ground-floor materials used in early twentieth-century mill buildings include concrete, asphalt, and wood.[22] Upper-floor materials include steel troughs, corrugated iron and brick arches, steel girder and timber, or slow-burning wood. Roof coverings include slate, asphalt, slag and gravel, corrugated iron, sheet steel, steel roll, tin or terne plate, metal shingle, rubber, and asbestos.[23]

Reinforced concrete was introduced in American industrial buildings at the beginning of the twentieth century. Most factory buildings constructed of reinforced concrete were monolithic structures. Among the earliest American industrial building that used this construction was Ernest L. Ransome's Pacific Coast Borax Refinery (189–1898) in Bayonne, New Jersey. Its 200-foot-long exterior walls were self-supporting and were pierced by only small window openings. The factory suffered a severe fire in 1902 and was subsequently rebuilt with fewer combustible materials.[24]

By 1902, Ransome had developed and patented a skeletal method of reinforced concrete construction in which the floor slab was extended beyond the face of the building to support brick panel walls and large windows. An early example of this construction was found in the Kelly & Jones Co. factory in Greensburg, Pennsylvania (ca. 1903). In this 60 by 300 foot, four-story building, the structural framework was constructed entirely of concrete reinforced by twisted steel rods.[25] In Ransome's buildings, the floor slab was poured as a unit, with the joists forming parallel ribs of rectangular section, the whole resting on deep girders spanning between columns in both directions. Beams and joists were reinforced with bars located near the undersurface.[26]

Figure 15.4. Interior of building showing concrete slab and mushroom column construction developed by C.A.P. Turner. Reprinted from C. K. Smoley, *Concrete Design* (Scranton, PA: International Textbook Company, 1910).

A second major development in reinforced concrete construction was column-and-slab framing. Developed by Minneapolis engineer C. A. P. Turner in 1905–1906, it combined concentric, radial, and continuous multiple-way reinforcement with the flared column capital (Figure 15.4). The first structure to use Turner's system was the Johnson-Bovey Building in Minneapolis (1906).[27]

MODERN INDUSTRIAL BUILDINGS

A thorough discussion of twentieth-century industrial building design could be the subject of an entire book. One architect whose designs influenced many modern industrial production sheds was Albert Kahn.[28] Kahn articulated his design philosophy in his 1939 article titled "Industrial Architecture":

> Industrial architecture must necessarily deal with the practical first, with proper functioning of the plant, with best work conditions, efficiency and flexibility, with economical and safe construction, and only last with external appearance.[29]

Elements common to many of his industrial buildings are long expanses of side walls filled with bands of concrete stucco and steel sash that conceal the interior framing bays, the disposition of conventional terminal elements such as corner piers and parapeted gable ends, and the later use of glazed sheds atop which roofs appear to float (Figure 15.5).[30]

INDUSTRIAL BUILDING TYPES

By the mid-nineteenth century, new manufacturing processes and new types of machinery led to the development of specialized industrial workspaces. Some of the more common building types are described below.

Blacksmith or Forge Shop

In this building, iron was heated in forges and beaten or hammered into desired shapes. A forge could usually be identified by several small chimneys, often constructed of sheet metal, rising through the roof marking the location of forges within (Figure 15.6). Sometimes a brick chimney, which drew a draft on the forges through an underground manifold, stood beside the structure. The

Figure 15.5. Example of an Albert Kahn-designed building. Ford Motor Company, Long Beach assembly plant. Long Beach, California. Northeast side. HAER CA-82A. Photograph by David G. DeVries, 1990.

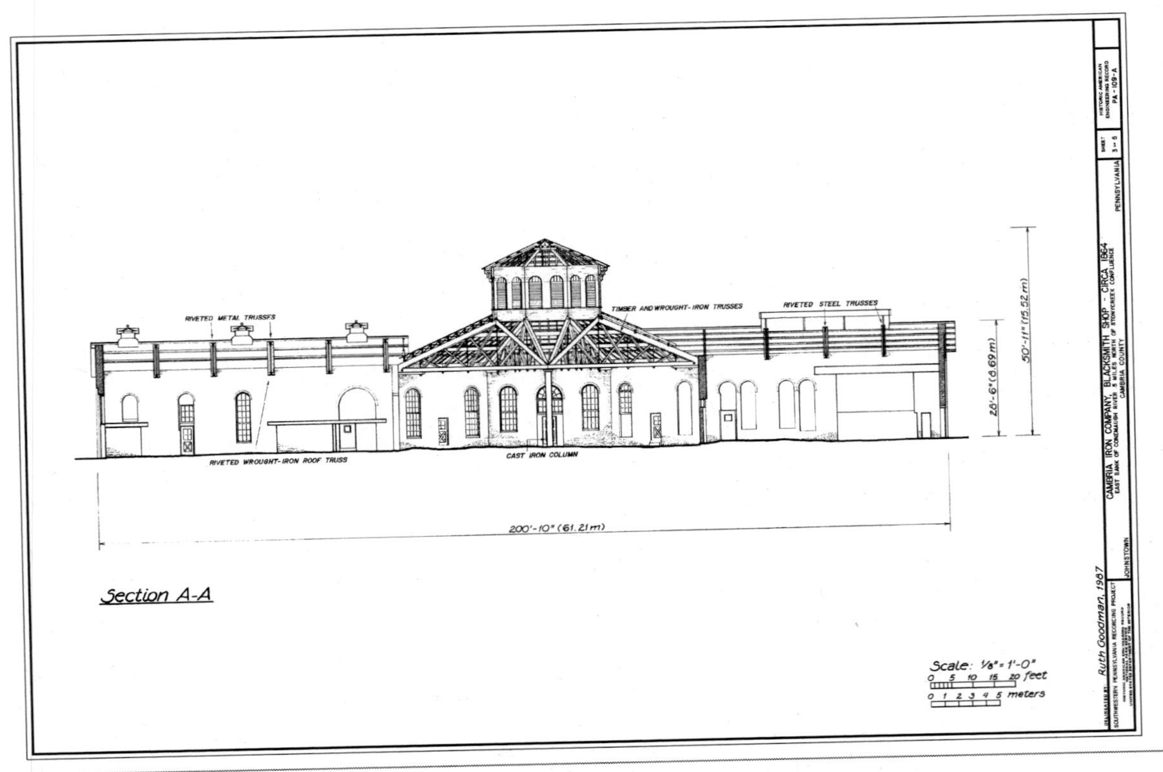

Figure 15.6. Blacksmith or forge. Cambria Iron Works, Johnstown, PA. HAER PA-109A, National Park Service. Delineated by Ruth Goodman, 1987.

forge was generally placed near an erecting shop or foundry.

The twentieth-century forge shop reflected changes in building design. Engineers recommended that the forge shop have a clear height under roof trusses of at least 14 feet and that side walls have 6-foot-tall bands of continuous window sash. By that time, the rows of small forge stacks had been largely placed by chimneys that could serve several forges.[31]

Boiler House

The boiler house is a freestanding structure, or half of a boiler and engine house complex, in which steam boilers are situated to provide power to an industrial complex. Depending on the energy needs of a facility, the boiler house may range in size from about 50 by 35 feet to 100 by 50 feet or more. The outside of the building is typically fenestrated with steel industrial sash windows, has little ornamentation, and is topped by a flat roof from which rise metal ventilators and discharge pipes. Rolldown garage doors are typically placed in walls. Commonly, the boiler house is divided into two rooms, separated by a concrete block fire wall. One room is typically used for general and fuel storage, while the other room contains the boiler or boilers. Overhead pipes typically transfer steam generated by the boiler(s) to consuming building(s).[32]

Erecting Shop

The erecting shop was a building or part of a building used for the final assembly of large products. In heavy industry, erecting shops were relatively large spaces kept as free as possible of columns and millwork. The height needed for crane operation and the assembly of large items might be equal to a two- or three-story space. Jibs and overhead traveling cranes lifted pieces into position, supported them as they were attached, and moved completed items out of the shop. Grates formed of I-beams were placed across pits in the floor so that the undersides of large and heavy items were accessible during assembly.[33]

Fitting Shop

A fitting shop was a building or portion of a building in which fitters assembled smaller parts or where parts were brought together for final assembly. An early example of this building type was the 1881 Rogers Locomotive Works Shop in Paterson, New Jersey (Figure 15.7). This four-story brick building had four floors used for machining operations. The lower floors were frame-fitting shops, while the upper floors housed lighter machining operations. Power for operations was provided from an adjoining wheelhouse via shafting that entered the building above the bearing walls.[34]

Foundry

The shed housing the foundry was sometimes the only building in a works to have a turret or a roof monitor for improved lighting and ventilation. The chimneys that stood near metal-heating cupolas were signature features of foundries.

In the mid-nineteenth century, the foundry was a freestanding structure situated as far as possible from the other buildings in the works because of the heat and smoke it generated. Foundries required yard room nearby for the storage of raw materials, flasks, and castings. The foundry housed a series of activities for making molds, heating iron, pouring molten iron, and processing castings.

Foundries as large as 100 feet in width and length were common by the mid-nineteenth century, though many were much smaller. Between 1880 and 1900, the foundry was transformed into a better lighted structure with a central craneway. Foundries were among the first industrial buildings to be erected with exterior walls filled with window sash. Cupolas were positioned just outside the building, with tapping spouts extending through the wall to free a central crane-served bay to serve as the molding and casting floor.[35]

Industrial Lofts

To accommodate industrial production in urban areas, companies erected multistory industrial buildings. Many such buildings remain and are well suited to adaptive use due to open floor areas, high ceiling heights, and large windows. Many cities now boast trendy loft residential districts, while other buildings have been successfully converted to office space.

The industrial loft was developed as a building that provided two or more stories of lofts, in which vertical circulation and service areas were grouped so as to intrude into the work space as little as possible. Industrial lofts were most typically constructed with an interior

ROGERS FITTING SHOP

THE ROGERS LOCOMOTIVE AND MACHINE WORKS FITTING SHOP SERVED AS ONE OF THE PRIME MANUFACTURING BUILDINGS WITHIN THE ROGERS WORKS COMPLEX. IT IS THE MOST RECENT OF THE EXTANT BUILDINGS OF THE COMPLEX, HAVING BEEN CONSTRUCTED IN 1881.

DURING ITS ORIGINAL USE, THE FITTING SHOP HAD FOUR FLOORS SERVING FOR MACHINING OPERATIONS, ON THE LOWER FLOORS WERE FRAME-FITTING SHOPS, WHILE THE UPPER FLOORS HAD LIGHTER MACHINING OPERATIONS. POWER WAS SUPPLIED FROM AN ADJOINING WHEEL HOUSE VIA SHAFTING ENTERING ABOVE THE BUILDING'S BEARING WALLS THROUGH THE MONITOR AREA.

THE BUILDING'S MASSIVE STRUCTURAL SYSTEM IS AN UNUSUAL ARRANGEMENT OF RIVETED WROUGHT IRON I-BEAMS, PLATES AND CHANNELS CHARACTERISTIC OF RAIROAD BRIDGES AND ELEVATED ROADWAYS OF THE PERIOD.

SOMETIME AFTER ITS CONSTRUCTION THE FOURTH FLOOR AND THE MONITOR BURNED. THE FIRE, POSSIBLY ON JANUARY 23, 1886, DEFORMED MANY OF THE BEAMS THAT SUPPORTED THE FOURTH FLOOR, PARTICULARLY ON THE NORTH SIDE OF THE BUILDING, AND THUS MADE IT IMPOSSIBLE TO REBUILD THE FOURTH FLOOR WITHOUT REPLACING THE MASSIVE, BUILT-UP BEAMS. THE DEFORMED BEAMS WERE LEFT INTACT AND THE FOURTH FLOOR AND MONITOR WERE REPLACED WITH A CONVENTIONAL SLOPED ROOF WITH ONLY A SHORT, ENCLOSED MONITOR AS AN INDICATION OF WHERE THE ORIGINAL MONITOR WAS LOCATED.

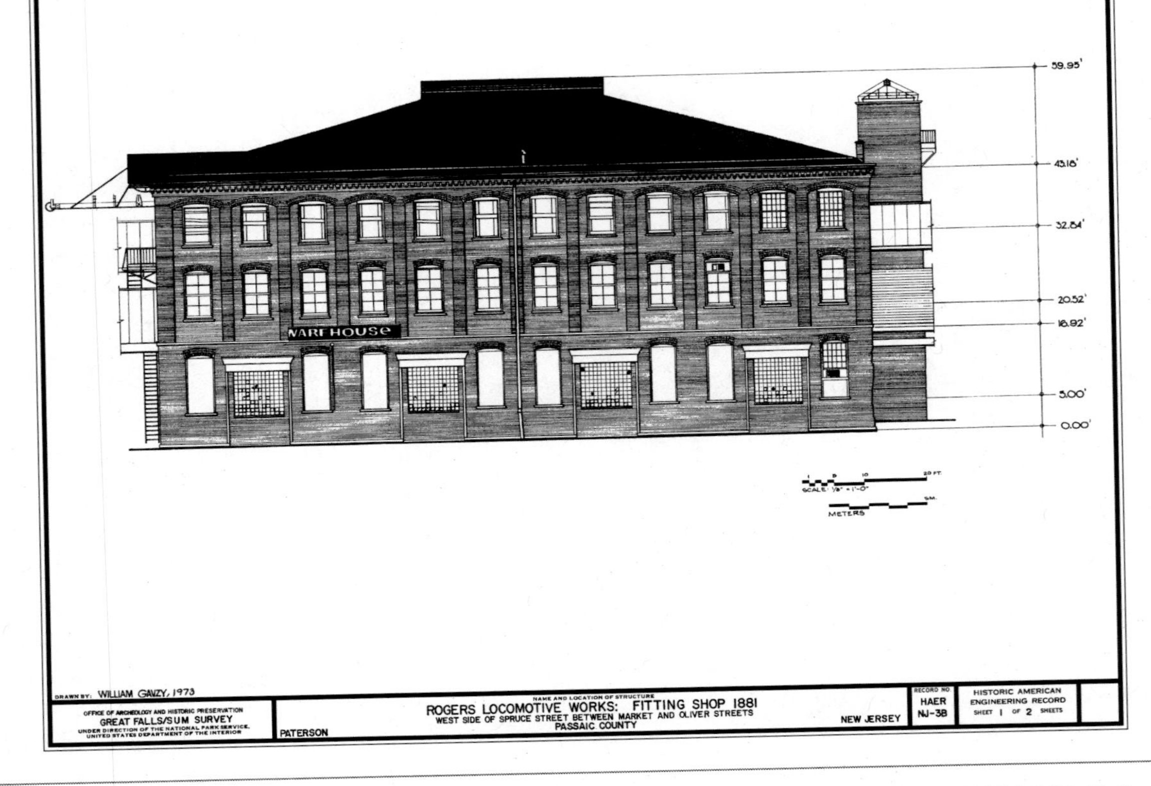

DRAWN BY: WILLIAM GAVZY, 1973

OFFICE OF ARCHEOLOGY AND HISTORIC PRESERVATION
GREAT FALLS/S.U.M. SURVEY
UNDER DIRECTION OF THE NATIONAL PARK SERVICE, UNITED STATES DEPARTMENT OF THE INTERIOR

PATERSON

NAME AND LOCATION OF STRUCTURE
ROGERS LOCOMOTIVE WORKS: FITTING SHOP 1881
WEST SIDE OF SPRUCE STREET BETWEEN MARKET AND OLIVER STREETS
PASSAIC COUNTY

NEW JERSEY

RECORD NO. HAER NJ-38

HISTORIC AMERICAN ENGINEERING RECORD
SHEET 1 OF 2 SHEETS

(ON FACING PAGES): **Figure 15.7.** Rogers Locomotive Works: Fitting Shop, Paterson, New Jersey. HAER NJ-38, National Park Service. Delineated by William Gavzy, 1973.

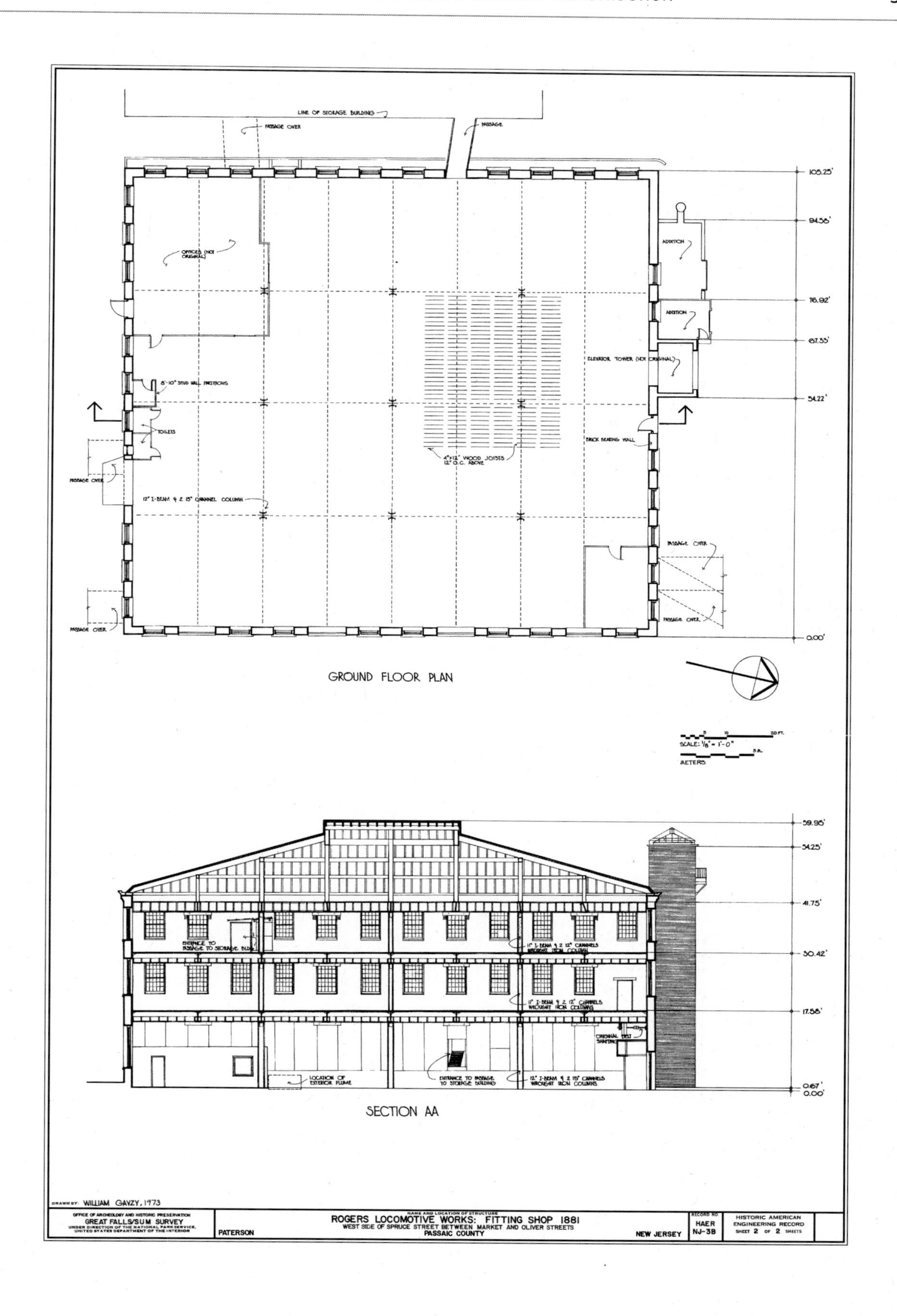
GROUND FLOOR PLAN
SECTION AA
DRAWN BY: WILLIAM GAVZY, 1973
OFFICE OF ARCHEOLOGY AND HISTORIC PRESERVATION
GREAT FALLS/SUM SURVEY
UNDER DIRECTION OF THE NATIONAL PARK SERVICE, UNITED STATES DEPARTMENT OF THE INTERIOR
ROGERS LOCOMOTIVE WORKS: FITTING SHOP 1881
WEST SIDE OF SPRUCE STREET BETWEEN MARKET AND OLIVER STREETS
PATERSON
PASSAIC COUNTY
NEW JERSEY
HAER NJ-38
HISTORIC AMERICAN ENGINEERING RECORD
SHEET 2 OF 2 SHEETS

Figure 15.8. The Ketterlinus Building, Fourth and Arch Streets, Philadelphia. Ballinger and Perrot, engineers and architects. Reprinted from Atlas Portland Cement Company, *Reinforced Concrete in Factory Construction* (New York: Atlas Portland Cement Company, 1907), 60.

wood frame and stone or brick exterior walls. For example, the Ketterlinus Building housed printing presses on the third through fifth floors (Figure 15.8).[36]

Detailing typically included a regular pattern of windows, raised loading platforms, loading bays with vehicular access doors on the street, and hoistways. Fire escapes and stair towers, a power transmission belt, and utilities sometimes projected from facades to keep floor areas unobstructed. Water tanks, perhaps enclosed in towers, and elevator bulkheads rose from flat roofs.

Machine Shop

A machine shop is a building or a portion of a building used for cold-shaping of iron and steel by means of machines. In the works, a machine shop might be a portion of a loft structure, an entire loft, or a one-story production shed. When possible, it constituted a separate building or a space partitioned off from other functions, to limit the presence of dust. Machine shops were engineered primarily for strength and resistance to vibration. The machine shop was often located near the works engine and boiler house to maintain efficient transfer of power, and it was also close to the erecting and forge shops.

The machine shop and the erecting shop were often combined after electric drive was introduced. Electric drive also permitted the use of wider machine shops, since roof lighting was more effective when no millwork obstructed the overhead area.[37]

Mill Building

A mill building was a wood-framed structure, a production shed, consisting of two rows of columns containing crane runways and supporting a roof and usually flanked by lean-tos at the sides. By the turn of the twentieth century, the types of wall construction most commonly employed in mill buildings included solid brick walls, iron columns with brick curtain walls, and iron columns and purlins covered with corrugated iron. Corrugated iron was the cheapest but could not be used for buildings that were to be heated, while brick walls supplied a rigid construction suited to withstand the action of cranes and heavy machinery.[38]

Office and Administration Buildings

The office of a nineteenth-century works was housed in either a freestanding building or a portion of a main loft building. During the mid-century, the office was generally a small two-story building of masonry or wood construction. General offices were located on the ground floor of these buildings. The upper floor, well-lighted and isolated from dust and vibration, was usually used as the drafting room. The office was placed in a central location or near the main gate to facilitate supervision of the works.[39]

Pattern Storehouses or Shops

Wood and metal patterns were used in metal casting and other molding operations and also served as models for various types of products. Master templates either supplemented or substituted for shop drawings as the design record. Specialized buildings or parts of buildings were erected for their safekeeping. These buildings, often two or three stories in height, were invariably of fireproof construction to protect the material stored there.[40]

An example of a small pattern storehouse is that of the Hardie-Tynes Manufacturing Company in Birmingham, Alabama, manufacturers of mine hoists (Figure 15.9). The pattern shop was a two-story building with carpentry tools on the upper floor, space for lumber storage, and racks for pattern storage.

Powerhouse

Powerhouses were constructed of noncombustible materials, usually stone or brick, and later reinforced concrete. They were two-part facilities, with separate rooms for boilers and engines, and sometimes a third area for coal storage. Each area was spanned with truss roofs to eliminate interior columns. The size of the structure was governed by the number of boilers and engines it housed and by the space needed to stoke boilers, maintain equipment, and replace boiler tubes.[41]

Production Shed

A production shed was a one-story industrial building with interior spaces of considerable height and wide bays, possessing the strength and stability to support overhead traveling cranes. After 1880, exterior walls were engineered curtain walls sheathed with a variety of materials. Roofs usually incorporated means for lighting and ventilation. By the end of the nineteenth century, the production shed often became a series of bays of various dimensions, some served by cranes.[42]

POWER GENERATION AND DISTRIBUTION

During the history of manufacturing in the United States, various prime movers have been used to generate industrial power. These have included water, steam, and electric power.

Waterpower

As noted elsewhere, waterpower was essential to the early development of the textile industry in the United

Figure 15.9. Pattern Shop. Hardie-Tynes Manufacturing Company, Birmingham, Alabama. HAER AL-13, National Park Service. Photograph by David Diesing, 1992.

Figure 15.10. Site plan of Monadnock Mill, Claremont, New Hampshire, showing the source of waterpower, power canal, dams, and wheelhouse. HAER NH-1. Delineated by Beth Cohen and Margaret Mook, 1978.

States and also played an important role in the early development of other industrial sectors. With a waterwheel or turbine, the ideal location for an industrial building was at the edge of a river or canal so that the power source could be incorporated into the base of the structure (Figure 15.10).

When this was not possible, water was brought to the site by a flume or canal, or power was brought mechanically from an external waterwheel, sometimes by a rope drive. These industrial enterprises were limited in size by the waterpower capacity.

Steam Power

With the development of the practical steam engine in the 1840s, manufacturers began to be liberated from siting constraints associated with waterpower. In addition, more power became available. Steam power influenced the design of manufacturing works. If a single steam engine powered the facility, it was centrally located for efficient power distribution. Decentralized steam power permitted buildings to be sited to facilitate the flow of production rather than for proximity to the engine house.[43]

In some factory complexes, such as Alabama's Prattsville Manufacturing Company, a textile mill, the use of steam power represented part of the evolution of the facility's power system. The first mill in the complex was powered by water turbines. When the second mill was built, the decision was made to operate it using steam power. A steam engine with a single boiler was housed in an auxiliary building west of the mill. The principal power belt from the steam engine entered the basement, and shafts, belts, and pulleys distributed power through the floor.[44]

Figure 15.11. Power transmission pulleys and shaft. Crown and Eagle Mills, South Uxbridge, Massachusetts. HABS MA-991, National Park Service. Photograph by R. Randolph Langenbach, 1967.

Millwork

Water and steam power were mechanically distributed through millwork that linked the building and mechanical equipment. Millwork consisted of shafts, gears, pulleys, and belting. Shafting was supported by metal hanger-plates that were attached to the bottoms of girders, beams, and joists. From a main shaft on each floor, power was distributed to several line shafts. Belts extended downward from a line shaft to machinery below it and also upward through the floor to run machinery on the story above (Figure 15.11).

This method of power distribution had a direct impact on the form of a textile mill. The vertical shafts extended up through the floors and connected to bevel gear wheels and horizontal shafts that extended the length of each floor of the mill. This system required mills to be tall and narrow because power could not be transferred effectively over 100 feet. Floor areas of limited length were stacked to provide the needed space. Using efficient water turbines and positioning the main drive shaft in a central location so it could operate two systems of millwork, each half the length of the building, made it possible to erect mills several hundred feet long and only three stories high.[45]

Electrical Power

Electric lighting and electric drive revolutionized industrial operations. The elimination of the mechanical distribution of power through millwork led to a new freedom in production layouts. Using electric drive freed up spaces formerly occupied by steam engines and millwork for the manufacturing process and offered the economies of a centralized power plant.

Small electric motors were designed specifically for use by manufacturing operations of limited size. The adoption of electric drive did not eliminate all millwork. Experts recommended that large machines use individual electric motors. Smaller tools and machines kept in nearly constant use, such as textile looms, could be grouped and driven by motor-driven shaft in an arrangement called a group drive. This arrangement used only short lengths of light shafting. Motors for the group drive were either supported by brackets on side walls or hung from the ceiling and were often connected through the floor to machinery on the floor above.[46]

Types of Construction

Factory fires were a frequent danger in many nineteenth-century industrial buildings, due to the character of industrial processes, the generation of sources of fire, and the presence of combustible materials. By the mid-nineteenth century, manufacturers had developed several approaches to prevent, detect, and suppress fires in their factories.

A building could be considered fireproof if it were erected primarily of noncombustible or fire-resistant materials and incorporated elements of fire-resistive construction. Fireproof construction usually referred to assemblies that eliminated wood and were of masonry, metal, or concrete construction. Nineteenth-century fireproof construction relied on the use of noncombustible materials, such as heavy timber framing, in sufficient quantity that a fire might damage the surfaces but not compromise structural integrity.

Other fireproofing elements included the use of intermediate fire walls extended above the roof level, and installation of automatic sprinkler systems. Early sprinkler systems designed for textile mills consisted of perforated pipes placed under ceilings and supplied with water from a tank or reservoir. The system was operated by controls on the exterior of the building. Isolation was a dominant principle in fire-resistive construction. Hazardous facilities such as furnaces and boilers, forges, lumber-drying kilns, and dust-laden operations were housed in freestanding structures.

Another element of this construction was the use of sets of fire doors, lined with tin or other metal sheets on both sides, and hung on hardware that caused them to close automatically after each opening. The doors, made of two thicknesses of matched boards, were 4 inches larger in each direction than the openings they protected (Figure 15.12). The doors were usually hung on sloping tracks, and some were equipped with fusible solder to close the door when the temperature reached a certain point.[47] Vertical flues, such as stairways, were isolated to the extent possible, in towers rising along the side of a factory building.

Elevators were placed in fireproof shafts, guarded by fire doors and covered overhead by skylights glazed with thin glass; alternatively, the hatchways in each floor could be guarded by automatic or self-closing latches. All belts or ropes were placed in incombustible vertical belt chambers. The most important feature in this construction was to make each floor continuous, avoiding belt holes and open ways, so that a fire originating in any one room could be confined to that room or story. As a rule, timbers were left unprotected.[48]

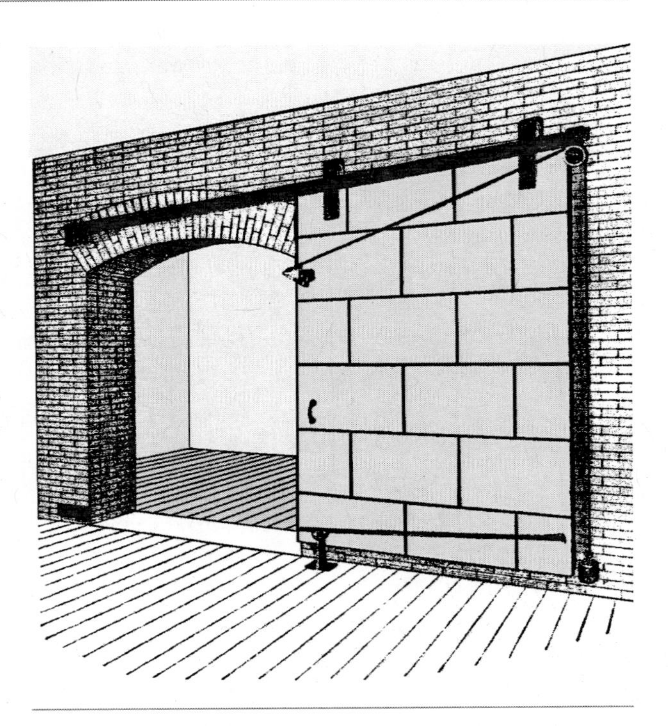

Figure 15.12. Victor Automatic Sliding Fire Door. Reprinted from Everett Uberto Crosby, *Hand-book of the Underwriters' Bureau of New England* (Boston: Press of the Standard Publishing Company, 1896), 20.

In addition, metal fire shutters were frequently installed. Early fire shutters were of thin iron sheet, unable to withstand heat. As a result, flames quickly penetrated the panes and ignited the interior woodwork. Later fire shutters were of more substantial construction and were fitted to the adjacent firebrick to reduce fire infiltration.

Factory or Mill Construction

Ordinary or non-fireproof construction was also modernized with modifications termed "mill" or "factory" construction. In these buildings, walls were masonry, usually brick, no less than 1 foot thick at the top story and increased in thickness at lower floors. Roof planks were at least 2¼ inches thick laid on timbers no less than 6 inches in either dimension. Floor planks were 3 or more inches thick, with 1-inch top floors laid at right angles resting on timbers no less than 8 inches in either dimension. All timbers were self-releasing by beveling the end of the beam that rested in the wall so that if the other end fell it would not rupture the wall.[49]

Slow-Burning Construction

Ordinary interior wood-framed construction used a large number of small pieces, such as floor timbers, studs, and braces, which burned rapidly, while the spaces between the beams and studs formed flues to give draft to the fire and spread it throughout the structure. When this construction was used in mills, a fire could rapidly consume the floors and allow heavy machinery to crash to the basement.[50]

Textile mills were erected with a particular type of fire-resistant construction, heavy-timber framing that became known as slow-burning construction or mill construction. This type of construction featured the use of wood for columns, floors, and roofs shaped and placed to produce the least favorable conditions for fire. James Montgomery wrote one of the early descriptions of this construction in 1840, when he noted that instead of joists, New England cotton mills used large beams with each end fastened to the side wall by a bolt and wall plate. The beams, about 5 feet apart, were supported in the center by wooden pillars with a double floor above.[51]

The beams and columns were so proportioned that they would retain enough strength not to fail even after one-third of their bulk head been charred or burned. Instead of a large number of small pieces, a small number of very large pieces were used (Figure 15.13).[52]

Praray Improved System of Mill Construction

In the late 1990s, HAER researchers studying southern textile mills rediscovered examples of a mill design system patented by Charles A. M. Praray, a Rhode Island mill engineer, in 1894. In the patent documentation, Praray describes the system as consisting of the combination, in a building, of a series of columns, resting upon piers; floor beams, wholly supported by said columns; and outer walls, independent of the floor beams and erected upon foundations of their own. The walls served as a shell or outer covering, provided with windows to furnish light. Praray's system was schematically illustrated in his patent drawing.[53]

According to HAER researchers, Praray built only five mills using the patented system, all of them located in the southern states. These include Dixie Mill in LaGrange, Georgia; the Thomas Holt Mill in Haw River, North Carolina; the Cora Cotton Mill, also in Haw River; the Georgia Western Cotton Mills at Douglasville, Georgia; and the Selma Cotton Mill in Alabama. The last mentioned mill, though later

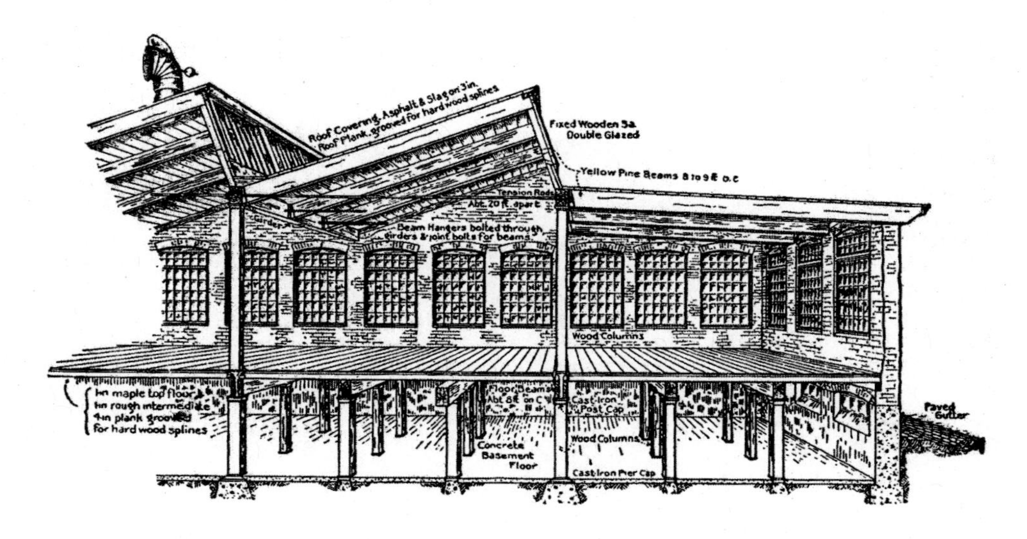

Figure 15.13. Mill fire-resistant construction. Reprinted from Insurance Institute of America, *Educational Committee, Building Construction: A Text-book Outline* (Boston: Insurance Press, 1916), 14.

converted to a cigar-making plant, is the most intact of his designs.[54]

Windows

During the mid-nineteenth century, most factories were fenestrated with multi-light wood sash, either double hung or fixed, similar to that used in other building types. Improvements in factory windows were discussed in the nineteenth-century architectural press. For example, an article in *The Manufacturer and Builder* suggested the use of wider windows supported by a thicker wall. Double-hung sash windows could be replaced with large planes of rolled glass, with the lower ones fixed and the upper one to open on hinges for ventilation.[55]

By the late nineteenth century, several types of glass were used in factory windows. Opaque glass admitted soft, diffused light. Rolled cathedral glass was translucent but not transparent, diffusing light and eliminating hard shadows.[56] Prismatic glass, with parallel ribs, bent rays of light and projected them deeper into the interior. It was recommended that clear glass be used at eye level.

Counterbalanced double-hung sash allowed large windows to be opened easily. The lowering of the top sash and the raising of the lower one facilitated the movement of air by providing an outlet for hot air. An alternative was the use of a double-hung sash window topped by an awning window. The introduction of mechanical sash-control devices in the late nineteenth century transformed window sashes into ventilating equipment. Fixed sash that had been used in roof monitors or clerestories could be replaced with horizontal pivoting sash easily operated from the shop floor (Figure 15.14).

The poor fire-resistance of glass was a problem solved in part by the introduction of wire glass. This shatter-proof material, consisting of wire mesh embedded between layers of glass, became available in the 1880s. At the turn of the twentieth century, the use of metal window frames and sash became more common.

Industrial Steel Sash

As the construction of fireproof reinforced concrete industrial lofts increased in the first part of the twentieth century, the need for rigid and fire-resistant metal sash to fill pier-to-pier window openings became more necessary. Around 1910, several American manufacturers began to offer lines of steel sash that soon became the standard in industrial buildings.

Industrial steel sash was made of rolled-steel sections (often T-shaped) designed to provide the greatest strength with the smallest dimension in order to offer the maximum glazing area. By the late 1910s, larger panes were preferred because windows were easier to keep clean when they didn't have so many dirt-catching horizontal muntins. From their initial introduction, pivoting sections of sash, which provided ventilation without admitting precipitation, were popular. Units of

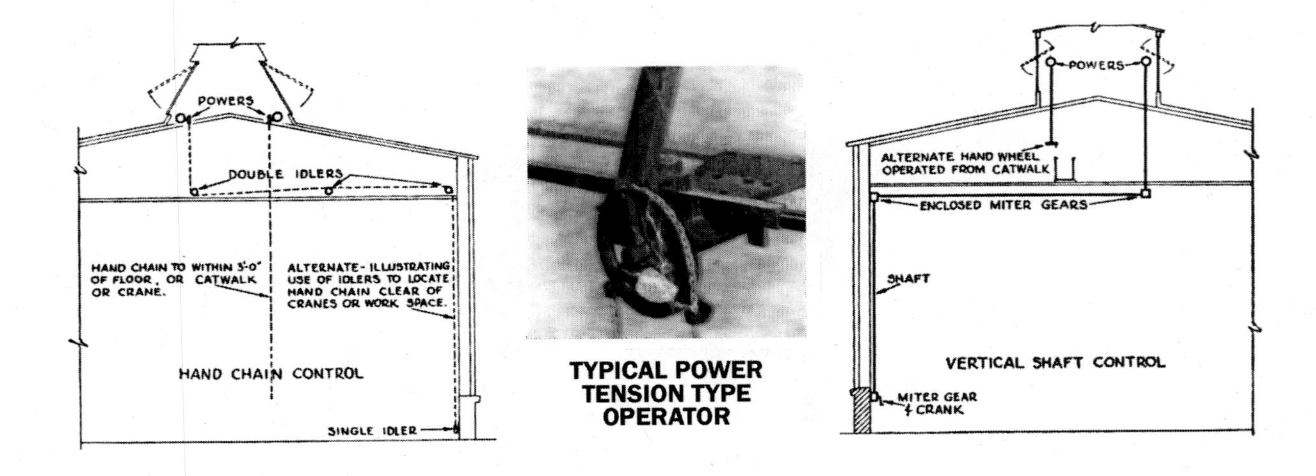

Figure 15.14. Diagram showing mechanical operation of roof monitor windows using miter gear and crank. Reprinted from Truscon Steel Company, *Truscon Steel Windows and Hangar Doors* (Youngstown, OH: Truscon Steel Company, 1945), 9.

steel sash, in various combinations of fixed and pivoting sash, came in standard sizes and were placed side by side to fill large openings or expanses of walls.

Top-hung continuous sash, operated in long sections by mechanical or electrical operating devices, was initially used in roof monitors and later in window openings. Continuous sash was placed on the exterior of some columns of loft buildings, creating expanses of sash that spanned two or three bays (Figure 15.15).

Monitor Roof

Mill construction experts advised the use of a monitor in the center of the roof to provide light and ventilation.

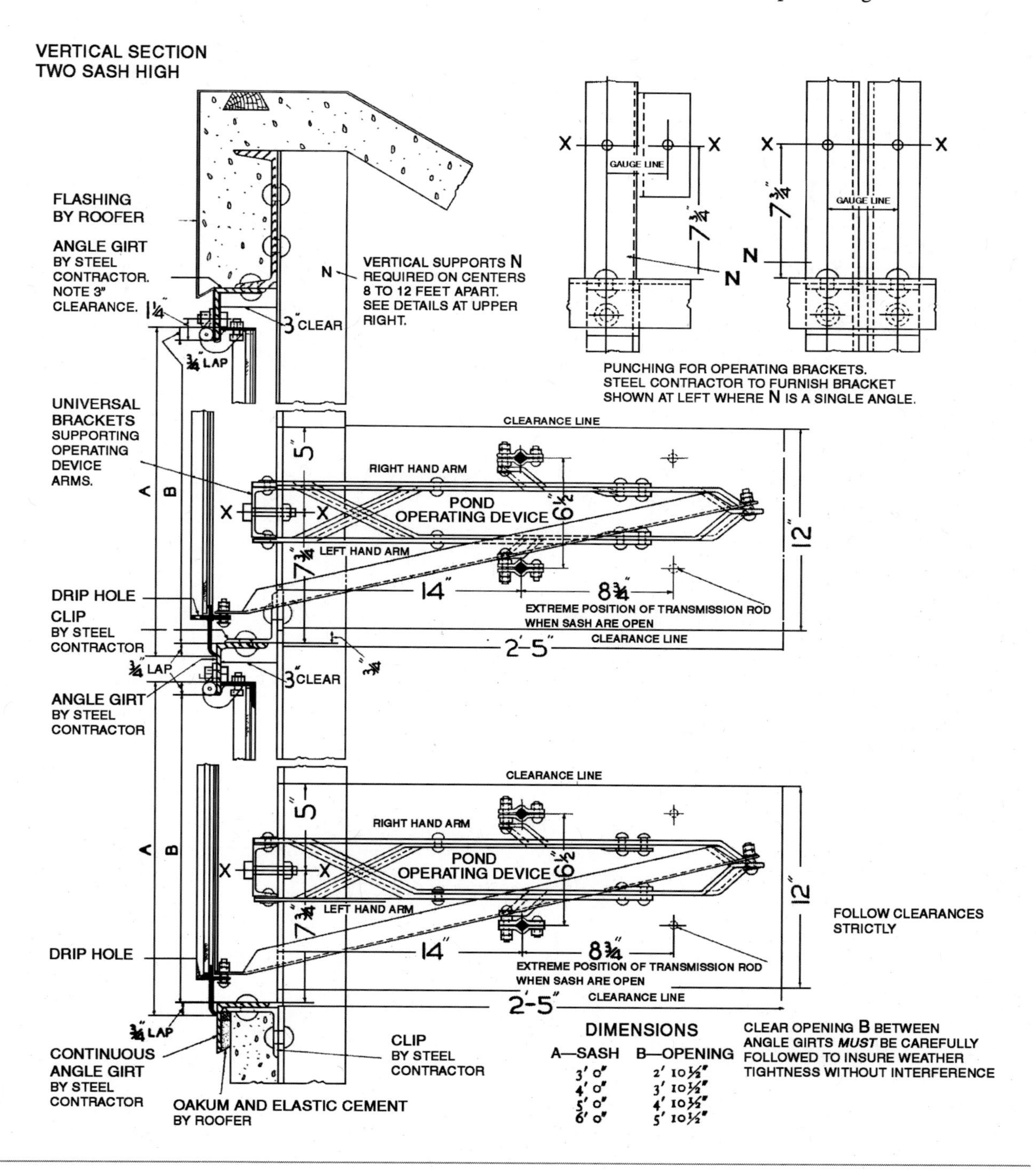

Figure 15.15. Installation and mechanism of continuous sash (Lupton 1928).

Because the roof planks did not extend under it, a substantial amount of light was thrown into the center of the upper story, making a very light room for work. The windows were usually swiveled on iron rods, so that they could be opened to allow heated air to pass out and ventilate the room. The pitch of the roof monitor was typically the same as the roof as a whole. Some monitors were built in small detached portions, and others were constructed with a continuous structure running almost the entire length of the mill.[57]

Sawtooth Roof

A sawtooth or "weaving shed" roof, a series of parallel one-sided skylights, was generally placed so that only northern light, not direct sunlight and its heat, was admitted to an industrial building. The short leg of the roof, either inclined or vertical, was glazed with glass or translucent fabric. The glazed leg of the roof typically faced north to provide a constant and agreeable light without need of window shades.

An improved version of the sawtooth roof, in which several ridges of the sawtooth structure were tied together by chords to reduce the number of interior support columns need, was patented in 1921 by Walter F. Ballinger and Clifford H. Shivers, Philadelphia-area architects (Figure 15.16). Their architecture and engineering firm used this construction in industrial commissions, most notably the Philadelphia factory of the Atwater Kent Company.

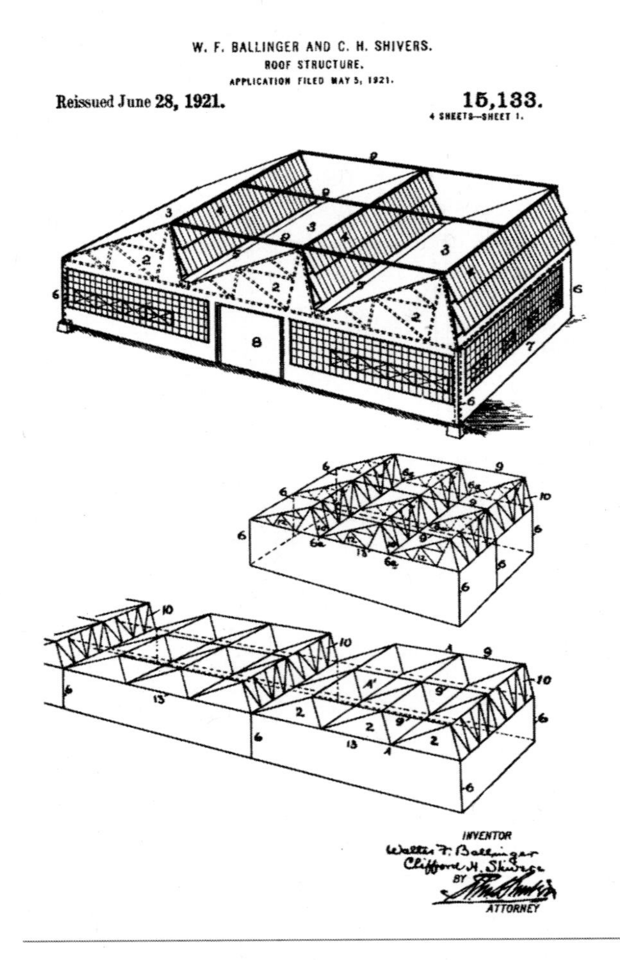

Figure 15.16. Improved sawtooth roof structure. Patent 15,133. Patented by Walter F. Ballinger and Clifford H. Shivers, June 28, 1921.

Building Framing

Frames of mill buildings are composed of a combination of trusses, monitors, rafters, purlins, chords, columns, and girders, properly braced together to form a shelter and enclosure for cranes and machinery. Framing may consist of a single-span roof resting on side walls or columns in the walls, or may have one or more rows of interior columns to support the roof and crane tracks with or without intermediate floors or galleries.

Rafters were typically constructed of two angles placed back to back with connection plates between them at the joints. The angles were riveted together at intervals of 2 to 4 feet. The bottom chords of roof trusses were also typically constructed of two angles placed back to back, but if weight had to be borne, a continuous plate was inserted between the angles.[58]

As Milo Ketchum noted, building frames for shops with heavy cranes could be more fittingly called covered craneways rather than shop buildings, for most of the framing material was in the crane supports. The crane girders and supporting columns were rigidly connected to the roof trusses.[59]

In early twentieth-century shops and mills, concrete members were often combined with steel members. Columns, sills, lintels, foundations, floors, and beams were constructed of reinforced concrete, while trusses and heavy girders were made of steel.[60]

Roof Forms

Historically, factories were constructed with gable, flat, or mansard roofs. Each form had its advantages and disadvantages.

A gable roof provided attic space and was common in nineteenth-century factories (Figure 15.17). Traditional rafters and purlins could be used to support the roof of a wide building if two rows of columns were used to divide the interior into thirds. The central bay of the building was then raised above the shed roof covering the side bays. This form became known as a monitor. The clerestory walls of the central bay were filled with windows to provide light and ventilation.

The introduction of flat roofs in the mid-nineteenth century was made possible by the introduction of built-up roofing, developed during the 1840s, that could be used on flat and very low-pitched roofs. In the following decades, flat roofs became standard for urban industrial and commercial buildings.

The construction of industrial buildings with mansard roofs was limited to a short period, generally the 1860s and 1870s. Mansard roofs allowed the introduction of light into the attic story through dormers that projected from the lower slope of the roof and, sometimes, by skylights in the upper, flatter portion of the roof.

Roof Trusses and Arches

In the middle of the nineteenth century, American industrial buildings grew substantially as technology improved, machinery grew in size, and output increased. Builders sought roof-support systems that would make possible long, clear spans. Borrowing from the bridge industry, they employed wood or iron

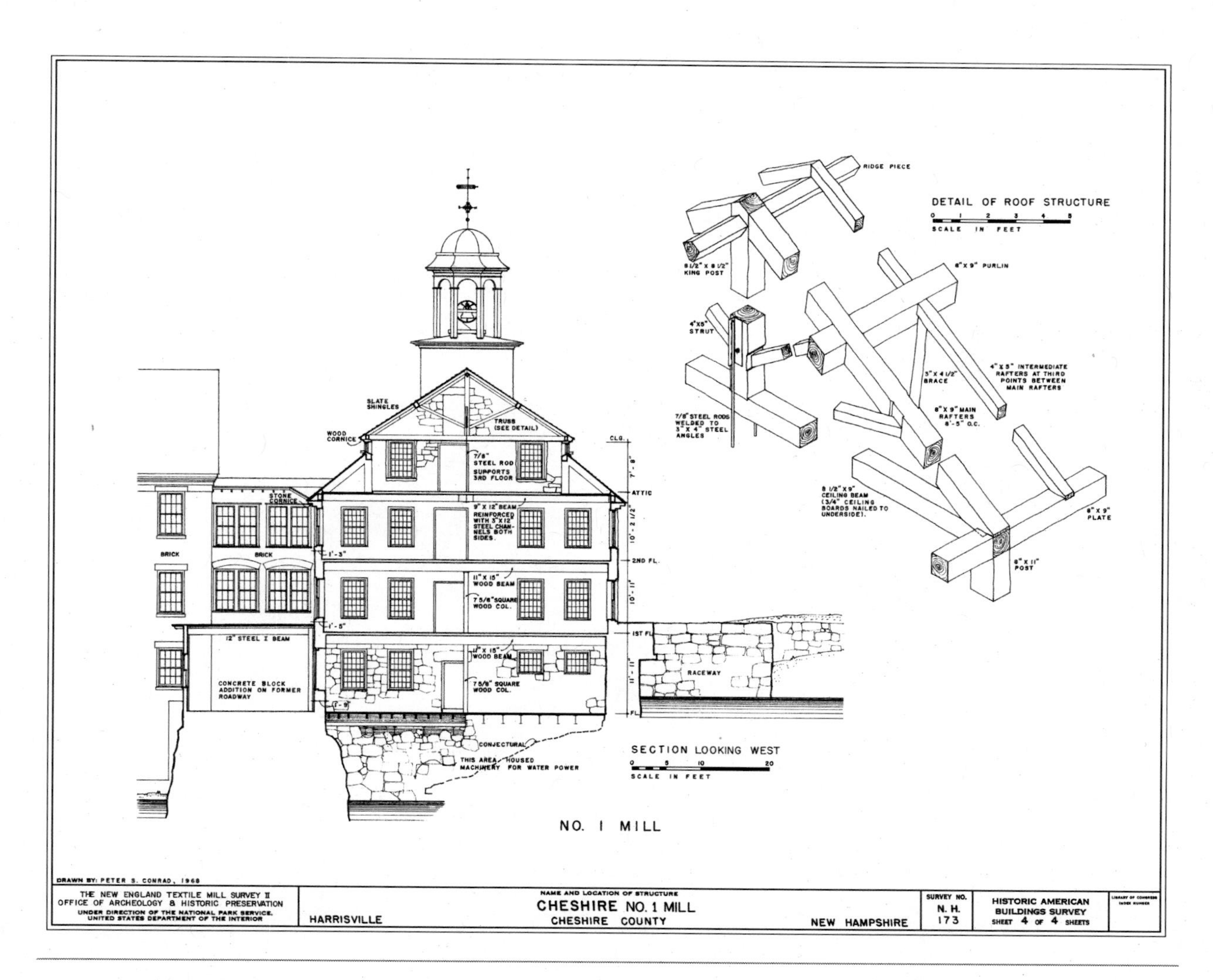

Figure 15.17. Detail of gable roof framing of Cheshire No. 1 Mill, Harrisville, New Hampshire. HABS NH-173, National Park Service. Delineated by Peter S. Conrad, 1968.

trusses to support the roof, thus eliminating the need for interior columns.

Roof trusses, like bridge trusses, may be composed of either a single web or a double web system, though a single web was often preferred since stresses are more readily determined. Roof trusses employ some of the same names as bridge trusses, including Howe, Pratt, Warren, and Fink.

The impetus for the development of long-span roof trusses came from the need for large, clear-span train sheds in metropolitan terminals. Truss systems for train sheds evolved in three recognizable stages. In the first stage, the roof truss rested atop rigid masonry walls. The next step was to eliminate the need for heavy, costly masonry walls, using instead columns of iron and steel. To make these columns more rigid, they were joined to the truss with an additional member called the knee brace. The knee brace forms a triangle at the corner made by the lower chord of the truss and the vertical column.

Span length was the principal consideration in the choice of a truss form. Roof trusses were designed so that there would be a panel point directly under each purlin. The maximum distance between panel points and purlins was 8 feet. A few truss configurations—Pratt, flat Pratt, Warren, Howe, and flat house—could be used for spans from 20 to 80 feet by varying the number of panels.

Simple trusses were viewed as practical for spans up to 100 feet. For larger clear spans, engineers turned to the arch. Roof arches are designated in the same way as bridge arches, by the number of pins or hinges they employ. A roof arch may be further classed as a braced or trussed arch, with an open framework or a ribbed arch, made of solid flat plates.[61]

By the early twentieth century, the Fink truss was the type most commonly used for the roofs of small buildings (Figure 15.18). It was economical because most of the members are in tension and the struts are short.

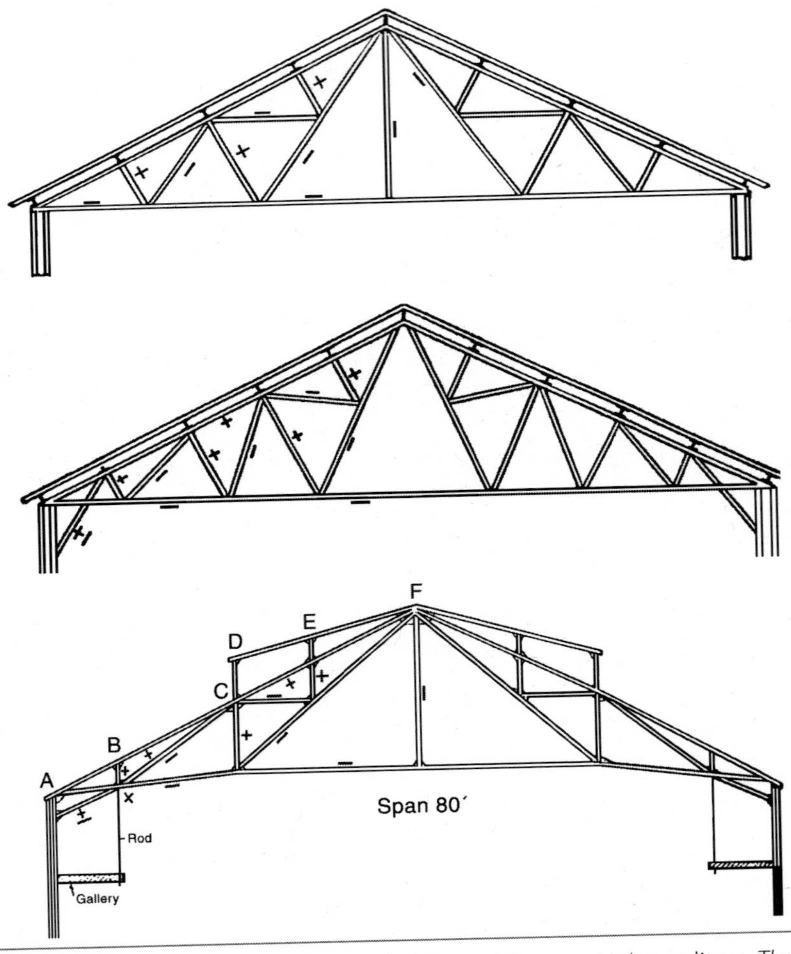

Figure 15.18. Fink roof trusses. Reprinted from Frank E. Kidder and Thomas Nolan, editors, *The Architects' and Builders' Handbook* (New York: John Wiley & Sons, Inc., 1921).

For small spans up to about 30 feet, sheets of corrugated iron could be curved and provided with a single tie-rod across the bottom.[62]

Roof Materials

The primary roof coverings in early twentieth-century mill buildings include corrugated steel or iron, slate, tile, tin, sheet metal, and tar and gravel. Corrugated steel or iron was usually fastened directly to the purlins by means of clips. Slate was usually nailed to sheeting boards with a layer of roofing felt between. Tile was usually fastened directly, the angle purlins spaced about 13 inches apart. Tin and sheet metal were usually laid on sheeting with roofing felt between, while tar and gravel roofs were either laid on wooden sheathing or reinforced concrete slabs.

Ventilation through the roof in the early twentieth century typically took the form of a monitor, cylindrical vents, or louvered vents. Typically a machine shop would employ cylindrical vents, a mill would employ louver vents, while a forge would employ louvers or open vents.[63]

GLOSSARY

Aiken roof. A roof consisting of alternating high and low bays formed by positioning roof purlins to span the top, and then the bottom, chords of trusses. The roof takes its name from Henry Aiken, an early twentieth-century Pittsburgh consulting engineer.

American bond. A brick bond formed by multiple rows of stretcher bricks, with a row of header brick placed between every three, five, or seven courses of stretchers. Also known as *running bond.*

Archivolt. An ornamental molding on the face of an arch that follows the curve of the opening.

Bat. A particular type of closer brick. Types of bats include queen closers, king closers, three-quarter bat, half bat, and quarter bat.

Bay. A regularly repeated interior space defined by beams and supporting columns. Mill construction was often termed, for example, an 8-foot bay or a 10-foot bay.

Bent. In steel-framed construction, a two-dimensional frame, usually consisting of columns and the truss they support.

Bollard. A post of metal or concrete positioned at a door frame or building corner to protect the lower wall or frame.

Bonded arch. A brick arch formed from two courses of brick ends in which the brick ends are bonded together.

Bowstring truss. A truss with a horizontal lower chord and arched upper curve.

Butterfly monitor or roof. A roof shape with two planes that rise from the center to the eaves with a valley in the center.

Clerestory. In industrial buildings, a series of window openings in the side walls of a monitor that provides light and ventilation.

Closer. Blocks varying in size from ordinary bricks used to finish or close the length of courses that have been adjusted to obtain the bond. Closers are made by cutting bricks with a blow with the edge of a steel trowel.

Continuous sash. Long sections of steel sash supported from the top and pivoted outward by a mechanical device.

Corbel. A projection on the face of a brick wall formed by one or more courses of brick, each projecting over the course below.

Counterbalanced sash. Double-hung sash connected by sash cords or chains over pulleys in a way that one automatically lowers when the other is raised.

Crane. A hoist that can also move the load in a lateral direction.

Craneway. A portion of a structure that supports and provides space for the operation of a traveling bridge crane.

Curtain wall. A nonbearing wall that encloses a building.

Double monitor. A roof that incorporates a narrow monitor at the roof edge (for ventilation) above a wider one (for light) that covers about a third of the building's width.

Elevator bulkhead. A structure on the roof that encloses the upper portion of the elevator shaft.

English bond. A brick bond formed by alternating header and stretcher courses. Longitudinal bond is provided by quarter or three-quarter bat closers placed in alternate courses.

Fire door. A fire-resistant door assembly, usually of metal-sheathed wood, fitted with an automatic closing mechanism.

Fire-resistant construction. The arrangement of a building to limit the spread of fire by isolating hazardous operations and vertical flues, by compartmentalizing a large building, and by using fire doors and fire shutters.

Fire shutter. A fire-resistant metal shutter, either mounted at the sides of the window opening or rolled down from a housing attached at the lintel.

Fire wall. An interior wall, usually 12 inches thick and extended above the roof 3 feet, with all openings fitted with self-closing fire doors or shutters, designed to stop the spread of fire.

Fireproof construction. Building construction in which structural members are made of noncombustible materials.

Flemish bond. A bond consisting of rows of alternating headers and stretchers with adjacent courses offset to create a series of cross patterns on the wall.

Framed structure. A building with a structure composed primarily of an arrangement of beams, girders, and columns, as opposed to masonry bearing walls.

Gallery. In production sheds, a side aisle, consisting of a ground floor and one or more upper levels, flanking a central crane-served bay; also referred to as a *lean-to*.

Half bat. A closer formed by cutting off one-half the length of a brick.

Head house. A relatively narrow multistory structure at one end of a lower, generally larger one-story building.

Header. A brick placed in a wall with only the ends exposed to view.

Header (or *heading*) *bond.* A brick bond composed entirely of headers, suitable only for sharp-curved walls.

Hoist. A pulley and rope or chains used to lift an object.

Hollow tile. Structural clay tile, hollow masonry building units made of terra-cotta or fire clay.

Industrial loft. A multistory building with relatively large, open floor areas in which various types of light manufacturing operations are housed.

Industrial steel sash. Window sash formed of rolled steel sections holding small panes of glass and usually combined into large expanses with units of operable and fixed sash.

Joists. One of a series of parallel beams used to support floor and ceiling loads.

King closer. A closer formed by cutting the corner off a whole brick at an angle.

Lean-to. A side bay of a one-story production shed, lower than the central bay, and featuring a single-pitched roof.

Loading bay. An internal loading platform located at the shipping doors of an industrial or commercial building.

Loading dock or *platform.* A platform, either timber-framed or of reinforced concrete construction, at the height of a cart, railroad car, or truck bed. A loading platform generally extends on the outside of the building along a railroad spur or adjacent to a yard area. A loading dock may refer to a platform positioned on the interior of a building adjacent to large door openings.

Loft. An upper level without partitions or elaborate finish intended to be used for manufacturing or storage.

Mezzanine. A partial floor or large balcony area inserted above the ground floor.

Mill. A building designed or fitted with machinery for a certain industry, such as a flour mill or textile mill.

Monitor. A section of a roof of an industrial building raised for light and ventilation, with side walls fitted with fixed or movable louvers or sash.

Purlin. A roof-framing member positioned horizontally between principal rafters or trusses. Roof purlins are often made of channels, I-beams or angled, fastened to the rafters of the roof trusses so that the roofing materials may rest upon one of their flat sides.

Quarter bat. A closer formed by cutting off three-quarters of the length of a brick.

Queen closer. A closer made from a whole or half brick cut lengthwise.

Rigid frame. A structural framework of welded steel or reinforced concrete in which all columns and beams are rigidly connected.

Rowlock arch. An arch built so that the ends of the bricks are visible and two or more concentric rings of brick ends frame the opening. In this type of arch, each course of brick is turned independently—that is, the headers of the two rows are offset from each other.

Running bond. A brick bond consisting of multiple rows of stretchers with rows of headers placed at regular intervals; also known as *American bond.*

Sag ties. Objects inserted to prevent sagging of purlins and to take the component of the load parallel to the roof. They are usually formed of round rods, threaded at the ends, run through holes in the purlin web with nuts to hold them in place.

Sawtooth roof. A series of parallel one-sided skylights or half-monitors placed so that only north light is admitted into a building.

Stack. A short steel chimney.

Store or *storehouse.* A storage building in a manufacturing works.

Stretcher. A brick laid lengthwise on the face of a wall.

Stretcher (or *stretching*) *bond.* A bond in which all the courses consist of brick stretchers, suitable only for partitions a maximum of 4 inches in thickness.

Three-quarters bat. A closer formed by cutting off one-quarter of the length of a brick.

Transept. In industrial buildings, transepts, often transportation corridors, cross shop areas at right angles.

Traveling crane. A crane in which a trolley moves across a bridge that spans overhead tracks. The assembly moves longitudinally as well.

Web. In steel-framed construction, the portion of a truss between the chords.

Window wall. Wall in which there are large, continuous expanses of sash instead of individual windows.

Wire glass. A shatterproof material in which wire mesh is embedded between two layers of glass.

REFERENCES

Anonymous. "The Construction of Mills." *The Manufacturer and Builder* 21:2 (February 1889): 40.

Atlas Portland Cement Company. *Reinforced Concrete in Factory Construction.* New York: Atlas Portland Cement Company, 1907.

Bradley, Betsy Hunter. *The Works: The Industrial Architecture of the United States.* New York: Oxford University Press, 1999.

Candee, Richard M. "The 1822 Allendale Mill and Slow-Burning Construction: A Case Study in the Transmission of an Architectural Technology." *IA: The Journal of the Society for Industrial Archeology* 15:1 (1989): 21–34.

Condit, Carl W. *American Building Art: The Nineteenth Century.* New York: Oxford University Press, 1960.

Crosby, Everett Uberto. *Hand-book of the Underwriters' Bureau of New England.* Boston: Press of the Standard Publishing Company, 1896.

Davidson, Lisa Pfueller. "Through the Mill: Documenting the Southern Textile Industry." *CRM* 4 (2000): 15–17.

Greenwood, Richard E. "Industrial Architecture of the Blackstone River Valley." In *The Early Architecture and Landscapes of the Narragansett Basin.* Prepared for the Annual Meeting and Conference of the Vernacular Architecture Form, Newport, Rhode Island, 2001.

Hunter, Louis C. *A History of Industrial Power in the United States, 1780–1930.* Volume 1: *Waterpower.* Charlottesville: University Press of Virginia, 1979.

International Correspondence Schools. *Yarns, Cloth Rooms, Mill Engineering, Reeling and Baling, Winding.* Scranton, PA: International Textbook Company, 1921.

Insurance Institute of America. Educational Committee. *Building Construction: A Text-Book Outline.* Boston: Insurance Press, 1916.

Ketchum, Milo Smith. *The Design of Steel Mill Buildings and the Calculation of Stresses in Framed Structures.* New York: McGraw-Hill Book Company, 1912.

Lewandoski, Jan Leo. "Traditional Timber Framing in Vermont, 1780–1850." *Association for Preservation Technology Bulletin* XXVI (2/3, 1995): 42–50.

Main, Charles Thomas. *Industrial Plants.* Boston: Caustic-Claflin Company, 1923.

Slaton, Amy. "Origins of a Modern Form: The Reinforced Concrete Factory Building in America, 1900–1930." Ph.D. dissertation, University of Pennsylvania, 1995.

Smoley, C. K. *Concrete Design. International Library of Technology 433B.* Scranton, PA: International Textbook Company, 1910.

Tolles, Bryant Franklin, Jr. "Textile Mill Architecture in East Central New England: An Analysis of Pre-Civil War Design." *Essex Institute Historical Collections* 107 (July 1971): 223–253.

Truscon Steel Company. *Truscon Steel Windows and Hangar Doors.* Youngstown, OH: Truscon Steel Company, 1945.

Tyrell, H. G. *Mill Building Construction.* New York: The Engineering News Publishing Company, 1901.

—. *A Treatise on the Design and Construction of Mill Buildings and Other Industrial Plants.* Chicago: Myron C. Clark, 1911.

—. *Engineering of Shops and Factories.* New York: McGraw-Hill, 1912.

Wermiel, Sara E. "Heavy Timber Framing in Late-Nineteenth-Century Commercial and Industrial Buildings." *Association for Preservation Technology Bulletin* XXXV (January 2004): 55–60.

NOTES

1 Richard E. Greenwood, "Industrial Architecture of the Blackstone River Valley," in *The Early Architecture and Landscapes of the Narragansett Basin* (Newport, RI, 2001), 21.

2 Martha and Murray Zimiles, *Early American Mills* (New York: Clarkson N. Potter, Inc., 1973), 112.

3 Greenwood, 21–22.

4 Bryant Franklin Tolles, Jr., "Textile Mill Architecture in East Central New England: An Analysis of Pre-Civil War Design," in *Essex Institute Historical Collections* 107 (July 1971): 229–230; Carl W. Condit, *American Building Art: The Nineteenth Century* (New York: Oxford University Press, 1960), 16–17.

5 Tolles, 232; Zimiles and Zimiles, 112–113.
6 Condit, 18.
7 Betsy Hunter Bradley, *The Works: The Industrial Architecture of the United States* (New York: Oxford University Press, 1999), 135.
8 Bradley, 135.
9 Zimiles and Zimiles, 113.
10 Tolles, 238.
11 Tolles, 240.
12 Condit, 19.
13 Bradley, 136.
14 Bradley, 139–140.
15 Bradley, 145–146.
16 "High vs. Low Factory Buildings," *The Manufacturer and Builder* (February 1880): 46.
17 Zimiles and Zimiles, 194–195.
18 "Specimen Modern Manufacturing Establishment," *The Manufacturer and Builder* (June 1893): 132.
19 Bradley, viii.
20 H. G. Tyrell, *Mill Building Construction* (New York: Engineering News Publishing Co., 1901), 9.
21 Henry Gratton Tyrrell, *A Treatise on the Design and Construction of Mill Buildings and Other Industrial Plants* (New York: Myron C. Clark Publishing Company, 1911), 62–63.
22 Tyrell 1911, 17–19.
23 Tyrell 1911, 20–29.
24 Atlas Portland Cement Company, *Reinforced Concrete in Factory Construction* (New York: Atlas Portland Cement Co., 1907), 46–59.
25 Bradley, 156–157; "The Kelly & Jones Company's Concrete-Steel Factory Building." *The Engineering Record* 49:6 (February 6, 1904): 153–154.
26 Carl W. Condit, *American Building Art: The Twentieth Century* (New York: Oxford University Press, 1961), 154.
27 Condit, 167–168.
28 An extended discussion of Kahn's industrial buildings is found in Grant Hildebrand's *Designing for Industry: The Architecture of Albert Kahn* (Cambridge: MIT Press, 1974).
29 Albert Kahn, "Industrial Architecture," *Michigan Society of Architects Weekly Bulletin* 13 (November 7, 1939): 5–9.
30 Bradley, 256–257.
31 Bradley, 42–43.
32 Middlesex Sampling Plant, Boiler House, HAER NJ-107-B, 2; United Engineering Company Shipyard, HAER CA-295-D, 4, 7, 9.
33 Bradley, 46–47.
34 William Gavzy, Roger Locomotive Works: Fitting Shop, 1881. HAER NJ-38. 1973.
35 Bradley, 40–42.
36 Atlas Portland Cement Company, 60–62.
37 Bradley, 44–46.
38 Bradley, 38.
39 Bradley, 35–37.
40 Bradley, 37–38.
41 Bradley, 49–51.
42 Bradley, 38–39.
43 Bradley, 89.
44 Lee Ann Bishop Land, Prattville Manufacturing Company, Number One, HAER No. AL-183, 1998: 17–18.
45 Bradley, 92–95.
46 Bradley, 95–97.
47 "Fire-Proof Construction of Mill Buildings," *The Manufacturer and Builder* (November 1890): 251.
48 "Fire-Proof Construction," 251–252.
49 Insurance Institute of America, *Building Construction: A Text-Book Outline* (Hartford: Insurance Institute of America, 1916), 11–12.
50 International Correspondence Schools, *Yarns, Cloth Rooms, Mill Engineering, Reeling and Baling, Winding* (Scranton, PA: International Textbook, 1921), 87:54.
51 Richard M. Candee, "The 1822 Allendale Mill and Slow-Burning Construction: A Case Study in the Transmission of an Architectural Technology," in *IA: The Journal of the Society for Industrial Archeology* 15:1 (1989): 21.
52 International Correspondence Schools, 87:55.
53 Charles A. Praray, "Building," Letters Patent 518,274, dated April 17, 1894.
54 Lisa Pfueller Davidson, "Through the Mill: Documenting the Southern Textile Industry," *CRM* 4 (2000): 15–17.
55 "The Construction of Mills," *The Manufacturer and Builder* (February 1889): 40.
56 "Fire-Proof Construction," 251.
57 International Correspondence Schools, 87:65.
58 Milo Smith Ketchum, *The Design of Steel Mill Buildings and the Calculation of Stresses in Framed Structures*. (New York: McGraw-Hill Book Company, 1912), 129–130.
59 Ketchum, 136–137.
60 Ketchum, 170.
61 David Weitzman, *Traces of the Past: A Field Guide to Industrial Archaeology* (New York: Charles Scribner's Sons, 1980),130.
62 Tyrell 1911, 9–10.
63 Tyrell 1911, 34–40.

Chapter 16

Industrial Landscapes

The previous chapters have primarily considered individual industrial archaeological resources, buildings, structures, objects, and sites as discrete subjects of investigation. However, many of these resources are also components in larger landscapes, landscapes that typically include industrial buildings and structures, pathways, residential buildings, commercial and institutional buildings, infrastructure, and associated public and private places. Frederic Quivik discusses the broadening of industrial archaeology to include larger landscapes:

> We now devote considerable attention to recording, analyzing, and interpreting whole landscapes comprised of complexes of industrial buildings, the linear systems that link buildings within those complexes to each other and connect them to the sources of supply, the neighborhoods that grew around industrial complexes to house workers, and the topographic features on the land, such as mine cuts or waste dumps, that have been caused by industrial activity.[1]

This chapter is intended as an introduction to some of these elements of industrial landscapes.

Iron Plantations

The principal characteristics of an iron plantation were its location in a remote landscape where the entrepreneur acquired a large tract and attracted to it numerous workers and their families. To produce iron commercially, the plantation had to be large. Hundreds of acres of hardwood forest were needed to produce the wood that would be converted to charcoal to fuel the furnaces and forges. Nearby deposits of iron ore and limestone were necessary, as well as a stream or river that could be used for waterpower. As Schallenberg and Ault note, "The name 'plantation' derives from the fact that these enterprises were often nearly self-sufficient communities, producing not only iron but also most of their own food, and were therefore as much agricultural as industrial operations."[2]

In addition, the entrepreneur had to acquire farmland where food could be raised to feed the workers at the plantation. Because the site would be remote from villages and towns when first established, the entrepreneur had also to provide housing for his workers, usually in the form of a cluster or village of small cottages near the ironworks. The center of the ironworks was usually a large, well-furnished "big house," in Pennsylvania often a Georgian or Federal stone house, surrounded by a thousand acres of woodland. The ironmaster residing there emulated the style of the English gentry.

The workforce at an iron plantation included about a dozen experienced furnace workers as well as woodcutters, colliers, and miners. In addition, if the works included a refinery forge, rolling mill, or slitting mill, workers experienced in those skills would also be needed. Some iron plantation villages included over 100 residents.

To maintain the plantation economy, the settlements also required a sawmill to furnish lumber and a gristmill to grind grain. An ironworks company store provided goods for residents. Larger iron plantations might also include a school or a church.[3]

The center of the colonial iron industry and of iron plantations was Pennsylvania. In about 1720, the first bloomery forge was erected in Pennsylvania, in Colebrookdale, near present Pottstown, Montgomery County. Other early furnaces include Chester County's Hopewell Furnace, as well as other iron plantations with names such as Reading, Warwick, Mount Pleasant, Durham, Cornwall, Elizabeth, Windsor, Rebecca, Pine Grove, and Mary Ann. Other iron plantations were established elsewhere on the eastern seaboard in locations convenient to raw materials for iron production. For example, William Hill's Ironworks in York County, South Carolina, established in 1779, occupied some 500 acres of land on Allison's Creek and included woodlands, iron ore banks, a sawmill, the ironworks itself, a gristmill, a tanyard, houses, and roads connecting the various functions.[4]

NEW ENGLAND MILL VILLAGES

Mill villages began as people started to congregate around custom gristmills, often established in the vicinity of a rural crossroads. These mills were frequently the most public place available in farming areas, and as such were popular gathering places. Before long, post offices and general stores were added, and people not engaged in farming would build their houses near the mill. Many of these villages remained small and died out as the fortunes of the mill waned.[5]

As Joseph Wood notes, the New England village, the epitome of a mill village, is a nineteenth-century creation.[6] He describes the character of these villages:

> Across New England after the Revolution, stores, shops, offices, and residences—material manifestations of a quickening rural economy—were gathered around the meetinghouse lots. . . . There was an opportunistic quality about these new villages. Few were planned, and speculation was rife as encirclement of the meeting house lots took place.[7]

Wood attributes the irregularity of New England village morphology to adaptations to site conditions and a road network that had been laid down for other reasons at an earlier time.[8]

While some of these villages declined as the flour industry moved west, others successfully transitioned from an economy based upon grist milling to one of multiple industries relying on waterpower. One such village was Spofford, New Hampshire. Spofford, like many other industrial villages, after a brief industrial heyday, soon lost most of its industry to communities with better transportation connections.

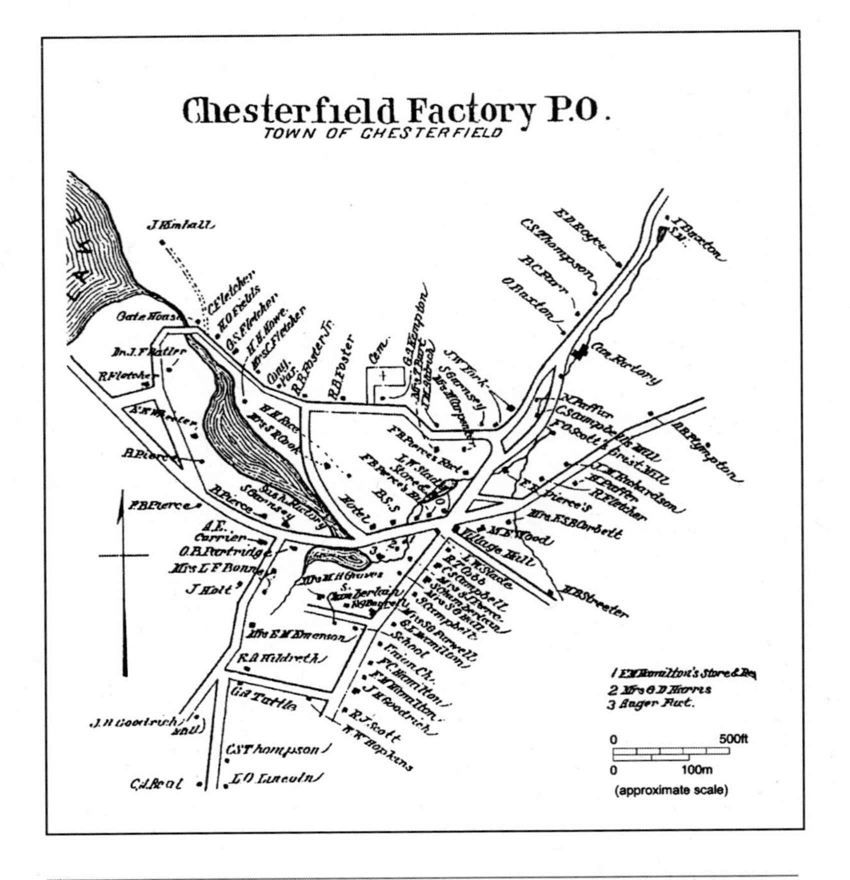

Figure 16.1. Map of Chesterfield Factory (Spofford), Cheshire County, New Hampshire. From *Town and City Atlas of the State of New Hampshire* (Boston: D. H. Hurd, 1892).

COMPANY TOWNS

The term company town is generally applied to communities that were built virtually overnight to house workers for a specific company, and in which all of the houses were built on a uniform plan and the company owned every house, renting them to their workers.[9]

James Allen characterizes common traits of company towns:

> If a person suddenly found himself in the middle of a company-owned town, he would have little difficulty identifying it as such, for certain general features usually stood out. First to be noted would be the standard uniform architecture of the company-owned houses. In a prominent location, however, would stand a larger, more imposing structure: the home of the company manager or superintendent. It would be observed that the town seemed to center around a focal point where a store, community hall, school, and other public buildings were located. The company store usually dominated the group. It would be noted that the settlement had no "suburbs," or no gradual building up from a few scattered homes to a center of population. Rather, one would note the complete isolation of the community and the definiteness of its boundaries.[10]

Early History

According to Leifur Magnusson, author of a pioneering study on company housing, company housing developed concurrently with the emergence of industry in the United States. The first New England mill town was established in 1791, while other early examples of company housing were found in Delaware by 1831 and in the southern states at about the same time. Brady's Bend, Pennsylvania, grew into an ironworker's town from land purchased in 1839 by Brady's Bend Iron Company. By the mid-nineteenth century, steel towns had been established in Michigan and Pennsylvania, while the first company towns appeared in Pennsylvania's anthracite coal region by 1840.[11]

Typology

Several scholars have developed typologies useful in characterizing company towns. Leland Roth chronicled four townscape plans that dominated company towns by the last decade of the nineteenth century. These plans can be further subdivided into grid versus curvilinear plans, town plans, and architect-designed and non-architect-designed houses.

The most basic type of company town consisted of housing arranged along a grid of streets, designed by the company engineer and built by company labor. The second type employed the same grid but incorporated a few houses designed by professional architects. These ostentatious houses were reserved for company managers and skilled laborers. The third type of company town distinguished by Roth used an overall design of curvilinear streets but with prosaic housing throughout. The addition of architect-designed dwellings to curvilinear streets defines Roth's fourth type.[12]

Several other typologies have also been postulated. Based upon his studies of West Virginia coal towns, Mack Gillenwater classified coal towns according to their relation to transportation corridors. The first type is a linear settlement: communities established along a road, trail, or valley stream. The next type was the cruciform plan that generally grew from earlier linear towns. The third type was the block or grid form, a type rarely suited to the mountainous terrain of West Virginia. A fourth type, fragmented settlements, derived from a lack of planning or combinations of the other plan types.

In his book *Company Towns in the American West*, James B. Allen concentrated more on the temporal sequence of town establishment than on its spatial dimensions. His first phase was a camp populated by bachelor miners, while his final phase was a conventional town of families.[13]

Textile Communities

Beginning in the late eighteenth century and into the early twentieth century, numerous cities in New England had an economy that revolved around textile production. Beginning in the late nineteenth century and continuing into the twentieth century, southeastern and southern cities and towns developed in response to the construction of textile factories. Studies have been made of the landscape of both community types.

New England Textile Communities

Richard M. Candee developed a typology of company towns based on his study of New England textile communities. This typology describes four kinds of company towns, which can be reduced to two types: the Rhode Island and the Waltham systems, based upon the presence or absence of landscape planning, the types and functions used, and the methods of wage distribution and product purchasing. The Rhode Island system refers to small-scale operations that offered little beyond providing for minimal needs: a factory, family housing, company store, and outlying agricultural fields. A subtype of the Rhode Island consisted of clusters of several villages lining a river and sharing the common power source. These towns relied on independent speculators to develop the non-company-owned land surrounding the village(s) for non-industrial uses.

Towns grouped under the Waltham system, named for the Massachusetts town established by the Francis Cabot Lowell enterprise, employed more entrepreneurial development. A corporation established these operations, with site plans guiding construction and with an array of amenities found in traditional cities. They were self-contained entities.

Examples of company-organized textile towns included Lawrence, Lowell, Ludlow, and Waltham, Massachusetts. The first houses established by the mill owners in each of these communites were boardinghouses, occupied primarily by unmarried employees. Such dwellings, emblematic of New England textile cities, are depicted in Figure 16.2.[14] Multiple tenements or barracks were also constructed. Workers' housing was constructed near the mill and was usually built by the company for rental to families.

Some communities, such as Fiskville and Harris, Rhode Island, had residential areas composed nearly entirely of small, single-family cottages set in rows near the mill. In other communities, full two-story houses were built similar to many traditional dwellings but partitioned inside for several families. Also common were tenements with stairs and entrances at each end or two doors in the front and rear. Especially during the later mill era, the classic New England triple decker became common in communities such as New Bedford and Fall River, Massachusetts.[15]

Figure 16.2. Textile city boardinghouses, South Bedford. Amoskeag Millyard, Manchester, New Hampshire. HABS NH-109, National Park Service. Photograph by R. Randolph Langenbach, 1967.

Magnusson, reporting on his study of a selection of textile communities in the early twentieth century, indicates there was considerable regard for the use of trees and vegetation as elements of beauty. In some cases, the streets were overly wide and lacked both gutters and sidewalks. The areas where the unskilled laborers lived were less neat and orderly than other areas. House types represented included similar numbers of semidetached, row, and flat or apartment dwellings. The average dwelling size was about five rooms.[16]

Most textile companies also operated a store either in a separate building or in the mill itself. The complex might also include blacksmith shop, while early spinning factories might also include a sawmill or gristmill.[17] Several New England textile communities, including Lowell, were known for their parks and promenades.[18]

Southern Textile Communities

As Magnusson notes in his study of company housing, most southern textile villages were established in isolated locations: "Situated with a view to the availability of power and the cheapness of land, mills generally are constructed at a distance from towns."[19]

Describing the southern textile mill village, Magnusson writes:

> Except for the dominating mill, the general appearance of the more common type of village is not very unlike that of a mining or steel town, with its rows of small houses of the same design and size, the unimproved streets and gutters and walks, the struggling young trees, the absence from the streets of vehicles, children and dogs. Almost invariable in the mill village, however, there are one or more churches, a substantial school building, [and] a good-size store.[20]

Margaret Crawford describes the elements common to an early twentieth-century Piedmont textile village:

> Lined up along the river, the long narrow buildings, three and four stories tall, dominated the landscape. Even in remote villages, the substantial red brick structures were embellished with decorative features that emulated those of civic and religious buildings. The rest of the village was casually laid out. Factory owners hired surveyors to plot roads and lots and allowed carpenters according to local custom. The "mill hill," as it name suggests, was generally located on a cleared site sloping up from the riverside mill. To save money, houses were all built at the same time with identical plans and laid out in roads leading from the mill. Houses followed a standard pattern, duplicating the most common and inexpensive type of rural dwelling in the Piedmont countryside.... These single-story frame structures . . . contained three or four rooms, a front porch, and often a rear kitchen extension. A small group of larger, better-built houses was reserved for mill managers or the town ministers.... Another group of smaller houses was often set apart on the outskirts of the village for black employees. The location of other community buildings did not form any particular pattern. A store, church, schoolhouse, and lodge hall might be clustered near the mill or interspersed with the houses without establishing a formal relationship to either.[21]

Coal Towns

Magnusson describes the coal towns that he observed in the bituminous regions of Pennsylvania and West Virginia in the 1910s. The towns were characterized by regularly laid-out streets and by uniformity in design and material of construction. Roads were dirt, generally lacked sidewalks, and rarely had gutters. The streets averaged about 40 feet in width. The predominant type of house in the soft-coal region was a semidetached wood-framed house usually of three or four rooms.[22]

Margaret Mulrooney compared company towns and housing forms in the coal regions of the eastern and central United States. She concluded that distinct regional variations existed within the bituminous industry, and these differences were most visible in the types of dwellings built for miners. Mulrooney identified four basic housing forms in the Appalachian coal belt: the shotgun, found in the southern portion of the eastern coal field; a one-story, detached or semidetached cottage built on post foundations with a hipped or pyramidal roof, found in the southern Appalachian region; a single-story, single-family, gabled roof bungalow with or without an attached ell that was common in the southern Appalachian region after 1910; and the Pennsylvania miner's dwelling, a two-story, semidetached, eaves-fronted, wood-framed building found throughout northern Appalachia.[23]

Coal communities, like other industrial landscapes, embodied a variety of types. The simplest landscapes were the earliest anthracite towns, known as patches, built beginning in the mid-nineteenth century. Patch towns were constructed in isolated valleys and were characterized by primitive living conditions. An example of a coal patch is Eckley, Pennsylvania, laid out adjacent to the coal breaker. The small community was arranged in a linear fashion on a single main street and two parallel streets, with the further division of two back alleys and three cross streets. The community had a company store, a hotel, churches, a school, and worker housing. Company officials were provided with single-family detached houses, while miners and laborers were provided with eaves-fronted, semidetached, wood-framed double houses, with a main block measuring 30 feet wide and 25 feet deep and a rear ell measuring 40 feet wide and 10 feet deep.[24]

The equivalent of the coal patch community in Pennsylvania's bituminous region were satellite communities of coal centers, such as Eureka No. 40, a satellite of Windber, home of the Berwind Coal Company.

The coal patch communities were replaced by anthracite-coal towns. These settlements were always situated in close proximity to the mine and along a highway through the region. Typically the settlement consisted only of groups of houses near the shafts and breakers. The principal street was frequently the principal township highway, but dirt roads sometimes covered in coal dust waste were the rule. Streets were frequently 45 feet in width. Lot width varied from 25 to 60 feet.

The average size of the miner's family quarters in the anthracite region was approximately five rooms. Most of the houses were wood-framed, weatherboarded, and were either semidetached or detached.[25]

Margaret Mulrooney describes the layout of Star Junction, a settlement in the bituminous region of southwestern Pennsylvania:

> Although Star Junction appears to lack a cohesive town plan, in fact, the arrangement of houses, coke ovens, streets, railroad tracks and tipples was carefully thought out. The location of each had to conform to the natural terrain, yet be organized in the most efficient manner possible. As in most mining towns, the company engineers gave first priority to locating the mine entries, tipples and coke ovens in the relatively flat land of the valley bottom. Next they located the railroad tracks along the valley floor, parallel to Washington Run and the long banks of ovens. Because it was a coke works, the engineers also had to make room for a coal ash dump. . . . It was located in the corner of the valley floor along the southeastern hillside. Then the engineers laid out streets.[26]

The center of Star Junction included public buildings and bosses' houses. The Junction House Hotel occupied one corner, the doctor's house the opposite corner, and the company store and store manager's house on the remaining two corners. Nearby were the two churches, the public school, and the theater building.[27]

While many of the elements of the anthracite landscape of northeastern Pennsylvania are gone, an enduring part of the landscape is the piles of culm or waste rock. Culm contains the slate and other rock that was picked out of the coal by mechanical sorting machines, dirt, coal pieces too small to process for sale, and dust created by the physical process of breaking rock.[28] As Going and Raymond note, the contours of culm banks are part of the vernacular landscape. Houses nestle against them, and their softened profiles mimic those of the surrounding natural hills.[29]

Southern Coal, Iron, and Steel Communities

Magnusson reported that with a few exceptions, the company towns of the southern region were not formally planned. Streets were not paved. A rectangular plan was typically used, although some principal streets were laid out according to land contours. No residential lot was under 25 feet in width, and most were between 50 and 75 feet. In each town surveyed, there was an African American district and a white district. The white district was typically further subdivided into ethnic neighborhoods.

Housing in this region consisted of single detached houses and semidetached houses. All were wood-framed, built in the form of a one-story cottage, square, with a hipped roof. Small southern towns frequently used "shotgun" houses, shaped like an oblong box and divided into three rooms in a row.[30]

Torok describes the layout of the coal towns of southern Appalachia:

> The narrow valleys and ravines of southern Appalachia were difficult places to build towns, and the natural landscape often dictated design and construction. Placed on the valley floor alongside the rivers and creeks, these towns usually had a linear design following the winding, meandering path the waterways had carved out over the years. Railroad tracks, house, and mine equipment were laid out. Stores, depots, machine shops, and mine sites were strung out along the railroad tracts. When all of the available land was used, houses were built on the surrounding hillsides.[31]

Exceptions to this ad hoc layout were found in model coal towns such as Stonega, Virginia, and Holden, West Virginia, which were constructed with beautifully landscaped streets, parks, recreation facilities, and modern conveniences. The houses were of better quality than those in other coal towns, sited on larger lots, manufactured with better wood, and generally boasted indoor plumbing.

Midwestern and Western Industrial Communities

Iron and Copper Mining Towns

As in other regions, the landscape ranged from functional communities designed by company employees to elaborate model villages designed by outside landscape architects. An example of the latter was the Cleveland-Cliffs Iron Company village of Gwinn, Michigan, designed by landscape architect Warren Manning. Manning made special efforts to identify and preserve the primary native plant species on the townsite and to plant additional trees and shrubs as a greenbelt. The town design was a compact grid at a 45-degree angle to range lines that featured a central town common. Lots for public buildings were selected so that the structure would terminate the view down a major street. The houses were duplexes of one and a half stories, with four to six rooms, built to one of 14 plans and painted with one of several exterior colors.[32]

Bituminous Coal Camps of the Mountain West

According to Magnusson, in the early days of coal production in the region, many employees constructed their own shelters or lived in discarded boxcars. Early company houses were simple wood-framed structures, unfenced and unscreened, and with primitive sanitary arrangements. The layout of the coal camps generally involved a main road that followed the contour of the site, with side roads extending off at a tangent or straggling up canyons. Most dwellings were wood-framed, with smaller numbers constructed of concrete blocks or finished in pebble-dash. Most houses were detached and consisted of four rooms.[33] Numerous camps also included boardinghouses in which men typically slept two to a room.

Western Mining Communities

The earliest communities in the western mining regions were camps. Described by Arnold R. Alanen as "company-built and controlled, the camps were built to house men who either were engaged in exploratory work or were seeking to assess the productivity of newly-opened mine sites." These camps were considered only temporary points of habitation.[34]

Once the economic potential of a new mine site had been determined, the companies developed small residential settlements or locations within walking distance of the mine. These more permanent settlements were of four types: unplatted/squatters' locations, with poorly built housing strung out along haphazardly arranged streets and pathway; company locations, laid out by mining engineers in undeviating orthogonal patterns; smaller numbers of model locations; and speculative developments, promoted by individuals and companies seeking to capitalize on the real estate potential of booming mining areas.[35]

Richard Francaviglia discusses the components of the historic mining landscape:

> The historic mining landscape consists of many visual elements—among them the natural and manmade topography, vegetation, structures and buildings of several major types (commercial, residential, institutional, and industrial), the way they are situated with regard to each other, the street pattern, the transportation lines, and the property parcels. Mining-related activities often concentrate many features into a relatively small space, and reading their landscapes usually requires that we consider the architecture of a particular building or structure in the context of other buildings and features.[36]

Using Tonapah, Nevada, as an example, Francaviglia points out that mining activities are closely related to topography. The town is almost surrounded by mines that are situated on the hillsides above the town. The railroad follows the contour lines as closely as possible in order to reduce the gradients while reaching the mines. Mine dumps are located very close to the mines.[37]

Francaviglia notes that the downtown business district of a mining town served both as a "functional marketplace and a stage setting for the drama of everyday life." On the main street, development of the mining district is apparent, as it may have been home to dozens of competing businesses or dominated by a single store of a large company. In western mining districts, the placement of buildings in the context of property lots and peculiar topographic situations is often distinctive, with parcels sloping steeply up or down on either side of the thoroughfare. Rear portions of commercial buildings are frequently either built into the hillside or, on a downhill slope, stand on substantial foundations.[38]

Other property types represented in mining towns include town halls, union halls, churches, and schools. As Francaviglia notes, mining communities may have had a relatively large number of union buildings or lodge halls due to their cosmopolitan but clannish population of relatively young working males. As noted elsewhere, three main types of houses were frequently seen in western mining towns: the rectangular miner's cabin, the T-plan cottage, and the pyramidal roofed cottage.[39]

Another physical landmark in most metal mining districts is the slag piles or heaps that result from the smelting of ores. In many cases they outlast the smelters themselves and are visible as dark, steep-sided hills or tablelands.[40]

Metal Mining Region of the Southwest

Donald L. Hardesty describes the early twentieth-century layout of the mining camp of Gold Bar, Nevada. In the fall of 1905, the settlement had a bunkhouse housing 14 miners, a boardinghouse, and the beginning a bungalow for the superintendent. The settlement was separated into two clusters. On the southwestern edge was the Gold Bar camp; at the opposite, northeastern edge of the site was the "Homestake camp," a larger cluster of house sites formed around roads to the Homestake Mine and Mill.[41]

Company Towns of the Pacific Northwest

As Linda Carlson indicates, the housing in Pacific Northwest company towns varied greatly. In some the lodging was more luxurious than anything available in nearby towns, whereas in others it was no better than a campsite. An example of the former was DuPont, Washington, with its street trees forming leafy canopies over broad streets lined with Craftsman bungalows. Other communities, such as Mason City, Washington, built for laborers on the Grand Coulee Dam, had 45 bunkhouses, each housing two dozen laborers.

In logging communities where buildings were designed to be moved, houses were small. In some, the earliest family houses were little more than boxcars without wheels, lacking both power and plumbing.

Most communities differentiated housing by professional and marital status, as well as by racial and ethnic group. In Potlatch, Washington, residential areas were laid out so that managers lived high on the hill, where smoke from the mill wouldn't blow into their houses. Laborers lived within walking distance of their jobs; Greek, Italian, and Japanese employees lived farthest down the hill in the smallest houses.[42]

DESIGNED COMPANY TOWNS

As Margaret Crawford writes, until 1900, most American company towns represented industrial landscapes, "direct translations of the technical and social necessities of a particular method of production into a settlement form."[43]

One early model company town was Leclaire, Illinois, described by John S. Garner. Founded by N. O. Nelson, the president of the N. O. Nelson Manufacturing Company, it was established on the principle of cooperation. Of the 125 acres purchased by the company, 15 were reserved for the factories, and the remainder was left for housing and community facilities. A fence formed by 30- foot trees separated the factories from the residential area. The principal streets were planned to follow the contour of the land to wind through the site in a picturesque manner. Trees lined the streets, and walks meandered through the site. Outdoor amusement and recreation were provided at a village green, a lake with picnic grounds, and a pavilion. Workers were offered the choice of four-, six-, and eight-room houses.[44]

After 1900, professional designers took over the design of many company towns, and visual separation from the factory was a key element in their design. One such town was Fairfield, Alabama, designed by planner George Miller and architect William Leslie Walton. The community had a range of bungalows in a park-like setting. Crawford notes, "the craftsman style, with its natural building materials and complex detailing, symbolically counteracted in the domestic sphere the realities of the workers' daily activity in the steel mill." Fairfield projected two separate images, an informal, welcoming residential area, and a commercial block indistinguishable from that of any other small town.[45]

By the end of the 1920s, the era of the new company town had begun to pass. In the first decades of the twentieth century, more than 40 new industrial towns had been built, but the availability of the inexpensive automobile gave workers mobility and lessened the need to live close to their place of work.[46]

BIBLIOGRAPHY

Alanen, Arnold R. "Documenting the Physical and Social Characteristics of Mining and Resource-Based Communities." *Bulletin of the Association for Preservation Technology* 11:4 (1979): 46–68.

Allen, James B. *The Company Town in the American West.* Norman: University of Oklahoma Press, 1966.

Beaudry, Mary C., and Stephen A. Mrozowski, "The Archeology of Work and Home Life in Lowell, Massachusetts: An Interdisciplinary Study of the Boott Cotton Mills Corporation."*IA, The Journal of the Society for Industrial Archeology* 14:2 (1988): 1–22.

Candee, Richard M. "New Towns of the Early New England Textile Industry." *Perspectives in Vernacular Architecture* 1 (1981): 31–51.

Carlson, Linda. *Company Towns of the Pacific Northwest.* Seattle: University of Washington Press, 2003.

Chidester, Robert C. "Analysis and Discussion of Labor Archaeology in Maryland." In *A Historic Context for the Archaeology of Industrial Labor in the State of Maryland.* www.heritage.umd.edu/CHRSWeb/ AssociatedProjects chidestreport/Chapter%201

Crawford, Margaret. "The 'New' Company Town." *Perspecta* 30 (1999): 48–57.

Eggert, Gerald G. *The Iron Industry in Pennsylvania.* Pennsylvania History Studies No. 25. Middletown, PA: The Pennsylvania Historical Association.

Ferguson, Terry A., and Thomas A. Cowan. "Iron Plantations and the Eighteenth- and Nineteenth-Century Landscape of the Northwestern South Carolina Piedmont." In *Carolina's Historical Landscapes: Archaeological Perspectives,* edited by Linda F. Stine, Martha Zierden, Lesley M. Drucker, and Christopher Judge. Knoxville: The University of Tennessee Press, 1997.

Francaviglia, Richard V. *Hard Places: Reading the Landscape of America's Historic Mining Districts.* Iowa City: University of Iowa Press, 1991.

Garner, John S., editor. *The Company Town: Architecture and Society in the Early Industrial Age.* New York: Oxford University Press, 1992.

Going, Peter, and Elizabeth Raymond. "Living in Anthracite: Mining Landscape and Sense of Place in Wyoming Valley, Pennsylvania." *The Public Historian* 23:2 (Spring 2001): 29–45.

Greenwood, Richard. "A Mechanic in the Garden: Landscape Design in Industrial Rhode Island." *IA: The Journal of the Society for Industrial Archeology* 24:1 (1998): 9–18.

Hardesty, Donald L. *The Archaeology of Mining and Miners: A View from the Silver State.* Special Publications Series No. 6. Society for Historic Archaeology, 1988.

Heiss, M. W. "The Cotton Mill Village: A Viewpoint." *Journal of Social Forces* 2:3 (March 1924): 345–350.

Magnusson, Leifur. *Housing by Employers in the United States.* Bulletin 263. Washington, DC: United States Department of Labor, Bureau of Labor Statistics, 1920.

Metheny, Karen Bescherer. *From the Miners' Doublehouse: Archaeology and Landscape in a Pennsylvania Coal Company Town.* Knoxville: University of Tennessee Press, 2007.

Mulrooney, Margaret M. *A Legacy of Coal: The Coal Company Towns of Southwestern Pennsylvania.* Washington, DC: Historic American Buildings Survey/Historic American Engineering Record, 1989.

Perry, Martin. Coal Company Towns in Eastern Kentucky, 1854–1941. On Coal Education website: www.coaleducation.org/CoalHistory/coaltowns/, 1991.

Quivik, Fredric. "Landscapes as Industrial Artifacts: Lessons from Environmental History." *IA: The Journal of the Society for Industrial Archeology* 26:2 (2000): 55–64.

Schallenberg, Richard H., and David A. Ault. "Raw Materials Supply and Technological Change in the American Charcoal Iron Industry." *Technology and Culture* 18:3 (July 1977): 436–466.

Wood, Joseph S. "The New England Village as an American Vernacular Form." *Perspectives in Vernacular Architecture* 2 (1986): 54–63.

Wyckoff, William. "Postindustrial Butte." *Geographic Review* 85:4 (October 1995): 478–496.

NOTES

1 Frederick Quivik, "Landscapes as Industrial Artifacts: Lessons from Environmental History," *IA: The Journal of the Society for Industrial Archeology* 26:2 (2000): 56.

2 Richard H. Schallenberg and David A. Ault, "Raw Materials Supply and Technological Change in the American Charcoal Iron Industry," *Technology and Culture* 18:3 (July 1977): 436.

3 Gerald G. Eggert, *The Iron Industry in Pennsylvania* (Middletown, PA: The Pennsylvania Historical Association, 1994), 16–17; W. David Lewis, "The Iron Plantations," in National Park Service, *Hopewell Furnace* (Washington, DC: National Park Service, 1983), 6–8.

4 Terry A. Ferguson and Thomas A. Cowan, "Iron Plantations and Eighteenth and Nineteenth Century Landscape of the Northwestern South Carolina Piedmont," in *Carolina's Historical Landscapes: Archaeological Perspectives*, edited by Linda F. Stine, Martha Zierden, Lesley M. Drucker, and Christopher Judge (Knoxville: University of Tennessee Press), 117–119.

5 Robert D. Chidester, "Analysis and Discussion of Labor Archaeology in Maryland," in *A Historic Context for the Archaeology of Industrial Labor in the State of Maryland*, www.heritage.umd.edu/chrsweb/associatedprojects/ChidesterReport.htm.

6 Joseph S. Wood, "The New England Village as an American Vernacular Form, *Perspectives in Vernacular Architecture* 2 (1986): 54.

7 J. S.Wood, 58.

8 Robert S. Wood, "'Build, Therefore, Your Own World': The New England Village as Settlement Ideal," *Annals of the Association of American Geographers* 81:1 (March 1991): 36.

9 Chidester, 4.

10 James B. Allen, *The Company Town in the American West* (Norman: University of Oklahoma Press, 1966), 79–80.

11 Leifur Magnusson, *Housing by Employers in the United States* (Washington, DC: U.S. Department of Labor, Bureau of Labor Statistics, 1920), 7–8.

12 Leland M. Roth, "Company Towns in the Western United States," in *The Company Town: Architecture and Society in the Early Industrial Age*, edited by John S. Garner (New York: Oxford University Press, 1992), 173–205.

13 Martin Perry, Coal Company Towns in Eastern Kentucky, 1854–1941: A Historic Context, www.coaleducation.org/CoalHistory/coaltowns/home.htm (1991).

14 For a study of boardinghouses in Lowell, see Mary C. Beaudry and Stephen A. Mrozowski's "The Archeology of Work and Home Life in Lowell, Massachusetts: An Interdisciplinary Study of the Boott Cotton Mills Corporation," *IA: The Journal of the Society for Industrial Archeology* 14:2 (1988): 1–22.

15 Richard M. Candee, "New Towns of the Early New England Textile Industry," *Perspectives in Vernacular Architecture* 1 (1981): 37.

16 Magnusson, 128–131.

17 Candee, 37.

18 See Patrick M. Malone and Charles A. Parrott, "Greenways in the Industrial City: Parks and Promenades along the Lowell Canals," *IA: The Journal of the Society for Industrial Archeology* 24:1 (1998): 19–40.

19 Magnusson, 142.

20 Magnusson, 143

21 Margaret Crawford, "Earle S. Draper and the Company Town in the American South," in *The Company Town: Architecture and Society in the Early Industrial Age*, edited by John S. Garner (New York: Oxford University Press, 1992), 148.

22 Magnusson, 58.

23 Karen Bescherer Metheny, *From the Miners' Doublehouse: Archaeology and Landscape in a Pennsylvania Coal Company* (Knoxville: University of Tennessee Press, 2007), 29.

24 Bescherer, 32–33.

25 Magnusson, 100–103.

26 Margaret M. Mulrooney, *A Legacy of Coal: The Coal Company Towns of Southwestern Pennsylvania* (Washington, DC: National Park Service, 1989), 38.

27 Mulrooney, 38.

28 Peter Going and Elizabeth Raymond, "Living in Anthracite: Mining Landscape and Sense of Place in Wyoming Valley, Pennsylvania," *The Public Historian* 23:2 (Spring 2001): 35.

29 Going and Raymond, 38.

30 Magnusson, 71–74.

31 George Torok, *A Guide to Historic Coal Towns of the Big Sandy River Valley* (Knoxville: University of Tennessee Press, 2004).

32 Arnold R. Alanen and Lynn Bjorkan, "Plats, Parks, Playgrounds, and Plants: Warren H. Manning's Landscape Designs for the Mining Districts of Michigan's Upper Peninsula, 1899–1932," *IA: The Journal of the Society for Industrial Archeology* 24:1 (1998): 46–47.

33 Magnusson, 91–92.

34 Arnold R. Alanon, "Documenting the Physical and Social Characteristics of Mining and Resource-Based Communities," *Bulletin of the Association for Preservation Technology* 11:4 (1979): 51.

35 Alanon, 51.

36 Richard Francaviglia, *Hard Places: Reading the Landscape of America's Historic Mining Districts* (Iowa City: University of Iowa Press, 1991), 12.

37 Francaviglia, 17.

38 Francaviglia, 40.

39 Francaviglia, 42–47.

40 Francaviglia, 27.

41 Donald L. Hardesty, *The Archaeology of Mining and Miners: A View from the Silver State*," The Society for Historical Archaeology, 70–71.

42 Linda Carlson, *Company Towns of the Pacific Northwest* (Seattle: University of Washington Press), 14–15, 29.

43 Margaret Crawford, "The 'New' Company Town," *Perspecta* 30 (1999): 49.

44 Leclaire is fully described in John S. Garner, "Leclaire, Illinois: A Model Company Town (1890–1934), *The Journal of the Society of Architectural Historians* 30:3 (October 1971): 219–227.

45 Crawford, 53.

46 Crawford, 55.

Index

A

B

C

Q

R

W

Z

About the Author

Douglas C. McVarish has been employed as an architectural historian by John Milner Associates, Inc. since 1992. He currently is Principal Architectural Historian and Associate and works in JMA's Philadelphia office. Born in Norwalk, Connecticut, he holds a B.A. degree in American studies from Hampshire College, a Master of Planning degree from the University of Virginia and a M.S. degree in historic preservation from the University of Vermont. He has surveyed and documented many industrial structures and landscapes including two slackwater navigation corridors, aluminum industry facilities of western Pennsylvania, a Maryland mine fan complex, many bridges, several electrical generating stations, a shipyard, a sanitary pottery, a forge, and several airports. He resides in Collingswood, New Jersey with his wife, Lois Maynard.